Lecture Notes on Data Engineering and Communications Technologies **161**

The aim of the book series is to present cutting edge engineering approaches to data technologies and communications. It will publish latest advances on the engineering task of building and deploying distributed, scalable and reliable data infrastructures and communication systems.

The series will have a prominent applied focus on data technologies and communications with aim to promote the bridging from fundamental research on data science and networking to data engineering and communications that lead to industry products, business knowledge and standardisation.

Indexed by SCOPUS, INSPEC, EI Compendex.

All books published in the series are submitted for consideration in Web of Science.

Leonard Barolli
Editor

Advances in Internet, Data & Web Technologies

The 11th International Conference on Emerging Internet, Data & Web Technologies (EIDWT-2023)

Editor
Leonard Barolli
Department of Information and Communication Engineering
Fukuoka Institute of Technology
Fukuoka, Japan

ISSN 2367-4512 ISSN 2367-4520 (electronic)
Lecture Notes on Data Engineering and Communications Technologies
ISBN 978-3-031-26280-7 ISBN 978-3-031-26281-4 (eBook)
https://doi.org/10.1007/978-3-031-26281-4

This Springer imprint is published by the registered company Springer Nature Switzerland AG
The registered company address is: Gewerbestrasse 11, 6330 Cham, Switzerland

Welcome Message of EIDWT-2023 International Conference Organizers

Welcome to the 11th International Conference on Emerging Internet, Data and Web Technologies (EIDWT-2023), which will be held from February 23 to February 25, 2023.

The EIDWT is dedicated to the dissemination of original contributions that are related to the theories, practices and concepts of emerging Internet and data technologies yet most importantly of their applicability in business and academia toward a collective intelligence approach.

In EIDWT-2023, topics related to Information Networking, Data Centers, Data Grids, Clouds, Crowds, Mashups, Social Networks, Security Issues and other Web implementations toward a collaborative and collective intelligence approach leading to advancements of virtual organizations and their user communities will be discussed. This is because Web implementations will store and continuously produce a vast amount of data, which if combined and analyzed through a collective intelligence manner will make a difference in the organizational settings and their user communities. Thus, the scope of EIDWT-2023 includes methods and practices which bring various emerging Internet and data technologies together to capture, integrate, analyze, mine, annotate and visualize data in a meaningful and collaborative manner. Finally, EIDWT-2023 aims to provide a forum for original discussion and prompt future directions in the area.

An international conference requires the support and help of many people. A lot of people have helped and worked hard for a successful EIDWT-2023 technical program and conference proceedings. First, we would like to thank all authors for submitting their papers. We are indebted to Program Area Chairs, Program Committee Members and Reviewers who carried out the most difficult work of carefully evaluating the submitted papers. We would like to give our special thanks to Honorary Chair of EIDWT-2023 Prof. Makoto Takizawa, Hosei University, Japan, for his guidance and support. We would like to express our appreciation to our Keynote Speakers for accepting our invitation and delivering very interesting keynotes at the conference.

EIDWT-2023 Organizing Committee

Honorary Chair

Makoto Takizawa	Hosei University, Japan

General Co-chairs

Olivia Fachrunnisa	UNISSULA, Indonesia
Juggapong Natwichai	Chiang Mai University, Thailand
Tomoya Enokido	Rissho University, Japan

Program Co-chairs

Ardian Adhiatma	UNISSULA, Indonesia
Elis Kulla	Fukuoka Institute of Technology, Japan
Admir Barolli	Alexander Moisiu University, Albania

International Advisory Committee

Janusz Kacprzyk	Polish Academy of Sciences, Poland
Arjan Durresi	IUPUI, USA
Wenny Rahayu	La Trobe University, Australia
Fang-Yie Leu	Tunghai University, Taiwan
Yoshihiro Okada	Kyushu University, Japan

Publicity Co-chairs

Naila Najihah	UNISSULA, Indonesia
Farookh Hussain	University of Technology Sydney, Australia
Keita Matsuo	Fukuoka Institute of Technology, Japan
Pruet Boonma	Chiang Mai University, Thailand
Flora Amato	Naples University Frederico II, Italy

International Liaison Co-chairs

Muthoharoh	UNISSULA, Indonesia
David Taniar	Monash University, Australia
Tetsuya Oda	Okayama University of Science, Japan
Omar Hussain	University of New South Wales, Australia
Nadeem Javaid	COMSATS University Islamabad, Pakistan

Local Organizing Committee Co-chairs

Agustina Fitrianingrum	UNISSULA, Indonesia
Andi Riansyah	UNISSULA, Indonesia

Web Administrators

Kevin Bylykbashi	Fukuoka Institute of Technology, Japan
Ermioni Qafzezi	Fukuoka Institute of Technology, Japan
Phudit Ampririt	Fukuoka Institute of Technology, Japan

Finance Chair

Makoto Ikeda	Fukuoka Institute of Technology, Japan

Steering Committee Chair

Leonard Barolli	Fukuoka Institute of Technology, Japan

PC Members

Akimitsu Kanzaki	Shimane University, Japan
Akira Uejima	Okayama University of Science, Japan
Alba Amato	National Research Council (CNR) - Institute for High-Performance Computing and Networking (ICAR), Italy
Alberto Scionti	LINKS, Italy
Antonella Di Stefano	University of Catania, Italy
Arcangelo Castiglione	University of Salerno, Italy

Beniamino Di Martino	University of Campania "Luigi Vanvitelli", Italy
Bhed Bista	Iwate Prefectural University, Japan
Carmen de Maio	University of Salerno, Italy
Chotipat Pornavalai	King Mongkut's Institute of Technology Ladkrabang, Thailand
Dana Petcu	West University of Timisoara, Romania
Danda B. Rawat	Howard University, USA
Elis Kulla	Fukuoka Institute of Technology, Japan
Eric Pardede	La Trobe University, Australia
Fabrizio Marozzo	University of Calabria, Italy
Fabrizio Messina	University of Catania, Italy
Farookh Hussain	University of Technology Sydney, Australia
Francesco Orciuoli	University of Salerno, Italy
Francesco Palmieri	University of Salerno, Italy
Gen Kitagata	Tohoku University, Japan
Giovanni Masala	Plymouth University, UK
Giovanni Morana	C3DNA, USA
Giuseppe Caragnano	LINKS, Italy
Giuseppe Fenza	University of Salerno, Italy
Harold Castro	Universidad de Los Andes, Bogotá, Colombia
Hiroaki Yamamoto	Shinshu University, Japan
Hiroshi Shigeno	Keio University, Japan
Isaac Woungang	Toronto Metropolitan University, Canada
Jiahong Wang	Iwate Prefectural University, Japan
Jugappong Natwichai	Chiang Mai University, Thailand
Kazuyoshi Kojima	Saitama University, Japan
Kenzi Watanabe	Hiroshima University, Japan
Kiyoshi Ueda	Nihon University, Japan
Klodiana Goga	LINKS, Italy
Lidia Fotia	Università Mediterranea di Reggio Calabria (DIIES), Italy
Lucian Prodan	Polytechnic University Timisoara, Romania
Makoto Fujimura	Nagasaki University, Japan
Makoto Nakashima	Oita University, Japan
Marcello Trovati	Edge Hill University, UK
Mauro Marcelo Mattos	FURB Universidade Regional de Blumenau, Brazil
Minoru Uehara	Toyo University, Japan
Mirang Park	Kanagawa Institute of Technology, Japan
Naohiro Hayashibara	Kyoto Sangyo University, Japan
Naonobu Okazaki	University of Miyazaki, Japan
Nobukazu Iguchi	Kindai University, Japan

Nobuo Funabiki	Okayama University, Japan
Olivier Terzo	LINKS, Italy
Omar Hussain	UNSW Canberra, Australia
Pruet Boonma	Chiang Mai University, Thailand
Raffaele Pizzolante	University of Salerno, Italy
Sajal Mukhopadhyay	National Institute of Technology, Durgapur, India
Salvatore Ventiqincue	University of Campania Luigi Vanvitelli, Italy
Shigetomo Kimura	University of Tsukuba, Japan
Shinji Sugawara	Chiba Institute of Technology, Japan
Shinji Sakamoto	Kanazawa Institute of Technology, Japan
Sotirios Kontogiannis	University of Ioannina, Greece
Teodor Florin Fortis	West University of Timisoara, Romania
Tomoki Yoshihisa	Osaka University, Japan
Tomoya Enokido	Rissho University, Japan
Tomoya Kawakami	NAIST, Japan
Toshihiro Yamauchi	Okayama University, Japan
Toshiya Takami	Oita University, Japan
Xu An Wang	Engineering University of CAPF, China
Yoshihiro Okada	Kyushu University, Japan
Jindan Zhang	Xianyang Vocational Technical College, China
Luca Davoli	University of Parma, Italy
Ricardo Rodriguez Jorge	Jan Evangelista Purkyně University, Czech Republic
Yusuke Gotoh	Okayama University, Japan

EIDWT-2023 Reviewers

Adhiatma Ardian
Amato Flora
Amato Alba
Barolli Admir
Barolli Leonard
Bista Bhed
Chellappan Sriram
Chen Hsing-Chung
Cui Baojiang
Di Martino Beniamino
Enokido Tomoya
Esposito Antonio
Faiz Iqbal Faiz Mohammad
Fachrunnisa Olivia
Fun Li Kin
Funabiki Nobuo
Gotoh Yusuke
Hussain Farookh
Hussain Omar
Javaid Nadeem
Ifada Luluk
Iio Jun
Ikeda Makoto
Ishida Tomoyuki
Kamada Masaru
Kato Shigeru
Kayem Anne
Kikuchi Hiroak
Kohana Masaki
Kulla Elis

Leu Fang-Yie
Leung Carson
Matsuo Keita
Ogiela Lidia
Ogiela Marek
Okada Yoshihiro
Pardede Eric
Paruchuri Vamsi Krishna
Rahayu Wenny
Spaho Evjola
Sugawara Shinji
Takizawa Makoto
Taniar David
Uehara Minoru
Yoshihisa Tomoki
Venticinque Salvatore
Wang Xu An
Woungang Isaac
Xhafa Fatos

EIDWT-2023 Keynote Talks

Fueling the Data Engine to Boost the Power of Analytics

Wenny Rahayu

La Trobe University, Melbourne, Australia

Abstract. Data analytics is often considered in isolation. The attractiveness of the problems that need to be solved, the sophistication of the solutions, and the usefulness of the results are certainly the significant strengths of work on data analytics. However, the input data is often too simplistic, or at least the assumption that the data is already readily prepared for data analytics often neglects the fact that preparing such an input data is in many cases, if not all, actually the major work in the data life cycle. The pipeline from the operational databases that keep the transactions and raw data to the input data for data analytics is very long; it often occupies as much as 80% (or sometimes even more) of the entire life cycle. Therefore, we need to put much effort to this preparation and transformation work in order to value the work and the results produced by data analytics algorithms. Having the correct input data for the data analytics algorithms, or in fact for any algorithms and processes, is critical, as the famous quote "garbage in garbage out" had said. Even when the original data is correct, but when it is presented inaccurately to a data analytics algorithm, it may consequently produce incorrect reasoning. This talk will present a systematic approach to build a data engine for effective analytics.

Impact of Uncertainty Analysis and Feature Selection on Data Science

Ricardo Rodriguez Jorge

Jan Evangelista Purkyně University, Ústí nad Labem, Czech Republic

Abstract. Data science applications usually need a previous preprocessing stage for feature extraction and data validation. The data needs to be preprocessed and analyzed to minimize the dataset while preserving variance and patterns in order to find the optimal feature vector configuration. The feature selection algorithm allows finding the feature vector configuration to ensure minimal uncertainty in mapping the corresponding outputs and feature vectors. In data science, feature vector designs can be performed by different techniques and the validation can be performed by uncertainty analysis. These considerations are timely because wearable devices are increasingly being used on a large scale in different scientific fields. This talk will contribute to recommendations for the use of signals and data as a means of informing the impact of different uncertainty analysis and feature selection methods for data science applications. Using this new knowledge together with machine learning, data science applications can be evaluated with more confidence.

Contents

Data Integration in Practice: Academic Finance Analytics Case Study

Kittayaporn Chantaranimi[1(✉)], Juggapong Natwichai[2], Pawat Pajsaranuwat[3], Anawat Wisetborisut[4], and Surapong Phosu[1]

[1] Data Science Consortium, Faculty of Engineering, Chiang Mai University, Chiang Mai, Thailand
{kittayaporn_c,surapong_ph}@cmu.ac.th
[2] Department of Computer Engineering, Faculty of Engineering, Chiang Mai University, Chiang Mai, Thailand
juggapong@eng.cmu.ac.th
[3] Department of Statistics, Faculty of Science, Chiang Mai University, Chiang Mai, Thailand
pawat.pak@cmu.ac.th
[4] Department of Family Medicine, Faculty of Medicine, Chiang Mai University , Chiang Mai, Thailand
anawat.w@cmu.ac.th

Abstract. Financial sustainability is one of the crucial operations of many higher education institutes. Though since late 2019, the inevitable disruption and significant changes in the higher education system have continued after the increasing in COVID-19 transmissions. These affect the operations of higher education institutions in numerous ways, such as students' admission, financial management and teaching strategies. The purpose of this study is to present a data integration aspect of the analysis of financial data from academic income. Such data integration relates to the data from enrollment, admission, and research from many heterogeneous sources within the institution. In addition, the *k*-mean clustering approach is applied to group academic programs for further analysis. In the future, the institution's financial and risk management, research enhancement, and reputation and positioning will employ this analytics to support and shape the institution's operations.

Keywords: Data integration · Finance analytics · Clustering · Higher education

1 Introduction

Chiang Mai University (CMU), which establishes in 1984 and is a university for knowledge acquisition and transfer in both the arts and the sciences, as well as social services, defines its organizational structure and functions into multiple divisions, e.g., university committees, administrative offices, registration office, schools, institutes, and information and technology office. CMU offers degrees

L. Barolli (Ed.): EIDWT 2023, LNDECT 161, pp. 1–11, 2023.
https://doi.org/10.1007/978-3-031-26281-4_1

at various levels including bachelor's degrees, master's degrees, postgraduate certificates and high diplomas, and doctorates.

However, like other institutions, its operations have been affected by the COVID-19 pandemic since late 2019. For example, the increased cost of a college education especially health and safety cost, the decrease in the number of new students, the lower enrollments, and the decline in university revenue because of ad hoc policy (e.g., discount in tuition fees, financial aids) [9,21]. With this situation and in the future, the entire university needs to significantly operational reassessment policies regarding academics and services: financial stability, mission, academic programs, program delivery, technology, library, and student services. Among these, financial planning is the top priority [25,27]. Additionally, the effective utilization of resources to many divisions has emerged as a major concern for university administrators [7].

This study presents the result of analytics from the integration of data produced by heterogeneous sources to review university's outcomes and to support policy planning. Therefore, the goal of division assessment, particularly for faculties and colleges, is to determine whether the university's and divisions' establishment objectives, and mission and vision are accomplished, and how to enhance the operation if necessary. However, while revenue is a critical concern for our university, the commitment to social responsibility for sustainable development through academic, innovation, research, and social service is also highly significant. At this point, our analysis is comprised of financial and research aspects of each division compared to those of other divisions in order to pinpoint each cluster-specific group's strengths and weaknesses.

Despite the fact that each division of CMU adheres to the policies established by the CMU board of directors, the directors of each division are allowed to implement their policies independently of other divisions so long as they do not conflict with the top policies. Because of the widespread adoption of several systems, data schemas, and semantic variations, the data analytics team is currently faced with a tremendous obstacle: data integration issues. Various techniques are used to deal with the issues, including data gathering and integration, exploratory data analysis, cluster analysis, and data visualization. Thus, this study presents the procedures for gathering, integrating, evaluating, and presenting analytic results in order to facilitate better decision-making.

The paper is organized as follows. In Sect. 2 related literatures, we review and discuss some concepts of data integration, k-mean clustering, and financial sustainability. In Sect. 3, we introduce the work of data integration for the operational analysis of an institution, in this work the CMU's case. Thereafter, the analytics result, conclusion and discussion are presented in Sect. 4.

2 Related Literatures

In this section, we review the concepts of data integration and data clustering, respectively.

2.1 Data Integration

2.1.1 Traditional Data Integration

From the perspective of database systems, database integration is referred to an approach or a process of synchronizing and integrating multiple autonomous and heterogeneous data sources, e.g., relational databases, XML documents, JSON files, legacy data systems, to provide a single uniform view or an integrated schema [1,6,17,20,26]. In [2], the two main reasons for integration: First, given existing information, an integrated view can be established to simplify information access and reuse. Second, given a certain information need, data from different sources is combined to satisfy that requirement. Moreover, either physical or virtual databases could be the outcome of an integration process. In the meantime, the traditional data integration pipeline tends to be a one-way, batch process, typically implemented using the extract-transform-load (ETL) methodology, where the bottleneck is formed by labor-intensive activity [8]. Example of general data integration approaches are Manual Integration, Common User Interface, Integration by Applications, Integration by Middleware, Uniform Data Access, and Common Data Storage [2].

2.1.2 Heterogeneity Problems

Integration problems are escalated even if the databases are implemented with the same architecture since data from different sources is heterogeneous in various meanings. According to [26], the following issues can exist: 1) structural differences, e.g., decomposition (horizontal and vertical) type and data type, 2) naming differences in metadata or data itself such as "name" versus "student name", 3) semantic differences that occur when meanings are similar but not precisely equivalent; and 4) content differences, which is the possibility that implicit, derivable, or missing contents occur across environments. Particularly for semantic issues, data models cannot describe all of the semantics of the real world since schemas differ depends on the designer's perspective [19].

2.1.3 Data Integration and Machine Learning

In the big data and data-driven environment, Data Integration (DI) goes beyond historical terms as it points to capabilities and usability of integrated data for data analytics and Business Intelligent (BI) rather than only focuses on operational database integration. In [22], DI includes data acquisition, analytics, and data warehousing, data synchronization between operational applications, data migrations and conversions, master data management, data sharing, and delivery of data services. Thus, the integrated data from the well-define integration solution enables an analytic team to provide insights of the organization's operations by implementing historical and current data to introduce better analytic outcomes, such as reporting, key performance indicators, and predictive models or analytics.

In [10], cutting-edge data integration solutions utilize machine learning (ML) to achieve accurate results and efficient human-in-the-loop pipelines and end-to-end machine learning applications rely on data integration to find clean, relevant data for analytics, are illustrated. In the meantime, there are various works utilizes ML and statistics for DI such as implementation of probabilistic model in DI [23], the use of ML and neural network in omics and clinical data integration [18], outlier detection and repair of mixed-type data [3]. Subsequently, in [24], an end-to-end data collection guideline is presented, which includes the processes of data acquisition, data labeling, and data improvement. They emphasize machine learning's contribution to data management discipline and vice versa by introducing the decision flow chart for data collection and technique selection.

2.2 Data Clustering

Data clustering is underline unsupervised classification where the purpose is to partition an unlabeled set of data points, such as observations, data items, or feature vectors, into meaningful groupings or clusters that are as similar as possible [4, 15]. In short, it is the method that the objects (data points) are clustered or grouped based on *"the principle of maximizing the intraclass similarity and minimizing the interclass similarity"* [13].

There are two types of clustering approaches: hierarchical and partitional. The difference is that hierarchical methods generate a nested series of partitions, whereas partitional methods generate only one shot. However, in this study, we do rely on partitional approaches. For partitional approaches, an initial set of seeds (or clusters) is required, which the representative points chosen at different iterations are thereafter improved iteratively [4]. However, among various algorithms of partitional methods, k-Means clustering which is distance-based algorithm is used because of its classical and simplicity.

Algorithm 1 illustrate the process of k-Means clustering.

Algorithm 1. k-Means clustering

1: Select k points as initial centroids.
2: **repeat**
3: Form K clusters by assigning each point to its closest centroid.
4: Recompute the centroid of each cluster.
5: **until** convergence criterion is met.

In the higher education sector, [7] conducts a cluster analysis using efficiency scores and aggregate efficiency decomposition of each department's teaching, publications, and external grants. As a result, departments are then categorized into three groups: Teaching, Research, and General. Furthermore, in [11], universities are divided into four groups according on their research and international collaboration characteristics. As a result, the author suggests that offering scholarships will strengthen academic quality. On the other hand, community learning and individual learning are the groups of higher education institutions presented in [12].

3 Case Study: Higher Education Academic Finance Analytics

This study presents a case study of a financial data analytics project for Chiang Mai University (CMU) focusing on academic revenue. Based on the concept of data collecting from data lake searching, in which many datasets are generated internally and are not easily discoverable by other teams, we follow a post-hoc approach [24] to gather data. Furthermore, we introduce our data analytics process by adapting the Cross-Industry Standard Process for Data Mining (CRISP-DM) [14] and the Analytics Solutions Unified Method for Data Mining/predictive analytics (ASUM-DM) [5], a refined CRISP-DM.

3.1 Data Analytics Process

Our data analytics process is illustrated in Fig. 1. We divide the process into five main stages: Determining Goals and Objectives, Data Gathering, Data Integration, Data Analytics, and Reporting. This workflow with brainstorming in every stage is provided for conducting analytics projects within the realistic context of CMU's workplace.

Determining Goals and Objectives. A review of the current university and educational system scenario and issues is the first step to understanding the state. In this condition, brainstorming, a method for eliciting ideas, requirements, potential solutions, and discussion, is used to determine the goals and objectives associated with the current situation precisely. Consequently, an initial proposal regarding analytics is submitted to the stakeholder committee. However, if the proposal is not approved, the team can move back and forth between tasks to refine another edition as necessary.

Data Gathering. In order to mitigate the Garbage in - garbage out problem [16], the attention is initially placed on the identification of data requirements, data sources, and data owners, as a result of the need for established analytics to ensure that high-quality data, or at least data from various appropriate sources, is collected. In addition, for internal data sources such as the Registration Office and Finance Division of University, the analytics team may encounter the heterogeneity problem. The data collection report, therefore, provides data owners with a description of the quantity and quality of data required. However, there is also the case where the information collection does not satisfy the requirement and the team have to return to the first task of this stage. Finally, the data for this study is mainly from internal sources, it consists of the student information and financial information.

Data Integration. This stage is a time-consuming part of the analytics process and is very prioritized. The challenge arises because of the heterogeneity problem. For example, there are two data sources storing the same identity card number in different data types (string versus integer). In fact, it is estimated that data preparation usually takes 50–70% of a project's time and effort [14]. Accordingly, our integration phase begins with a brainstorming session, followed by an investigation of the collected data's type, structure, and context. Therefore, manual integration is performed via hand-coding and several Excel and Python libraries. The ETL process, which combines the extract, transform, and load operations, is a crucial part. Meanwhile, the data preparation technique consists of the following sub-tasks: data selection, data cleaning, data transformation, data merging, variables deriving, and data storing.

Data Analytics. The well-prepared data from the previous stage is an input for this stage. This study uses Microsoft Excel, Python, and R programming to analyze the data. First, we use many statistical methods, e.g., descriptive statistic, forecasting technique, and correlation analysis, to investigate trends, patterns, and relationships between the university's income and other variables. Secondly, this paper applies k-Means cluster analysis to group the curriculum at CMU regarding 1) the number of enrolled students in each curriculum, and 2) the correlation coefficient between the university's income and the number of enrolled students. After that, we interpret our result. However, if the output from this stage is not acceptable, the analytic team can go back to the earlier stage.

Reporting. Three main tools to present the insight and result include an in-detail document, a presentation showing summarized text, graph, and visualization, and Power BI dashboard.

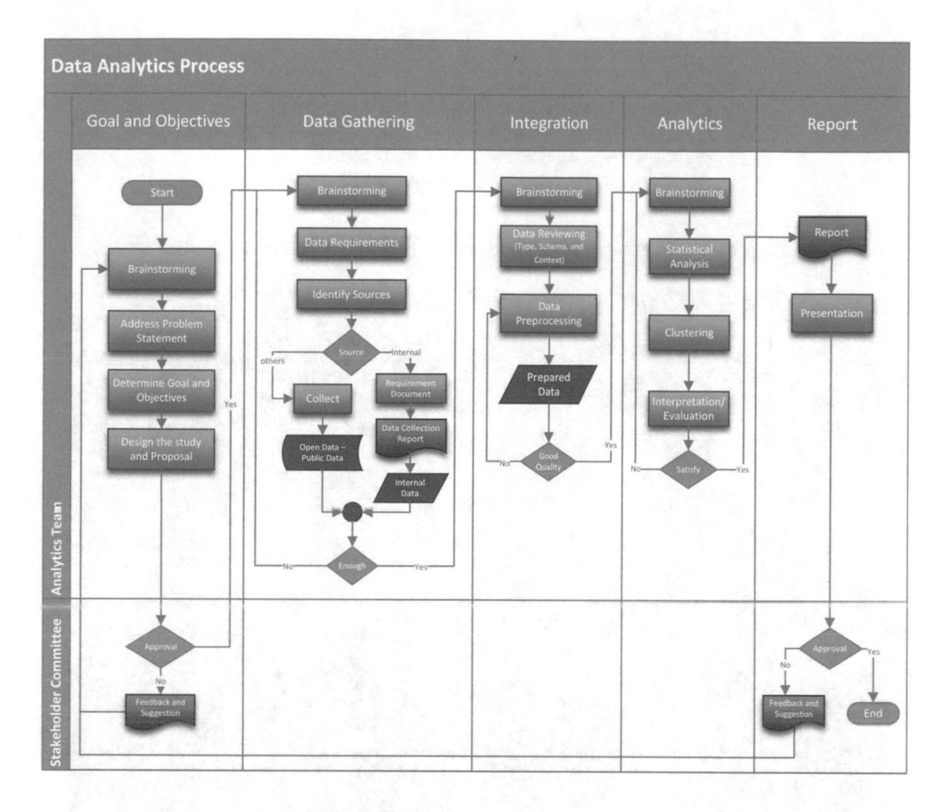

Fig. 1. Data analytics process

4 Result and Discussion

4.1 Result

In terms of clustering results, the k-Means technique produces clusters that divide each curriculum into four groups. Figure 2 depicts a simple visualization of four groups.

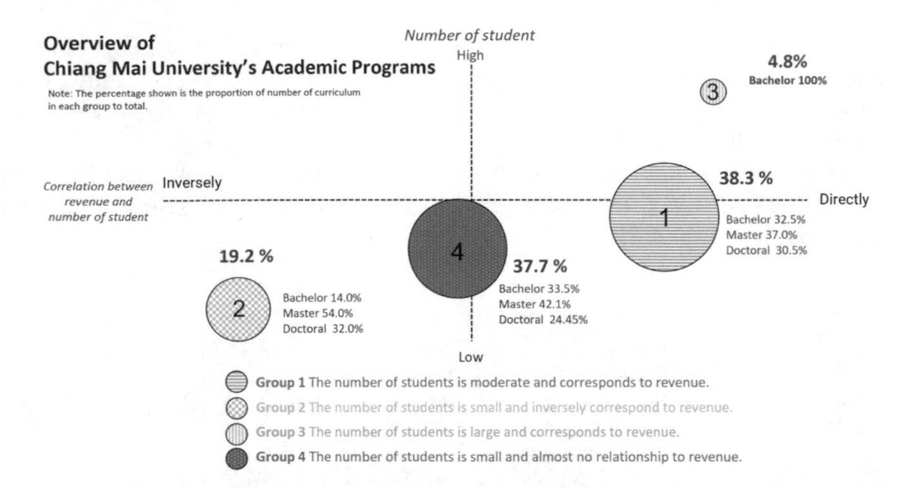

Fig. 2. Four group of Chiang Mai University's curriculum

Group	%	Characteristic	Suggestion for Improvement
1	38.3	Each unit has a moderate number of enrolled students, with a positive correlation to income received. Thus, this number of students is in line with the trend of income.	• Review of marketing strategies and plans.
2	19.2	The number of students is small and inversely correspond to revenue. There is a tendency for the number of enrolled students to decrease, in contrast to the increasing of income.	Review and redefine curriculum details, especially cost management and resource utilization.
3	4.8	The number of students is large and corresponds to the high revenue.	Design and the Implementation of Knowledge Management
4	37.7	The number of students is small and almost no relationship to revenue. Thus, the number of enrolled students is relatively stable.	• Review and redefine curriculum detail. • Review of marketing strategies and plans. • Review of cost management.

Fig. 3. The characteristic of each group

From the result, it is demonstrated that the majority is in groups 1 and 4. These two groups have a small to moderate number of students enrolled. In contrast, the number of enrolled students in group 4 is relatively stable. In addition, the effective curriculum (group 3) where a large number of students corresponds to the high revenue has a small fraction. On the other hand, group 2 is unsuccessful since there is a tendency for the number of enrolled students to decline, despite an increase in total income. Figure 3 summarizes the characteristics of each group as well as suggestions for improvement.

4.2 Discussion

In this study, Chiang Mai University serves as a case study for the introduction of the finance analytics procedure from problem statement to insight presentation. The workflow process consists of five stages: determining goals and objectives, data gathering, data integration, data analytics, and reporting. As the data utilized in our study are gathered from a variety of sources, we focus on data integration to assure data quality prior to conducting analysis. In addition, we believe that open and continuous collaboration to solve problems and develop innovative ideas is developed when brainstorming is assigned as the first task at each stage.

Additionally, clustering would be very useful when grouping different types of academic programs, as it can reveal the characteristics of each group, allowing university administrators to use the result or insight for decision-making and to tailor policies to each group. For example, the effective curriculum, which is a very small proportion should establish and share their best practice among other divisions in the organization. Moreover,

For future study, in supplement to manual data integration, modernized technology like high-performance computing, machine learning, and deep learning should be studied to fulfill the gap of heterogeneous data integration issue. Moreover, focus group methods, when combined with quantitative analytics, will be conducted, in wihch it can yield insights that would be impossible to discover in a computational way.

Acknowledgements. We are most thankful for the Faculty of Engineering, Finance Division of University, Planning Division of Office Of the University, Registration Office Chiang Mai University, Graduate School Chiang Mai University, Office of Educational Quality Development, for supporting us in this study.

References

1. Halevy, A.: Answering queries using views: A survey (2001). https://doi.org/10.1007/s007780100054
2. Ziegler, P., Dittrich, K.R.: Three decades of data intecration— all problems solved? In: Jacquart, R. (ed.) Building the Information Society. IIFIP, vol. 156, pp. 3–12. Springer, Boston, MA (2004). https://doi.org/10.1007/978-1-4020-8157-6_1
3. Eduardo, S., Nazabal, A., Williams, C.K.I.: Robust variational autoencoders for outlier detection and repair of mixed-type data. In: International Conference on Artificial Intelligence and Statistics, pp. 4056-4066. PMLR (2020)
4. Aggarwal, C.C., Reddy, C.K.: Data Clustering: Algorithms and Applications, 1st edn. Chapman & Hall/CRC (2013)
5. Angée, S., Lozano-Argel, S.I., Montoya-Munera, E.N., Ospina-Arango, J.D., Tabares-Betancur, M.S.: Towards an improved asum-dm process methodology for cross-disciplinary multi-organization big data & analytics projects. In: Uden, L., Hadzima, B., Ting, I.H. (eds.) Knowledge Management in Organizations, pp. 613–624. Springer International Publishing, Cham (2018)
6. Batini, C., Lenzerini, M., Navathe, S.B.: A comparative analysis of methodologies for database schema integration. ACM Comput. Surv. **18**(4), 323–364 (1986)

7. Chen, S.-P., Chang, C.-W.: Measuring the efficiency of university departments: an empirical study using data envelopment analysis and cluster analysis. Scientometrics **126**(6), 5263–5284 (2021). https://doi.org/10.1007/s11192-021-03982-3
8. Dayal, U., Castellanos, M., Simitsis, A., Wilkinson, K.: Data integration flows for business intelligence. In: Association for Computing Machinery, pp. 1–11 (2009). https://doi.org/10.1145/1516360.1516362
9. Deloitte Touche Tohmatsu Limited (2020) Covid-19 impact on higher education. https://www2.deloitte.com/us/en/pages/public-sector/articles/covid-19-impact-on-higher-education.html Accessed 11 August 2022
10. Dong, X.L., Rekatsinas, T.: Data integration and machine learning: A natural synergy. Proc VLDB Endow **11**(12), 2094–2097 (2018).https://doi.org/10.14778/3229863.3229876
11. Elbawab, R.: University rankings and goals: A cluster analysis. Economies **10**(9), 209 (2022) https://doi.org/10.3390/economies10090209, https://www.mdpi.com/2227-7099/10/9/209
12. Guzman, J.H.E., Zuluaga-Ortiz, R.A., Donado, L.E.G., Delahoz-Dominguez, E.J., Marquez-Castillo, A., Suarez-Sánchez, M.: Cluster analysis in higher education institutions' knowledge identification and production processes. Procedia Computer Science 203:570–574 (2022). https://doi.org/10.1016/j.procs.2022.07.081, https://www.sciencedirect.com/science/article/pii/S187705092200686X In: 17th International Conference on Future Networks and Communications/19th International Conference on Mobile Systems and Pervasive Computing/12th International Conference on Sustainable Energy Information Technology (FNC/MobiSPC/SEIT 2022), August 9-11, 2022, Niagara Falls, Ontario, Canada
13. Han, J., Kamber, M., Pei, J.: Data Mining Concepts and Techniques, third edition (2012). www.amazon.de/Data-Mining-Concepts-Techniques-Management/dp/0123814790/ref=tmm_hrd_title_0?ie=UTF8&qid=1366039033&sr=1-1
14. IBM Corporation: Ibm spss modeler crisp-dm guide (2021). https://www.ibm.com/docs/en/spss-modeler/18.1.1?topic=spss-modeler-crisp-dm-guide Accessed 11 August 2022
15. Jain, A.K., Murty, M.N., Flynn, P.J.: Data clustering: A review **31**(3), 264–323 (2000)
16. Kilkenny, M.F., Robinson, K.M.: Data quality: "garbage in -garbage out". Health Information Management Journal **47**(3), 103–105 (2018). https://doi.org/10.1177/1833358318774357
17. Lenzerini, M.: Data integration: A theoretical perspective. Association for Computing Machinery, New York, NY, USA, PODS '02, pp. 233–246 (2002). https://doi.org/10.1145/543613.543644
18. Li, Y., Wu, F.X., Ngom, A.: A review on machine learning principles for multi-view biological data integration. (2018). https://doi.org/10.1093/bib/bbw113
19. Parent, C., Spaccapietra, S.: Issues and approaches of database integration. Commun. ACM **41**, 166–178 (1998). https://doi.org/10.1145/276404.276408
20. Parent, C., Spaccapietra, S.: Database integration: The key to data interoperability. In: Advances in Object-Oriented Data Modeling, The MIT Press (2000)
21. Pavlov, O.V., Katsamakas, E.: Covid-19 and financial sustainability of academic institutions. Sustainability (Switzerland) **13**(7), 3903 (2021). https://doi.org/10.3390/su13073903
22. Poess, M., Rabl, T., Jacobsen, H.A., Caufield, B.: Tpc-di: The first industry benchmark for data integration. Proc. VLDB Endow **7**(13), 1367–1378 (2014). https://doi.org/10.14778/2733004.2733009

23. Rekatsinas, T., Chu, X., Ilyas, I.F., Ré, C.: Holoclean: Holistic data repairs with probabilistic inference. Proc VLDB Endow **10**(11), 1190–1201 (2017). https://doi.org/10.14778/3137628.3137631
24. Roh, Y., Heo, G., Whang, S.E.: A survey on data collection for machine learning: a big data-ai integration perspective. Trans. Knowl. Data Mach. Learn. **33**(4), 1328–1347 (2021). https://doi.org/10.1109/TKDE.2019.2946162
25. Rowley, W.J.: Higher education in the midst of a pandemic: a dean's perspective. Int. Dialogues Educ. **7**, 108–115 (2020)
26. Sujansky, W.: Heterogeneous database integration in biomedicine. J. Biomed. Inform. **34**, 285–298 (2001). https://doi.org/10.1006/jbin.2001.1024, http://www.sciencedirect.com/science/article/pii/S153204640191024X
27. Witze, A.: Universities will never be the same after the coronavirus crisis (2020). www.nature.com/articles/d41586-020-01518-y Accessed 11 August 2022

Proposal of an Aquarium Design Support Virtual Reality System

Fumitaka Matsubara and Tomoyuki Ishida(✉)

Fukuoka Institute of Technology, Fukuoka 811-0295, Fukuoka, Japan
s19b2039@bene.fit.ac.jp, t-ishida@fit.ac.jp

Abstract. In this paper, we propose an aquarium design that supports a virtual reality system that can easily simulate aquarium designs. When a user designs a san aquarium in a physical location, they must repeatedly rearrange the aquarium's plants and driftwood to get close to the ideal aquarium. As a result, designing the aquarium requires a lot of time and work. The user first selects the aquarium size through this system. The perfect aquarium is then constructed in the virtual environment by freely placing aquarium plants, driftwood, stones, etc. This system aims to acquire new users by lowering the hurdles of aquarium design.

1 Introduction

As people spend more time at home as a result of the new coronavirus infection, more people are beginning to keep pets. One of them, the killifish, is expected to experience a boom due to its ease of raising in a confined space at home as a therapeutic object [1]. Aquariums have relaxing and stress-relieving effects [2]. As a result, more people are starting to keep aquariums as a new hobby and source of comfort.

Keeping an aquarium is not an easy task. The user must set up the necessary equipment for the fish they intend to keep, feed them regularly, and maintain the aquarium. Some new users feel that keeping is too difficult. Aquarium manufacturers are creating tools that little maintenance work to lower the barrier.

On the other hand, when building the aquarium, the user chooses and buys the fish, aquarium plants, driftwood, stones, etc. by himself. At that time, the user purchases them while considering the ideal aquarium design. When an aquarium is designed, it is not always possible to design the perfect aquarium. In practice, the user must redesign the aquarium several times while taking placement considerations into account, as well as numerous times rearranging the placement of stones and driftwood. We, therefore, reasoned that one of the challenges of aquarium design is the difficulty of repeatedly designing using actual aquariums, aquarium plants, driftwood, etc.

As a result, in this study, we build a design simulation system for the aquarium and a 3D model based on real objects (aquarium plants, driftwood, and stones).

The rest of the paper is organized as follows. The related works are presented in Sect. 2. The objective of our study is described in Sect. 3. Section 4 explains the system configuration and architecture of our suggested aquarium design support virtual reality system. The prototype system is described in Sect. 5. Finally, we conclude our findings in Sect. 6.

L. Barolli (Ed.): EIDWT 2023, LNDECT 161, pp. 12–21, 2023.
https://doi.org/10.1007/978-3-031-26281-4_2

2 Related Works

A virtual reality aquarium display that enables multiple users to view an aquarium's world in a spherical display was created by Fafard et al. [3]. The user can observe the aquarium display from any angle. By using a laser pointer to direct a fish or shellfish onto the screen, a user can encourage user interaction and communication. Since this system aims to share communication and experiences among multiple users, it is considered to emphasize entertainment.

Multiple users can view, share, and transfer an interactive 3D fish simultaneously thanks to the creation of gCubik+i by Roberto et al. [4]. gCubik+i consists of two displays, a small 3D display connected to the tabletop display. Multiple users can surround a table with a static 2D image of a fish displayed and pick up the fish in their hands using a 3D display. After picking it up, you can help others collaborate by observing the fish on the 3D display and handing it to them.

Tanaka et al. [5] used mixed reality to create the interactive aquarium system "MR Coral Sea." Users can play with virtual fish through the MR video experience and coral display. A virtual fish swims in response to the detected position of the user's hand in this system. Users can directly experience events in the virtual space without wearing a head-mounted display.

Matsuo et al. [6] proposed an aquarium simulation system for augmented reality entertainment. This system superimposes 3D models such as fish created by computer graphics on the environment photographed in real-time by a web camera. This allows the user to feel as if they are close to the fish.

3 Research Objective

In this study, we propose a system that can easily simulate the layout of objects in an aquarium. This proposed system includes a feature that allows users to freely place 3D objects like driftwood and stones by selecting their preferred aquarium size. Users do not need to manually move the actual aquarium, aquarium plants, driftwood, and so on, so they can simulate the aquarium design many times using this system. This study lowers the barriers to aquarium design for users by providing a setting in which they can create their ideal aquarium.

4 System Configuration and System Architecture

Figure 1 and 2 show the system configuration and system architecture of this study. A user agent, an aquarium management agent, a VR space control function, and an aquarium design virtual reality space comprise this system.

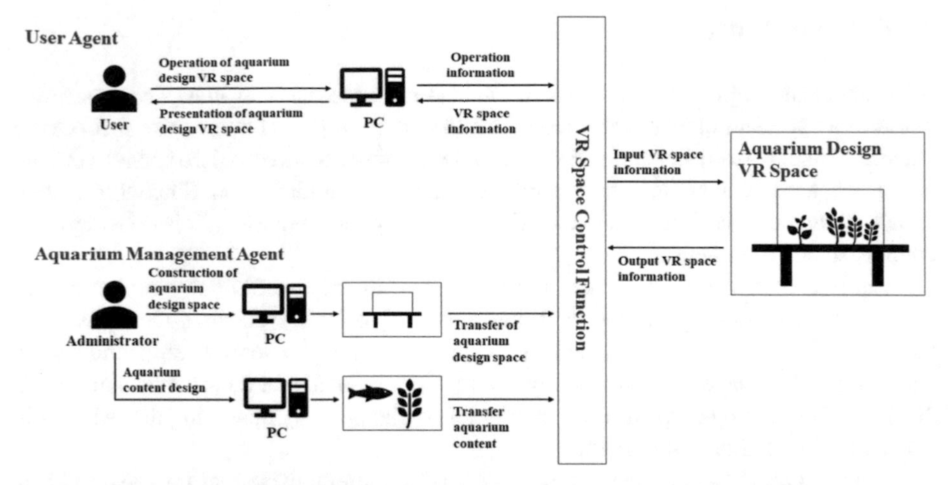

Fig. 1. System configuration of the aquarium design supports a virtual reality system.

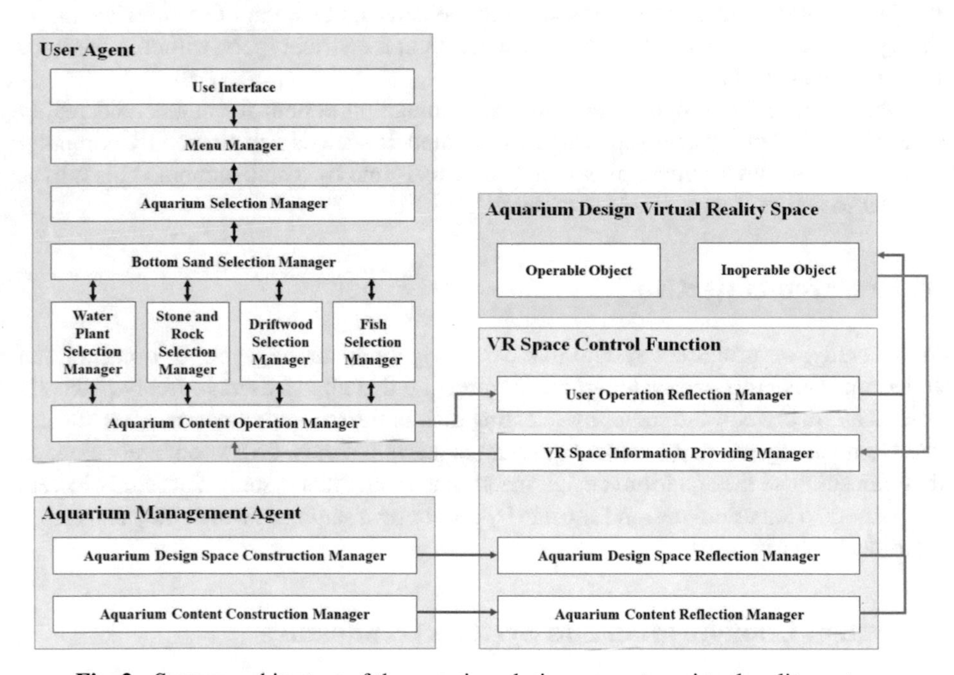

Fig. 2. System architecture of the aquarium design supports a virtual reality system.

4.1 User Agent

The user agent can experience the aquarium design VR space through the PC. The aquarium design VR space is operated by the user using the mouse, and the operation information is reflected in the aquarium design VR space via the VR space control function.

4.2 Aquarium Management Agent

The aquarium management agent creates and designs the aquarium design space as well as the aquarium contents (3D objects such as aquariums, aquarium plants, driftwood, and stones), which are then reflected in the aquarium design VR space via the VR space control function.

4.3 VR Space Control Function

The VR space control function provides the aquarium design VR space in response to the user agent's request, as well as displaying the user's operation information in the aquarium design VR space. Furthermore, the aquarium design space and aquarium contents are reflected in the aquarium design VR space in response to a request from the aquarium management agent.

4.4 Aquarium Design Virtual Reality Space

The aquarium layout the VR space is a VR spacc in which the user can design the aquarium, and the user is given a user interface to choose the size of the aquarium first. Following the selection of the aquarium, the user interface is presented in thc order of selecting the substrate and objects (aquarium plants, driftwood, stones), and the user can design an arbitrary aquarium.

5 Aquarium Design Support Virtual Reality System

5.1 Organization of Objects in VR Space

The aquarium stand object, the floor object, the aquarium objects, the aquarium floor objects, the aquarium plant objects, the driftwood objects, and the stone objects comprise the 3D model of this system's VR space. The aquarium objects, aquarium floor objects, aquarium plant objects, driftwood objects, and stone objects were created using Blender [7]. Blender objects were exported in fbx format and imported into Unity [8]. We created four types of aquarium objects. The four types of aquarium objects are 20-cm aquarium (width 20 cm × depth 20 cm × height 20 cm), 30-cm aquarium (width 30cm × depth 30 cm × height 30 cm), 45-cm aquarium (width 45 cm × depth 24 cm × height 30 cm), and 60-cm aquarium (width 45 cm × depth 24 cm × height 30 cm) (width 60 cm × depth 30 cm × height 30 cm) as shown in Fig. 3.

Fig. 3. Aquarium objects (20, 30, 45, and 60 cm aquariums from left to right).

Fig. 4. Aquarium floor objects (soil, gravel, white sand from left to right)).

Next, we created three types of aquarium floors. As shown in Fig. 4, the aquarium floor consists of soil, gravel, and white sand.

Finally, we created the aquarium plant object, driftwood object, and stone object. We prepared two types of aquarium plants, one type of driftwood, and four types of stones for the prototype system. Pearl grass (Fig. 5, left) and Rotala rotundifolia high red are the two types of aquarium plants (Fig. 5, right). The 3D models of the created stone and driftwood are shown in Fig. 6.

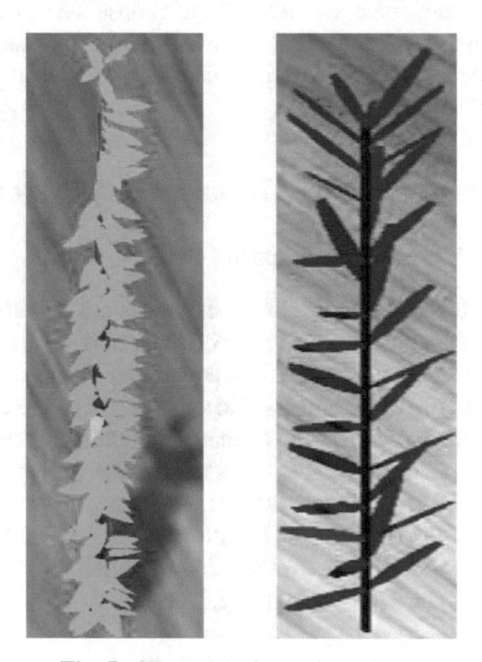

Fig. 5. 3D model of aquarium plants.

Fig. 6. 3D model of stone and driftwood.

5.2 System Functions

The user begins by selecting the aquarium to be designed from the aquarium selection UI when using this system (Fig. 7). Next, the user selects the aquarium floor to be laid on the aquarium floor from the aquarium floor selection UI. Finally, the user selects objects (aquatic plants, driftwood, stones) from the object selection UI and drags them into the aquarium design.

Fig. 7. Aquarium selection UI.

The user selects the aquarium from four aquarium selection UI. When the mouse cursor is over each UI, the name and size of the aquarium are displayed as shown in Fig. 8. Figure 9 depicts the user's position with the mouse cursor over the UI of the 30-cm aquarium. After selecting an aquarium, the user presses the "Confirm Aquarium" button, which displays the aquarium floor selection UI.

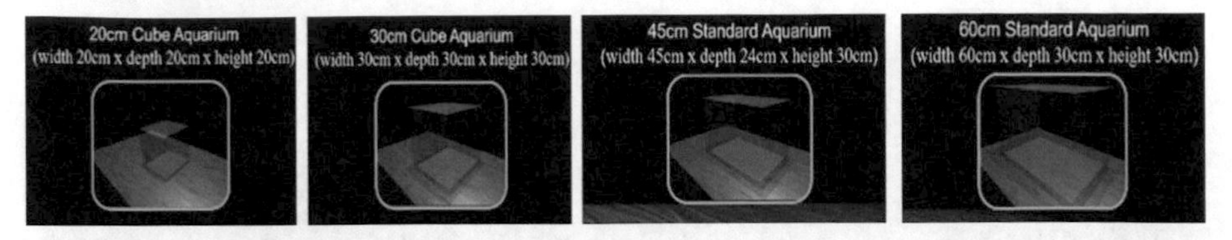

Fig. 8. Display of aquarium name and size when hovering over each UI.

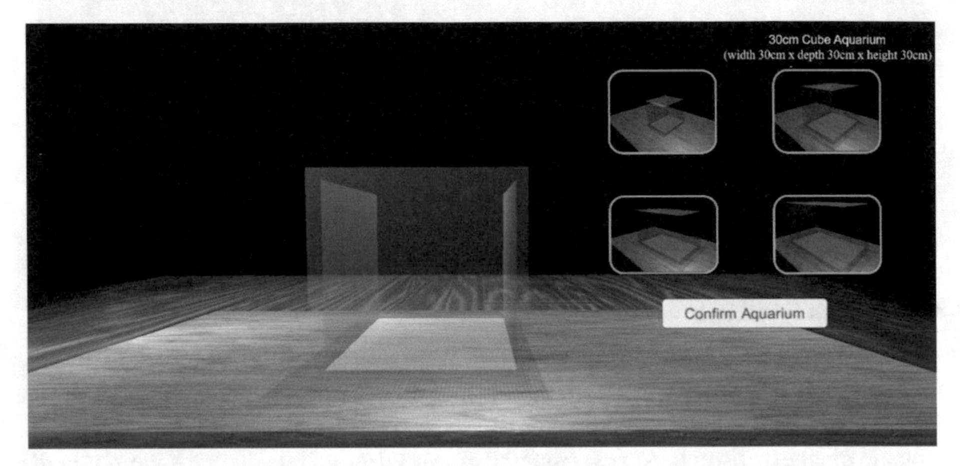

Fig. 9. Aquarium selection UI when selecting a 30-cm cube aquarium.

The user then selects an aquarium floor from one of three aquarium floor selection UI. Figure 10 depicts the user interface when the user selects soil. The object selection UI is displayed when the user selects the aquarium floor and presses "Confirm Aquarium Floor."

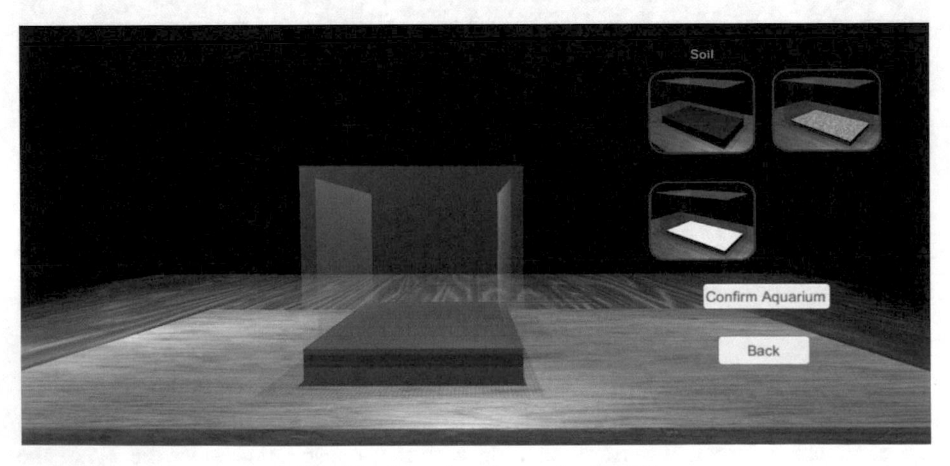

Fig. 10. Aquarium floor selection UI.

Finally, the user selects objects (aquarium plants, driftwood, stones) for designing the aquarium's interior and drags them into place. When the user selects the aquarium plant from the object selection UI shown in Fig. 11, the aquarium plant selection UI shown in Fig. 12 is displayed. When the user chooses an aquarium plant from the aquarium plant selection UI, it is displayed on the right side of the aquarium object, as shown in Fig. 13. The aquarium plant object, which is displayed on the right side of the aquarium, can be dragged and moved, and the user can freely place it in the aquarium. Figure 14 depicts an example of a user designing an aquarium.

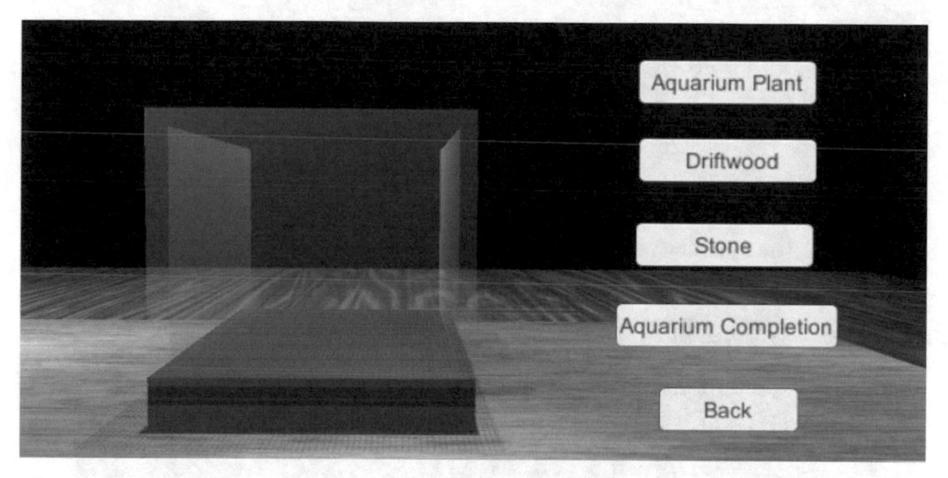

Fig. 11. Object selection UI.

Fig. 12. Aquarium plant selection UI.

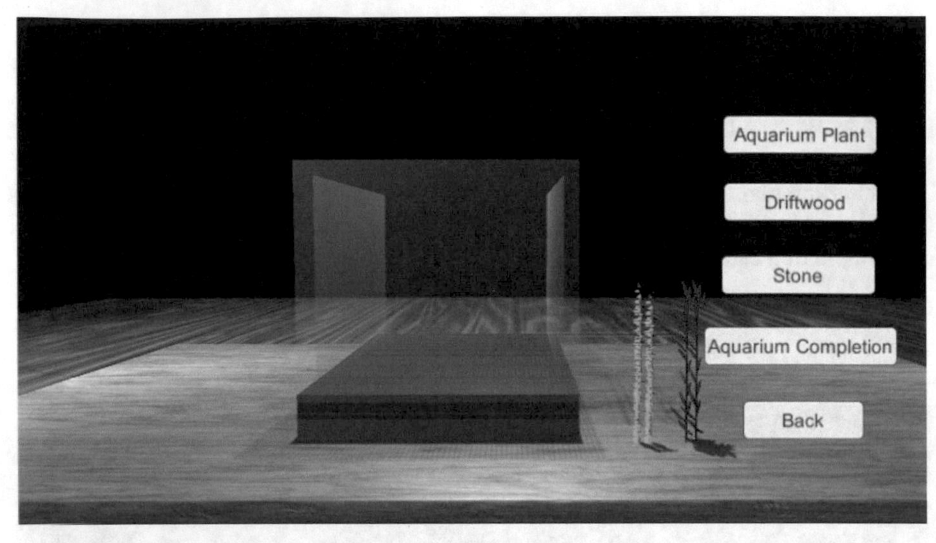

Fig. 13. The aquarium plant objects are displayed to the right of the aquarium object.

Fig. 14. Aquarium design example.

6 Conclusion

In this paper, we proposed the aquarium design supports a virtual reality system that can easily simulate aquarium designs. After selecting the size of the aquarium, the user can freely arrange aquarium plants, driftwood, stones, and so on to design the aquarium using this system. We intend to conduct a questionnaire survey of the subjects in the future to assess the effectiveness and operability of this system.

References

1. Japan Pet Food Association: 2021 Nationwide Dog and Cat Breeding Survey (2021). https://petfood.or.jp/data/chart2021/index.html. Accessed Dec 2022
2. GEX.CO.,LTD.: Joint research between Gifu University and Gex Co., Ltd. (2020). https://www.gex-fp.co.jp/fish/blog/labo/gex-lab-20200505/. Accessed Dec 2022
3. Fafard, D.B., Zhou, Q., Chamberlain, C., Hagemann, G., Fels, S., Stavness, I.: Design and implementation of a multi-person fish-tank virtual reality display. In: Proceedings of the 24th ACM Symposium on Virtual Reality Software and Technology, pp. 1–9 (2018)
4. Roberto, L., Yoshida, S., Makino, M., Yano, S., Ando, H.: gCubik +i virtual 3D aquarium : natural interface between a graspable 3D display and a tabletop display. Trans. Virtual Reality Soc. Japan **15**(2), 147–155 (2010)
5. Tanaka, Y., Tanaka, C., Ohshima, T.: MR coral sea - mixed reality aquarium with physical MR display. In: Proceedings of the Entertainment Computing Symposium 2014, pp. 22–25 (2014)
6. Matsuo, K., Hagiwara, M.: Entertainment AR aquarium. J. Soc. Art Sci. **10**(4), 226–233 (2011)
7. blender.org, Features - blender.org (2022). https://www.blender.org/features/. Accessed Dec 2022
8. Unity Technologies: Unity Real-Time Development Platform I 3D, 2D VR & AR Engine, https://unity.com/. Accessed Dec 2022

The Source Code Maintenance Time Classifications from Code Smell

Patcharaprapa Khamkhiaw(✉), Chartchai Doungsa-ard, and Passakorn Phannachitta

College of Arts, Media and Technology, Chiang Mai University, Chiang Mai, Thailand
{patcharaprapak,chartchai.d,passakorn.p}@cmu.ac.th

Abstract. Software maintenance is necessary for the software development process after the software deployment. The maintenance estimation time is required for repair maintenance. Many metrics have been used to estimate the maintenance time. In this paper, a maintenance estimation method is proposed to explore the possibility to accurately estimate the maintenance time using the amount of code smells. In the experiment, the amount and types of code smells were extracted by using SonarQube and the maintenance time was extracted from Github's issue trackers. As a preliminary study, two machine learning techniques, i.e., Support Vector Machines (SVM) and Gradient Boosting Classifier (GBC) were applied to discover the relationship between three types of code smells, i.e., Bloaters, Object Orientation Abusers, and Dispensable and the median of maintenance time over Apache Flink and Apache Hive projects. The results showed that SVM and GBC could achieve 61% and 65% accuracy in terms of measurement of test accuracy calculated based on test accuracy and recall respectively. The results were shown as promising as the experiment was carried out only with small dataset and did not take any parameters. Optimization into an account, where the maximum f1 achieved from the two projects was above 0.60%. Thus, there is a possibility for develop a model better performance.

Keywords: Code smell · Software maintenance · Estimation · Classification · Machine learning

1 Introduction

Software maintenance is the process that aims to correct bugs, improve performance, enhance software capability, and remove outdated functions after the project is delivered to improve functionality and increase efficiency [1] because the software requirements can be changed at any time. Software maintenance also occurred frequently. The goal of the software maintenance is to evolve the software to be consistent with the changed requirements and also improve performance even if no error occurs [2].

However, when the changes occur, the time is required for change the software planning, the estimation of the fix time is important for software project managers as it can help them to prioritize the tasks to make the project complete in time at the desired quality and within budget. There are many factors related to maintenance estimation, such as complexity of code, developer resources, term experience in which can be analyzed

L. Barolli (Ed.): EIDWT 2023, LNDECT 161, pp. 22–32, 2023.
https://doi.org/10.1007/978-3-031-26281-4_3

from human assessment, job size assessment, budget funds assessment, and the estimated time of work [3]. It is widely observed that [4] time management has relied on the experience of the software maintainer. The quicker the software maintenance process may complete with desirable outcomes. Thus, the level of comprehension of the written codes in an important factor determining source code quality. In practice, code smell can imply source code quality by the measure of existence in the source code is a lot of code smell that shows highly complex [5].

The code smell is a source code that is prone to future errors or hard to understand how the system works [6]. There are many types of code smells which are grouped by several studies. For example, Bloaters are caused by increasing numbers between classes and methods, Object Orientation Abusers are the code development that violates the object-oriented programming principles, and Dispensable means that the source code is not executed when the system runs. Typically, these unnecessary source codes are eliminated by refactoring to improve the source code quality and help software maintainers better understand the source code.

It must also be noted that eliminating code smells is costly and not every developers makes code smells by their intention but mostly due to the time limit causing them to take short cut and left code smell in the source code to get the task done promptly, regardless of quality and can lead to the accumulation of technical debt making the software development and maintenance more difficult in the future [7, 8].

As a preliminary study, two machine learning techniques are applied to discover relationships between three types of code smell and the median of maintenance time. The three types of code smell to are Bloaters, Object Orientation Abusers, and Dispensable, and the two machine learning techniques are Support vector machines (SVM) [9] and Gradient boosting classifier (GBC) [10] Evaluated over Apache Flink and Apache Hive projects.

The paper is organized as follows. Section 2 provides the related works. The technique we used to provide the classification models is explained in Sect. 3. In Sect. 4, our experimental result and the discussion are provided. Lastly, the conclusion and future work is addressed in Sect. 5.

2 Literature Review

2.1 Maintenance Time Estimation

One problem of the software project management is to estimate the time for software maintenance because the software time maintenance is hard to estimated due to many factors of the source code.

Luo et al. [2] presented a maintenance time estimation in the design phase based on a Fuzzy v-support vector machine (Fv-SVM). They analyzed the influencing factors of maintenance time in the software designed, then proposed the estimation time methods using Fv-SVM. The result showed that the factors on the maintenance processes can be used to model the maintenance procedure, but the different factors may returns the different accuracy.

The other factors which can be used as the source for software maintenace model estimation is the UseCase Point. PointSima [3] studied the software project from five companies. The software process for the projects has been analyzed to developed the Risk Extended Use Case Points (RUCP) method. The result was the influence of each factor in the software development process such as technology, user or process is difference in the different software processes. As the consequence, each factor required more investigated influence.

And in addition, Elmidaoui [11] presented predicting software maintainability using experimental results suggesting that GBC are the best ensemble for all datasets, since they ranked first and second, respectively. They are analyzing influencing factors of maintenance time and putting forward a novel maintenance time estimation method based on SVM and GBC.

2.2 Code Smell

Code smell is the complex source code that is hard for understanding the operation of the system. Code smell may cause errors in the future. According to Sae-Lim's [12] suggestion, Code smell can be divided into 5 categories.

- Bloaters are code, methods, and classes that have increased to such enormous balances that they're hard to work with.
- Object-Orientation Abusers are the source code which is incomplete or incorrect applications of object-oriented programming principles.
- Change Preventers are a complex set of code smell that cannot be easily maintained by software it will takes time and many resources to correct the software.
- Dispensables is the source code that is pointless and unneeded.
- Couplers is group contribute to the excessive coupling between classes which makes it difficult to validate the execution once the code has been modified.

Sae-Lim et al. [6] proposed the importance of each code smell to be removed. The Selecting and Prioritizing code smell has been proposed to help the developers to select each code smell should be removed first, but they did not estimate the time for removing the smells. Although, the SonarQube which is the tool to analyze code smell can estimate the code smell repairing time,

Palomba et al. [13] validated the accuracy of the SonarQube estimation time. They asked 65 novice developers to remove rule violations and reduce the technical debt in 15 Java open-source projects. The results showed that 52% of estimation has the accuracy of less than or equal 0.25. However, the accuracy is calculated per SonarQube issued, not by the real world practice.

In our paper, the code smell is used to estimate the maintenance time as it is the information which we can extracted from the tools, and affect the software maintenance process. In addition, the practical software development information will be used to analyze. The machine learning model will be used due to the complication of the problem.

3 The Experimental Design

In this work, the relationship between the amount of code smell and the total software maintenance time is investigated. The Fig. 1 illustrates the experimental setup comprising data source selection, code smell extraction, maintenance time extraction, classification, and evaluation respectively.

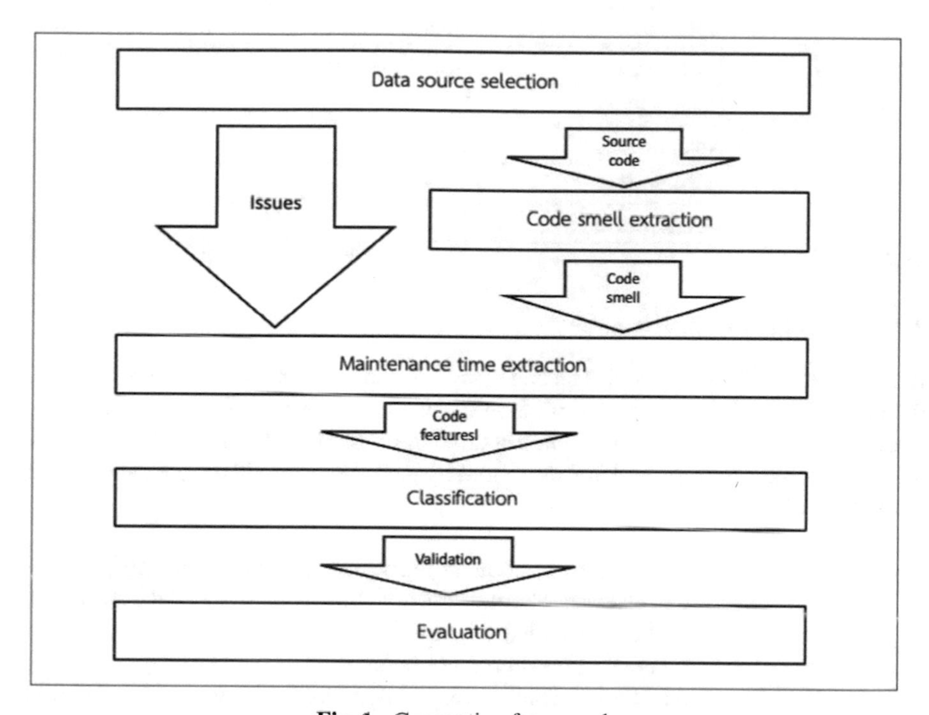

Fig. 1. Conception framework

3.1 Data Sources Selection

The first process of our work is to select the repository which provides the required data. The Apache Hive and Apache Flink projects [6, 10] are selected because the software development process details and the updated source code details. The example of the data which can be extracted from the issues are shown in Table 1.

Table 1. An example of information obtained from an open-source project problem report.

Data features	Valuable of hive
URL https://github.com/apache/hive/commit/dcfc31b	
Commit end	dcfc31b83
Commit begin	30b40c4
Open time	1/22/2010 5:01
Closed time	1/25/2010 18:54
Change flies	288
Addition files	5145
Deletion files	5142
Developers	1

3.2 Code Smell Extraction

After extracting the commit id, The code smell is extracted from the starting commit of the issue and the finished commit of the issue. The SonarQube [13] is an automatic tool to detect code smell in the static source codes. The code smells extracted by SonarQube are in the format as of rules defined by SonarQube. The example of rules are provided in the following list.

- java: S119 Type parameter names should comply with a naming convention
- java: S1123 Deprecated elements should have both the annotation and the Javadoc tag
- java: S1133 Deprecated code should be removed
- java: S1874 "@Deprecated" code should not be used
- java: S1948 Fields in a "Serializable" class should either be transient or serializable
- java: S2293 The diamond operator ("<>") should be used
- java: S 6212 Local-Variable Type Inference should be used
- java: S6213 Restricted Identifiers should not be used as Identifiers
- java: S3740 Raw types should not be used)
- java: S3776 Cognitive Complexity of methods should not be too high

The amount of code smell in a source code snapshot is collected for each group by executing SonarQube at the first commit and the last commit of a source code version, respectively. Table 2 shows an example of these data preprocessed from Apache Hive project, where postfix "begin" and "end" denotes the amount of code smells at the first commit and the last commit, respectively.

Table 2. Hive project data collection example

data features.	valuable of Hive commit
end dcfc31b83	
commit begin	30b40c4
s106 begin	5
s106 _end	9
s112 begin	0
s112 end	2
...	...
s125 begin	0
s125-end	1
s135 begin	3
s135-end	0

However, if there are too many groups of data, it is difficult to define the classification model. The code smells generated from the SonarQube are grouped into three code smell groups which are Bloaters(B), Object Orientation Abusers(OOA), and Dispensable(D) shown in Table 3.

Table 3. Group of code smells based on SonarQube

Group of code smell						
Bloaters	Object orientation abusers			Dispensable		
s1066	s135	s1197	s3358	s106	s1141	s1874
s1199	s1104	s1610	s5803	s116	s1155	s2093
s2293	s1117	s2139	s5998	s1123	s1172	s3626
s3358	s1124	s2986	s6212	s1133	s1182	s5411
s3776	s1181	s3011		s1134	s1854	s6212

These rules can also be categorized into a more well-known groups based on Sae-Lim's [12] which are Bloaters(B), Object Orientation Abusers(OOA), and Dispensable(D) for dispart categories of code smells that may cause errors in the future. Where a subset of rules are shown in Table 3. Thus, Table 2 was processed further by concatenating the rules themselves as shown in Table 2 and their categories and an example of the final product is presented in Table 4.

3.3 Maintenance Time Extraction

Maintenance time is the target variable of the present study where ideally it should be determined directly by the extracted code smells. After collecting all data, the features for creating the model should be prepared. The output value is classified. In the proposed approach, the fix time is classified in two categories. Easy to fix, and hard to fix.

Table 4. Hive project issues information

data features.	valuable of Hive	
Commit end	dcfc31b83	a3e2ae47
Commit begin	30b40c4	44797e96b9
s106 begin	5	0
s106 end	9	4
...	0	2
s135 begin	1	7
s135 end	3	3
Opened time	1/22/2010 5:01	9/24/2014 20:00
Closed time	1/25/2010 18:54	10/6/2014 21:20
Change flies	288	60
Addition files	5145	3643
Deletion files	5142	1652
Developers	1	2
B begin	20	17
B end	11	20
OOA begin	27	22
OOA end	31	13
D begin	7	12
D end	11	6

The fix time is calculated by the open time, and the close time as shown in Eq. 1 [4]. The definition in Eq. 1 is taken from Table 1. The fix time from every issue are calculated for the median value of the fix time. The median value is used as the threshold for the output class.

$$Fix\ time = Close\ time - Open\ \ time \quad (1)$$

The Eq. 2 provides how to define the class. Where the two are quantified by subtracting the time when the related issue of the particular committed is close by the time when the issue is open and calculate the median of the product across the entire dataset. Then, for the cases taking the fix time higher than the median overall, these cases are considered as hard to fix and labelled as 1 used justified by a classifier. For the otherwise, they are denoted as easy to fix cases and labelled as 0.

$$Fix\ time\ class = \begin{cases} 1, & Fix\ time > median(Fix\ Time) \\ (2)\ 0, & Fix\ time < median(Fix\ Time) \end{cases} \quad (2)$$

However, since this preliminary study target only whether the relationship between the amount of code smell and that of the maintenance time are existed or not. For the current study, the 30 issues have been randomly picked for the experiment, and the median fix time for the two selected repositories are presented in Table 5.

Table 5. The median of fix time in each project

Projects	Median of fix time
Hive	15 days 00:43:00 h
Flink	7 days 05:57:00 h

3.4 Classification

The classification techniques is used in order to classified the class of fix time based on the amount of smells to be fixed. The classification techniques such as Gradient Boosting Classifier (GBC) and Support Vector Machines (SVM) are selected because Luo et al. [2] present a maintenance time estimation in the design phase based on SVM, Assessing the issues of finite samples and uncertain data in the maintenance time estimation. A classification model is constructed to predict the class of the fix time in each issue by code smell illustrated in Table 4. The feature sets are selected in these case study the selected feature are the amount of code smell in each group at the beginning and the end of the issue (Commit end, Commit begin, s106 begin, s106 end,…, Opened time, Closed time, Change flies, Addition files, Deletion files, Developers, B begin, B end, OOA begin, OOA end, D begin, D end). The issues are separated to the training data and testing data for 7:3 ratio. In other words, the 21 issues are random selected to be training data, and the remaining 9 issues are the testing data in each iteration. The cross validation has been tested for 3 times to measure the performance of the model. Table 6 shows the parameter configured for the two classification techniques when building them.

Table 6. Modeling for prediction

Models	Parameter
Support Vector Machines (SVM)	gamma = 'auto'
Gradient Boosting Classifier (GBC)	n estimators = 1 learning rate = 1.0, max depth = 1, random state = 2

3.5 Evaluation

The classification performance was evaluated using three conventional metrics, which are precision, recall, and f-measure (f1 Score) [14, 15] because they are standards of accuracy criteria. Precision is a measure of how many of the optimistic predictions made are correct (true positives). Recall is a measure of how many of the positive cases the classifier correctly expected over all the positive cases in the data. F1-Score is a measure incorporating both precision and recall. It is normally explained as the harmonic mean of the two. Harmonic mean is another way to calculate an average of values, generally described as more suitable for ratios such as precision and recall than the traditional arithmetic mean.

4 Experimental Results

To experiment with our model, the Apache Flink and Apache Hive project are selected because there are many issues provided in the project which match our requirements. The data has been acquired from the project issues and set up the data frame to create the model. The Support Vector Machine (SVM) and Gradient Boosting Classifier techniques are used as the classification technique in terms of the Precision recall and F1 score are provided in Table 7.

Table 7. Compare the results of projects Flink and Hive.

Models	Flink			Hive		
	Precision	Recall	F 1	Precision	Recall	F 1
SVM	0.56	0.77	0.65	0.62	0.61	0.61
GBC	0.52	0.52	0.49	0.56	0.77	0.65

Comparing between the results obtained from SVM and GBC, SVM returned slightly litter better performance for the Apache Flink project for all three metrics; however, it is remaining inconclusive for the Apache Hive project where GBC won in terms of the recall and f1 score. Overall, even if these results are generated by small datasets and did not have any parameter optimization performed, the returned results are better than random. This shows that with the default configuration as a consequence. Thus, there is a potential to further improve the accuracy. However, the proposed work aims to assist the project manager for their task planning. Unfortunately, the features that can only be obtained at the last commit are not available at the time of taking planning. As a consequence in the practical problem, only the feature is available at the beginning which we can archive at the beginning of the issue. Thus, this experiment further explores if it is possible to make the prediction only with the available information at the beginning of each issue.

5 Conclusion

Software maintenance is necessary for the software development process after the software deployment. The maintenance estimation time is required for repair maintenance. Many metrics have been used to estimate the maintenance time. In this paper, the maintenance estimation prediction classification has been proposed. The source code is analyzed for detecting code smell which may lead to technical debt. This study examined the relationship between the amount of code smell and the amount of time to maintain the software to explore the effect of the code smell to include the data at the begin of the issue, and at the end of the issue. The maintenance time was extracted from the change request issues in the project. Specifically, the amount and types of code smells are extracted by using SonarQube. As a preliminary study, two machine learning techniques are applied to discover relationships between three types of code smell and the median of maintenance time the three types of code smell to are Bloaters, Object Orientation Abusers, and Dispensable and the two machine learning techniques are SVM and GBC evaluated over Apache Flink and Apache Hive projects. The result shows that SVM and GBC could achieve 61% and 65% accuracy in terms of f1 score respectively. This work explores the possibility to identify the relationship between the amount of code smells and the maintenance time, where the amount of code smells was extracted using SonarQube, and the maintenance time was extracted from the time between an issue being opened until it is closed. As a preliminary study, the maintenance time was further simplified into two categories: easy to fix and hard to fix quantified by being above or below the median of the time between an issue being opened until it is closed calculated from the entire dataset. The dataset comprises 30 records preprocessed from two GitHub projects namely Apache Hive and Apache Flink were used for building classifiers based on SVM and GBC techniques. The results were shown as promising as the experiment was carried out only with small dataset and did not take any parameter optimization into an account, where the maximum f1 achieved from the two projects was above 0.60%. Thus, there is a potential to further improve the accuracy. In our future work, the researchers plan to carry out the experiments of the present study over sufficiently large dataset where statistically significant test can be applied. Furthermore, parameter optimization using sophisticated techniques such as bayesian optimization will be taken into an account. The researchers believe this future direction will lead to the conclusion that the relationship between the amount of code smells and the maintenance time are existed.

References

1. Dai, Z., Yao, L., Qin, J.: Research on equipment maintenance time management based on risk evaluation, pp. 1576–1580 (2018)
2. Luo, X., Yang, Y., Ge, Z., Guan, F.: Maintenance time estimation method in the design phase based on fuzzy v-support vector machine, pp. 649–653 (2013)
3. Bagheri, S., Shameli-Sendi, A.: Software project estimation using improved use case point, pp. 143–150 (2018)
4. Sawarkar, R., Nagwani, N.K., Kumar, S.: Predicting bug estimation time for newly reported bug using machine learning algorithms, pp. 1–4 (2019)
5. Gupta, R., Singh, S.K.: Using software metrics to detect temporary field code smell, pp. 45–49 (2020)
6. Sae-Lim, N., Hayashi, S., Saeki, M.: How do developers select and prioritize code smells? a preliminary study, pp. 484–488 (2017)
7. Arif, A., Rana, Z.A.: Refactoring of code to remove technical debt and reduce maintenance effort, pp. 1–7 (2020)
8. Fontana, F.A., Ferme, V., Spinelli, S.: Investigating the impact of code smells debt on quality code evaluation, pp. 15–22 (2012)
9. Jesudoss, A., Maneesha, S., Naga Durga, T.: Identification of code smell using machine learning, pp. 54–58 (2019)
10. Tufano, M., et al.: When and why your code starts to smell bad (and whether the smells go away). IEEE Trans. Software Eng. **43**, 1063–1088 (2017)
11. Elmidaoui, S., Cheikhi, L., Idri, A., Abran, A.: Predicting software maintainability using ensemble techniques and stacked generalization (2020)
12. Singh, S., Kaur, S.: A systematic literature review: Refactoring for disclosing code smells in object oriented software. Ain Shams Eng. J. **9**, 2129–2151 (2018). https://doi.org/10.1016/j.asej.2017.03.002
13. Palomba, F., Zaidman, A., Lucia, A.D.: On the accuracy of sonarqube technical debt remediation time, pp. 317–324 (2019)
14. Kala, M., Lali´ s, A., Vittek, P.: Optimizing calculation of maintenance revisionˇ times in maintenance repair organizations, pp. 1–6 (2019)
15. Nagwani, N.K., Verma, S.: Predictive data mining model for software bug estimation using average weighted similarity, pp. 373–378 (2010)

Evolution Analysis of R&D Jobs Based on Patents' Technology Efficacy Labeling

Cui Ruiyi, Deng Na(✉), and Zheng Cheng

School of Computer Science, Hubei University of Technology, Wuhan 430068, China
iamdengna@163.com

Abstract. It is very important for companies and individuals to be able to judge the trend of change in a class of jobs. The number of patent applications and disclosures has been growing in recent years. In order to obtain the evolution of job trends in a field, the necessary reference objects are indispensable. Patents represent the latest technology in a field, and changes in the number of patents are a good reference for the development of a specific filed. In this paper, mechanical R&D job is mapped to mechanical patents, and trends in job change are inferred from trends in patents. The trend of job evolution in a field can be effectively inferred by comparing the line graph of patent development with the line graph of the number of jobs employed in the mechanical R&D category.

1 Introduction

With the continuous development of society, jobs are in a constant state of change. Some jobs emerge with new technologies and gradually become more and more important, and some jobs evolve with technology and gradually disappear. For example, with the development of computer software and hardware over the years, the changes in computer-related jobs have slowly become saturated with talent from the urgent need for a large pool of talent. How to predict job changes has become an indispensable tool for companies or countries. Each company studies changes in demand for jobs to determine the amount of talent to bring in. It is becoming more and more important to study how to find a reference for job changes.

Traditionally, the number of employees need to hire for a company generally depends on the analysis of experts, who artificially judge the demand for jobs based on the needs of their own companies. Current approaches to the evolution of jobs are generally based on statistical methods. A job evolution result is given by counting the number of current jobs and combining it with the individual experience of the expert. However, this statistical approach lacks information on technological changes and ignores the core technological changes needed for the development of a company, lacking a reference on technological changes.

Innovation is the first driving force that leads development, and protecting Intellectual Property Rights (IPR) is actually protecting innovation. As China has placed more emphasis on innovation, more attention is paid on the protection of IPR, and the number of patents' application has increased year by year.

L. Barolli (Ed.): EIDWT 2023, LNDECT 161, pp. 33–43, 2023.
https://doi.org/10.1007/978-3-031-26281-4_4

For a company, patent represents its core competitiveness, and possessing high-quality patents can make it stand out in the face of intense competition; furthermore, the need for new technology means the need for new talents. Therefore, the evolution of a company's position is closely related to the development of patents. Patent information, on the other hand, represents the technological evolution of a company or an industry. The evolution of an industry can be judged by changes in patents in an industry and whether the industry still has growth potential. For example, Lin [1] et al. analyzed that Kodak's bankruptcy was related to the development of Apple's camera industry by studying the changes in Kodak's and Apple's camera-related patents. Apple's camera patents soared after 2007, while Kodak's were relatively slow, leading to Kodak's gradual withdrawal from competition in the camera industry. For companies, the emphasis on innovation, patent evolution and talent acquisition can sustain the company's vitality.

In this paper, we study the mapping of jobs to patents to represent the technological change of a job and predict the evolutionary trend of jobs through technological change.

2 Related Work

Predicting the evolution of jobs has been an important problem for research in countries and companies. In particular, the evolution of technological jobs depends on technological development. While a company cannot grow without service jobs, technological innovation and technical jobs are fundamental for a company to maintain sustainable growth. The way most people look for work today is via the Internet. A Google search for the term "job" implies that there are a portion of unemployed population are looking for jobs. The Google search for jobs covers macroeconomics and can be used as a reference for trends in the evolution of jobs to some extent [2]. However, quantitative statistics of the search for jobs through the willingness of individuals can only reflect the demand for jobs, but can not effectively reflect the talent positions required by a company.

When considering the impact of technology on job evolution trends, Artificial intelligence has been successfully applied in many fields. Employee's job-hopping also affects the change of jobs in a company. Nathan Kosylo [3] proposed a novel artificial intelligence technique, i.e., sequential optimization with Naive Bayes. It uses several important job-hopping characteristics as predictive data and successfully predicts the job-hopping patterns of employees based on their profiles. To study the future job needs and requirements, Shravni Satish Kumar [4] et al. tried to develop an application program to predict the future jobs which can be used to understand the human resources (HR) companies, partners and future training as a way to meet the future job needs in advance.

TAEHYUN HA [5] et al. used the description of IPC (International patent classification) to map different jobs to patents and infer the trend of job change through the trend of patents. This approach takes into account the effect of technology on job changes, but the description of patent classification numbers is relatively abstract and vague and can not represent all patents well.

To solve the insufficiency in [5], this paper proposes a method for job evolution analysis. It uses technology and efficacy extracted from patent abstract texts to represent the category of patents, builds links between job description and patents, and infer job development trends based on patent trends.

2.1 BiLSTM-CRF

In this paper, we take the example of patents and technical jobs in the area of machinery. In order to investigate how jobs changes with technology, we establish relationships between jobs and technology and efficacy entities of patents.

Technology and efficacy extraction is a task in named entity recognition (NER) and NER is a basic task in natural language processing. Named entity recognition is to extract entities from unstructured text such as technologies and efficacy. Technology and efficacy extraction can be used for patent retrieval, patent infringement detection [6] and patent similarity calculation.

The technology, efficacy and mechanical components represent the core of a mechanical patent. In this paper, a bidirectional long short-term memory model paired with conditional random field model, namely, BiLSTM-CRF is used to achieve technology, efficacy and composition extraction from mechanical patents.

In this paper, the bidirectional short-term memory model and conditional random field (BiLSTM-CRF) model are used to extract the technology, efficacy and components of mechanical patents. Hochreiter and Schmidhuber [7] proposed LSTM model, which is a special neural network model and can more effectively obtain context information. A inverse LSTM is added to the LSTM model, and a bidirectional LSTM model is used to obtain contextual information before and after. The BiLSTM model can obtain information about two sequences: front-to-back and back-to-front. The combination of the two sequences information allows better access to semantic information and improves the accuracy and efficiency of mechanical patent work extraction.

Lafferty [8] et al. proposed the conditional random field, CRF, in 2001. Conditional stochastic domain is a discriminative model suitable for prediction-type tasks and has been successfully applied in NER, lexical annotation and other domains with promising results in recent years. CRF can optimize the label score of a word at the sentence level to select the most reasonable score and improve the accuracy of efficacy extraction across the technique.

2.2 Word2vec

The transformation of an incomputable and unstructured word into a computable and structured vector is called word embedding. Word2vec is a new word embedding method proposed by Google [9] in 2013. Word2vec addresses the explosion of vector dimensionality that occurs in traditional one-hot coding. The model is able to obtain word vectors by learning contextual semantic information. It maps each word from a high-dimensional word vector space to a low-dimensional space by learning, and preserve the positional order relation between word vectors, thus solving the data sparsity and semantic missingness problems. Word2vec contains two training methods, a continuous bag-of-words model, namely CBOW, and a skip-word model, namely skip-word model. As shown in Fig. 1.

Word2vec is a lightweight neural network model, which is divided into three layers, that is, input layer, prediction layer and output layer. CBOW and skip-gram are distinguished by the difference between input and output. CBOW model predict the current word w_t by inputting the context $w_{t-2}, w_{t-1}, w_{t+1}, w_{t+2}$. Conversely Skip-gram model is used by entering the current word w_t to predict the context $w_{t-2}, w_{t-1}, w_{t+1}, w_{t+2}$.

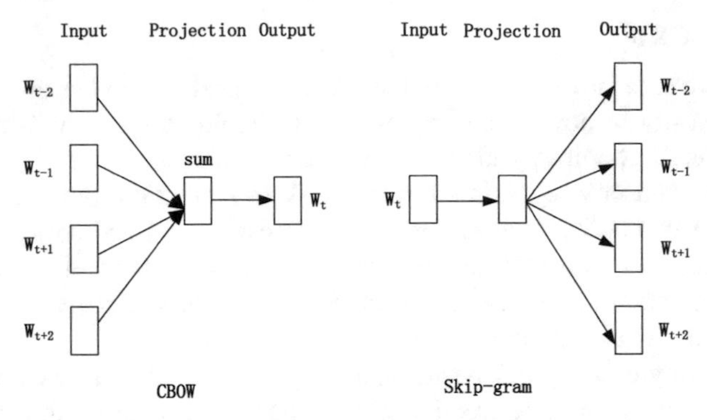

Fig. 1. Word2vec training model diagram

3 Job Prediction Model

In this paper, we build the R&D job evolution model BiLSTM-CRF+Word2vec with the structure of Fig. 2 based on the related studies mentioned above. Firstly, texts are labeled and the data is pre-processed. Then, the BiLSTM-CRF model is used as input to the Word2vec model to extract the technology and efficacy entities of the mechanical patents, and the word vectors of the technology and efficacy entities are obtained. Finally, jobs are mapped to patents by similarity calculation.

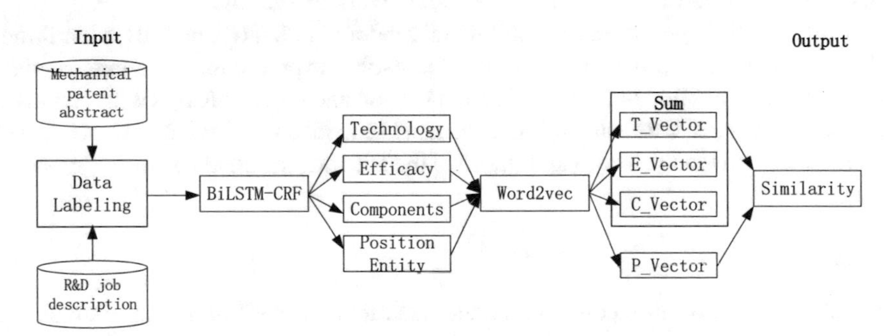

Fig. 2. Model of the evolution of R&D jobs

3.1 Data Pre-processing

(1) Data labeling. We use Baidu AI [10] to label the technology, efficacy and components of the patent abstract and job description. The labeling tags are technology, efficacy and component respectively.

(2) Data parsing and label building. Parse the JSON file generated by Baidu AI after the data labeling is completed. The labeled technology and efficacy entities are extracted from the JSON file, and the location information is resolved at the same time. Finally, a label information file corresponding to each word is created as a training and testing dataset for the model.

3.2 BiLSTM-CRF Neural Network Model Construction

The model consists of an input layer, an embedding layer, a BiLSTM layer and a CRF layer. The output of the previous layer is used as the input of the next layer. The input data to the input layer is the training and testing data sets after data preprocessing. The embedding layer transforms the incoming data from the input layer into a low-dimensional vector that can be recognized and computed by the computer. The BiLSTM layer extracts sentence information, and BiLSTM can record sentence semantic information from front-to-back and back-to-front, thereby improving the accuracy of technical efficacy extraction. The last layer is the CRF layer, whose input is the label score of each word output by the BiLSTM layer. The CRF layer can label the entire sentence information and address the adjacent label dependency problem, thus correcting the error label score for each word. Finally, the BiLSTM-CRF model is used to extract the technology, efficacy and component entities of patent abstracts and job descriptions in the dataset. As shown in Fig. 3.

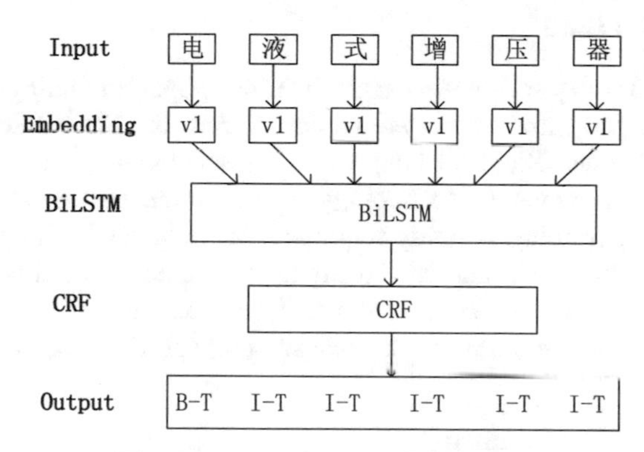

Fig. 3. Model of the BiLSTM-CRF.

3.3 Text Vectorization Representation

We use the CBOW in the Word2vec model to transform the technical and efficacy entities obtained from the upper layer into low-dimensional vectors. The technology and efficacy entities extracted from the BiLSTM-CRF model are trained as vectors with Word2vec. In order to expand the training sample and increase the training effect, the training set contains repeated technology and efficacy entities. After training, the de-duplicated technology, efficacy and component entities, and job description entities are fed into the Word2vec module again to obtain vectors for each technology efficacy, component, and job description. The output technology and efficacy vectors are averaged to represent vectors of a category of patents, and the job description vectors are averaged to represent vectors of a category of R&D jobs.

3.4 Similarity Calculation

Technology and efficacy vectors and job vectors are used as input of the similarity calculation. Suppose the technology efficacy vector is $A = (A_1, A_2, \ldots, A_n)$, and the job vector is $B = (B_1, B_2, \ldots, B_n)$. Then the similarity between patent and job is calculated using Eq. (1).

$$COS(\theta) = \frac{\sum_{i=1}^{n}(A_i \times B_i)}{\sqrt{\sum_{i=1}^{n}(A_i)^2} \times \sqrt{\sum_{i=1}^{n}(B_i)^2}} \quad (1)$$

By mapping the R&D jobs to patents through Eq. (1), the future evolution trend of overall R&D jobs is obtained by analyzing the evolution trend of patents in recent years.

4 Experimental Results and Analysis

4.1 Experimental Data

The data source for the job evolution analysis in this paper, including 1077 mechanical patents and R&D job information, are captured from the State Intellectual Property Office of China [11] and XiaoMuChong website respectively.

To ensure the accuracy of the technology and the efficacy extraction of the BiLSTM-CRF model, model training is firstly required. The data set in the patent technology efficacy extraction module is annotated using the BIO labeling method, which indicates the beginning, middle and non-entity of the entity, respectively. After that, the patent and job datasets are divided into training set, test set, and validation set according to 8:1:1.

4.2 Experimental Environment

This paper experiments with a machine with Windows operating system, Intel Core i7 CPU model, RTX 2060 graphics card model, and 16G video memory. Python version 3.8 and Torch 1.12.0 frameworks were built on the machine.

The technology efficacy extraction works best when the BiLSTM-CRF model training parameters are listed in Table (1) below.

Table 1. Experimental parameters.

Parameters	Value
Batch_size	64
Learning rate	0.001
Epoches	100
Hidden size	128

4.3 Technical Efficacy Extraction Results

The accuracy of technology and efficacy extraction accuracy affects the accuracy of the subsequent vectorization and similarity calculations. The accuracy of technology and efficacy extraction can more accurately represent a category of patents and map R&D jobs to patents more effectively, which is beneficial for inference studies of subsequent job evolution. The best accuracy of technology and efficacy extraction is shown in Table 2.

Table 2. Technical and efficacy extraction accuracy.

Tag	Accuracy	Recall rate	F1
I-T	79.36	86.13	82.61
B-T	81.52	64.10	71.77
I-C	84.77	85.02	84.89
B-C	84.25	81.64	82.92
I-E	75.76	67.16	71.20
B-E	82.81	60.23	69.74
O	96.58	96.80	96.69
Avg	94.00	94.03	93.99

Where the precision is calculated as follows:

$$P = \frac{T_p}{T_p + F_p} \quad (2)$$

The recall is calculated as:

$$R = \frac{T_p}{T_p + F_n} \quad (3)$$

The formula for F1 is:

$$F_1 = \frac{2PR}{P + R} \times 100\% \quad (4)$$

Here T_p is the number of technology and efficacy entities correctly identified by the model, F_p is the number of technology and efficacy entities identified incorrectly, and F_n is the number of technology and efficacy entities not identified by the model.

4.4 Word Embedding

The Word2vec model training parameters are listed below (Table 3):

Here, sg is 0, which means that the word2vec model is trained using the CBOW algorithm for data vectorization, and vector_size of 100 indicates that the training word vector dimension is 100. A window of 3 means that the maximum distance between the

Table 3. Word2vec model parameters.

Parameters	Value
sg	0
vector_size	100
window	3
min_count	1
workers	4
hs	0

current word and the predicted word in a sentence is 3, and a min_count of 1 means that words with word frequency less than 1 will not participate in the training. The number of workers represents the number of parallel training. If hs is 0, the negative sampling technique is used.

To verify the accuracy of the word vectors trained using the word2vec model, the similarity between the extracted patent components was calculated. The similarity between "piston" and "piston rod" is 99.7%, which indicates that the word2vec word vector training is relatively accurate and can be used as the next input to the model in this paper.

4.5 Similarity Calculation

In order to map R&D jobs to the corresponding patents, the similarity between patents and jobs needs to be calculated. In this paper, 1077 mechanical patents are selected and the BiLSTM-CRF model is used to extract the technology, efficacy, and component entities of each patent. These entities represent each patent. This mechanistic patent is represented by averaging over extracted technology and efficacy vectors. Meanwhile, 20 job descriptions of the mechanical R&D category are selected and the description entities of the jobs that can represent the jobs are extracted. The average vector of job entities is then obtained to represent this category of jobs. The average vector similarity between patents and jobs is obtained as follows.

Table 4. Similarity results.

Models	Similarity
Cosine similarity	96.8%
Pearson correlation coeffic	93.8%

From Table 4, it can be concluded that there is a high match between the patents in the mechanical domain and the job in the mechanical R&D category, and the mechanical patents can be used as a technical reference for the jobs. The evolution of mechanical R&D jobs can be inferred from the evolution of patents.

4.6 Patent Development

In this paper, we count the number of patents in the field of machinery disclosed from 2000 to 2021 from the website of the State Intellectual Property Office of China, as shown in the following table.

Table 5. Number of patent disclosures.

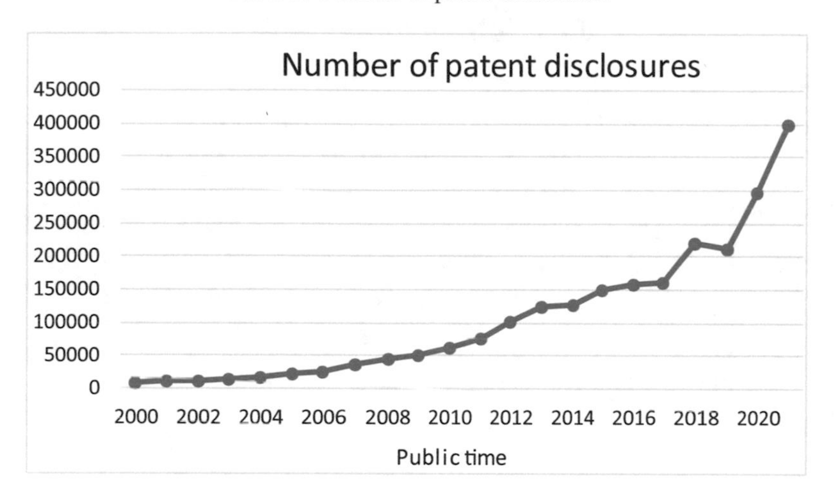

According to Table 5, the number of patent disclosures in the mechanical sector in China grew slowly before 2017, while the growth rate of patents in the mechanical sector increased significantly after 2017. Starting from 2017, China has experienced great development of new energy vehicles. Since the development of the auto industry cannot be separated from the development of machinery, it leads to a significant increase in the number of patents in the machinery sector. An increase in patents indicates an increase in demand for talents.

4.7 Evolution of Mechanical R&D Jobs

With the development of new energy vehicles in China, the demand for highly skilled personnel is increasing. 450 companies are hiring mechanical engineers in the R&D category above graduate level in 2022 on Boss [12]. The statistics of job postings for R&D PhDs for the last 3 years on the official website of XiaoMuChong are shown in Table 6. The demand for R&D jobs is 198 in 2019, 278 in 2020 and 389 in 2021.

Comparing Tables 5 and 6, it can be concluded that when the number of patents in the machinery field increases, the demand for R&D jobs in the machinery industry becomes larger. Patents represent the latest technology in an area, new technologies represent new vitality, business development cannot be separated from new technologies, and the emergence of new technologies cannot be separated from R&D talent. The number of patents in China's machinery field has exploded in the past three years, while the recruitment of R&D personnel in machinery has also increased. Therefore, it can be assumed that when the number of patents increases, the corresponding demand for R&D talent also increases. It can be seen through Table 5 that an increase in the number of patents in the mechanical sector means companies can recruit more mechanical engineers.

Table 6. Job evolution.

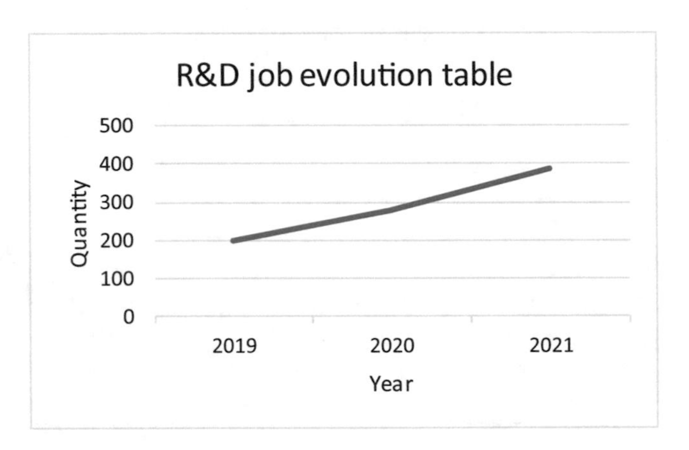

5 Conclusion

A significant increase in the number of patents means that an industry will flourish, more jobs will be created and more talent will be needed. The research in this paper maps R&D jobs to patents and infers the future evolution of jobs by analyzing the trends of patents. While this paper is only able to analyze changes in job development in one field, the next study will focus on the evolution of different jobs by analyzing patents in different fields.

References

1. Lin, Y., Sung, H.C., Yang, W.A., et al.: Identifying future competitors with patent text mining. In: Pacific Asia Conference on Information Systems 2022 (2022)
2. Borup, D., Cheistian, E.C.M.: In search of a job: forecasting employment growth using google trends. J. Bus. Econ. Statist. 40(1), 186–200 (2022)
3. Kosylo, N., Smith, J., Conover, M., et al.: Artificial intelligence on job-hopping forecasting: AI on job-hopping. In: 2018 Portland International Conference on Management of Engineering and Technology (PICMET), pp. 1–5. IEEE (2018)
4. Kumar, S.S., Adeoye, S., Bokhare, A., et al.: Predicting the work opportunities with proposed model: HR FUTURE of WORK 2030. Int. J. Innov. Sci. Res. Technol. 1362–1368 (2020)
5. Ha, T., Lee, M., Yun, B., et al.: Job forecasting based on the patent information: a word embedding-based approach. IEEE Access 7223–7233 (2022)
6. Lv, X.Q., Luo, Y.X., et al.: A review of chinese patent infringement detection. Data Anal. Knowl. Discov. **5**(03), 60–68 (2021)

7. Hochreiter, S., Schmidhuber, J.: Long short-term memory. Neural Comput. **9**(8), 1735–1780 (1997)
8. lafferty, J., Mccallum, A., Pereira, F.: Conditional random fields: probabilistic models for segmenting and labeling sequence data. In: Proceedings of the 18th International Conference on Machine Learning, pp. 282–289 (2001)
9. Kolesnikova, O., Gelbukh, A.: A study of lexical function detection with word2vec and supervised machine learning. J. Intell. Fuzzy Syst. **39**(2), 1993–2001 (2020)
10. Xu, J.: Baidu AI Cloud leads the intelligent upgrading of AI industry. Nan Fang Daily **24**(T09) (2022)
11. Zhang, Y.X.: State Intellectual Property Office: more than 1 million intellectual property talents will be available in 2025. Guangming Daily **31**(008) (2022)
12. Li, N.Y.: New opportunities for young people to find jobs. China Youth News **12**(001) (2022)

The Models of Improving the Quality of Government Financial Reporting

Edy Suprianto(✉), Dedi Rusdi, and Ahmad Salim

Universitas Islam Sultan Agung, Semarang, Indonesia
edysuprianto@unissula.ac.id

Abstract. This study will examine the factors that influence the success of the Semarang City Government in applying big data-based technology to increase accountability and transparency of local government financial reports. The results of the Audit Board of the Republic of Indonesia (henceforth, BPK) in the last five years show that the Semarang City Government has experienced significant developments. Even the Semarang City Government is the only local government that has been able to apply the accrual accounting basis in the province of Central Java, Indonesia. This study used 50 Regional Apparatus Organizations (OPD) in the Semarang City Government. The results of the study indicate that government accounting standards and internal control systems have a positive effect on Big Data. In addition, government accounting standard, internal control system, and Big Data have a positive effect on quality of financial report.

Keywords: Government accounting standard · Quality of Government Apparatus · Internal control system · Performance of local government financial report

1 Introduction

Government Regulation No. 71 of 2010 concerning Government Accounting Standards has mandated that all local governments must be able to implement these government accounting standards properly and correctly. This regulation has amended the previous Government Regulation No. 24 of 2005 which was based on accounting. Initially, local governments in the preparation of financial reports used a cash accounting basis. However, with the emergence of the new regulation, local governments must implement the accrual basis starting in 2015.

Based on the results of the observation of the Semarang City Government Financial Reports for the Fiscal Year 2007 to 2019 listed in Table 1 shows a fairly good trend. Starting with a fair opinion with an exception in 2007 to obtaining an unqualified opinion in 2019. These results show the hard work of all OPD and Regional Financial Management Officials (PPKD) in the Semarang City Government in understanding and implementing government accounting standards in preparation of local government financial reports (LKPD). This condition is of course a good example and can be an inspiration for other local governments in Indonesia in general or in the province of Central Java in particular.

L. Barolli (Ed.): EIDWT 2023, LNDECT 161, pp. 44–51, 2023.
https://doi.org/10.1007/978-3-031-26281-4_5

One of the factors that play a role in the success of the Semarang City Government in obtaining the current Unqualified (WTP) is the information technology factor. For more than 10 years, the Semarang City Government has implemented big data-based technology, namely the E-Reporting application. According to Effendy (2020), Big Data in the government system can create various policies that are faster, more accurate and cheaper with various institutions in the government. Technology plays a very important role in improving the quality of local government financial reports (Suprianto 2017). However, not all local governments are able to apply this big data system well. The government needs management and governance of big data-based system.

There are several factors that affect governance and management in using Big Data, such as: 1) Principles, policies and frameworks; 2) Process; 3) Organizational structure; 4) Culture, ethics and behavior; 5) Information; 6) Services, infrastructure and applications; and 7) HR, capabilities and competencies. There needs to be a synergy and integration between the seven factors to be able to create an application system that is beneficial for the institution.

Another important factor is the implementation of SAP itself. BPK as an auditor for local government financial reports makes this SAP as the basis of reference in verifying the fairness of financial reports. Nugraheni & Subaweh (2020) researched the implementation of SAP at the Inspectorate General in UAPPA E1 and UAPPB. He concluded that the application of SAP had a significant positive effect on the quality of financial reports. These results are also supported by research conducted by Suprianto (2015), that the application of SAP has a significant effect on the quality of financial reports.

Government Regulation Number 60 of 2008 concerning the Government's Internal Control System states that the Financial Reports presented by local governments must be based on an adequate internal control system and in accordance with Government Accounting Standards. The control function is carried out by the Regional Head through the Internal Control System. The internal control system is an integral process for actions and activities carried out continuously by the leadership and all employees to provide adequate confidence in the achievement of organizational goals through effective and efficient activities, reliability of financial reporting, safeguarding state assets, and compliance with laws and regulations. Suprianto (2014) conclude that the internal control system has a significant positive effect on the quality of local government financial reports.

This study will examine the factors that influence the success of the Semarang City Government in applying big data-based technology to increase accountability and transparency of local government financial reports. This research is important to study because it can provide input to local governments in Indonesia in implementing big data-based technology systems to be more efficient, effective and optimal.

2 Theory and Hypothesis

Big Data is a large collection of data, whether structured, semi, or unstructured so that it cannot be processed using ordinary relational database tools (Syamsurizal 2016). The emerging data has the opportunity to provide a policy guide without ever realizing it. Big Data is a technology trend to take a new approach to understanding the world and making business decisions.

Technology has an important role in the continuity of information, starting from information being created to being destroyed. Successful enterprises treat IT as a significant part of executing business processes. Business processes and IT must collaborate and work together so that IT can enter into governance and management. Factors that affect the governance and management of enterprise IT, namely: 1) Principles, policies and frameworks; 2) Process; 3) Organizational structure; 4) Culture, ethics and behavior; 5) Information; 6) Services, infrastructure and applications; and 7) HR, capabilities and competencies.

Big Data harvesting is expected to produce quality information that supports decision making. Good information quality will result in good organizational decision results so that it will increase profits for the enterprise. Big Data is a process of collecting data to find patterns and correlations that may not be obvious at first, but have the potential to be useful in business decision making. Hakim & Maulana (2019) concluded that the use of information technology can support the quality of financial reports. This statement is supported by Eveline (2017), Wulandari & Bandi (2015), and Kamela & Setyaningrum (2020).

Government accounting standards (SAP) is a reference for compilers of financial report in tackling accounting problems that have not been regulated in the standard. SAP are accounting principles applied in preparing and presenting government financial report. Thus, SAP is a requirement that has legal force in an effort to improve the quality of government financial reports in Indonesia. With the implementation of good SAP will produce local government financial reports that are better and orderly in administration and principles. Suprianto (2014) shows that the good application of government accounting standards can increase quality of local government financial reports. Juwita (2013) also concluded that SAP has a positive effect on the quality of financial reports.

The internal control system is an integral process for actions and activities that are carried out continuously to provide adequate assurance for the achievement of effectiveness and efficiency in achieving the objectives of state government administration, reliability of financial reporting, safeguarding state assets, and compliance with laws and regulations. Inayattulloh & Siswantoro (2020) also concluded that SPI has a positive effect on the quality of financial report. Based on the explanation above, the hypotheses in this study are:

H1: The application of Government Accounting Standard has a positive effect on the quality of financial report.

H2: The application of Government Accounting Standard has a positive effect on Big Data-based systems.

H3: The Internal Control System has a positive effect on the quality of financial report.

H4: The Internal Control System has a positive effect on Big Data based system.

H5: The application of Big Data-based system has a positive effect on the quality of financial reports.

We have conducted research on the quality of government financial reports starting in 2010 on the Model for Empowerment of Strategic Human Resource Management, Information Technology, and Implementation of Good Governance as Added Values in Increasing the Effectiveness of the Performance of the Urban Independent Community

Empowerment National Program (Pnpm) for Alleviating Poverty. Suprianto & Nugroho (2010) believed that it is necessary to pay attention to the factors of Strategic Human Resource Management, Information Technology, and Implementation of Good Governance to succeed the national program of empowering independent communities in urban areas in alleviating poverty. Furthermore, Suprianto (2014) found the importance of technology in the implementation of governance to improve local government performance. In 2015, Suprianto (2015) examined the effect of internal control in improving the quality of local government financial reports. Based on the study, it recommends that improving the quality of local government financial reports need to pay attention to the role of the internal control system. Rusdi & Suprianto (2016, 2017) also added the need for the use of information technology in improving the quality of local government financial reports. In 2021, it is important to research further on the factors that influence the use of big data-based systems to increase the accountability and transparency of the Semarang City Government Financial Reports (Fig. 1).

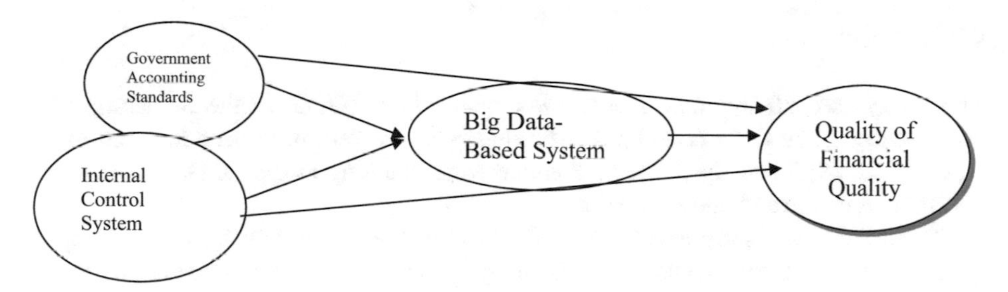

Fig. 1. Research model

3 Methodology

This study used 50 Regional Apparatus Organizations (OPD) in the Semarang City Government. The sample was obtained using the census method so that it could find out all relevant information about the role of each OPD. The unit of analysis is the Head of OPD and Head of Finance & Accounting at OPD.

This study used primary and secondary data. Primary data is in the form of questionnaires distributed directly to Government Accounting Standards (SAP), Quality of Government Apparatus, Internal Control System (SPI) and Quality of Financial Reports. While secondary data in the form of financial reports, audit results, organizational data etc. obtained from the OPD accounting/bookkeeping/finance staff, the head of the OPD finance sub-section, the head of the OPD Administration/Secretariat.

The quality of financial reports variable is measured by relevant, reliable, comparable and understandable indicators (SAP, 2010). While the internal control system variable is measured by indicators of the control environment, risk assessment, control activities, information and communication, and monitoring. Government Accounting Standards Variables measured by indicators of Accounting Base, Historical Value, Realization,

Substance Outperforms Formal Form, Periodicity, Consistency, Complete Disclosure, and Report Presentation.

The technique used to find the value of the regression equation is Least Square analysis by minimizing the sum of the squares of errors. The regression equation is formulated as follows:

$y = a + b_1x_1 + b_2x_2 + b_ex_e + e$

Which:

$y =$ Quality of Financial Report.

$x_1 =$ Government Accounting Standards.

$x_2 =$ Big Data-Based System.

$x_3 =$ Internal Control System.

$a =$ Regression Constant.

$b_1b_2b_3 =$ Regression Coefficient Variables.

e = error.

4 Results

This study used 50 Regional Apparatus Organizations (OPD) in the Semarang City Government. Each OPD is sent 3 questionnaires to answer questions addressed to the Head of the OPD and the Head of Finance & Accounting at the OPD. The returned questionnaires were 85 questionnaires.

Table 1 shows descriptive statistical data for each variable. For the financial report quality variable, the average is 4.814, meaning that the average respondent has a perception that the OPD financial reports have been prepared in an accountable and transparent manner. The average value of Big Data variable is 4,140, meaning that the OPD has applied the financial reporting system with big data optimally. The average SAP variable is 4.526, meaning that OPD has understood the current SAP, while the SPI variable is 4.550, meaning that OPD has an adequate internal control system.

Table 1. Descriptive statistics

Information	Mean Min	Max	Standard deviation
Quality_FR	4,814 1	5	0.390
Big Data	4,140 1	5	0.347
SAP	4,526 1	5	0.154
SPI	4.550 1	5	0.187

Based on the results of the regression test in Table 2, it is found that the coefficient of the SAP variable is 0.12 with a significant level of at the level of 0.01, while the SPI variable also shows a coefficient value of 0.23 with a significant level of 0.02. Thus, SAP and SPI have a positive effect on Big Data, so H1 and H2 are accepted. The results of this study support previous research conducted by Suprianto (2014), that the application

Table 2. First regression test

Independent variable	Predict sign	Dependent variable (Big Data)	
		(β)	(Prob)
Constant	?	0.02	0.00
SAP	+	0.12	0.01
SPI	+	0.23	0.02

of an information system-based accounting system must be supported by the maximum application of SPI and SAP. Besides, it is supported by Divine & Alia (2017), that good financial management requires cooperation between three state audit institutions, namely BPK, KPK and BPKP in controlling the government control system. (Surepno 2015) also added that the success of the Semarang City Government in implementing accrual-based accounting standards can help the quality of the Semarang City Government's financial reports which are based on integrated big data.

Big Data is expected to produce quality information that supports decision making. Good information quality will result in good organizational decision results so that it will increase profits for the enterprise. Big Data is a process of collecting data to find patterns and correlations that may not be obvious at first, but have the potential to be useful in business decision making. Hakim & Maulana (2019) concluded that the use of information technology can support the quality of financial reports. This statement is supported by Eveline (2017), Wulandari & Bandi (2015), and Kamela & Setyaningrum (2020).

Table 3. Second regression test

Independent variable	Predict sign	Dependent variable (Quality_FR)	
		(β)	(Prob)
Constant	?	0.08	0.00
SAP	+	0.90	0.01
SPI	+	0.12	0.00
Big Data+		0.11 0.01	

Based on the results of the regression test in Table 3, the results show that the SAP variable coefficient is 0.90 with a significant level of 0.01, while the SPI variable also shows a coefficient value of 0.12 with a significant level of 0.00, while the Big Data variable shows a coefficient value of 0.11 with a significant level of 0.01. Thus, SAP, SPI, and Big Data have a positive effect on the quality of financial reports, so it indicates H2, H4, and H5 are accepted. The results of this study support previous research conducted Suprianto (2014), that the application of good government accounting standards and good SPI can increase quality of local government financial reports. This result is also supported by Juwita (2013), Inayattulloh & Siswantoro (2020), Eveline (2017),Wulandari & Bandi (2015), and Kamela & Setyaningrum (2020).

Government accounting standards (SAP) is a reference for compilers of financial reports in tackling accounting problems that have not been regulated in standards. SAP is accounting principles applied in preparing and presenting government financial reports. Thus, SAP is a requirement that has legal force in an effort to improve the quality of government financial reports in Indonesia. With the implementation of good SAP will produce local government financial reports that are better and orderly in administration and principles. Suprianto (2014) shows that the good application of government accounting standards can increase quality of local government financial reports. Juwita (2013) also concluded that SAP has a positive effect on the quality of financial reports.

Internal control System is an integral process for actions and activities that are carried out continuously providing adequate assurance for the achievement of effectiveness and efficiency in achieving the objectives of state government administration, reliability of financial reporting, safeguarding state assets, and compliance with laws and regulations. The implementation of good SPI will produce better financial reports. Suprianto (2014) show that internal control system has a significant positive effect on the quality of local government financial reports. Inayattulloh & Siswantoro (2020) also concluded that SPI has a positive effect on the quality of financial reports.

5 Conclusion

Based on the results of the analysis, it can be concluded that government accounting standards and internal control systems have a positive effect on Big Data. In addition, government accounting standards, internal control systems, and Big Data have a positive effect on the quality of financial reports. The limitations of this study are *first*, the sample of this study is only limited to the Semarang City Government, future research needs to analyze other local governments. *Second*, this research only used a quantitative approach with online questionnaires without direct interviews with respondents, further research needs to analyze qualitatively more deeply with FGD methods and so forth.

References

Almunawwaroh, Marliana: The influence of CAR, NPF and FDR on profitability of islamic banks in Indonesia. J. Islamic Econ. Finan. **2**(1), 1–18 (2018). https://doi.org/10.31289/jab.v6i1.3010

Effendy, T.: Implementation of big data in government agencies (2020)

Eveline, F.: The influence of accrual-based sap, accounting information systems, quality of human resources, internal control and organizational commitment to the quality of financial reports at the national disaster management agency. Media Res. Account. Audit. Inf. **16**(1), 1 (2017). https://doi.org/10.25105/mraai.v16i1.2004

Hakim, L., Maulana, T.I.: Effectiveness of Integrated Financial Management and Information System (IFMIS) in central government financial reporting process. J. Account. Bus. **19**(1), 80 (2019). https://doi.org/10.20961/jab.v19i1.328

Illahi, B.K., Alia, M.I.: Accountability for state financial management through BPK and KPK cooperation. Integrity **3**(2), 37 (2017). https://doi.org/10.32697/integrity.v3i2.102

Inayattulloh, M.R., Siswantoro, D.: The effect of budget management quality and internal audit to financial statement quality in the ministries and agencies. J. Account. Bus. **19**(2), 218 (2020). https://doi.org/10.20961/jab.v19i2.431

Juwita, R.: The effect of implementation of government accounting standards and accounting information systems on the quality of financial statements. Trikonomika **12**(2), 201 (2013). https://doi.org/10.23969/trikonomika.v12i2.480

Kamela, H., Setyaningrum, D.: Do political factors affect financial performance in public sector? Indonesian Account. Finan. Res. **5**(2), 202–209 (2020). https://doi.org/10.23917/reaction.v5i2.11002

Suprianto, E.: The role of e-governance in increasing of local government performance (Case Study In Demak Province, Central Java, Indonesia). Int. J. Bus. Econ. Law **4**(1), 49 (2014). https://doi.org/10.28992/ijsam.v1i2.16

Suprianto, E.: The problem of Islamic banking. J. Bus. Econ. **12**(3), 1–9 (2017)

Suprianto, E.: Factors that affect non-performing loans in Islamic banking. Indonesian J. Bank. Account. **9**(1), 1–12 (2018). https://doi.org/10.20885/jaai.vol19.iss1.art1

Suprianto, E.: The role of the sharia supervisory board in improving the performance of sharia banking in Indonesia. Indonesian J. Account. Audit. **24**(1), 33–42 (2019). https://doi.org/10.20885/jaai.vol24.iss1.art4

Suprianto, E., Nugroho, M.: Empowerment Model of Strategic Human Resource Management, Information Technology, and Implementation of Good Governance as Added Values in Increasing the Effectiveness of the Performance of the Urban Independent Community Empowerment National Program (Pnpm) for M (2010)

Surepno: Key to success and role of accrual-based accounting implementation strategy. J. Account. Dyn. **7**(2), 119–128 (2015)

Syamsurizal: The effect of CAR (Capital Adequacy Ratio), NPF (Non Performing Financing), and BOPO (Operational Cost of Operating Income) on ROA (Return On Assets) at BUS (Islamic Commercial Banks) Registered with BI (Bank Indonesia). J. Socio-Religious Res. **19**(2), 151–173 (2016)

Wulandari, I., Bandi, B.: The influence of e-government, Apip capability and percentage of completion of follow-ups on audit opinions of local government financial statements in Indonesia. J. Account. Bus. **15**(2), 148 (2015). https://doi.org/10.20961/jab.v15i2.184

Fuzzy Mean Clustering Analysis Based on Glutamic Acid Fermentation Failure

Chunming Zhang(✉)

Engineering University of PAP, Xian 710086, China
1085798666@qq.com

Abstract. In the process of glutamic acid fermentation, the quality of the products fluctuates greatly, and faults and errors are not easy to be found in the early stage, which easily leads to waste of raw materials and idling of equipment. This paper takes glutamate fermentation as the research object, according to the actual industrial production situation, puts forward a new state identification method, including the off-line state division of the system and the on-line state condition identification. Through the simulation study, the stability, reliability and effectiveness of the proposed method are proved, which has certain theoretical significance and practical application value.

1 Introduction

Because of the nonlinearity, uncertainty and complexity of modern industrial production process, in the actual industrial production process, the parameters or structure of the controlled object in the control system may change, which makes it difficult to achieve the expected results. Therefore, the safety and reliability of the production process become very important. At the same time, in order to ensure personal and production safety and improve the economic benefits of production, online real-time monitoring of the production process has become a very important research direction in the field of process control. At this time, if the system can automatically and effectively identify the operation status in the production process, people can accurately and quickly control the system, thus effectively preventing the occurrence of various accidents; At the same time, it also provides a good foundation for the research of adaptive optimization control of production process, process performance evaluation, process monitoring, process fault diagnosis, etc.

2 Research on the Running State of Glutamic Acid Fermentation Process

2.1 The Main Control Variables of Glutamate Fermentation

In the process variables of glutamate fermentation, temperature (T), pH value, revolution per minute (RPM), dissolved oxygen (DO), oxygen uptake rate (OUR), carbon dioxide generation rate (CER), ammonia flow addition, optical density (OD) were measured

L. Barolli (Ed.): EIDWT 2023, LNDECT 161, pp. 52–63, 2023.
https://doi.org/10.1007/978-3-031-26281-4_6

online. During the fermentation process, the feed sugar concentration, the residual sugar concentration of the fermentation liquor, the dry cell weight (DCW) and the glutamate concentration were measured offline. Among the measured variables in the glutamic acid fermentation process, the temperature and pH value are related to the cell growth in the glutamic acid fermentation, and are process variables used to control the normal growth of glutamic acid. The information contained can be expressed by the cell concentration, and the cell concentration and the cell dry weight have a linear relationship [1].

This paper analyzes and studies several important parameters that can be detected online at present. Temperature (T), pH value, dissolved oxygen, fed sugar concentration, residual sugar concentration of the fermentation liquor, optical density (OD) are the fault variables set in this experiment. Selecting a variable of glutamic acid bacterial concentration can fully express the effective information of the above measured variables. Therefore, glutamic acid concentration alone can be used as the basis for the division of working conditions.

2.2 The Steps of State Diagnosis in Glutamic Acid Fermentation Process

Step1. K-means clustering analysis is conducted on the data collected on site to distinguish the data of normal fermentation group from the data of abnormal fermentation group, which provides a basis for the following work condition identification.

Step2. The data of the divided abnormal fermentation group is processed by principal component analysis (PCA) to reduce dimension and complete feature extraction. Feature data is extracted from the glutamate fermentation process and preprocessed. In this simulation, we conduct normalization preprocessing on the data.

Step3. Establish the membership function, initialize the membership matrix and cluster center, and then conduct iterative calculation until convergence.

Step4. Identify the judgment, get a stable membership matrix and the best clustering center, and output the clustering results.

2.3 State Diagnosis of Glutamic Acid Fermentation Process

In the process of glutamic acid fermentation, the condition of fermentation liquor needs to be monitored in real time. Ammonia consumption will affect the change of pH value of fermentation liquor, causing the change of the whole fermentation environment and directly affecting the fermentation results [2]. While sugar depletion will affect the fermentation yield and reduce the production efficiency, and even cause the operation failure that fermentation cannot be carried out normally. Therefore, it is very important to collect and analyze the sugar consumption and ammonia consumption in real time and draw accurate conclusions for the safety production of the whole plant.However, the dissolved oxygen value of fermentation liquid can affect the consumption of both bacteria. The failure occurred in fermentation batch (1) is the dissolved oxygen failure caused by mechanical failure. However, it is difficult to directly measure the oxygen solubility in fermentation broth. For this reason, we selected five characteristic variables: residual sugar concentration, flow sugar addition, temperature, PH value and OD value, and collected 45 groups of data for simulation experiment, including 30 groups of normal data and 15 groups of fault data. Input 45 groups of data into the K-means clustering model to see whether the model can distinguish the two groups of data, so as to achieve the purpose of judging the fault state of fermentation process. Figure 1 shows the classification results under a fault condition. This is an experiment to collect data in case of batch (1) failure.

Input the data into the K-means clustering model to comprehensively analyze the fermentation state. The original normal data and fault data are shown in Tables 1 and 2 below.

Table 1. 30 groups of normal data of fermentation batch (1)

Test data group	Residual sugar (g/dL)	Flow sugar (KL)	Temperature (°C)	Optical Density (OD)	PH value
1	1.5	11.8	35.3	0.82	7.1
2	1.7	11.4	35.3	0.96	7.1
3	1.5	11.7	35.3	0.94	7.1
4	2	9.9	35.3	1.03	7.1
5	1.6	11	35.3	1	7.1
6	2	9.6	35.3	0.98	7.1
7	1.7	11.2	35.3	0.9	7.1
8	1.9	10.6	35.3	1.01	7.1
9	1.3	10.6	35.5	0.87	7.1
10	1.6	10.8	35.5	0.92	7.1
11	1.7	11.8	35.3	1	7.1
12	2	11	35.4	1.02	7.1
13	2.5	9	35.3	1.06	7.1
14	2.4	9.8	35.2	0.95	7.1
15	1.5	10.8	35.3	0.92	7.1
16	1.8	10.3	35.3	0.9	7.1
17	2	10.4	35.3	1.02	7.1
18	2.1	10	35.3	1	7.1
19	2	10	35.5	1.02	7.1
20	2.7	9.9	35.3	0.95	7.1
21	1.9	11.6	35.3	0.99	7.1
22	1.6	11.8	35.3	0.97	7.1
23	1.3	10.2	35.3	0.88	7.1
24	1.6	9.7	35.2	0.96	7.1
25	1.7	9.9	35.3	0.95	7.1
26	1.7	10.8	35.3	0.9	7.1

(*continued*)

Table 1. (*continued*)

Test data group	Residual sugar (g/dL)	Flow sugar (KL)	Temperature (°C)	Optical Density (OD)	PH value
27	1.8	12.1	35.4	0.82	7.1
28	1.6	11.9	35.3	0.82	7.1
29	1.9	11.7	35.4	0.96	7.1
30	1.5	12	35.3	0.98	7.1

Table 2. 15 sets of fault data of batch (1)

Test data group	Residual sugar (g/L)	Flow sugar (KL)	Temperature (°C)	Optical Density (OD)	PH value
1	0.7	13.8	32.1	0.01	8
2	0.8	13.5	32.1	0.03	8.2
3	0.6	13.2	31	0	8.4
4	0.9	12.9	30.5	0.02	8.3
5	0.5	13.1	30.8	1	8.5
6	0.3	13.8	31.2	0.35	8.4
7	0.7	13.3	31.3	0.01	8.6
8	0.7	13.5	30.8	0.05	8.3
9	0.7	13.3	32	0.13	8.5
10	0.6	13.7	31.5	0.13	8.8
11	0.3	12.9	30.8	0.2	8.7
12	0.5	12.6	31.3	0.31	8.7
13	0.5	12.8	32	0.21	8.6
14	0.6	13.1	31.3	0.13	8.9
15	0.7	12.7	32	0.15	8.5

Any group of data to be tested can be preliminarily judged by comparing the fault center. In the iteration process, the number of recursive cycles is 8, and the final cost function value is 31.9. It can be seen that without any prior knowledge, unlabeled data groups are divided into two groups with different characteristics. The grouping is based on the membership degree of the feature points to the cluster center mentioned above. The degree of membership of a data point to the cluster center indicates the degree of similarity between the data point and the state features represented by the cluster center. The membership relationship between the data points obtained after clustering and different state classes is shown in Table 3 below.

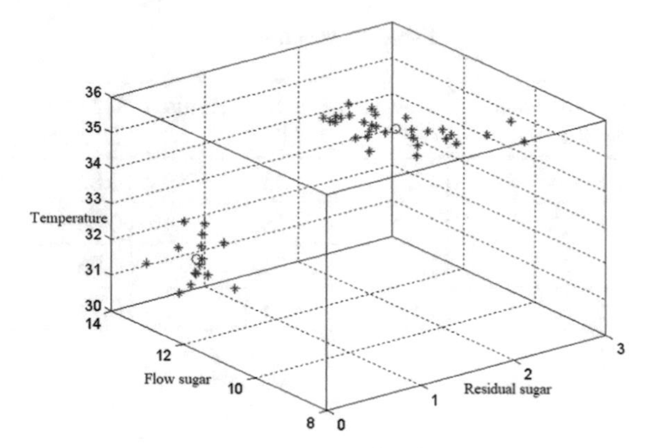

Fig. 1. Classification results of batch (1) under fault

Table 3. Membership of sample data to taxonomy

Data group \ Status class	Normal state	Sugar run out fault state	Data group \ Status class	Normal state	Sugar run out fault state
1	0.05	0.95	24	0.04	0.96
2	0.02	0.98	25	0.02	0.98
3	0.04	0.96	26	0.00	1.00
4	0.03	0.97	27	0.08	0.92
5	0.00	1.00	28	0.06	0.94
6	0.04	0.96	29	0.04	0.96
7	0.01	0.99	30	0.07	0.93
8	0.00	1.00	31	0.95	0.05
9	0.01	0.99	32	0.97	0.03
10	0.00	1.00	33	0.99	0.01
11	0.05	0.95	34	0.97	0.03
12	0.00	1.00	35	0.97	0.03
13	0.08	0.92	36	0.98	0.02
14	0.04	0.96	37	1.00	0.00
15	0.00	1.00	38	0.98	0.02
16	0.01	0.99	39	0.98	0.02
17	0.01	0.99	40	0.99	0.01
18	0.02	0.98	41	0.98	0.02
19	0.02	0.98	42	0.98	0.02
20	0.05	0.95	43	0.97	0.03
21	0.03	0.97	44	0.99	0.01
22	0.05	0.95	45	0.97	0.03
23	0.02	0.98			

From the membership degree table, we can see that 30 groups of data are classified under normal conditions and 15 groups are classified under fault. Obviously, the membership degree of the data points to the cluster center is above 0.9, and the accuracy is 100%.On the one hand, it shows that the model can well distinguish the data points collected in these two states; On the other hand, it also shows that these two data sets can better contain the normal and fault state characteristics. It can be used as the standard comparison data for judging the fault type in the future. According to the value of the fault center and the membership degree of the fault to the center, we can directly compare and judge the data to be tested, and draw a conclusion. Of course, when there are many data, it is necessary to input the model for analysis, and the identification is more rapid and accurate [3].

In the actual production process, there is not only the possibility of one kind of failure, but also the possibility of many kinds of failures. It is more difficult to distinguish. On the one hand, the collected data is numerous and shows a slow change trend, and the change value between the data may be very small [4]. On the other hand, the relationship between the data is non-linear, and it is more difficult to judge multidimensional data. Therefore, the combination of principal component analysis (PCA) and fuzzy C-means clustering is used to identify working conditions. If multiple fault conditions can be identified at the same time, it will be more helpful for practical application. Next, we input multiple sets of fault data into the model to view the experimental results. After the principal component analysis, the principal component contribution rate of each control parameter is shown in Fig. 2 below.

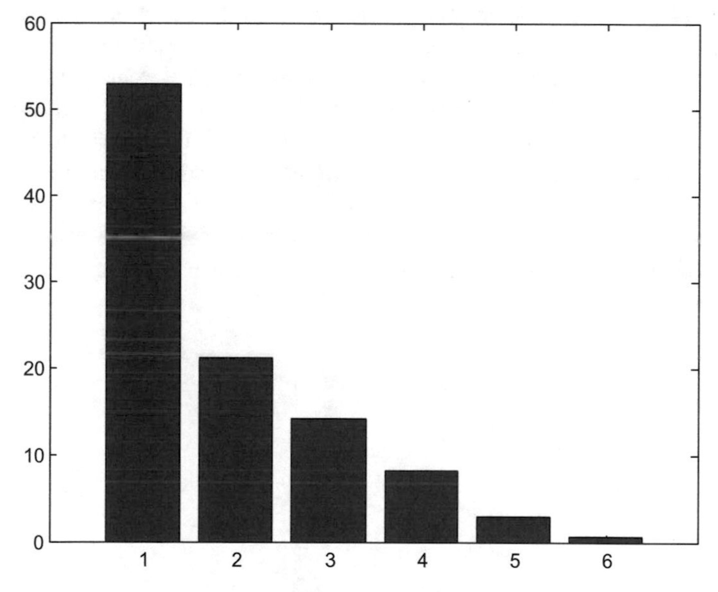

Fig. 2. Pivots contribution rate

Individual contribution rate:
0.52970.21080.1415 0.0624 0.0485 0.0071
Sum contribution rate:
0.5297 0.7405 0.8820 0.9444 0.9929 1.0000

From the above analysis, it can be concluded that the sum of the contribution rates of the first five pivots is 99.3%, so the first five pivots can be taken for analysis, and the coefficient matrix of the first five pivots is:

$$A = \begin{bmatrix} 0.4990 & -0.6107 & 0.2725 & -0.3891 & -0.0057 \\ 0.5174 & 0.5727 & 0.1292 & -0.5049 & -0.0582 \\ -0.3545 & 0.3631 & 0.7031 & -0.1041 & 0.2290 \\ -0.2312 & -0.2816 & 0.0189 & -0.3116 & 0.7484 \\ 0.4755 & -0.0223 & 0.4022 & 0.6946 & 0.2786 \\ -0.2795 & -0.2956 & 0.5025 & -0.0578 & -0.5535 \end{bmatrix}$$

It can be seen from the coefficient matrix that the coefficient of PH value in the first pivot is the largest, that is, PH value has the largest impact on the first pivot; In the second pivot, the coefficient of temperature is the largest, that is, the temperature has the largest influence on the second pivot; In the third pivot, the coefficient of optical density is the largest, that is, the optical density has the largest influence on the third pivot; In the fourth pivot, the coefficient of ventilation volume is the largest, that is, the ventilation volume has the largest impact on the fourth pivot; In the fifth pivot, the coefficient of feeding sugar is the largest, that is, feeding sugar has the largest impact on the fifth pivot [5]. Therefore, five control parameters, temperature, PH value, optical density, flow sugar and ventilation volume, were selected for fuzzy C-means cluster analysis.

A total of 100 groups of data were taken from the four failures and input into the fuzzy C-means clustering model at the same time. Since it is necessary to draw a three-dimensional clustering graph, it can be seen from the above principal component analysis that the sum of the contributions of the first three principal components is 88.2% (more than 73% that we often refer to). Therefore, we selected three variables--temperature, PH value, and optical density, to map and view the classification results. The results are shown in Fig. 3.

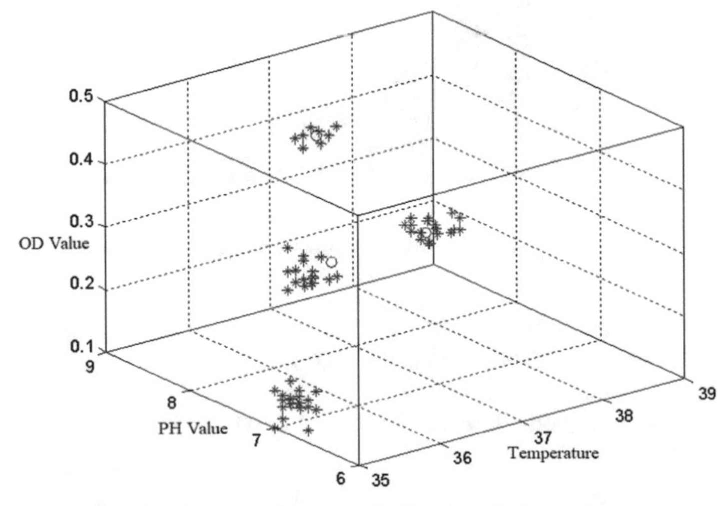

Fig. 3. Classification results under three fault conditions

The model divides the test data into four categories, and the final group center value is:

37.4 8.8 0.4 16.1 4000
37.3 7.4 0.3 14.3 3400
35.3 7.0 0.1 15.4 2800
37.3 8.6 0.2 12.1 3400

Table 4. Membership degree matrix of experimental results of multi-fault data

Data \ State	Failure Status of Ammonia Addition	Sugar Depletion Failure Status	Normal State	Contamination State
1	1.00	0.00	0.00	0.00
2	0.99	0.00	0.00	0.00
3	0.10	0.32	0.06	0.52
4	0.10	0.32	0.06	0.53
5	0.11	0.27	0.06	0.56
6	0.11	0.30	0.06	0.53
7	0.12	0.31	0.06	0.51
8	0.09	0.21	0.05	0.66
9	0.11	0.32	0.05	0.52
10	0.12	0.36	0.06	0.46
11	0.12	0.44	0.06	0.38
12	0.12	0.40	0.06	0.42
13	0.10	0.28	0.05	0.57
14	0.03	0.05	0.02	0.89
15	0.03	0.05	0.03	0.89
16	0.03	0.06	0.03	0.88
17	0.02	0.03	0.01	0.94
18	0.01	0.02	0.01	0.95
19	0.02	0.03	0.02	0.93
20	0.01	0.03	0.01	0.95
21	0.02	0.04	0.02	0.91
22	0.01	0.03	0.01	0.94
23	0.02	0.04	0.02	0.93
24	0.02	0.03	0.02	0.94
25	0.03	0.04	0.02	0.90
26	0.11	0.21	0.09	0.59
27	0.13	0.22	0.10	0.56
28	0.11	0.20	0.10	0.59
29	0.12	0.20	0.11	0.58
30	0.13	0.23	0.11	0.53
31	0.12	0.20	0.10	0.58

The number of recursive cycles in the iterative process is 19, and the cost function value is 2.38. The data are accurately divided into four categories. The obtained cluster center is considered as the standard feature point of a certain state. The final membership degree matrix is a 4×100 matix. We intercepted 31 groups of data for analysis. The membership degree matrix is shown in Table 4 above.

In the following, we will analyze the problem of clustering large state data sets into multiple cluster centers through the diagnosis of bacterial contamination.

The most common failure form in the fermentation process, which also causes the greatest loss, is bacterial contamination. The occurrence of bacteria infection is different according to the degree of visual symptoms, and may occur in various stages of fermentation, with a large degree of harm affecting a long time. Now let's focus on the analysis of bacterial contamination failure. We still select the above five characteristic variables obtained through principal component analysis, which are temperature, PH value, optical density, flow sugar and ventilation [7].

We randomly selected 100 groups of data from the bacterial contamination fault data in the production process, and added 50 groups of normal data to the fault diagnosis model at the same time. See what results can be obtained. In the previous examples, we analyzed the problem of mixed fuzzy classification of normal data and fault data. This time, we focus on the analysis of bacterial contamination faults. Due to the large number of input data, we do not know the optimal number of categories, and the tentative number of categories is 2. First, divide the data into two categories. The following results are obtained. The clustering diagram is shown in Fig. 4. We selected 40 groups of data from the membership matrix for analysis, as shown in Table 5.

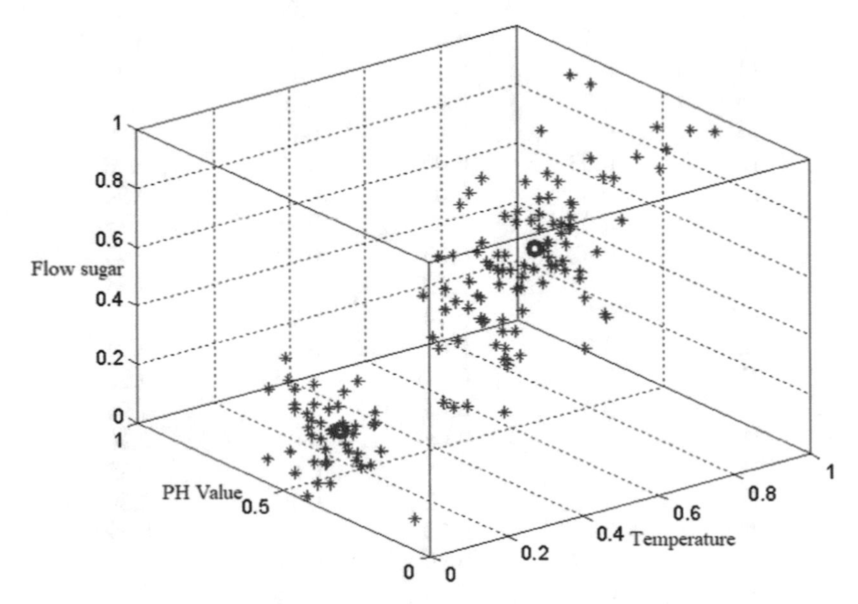

Fig. 4. Analysis results of bacteria contamination failure

Table 5. Membership degree matrix of bacterial contaminated sample data when fuzzy clustering number is 2

State / Data	Normal state	contamination state	State / Data	Normal state	contamination state
1	0.98	0.02	21	0.08	0.92
2	0.92	0.08	22	0.28	0.72
3	0.89	0.11	23	0.04	0.96
4	0.98	0.02	24	0.27	0.73
5	0.99	0.01	25	0.06	0.94
6	0.97	0.03	26	0.10	0.90
7	0.98	0.02	27	0.27	0.73
8	0.95	0.05	28	0.11	0.89
9	1.00	0.00	29	0.29	0.71
10	0.99	0.01	30	0.04	0.96
11	0.83	0.17	31	0.12	0.88
12	0.97	0.03	32	0.04	0.96
13	1.00	0.00	33	0.39	0.61
14	0.08	0.92	34	0.33	0.67
15	0.05	0.95	35	0.38	0.62
16	0.05	0.95	36	0.21	0.79
17	0.24	0.76	37	0.02	0.98
18	0.13	0.87	38	0.14	0.86
19	0.53	0.47	39	0.27	0.73
20	0.06	0.94	40	0.46	0.54

The cluster center obtained after clustering is:

$$U = \begin{bmatrix} 36.2\ 7.1\ \ 0.3\ \ 10.3\ 3400 \\ 38.7\ 8.7\ 0.21\ 11.3\ 3400 \end{bmatrix}$$

The number of recursive cycles in the iterative process is 26, and the value of the cost function is 21.6. From the clustering results, it can be seen that the value in the membership matrix clearly reflects the relationship between the data point and the cluster center. The higher the membership degree, the higher the similarity between the data point and this data class, and the greater the probability of occurrence of the corresponding state. On the contrary, it is smaller. From the table, we can see that most of the data points can be classified into a certain category with a higher degree of membership, but some data points have no obvious characteristics, and their membership values for the two categories are very close. For example, the membership degree of the 19th group of data in the above table is 0.53 for the fault class and 0.47 for the normal class. This result shows that it belongs to the middle of the two classes and does not have the obvious characteristics of a certain state. Therefore, we try to change the number of clusters to 3 to see the clustering results again. As shown in Fig. 5 below.

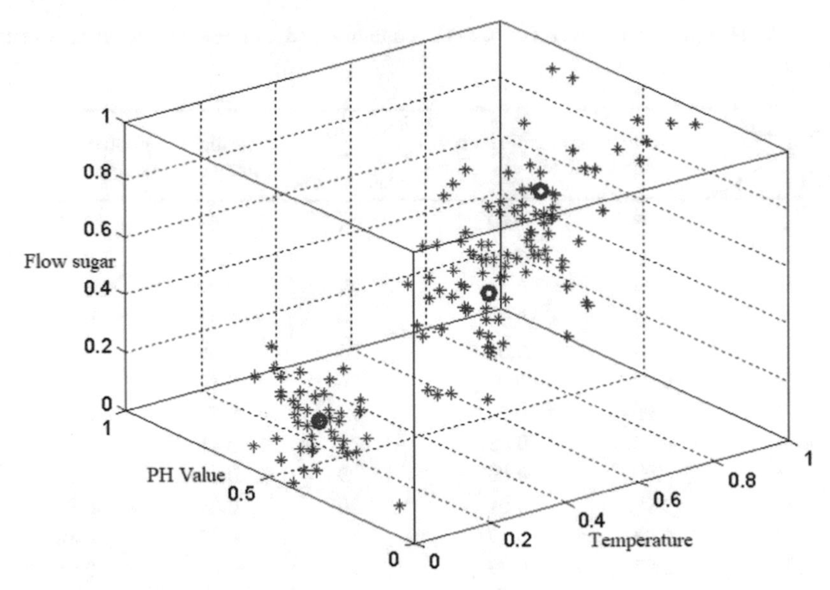

Fig. 5. Clustering results of contaminated fault samples when the number of cluster centers is 3

The cluster center obtained after clustering is:

$$U = \begin{bmatrix} 36.2 & 7.1 & 0.3 & 10.3 & 3400 \\ 38.1 & 8.3 & 0.25 & 10.8 & 3400 \\ 38.9 & 8.9 & 0.19 & 11.5 & 3400 \end{bmatrix}$$

The number of recursive cycles of the iterative process is 78, and the cost function value is 45.7. It can be seen from the clustering results that the clustering data points obtained more reasonable clustering. The membership of data belonging to a certain category is quite different from that of the other two groups. For example, the membership of the 30th group of data relative to State I is 0.80, while the membership of the other two groups is 0.03/0.16. It indicates that the data point has a high similarity with the feature of the cluster, and the probability of occurrence is high. On the contrary, it is less likely to happen. Based on this, we can propose a diagnosis processing model. In the production, the fault source data group is extracted, and then the input data to be tested and the data in the fault source are classified, so as to obtain which cluster center is closest to the result, and which fault is more likely to occur. The membership matrix obtained by dividing the source data into two categories is not very reasonable. A more reasonable classification is obtained by changing the number of clusters to 3. Therefore, we get that the information contained in the fault data is also divided into two categories, and the corresponding reality is the same. The occurrence of bacterial contamination failure is not sudden, but a gradual transition process.

Here, the two fault categories represent two kinds of situations--the early stage and late stage of bacterial infection, the situation in the early stage of bacterial infection is not obvious, and the data reflection is not drastic, so it is neither completely normal nor completely standard fault type in the late stage of infection. The result of the experiment is also consistent with the actual situation.

3 Conclusion

According to the clustering results, the center value of fuzzy clustering is different for different fault categories, which can be used as the evaluation basis for different faults in fermentation process. Through three simulation comparisons, we can see that without any prior knowledge, K-means clustering state diagnosis model can distinguish fault data and normal data well, and the hybrid algorithm of principal component analysis (PCA) and fuzzy C-means clustering can distinguish multiple faults at the same time. In the analysis of bacterial infection fault, multi-level state diagnosis can be preliminarily realized, and the conclusion consistent with the actual situation is obtained.

References

1. Chen, X., Li, X., Fu, G., et al.: Application of enzymatic corn syrup in glutamic acid fermentation. Ferment. Sci. Technol. Commun. **50**(3), 141–144 (2021)
2. Zheng, R., Pan, F.: Prediction modeling of glutamate fermentation product concentration based on PLS-LSSVM. J. Chem. Eng. **68**(3), 976–983 (2017)
3. Cai, Y., Wang, Q., Wang, W., et al.: Bacillus siamese LW-1 production γ- Optimization of polyglutamic acid fermentation medium. Sci. Technol. Food Indust. **42**(16), 163–170 (2021)
4. Zhou, J.: Management status and development trend of glutamic acid fermentation equipment. Ferment. Sci. Technol. Commun. **51**(2), 110–115 (2022)
5. He, N., Shan, P., He, Z., et al.: Fractional baseline correction of ATR-FTIR spectral signal in polyglutamic acid fermentation. Spectroscopy Spectral Anal. **42**(6), 1848–1854 (2022)
6. Ding, J., Cao, Y., Shi, Z.: Automatic glucose feeding system in glutamic acid fermentation process. J. Biol. **27**(2), 40–42 (2010)
7. Hu, H., Xu, D., Xu, Q., et al.: Study on membrane coupling batch dialysis fermentation process in glutamic acid fermentation. Food Ferment. Technol. **54**(1), 9–13, 23 (2018)

Mustahik Micro Business Incubation in Poverty Alleviation

Zainal Alim Adiwijaya(✉), Edy Suprianto, and Dedi Rusdi

Universitas Islam Sultan Agung, Semarang, Indonesia
edysuprianto@unissula.ac.id

Abstract. Zakat is an Islamic economic instrument that is relevant in poverty alleviation efforts. Several studies had been conducted to suggest that zakat be distributed productively will more effective in reducing poverty. This paper aims is to propose a micro business incubation model for mustahik to be financially independent. The research approach used was qualitative descriptive with semi-structured in-depth interviews with muzakki, mustahik, and district/city BAZNAS administrators in Central Java. The results of the interview were then confirmed based on agency theory and sharia enterprise theory. This research has identified the things needed in business incubation and proposed a series of microenterprise incubation models for mustahik. This research contributes to the novelty of the micro business incubation model which includes the provision of capital assistance, management skills training, strengthening innovation, and assistance from the zakat institution. The zakat institution provides capital assistance from zakat funds to mustahik which can be used to start a business, then zakat institution provides management skills training that helps mustahik in determining market segments, and improving product quality through innovation so that the business run by mustahik can develop. Assistance is also needed so that mustahik can optimize capital assistance that has been given.

Keywords: Innovation · Management skill · Financing · Assistance · Microbusiness

1 Introduction

Sustainable Millennium Development Goals (SMDGs) are a continuation of the Millennium Development Goals (MDGs) which is a development initiative formed in 2000 by representatives from 189 countries by signing a declaration known as the millennium declaration (Funds 2021). The declaration contains eight points that must be achieved by 2015, which include (1) eradicating poverty, (2) education for all humanity, (3) gender equality and empowering women, (4) fighting against HIV/AIDS, malaria, and so forth, (5) reducing child mortality, (6) improving reproductive health, (7) preserving the environment and protecting the environment, and (8) global cooperation for development. As one of the countries that ratified the global agreement, Indonesia is seriously making various efforts so that the eight goals can be achieved. However, the signs of achieving

L. Barolli (Ed.): EIDWT 2023, LNDECT 161, pp. 64–75, 2023.
https://doi.org/10.1007/978-3-031-26281-4_7

eight points in the MDGs are still not visible. It takes acceleration from all aspects to achieve the MDGs targets.

The first point of the eight points is the problem of poverty alleviation. Poverty is seen as a result of unemployment, low investment, and neglect of the government (Oyekale et al. 2012). Poverty is a phenomenon that poses a serious threat to development and national security (Abdurraheem and Suraju 2018). The importance of financial capital, human capital, and natural capital for development and for solving poverty is a popular issue (Hassan 2014). Zakat, infaq, and Alms are economic instruments in Islam that are attractive to be used as instruments for alleviating poverty (Abdullah et al. 2015; Hayati 2015; Razak 2020; Saputro and Sidiq 2020). Zakat is a fundamental pillar of Islam (Febianto and Ashany 2012; Sohag et al. 2015), which can increase the authorized capital of microfinance (Yumna and Clarke 2011; Zauro et al. 2020), improve the welfare of the poor (Yumna and Clarke 2011), and can help reduce poverty (Ayuniyyah et al. 2022; Zauro et al. 2020).

Zakat as an effective financial instrument to help economic development (Adnan et al. 2019) and poverty alleviation instruments have many advantages over other existing conventional fiscal instruments. First, the use of zakat has been determined in religious law (HQ. At Taubah: 60), where zakat is only intended for 8 groups (called as *ashnaf* namely: the indigent, the poor, zakat amil, *mu'allaf* (new convert to Islam), slaves, debtors, *jihad fi sabilillah* (people that fights in the way of Allah), and *ibn sabil* (people in the journey). *Jumhur Fuqaha* (Islamic Scholars) agree that apart from these 8 groups, it is not lawful to receive zakat. Neither party has the right to change these terms. This characteristic makes zakat inherently in favor of the poor. There is no conventional fiscal instrument that has this unique characteristic. Therefore, zakat will be more effective in alleviating poverty because the allocation of funds is certain and believed to be right on target.

Second, zakat has a low and fixed rate because it was regulated in Islam, that is 2.5% of the assets owned by the *muzakki* (the person who is obliged to pay zakat). Third, zakat is imposed on a broad basis and covers various economic activities. Fourth, zakat is a spiritual tax that must be paid by every Muslim who can afford it under any circumstances. Therefore, zakat receipts tend to be stable.

Zakat has a very strategic role in poverty alleviation efforts or economic development. If applied comprehensively, then a country will not experience poverty and unemployment issues (Yaacob and Azmi 2012). Ridwan (2005) stated that the strategic value of zakat can be seen through some items: (1) zakat is a religious call that reflects one's faith; (2) the source of zakat financing will never stop, meaning that people who pay zakat will pay annually or more; (3) zakat can empirically erase social inequality and can create asset redistribution and equitable development. This will ensure the sustainability of the poverty alleviation program in the long term.

Zakat is essentially an obligation for all Muslims who are able or have reached the *nisab* (the limit for being obliged in zakat) (Rehman and Aslam 2020; Nomran and Haron 2022). Conceptually, zakat is a vertical and horizontal relationship. In a horizontal relationship, the purpose of zakat is not only to support the poor consumptively but has a more permanent goal, namely alleviating poverty (Manurung 2014; Nurjanah et al. 2019; Ahmad 2019). The distribution of zakat funds has now developed, from being oriented only to meeting needs (consumptive) into being a source of productive funds that can boost the economy even further. In Indonesia, productive zakat was approved by the MUI in 1982, this is also strengthened by the existence of information about zakat collected by the Amil Zakat Institution and the Amil Zakat Agency which can be given consumptively for meeting the needs of daily life and can also productively increase the efforts of *mustahik*.

With all the potential of zakat as an instrument of poverty alleviation, the mechanism for the management of zakat agencies and the management of zakat funds must receive attention (Ali and Hatta 2014). Zakat management agencies are one of the tools introduced in Islam to fight poverty and improve welfare in society (Alaro and Alalubosa 2019). BAZ is a amil zakat agency established by the government to manage community zakat funds from the central (national) level to the district/city level. Furthermore, BAZ should have received full support from the government in its process, both from operational financing, as well as technical management of zakat funds.

The role of the government is very necessary to be able to optimize the role of BAZ. BAZ at the regional level (provincial and district/city) is an organized part of BAZNAS to carry out zakat management functions in the regions. Meanwhile, on the other hand, there is LAZ, which in the legislation is a zakat service institution formed by the community independently (apart from government intervention). The existence of BAZ and LAZ creates dualism because there is no clear coordination between both. The government as the policy maker wants zakat management institutions through one door, namely BAZ, but LAZ as a form of community self-help in managing zakat still wants to carry out its functions.

Amid the disagreements between LAZ and BAZ, the fact is that the role of institutions in collecting zakat funds is still very small from the overall proportion of existing zakat. The tendency is that people distribute their zakat personally. In this pattern, the zakat received by the community is only intended for temporary consumption. This is unable to remove the poor from the cycle of poverty. Therefore, it is important to distribute zakat as a productive fund, so that it is expected to bring added value to the welfare of communities. Zakat for productive purposes is more useful than consumptive, especially for empowerment purposes (Alim 2015).

Zakat is one of the most influential wealth distribution systems and ensures an even and trustworthy distribution of wealth (Djaghballou et al. 2018). Law Number 23 of 2011 article 27 concerning Zakat Management stated that zakat can be utilized for productive businesses, in the sense of productive zakat to handle the poor and improve the quality of the people, which can be done if the basic needs of *mustahik* have been fulfilled. This goal can be achieved if it has the support of all parties, and the management and distribution of zakat funds by zakat institutions are carried out professionally with good governance and following Islamic rules. Adiwijaya (2010) stated that good BAZNAS governance is transparency and accountability of BAZNAS which plays an important role as stakeholders, especially *muzaki* as funders. In addition, the management of BAZNAS is also needed for alleviating poverty. This means that the eradication of poverty must be felt by the public.

Productive zakat as capital for *mustahik* to carry out economic activities through the development of economic conditions and productivity potential of *mustahik*, because the majority of *mustahik* are constrained by capital from banking institutions (Sinaga et al. 2020). Making the poor more productive can be done by creating new jobs that can be realized if the community's micro-enterprises are supported to grow and develop. Empowered micro-enterprises have a major influence on several important aspects, such as more economic activity, employment opportunities, and an increase in income (Abdullah 2010). Micro-enterprises are entities that can survive in an underprivileged socio-economic environment with unique skills possessed, oriented to market sales and network activities that can increase entrepreneurial competence and company performance (Al Mamun et al. 2019).

The poor can be free from poverty if they are empowered through coaching and mentoring to become independent. The forms of empowering the poor who can be self-reliant and escape poverty are: (1) being employed, (2) being facilitated for business capital and business training, and (3) being facilitated for business assistance with a larger scope (Ghoniyah et al. 2019). The assistance is not only to be more productive but also to meet the needs of the family. If there is no assistance, likely, the zakat funds given as business capital will only be used for family needs. BAZNAS as the manager of zakat funds has several programs, including the independent partner development program, namely the empowerment program for the poor who have micro businesses through interest-free financing to support the sustainability of their business to achieve prosperity through productive economic activities. In other words, poverty can be reduced with this financing. In addition to solving the problem of poverty, interest-free financing and business assistance can foster the entrepreneurial spirit of the *mustahik* if they are seriously and patiently empowered.

Jumaizi and Sudarti (2018) researched 62 *mustahik* in BAZDA (Zakat institution established by the Regional Government in Central Java) using a mixed method approach and interview and questionnaire techniques. The results of that research present a series of micro-business incubation models. The respondents agreed and hoped for a capital loan, business management training, business quality improvement, business spiritual training, increasing self-confidence, and providing assistance if they experienced difficulties in running a business. Hassan and Noor (2015) used a qualitative approach and involved zakat recipients from the Selangor Zakat Board. The results show that the entrepreneurial capital assistance program provided by the Selangor Zakat Board to *mustahik* has effectively helped increase the income of *mustahik* and succeeded in changing the status of *mustahik* to *muzakki*. Olowu and Aliyu (2015) researched SMEs in Nigeria using a quantitative approach and a questionnaire to collect data. There are many financing policies for SMEs but the results are not as expected. The results show that management skills are the main determinants of micro business performance, especially in sales growth. Management skills that are still low hinder the performance of SMEs. Karabulut (2015) also researched to determine the role of innovation on company performance. As a result, product, process, and organizational innovation play an important role in company performance, which includes financial performance, customers, internal business processes, and growth. Meanwhile, innovations created as marketing techniques do not determine growth performance. Broadly, business actors who have innovations can improve business performance (Cravo and Pizza 2016; Karabulut 2015). Gstraunthaler (2010) proved that based on the aspect of sustainability, businesses in European countries that have been fostered in business incubation last longer than businesses without the support of an incubator. This study aims to present a mustahik empowerment program that can be carried out by zakat management institutions so that poverty alleviation efforts through zakat funds are more effective and efficient. As a novelty, this study propose an incubation model for micro-enterprises run by *mustahik*, which includes capital assistance in the form of interest-free financing, training to improve management skills, strengthening innovation to provide higher quality products and assistance for zakat institutions.

2 Theory and Hypothesis

2.1 Capital Assistance

Limited capital and difficult access to funding sources are obstacles for business actors (Huda 2012; Muhamat et al. 2013) that want to start or develop a business because the source of funds to start a business is important (Zin and Ibrahim 2020). A business will not be able to run or difficult to develop without capital. The distribution of zakat funds is considered a sustainable and beneficial approach, namely by using an interest-free financing scheme which is an opportunity for *muzakki* to improve welfare as a result of the expansion of micro-enterprises (Aderemi and Ishak 2022). The distribution of zakat to *mustahik* is useful as capital to start a business, which is generally a small-scale business that is not affordable by banking institutions. Utilization of zakat funds by providing business capital assistance to *mustahik* aims to develop and utilize zakat funds, so that the goals of zakat can be achieved. The provision of business capital assistance must be right on target, namely given to *mustahik* who are weak in the economy or can be given to *mustahik* who will develop a business. The purpose of providing capital assistance is to increase *mustahik's* income through the business they run (Hassan and Noor 2015), and can develop capital by running business activities, which can then change the status of *mustahik* to *muzakki*.

2.2 Management Skill

Low management skill can be one of the reasons why businesses can lose to compete in the market. Management skill can be measured by business planning, organizing, leading, and controlling the business (Olowu and Aliyu 2015). Financial management, strategic planning, and human resource management skill are also needed by business actors in managing their business (Asah et al. 2015) to grow and develop. Improper planning and selection of strategies can have an impact on not achieving the expected financial performance. Management skills will affect business performance. If the management skill in managing the business is good, then the business performance is also good and the success of the business can be achieved. Management skill becomes a necessary factor because businesses that focus on technical capabilities cannot develop their business properly. Management skill is very much needed in advancing micro-enterprises because, within management skill, it includes an analysis of market segments, innovation, and role organizing. Management skill is very different from product manufacturing technicality. Management skills can be improved through training programs (Al-Madhoun and Analoui 2003).

2.3 Innovation

A business can continue to grow and last long if business actors can innovate products continuously. On the other hand, a business can collapse if it is not able to adapt to the social developments of society and micro and small businesses that innovate, survive, and grow. Innovation is a strategic tool to sustain business and gain a competitive advantage in the global market (Karabulut 2015; Sumiati 2020), which can then improve business performance and beat competitors (Karabulut 2015), as well as expand available resources in unexpected ways (Codini et al. 2022). Innovation can be triggered by changes in the company's internal and external environment, which will then provide opportunities for companies to create systematic new procedures or processes to improve company performance (Sumiati 2020). While innovating, businesses create and deliver value mechanisms and begin to develop new value-capturing dynamics (Codini et al. 2022). The more innovation, the better the business performance. Innovation will increase productivity and business growth (Cravo and Piza 2016).

2.4 Assistance

Assistance provided by zakat institutions is needed to improve *mustahik's* ability to manage, improve, and develop businesses, as well as motivate them to do better in running their business. In addition, assistance is also useful for improving the quality of *mustahik's* faith. Without assistance, the productive zakat program will be inefficient. Business assistance can include the development of *mustahik's* skills in running a business such as production, marketing, supply chain management, customer relationship management, and others, as well as monitoring and evaluation of the *mustahik's* business that will encourage business growth and will lead to of the *mustahik's* welfare (Widiastuti et al. 2021). Training and guidance are also needed to help develop *mustahik's* ability to have a better mindset and be more financially independent. The more intensive the assistance provided, the business run by *mustahik* can develop.

2.5 Agency Theory

In agency theory, ideally, agents can be trusted to carry out their duties and responsibilities in maximizing prosperity (Jensen and Meckling 1976). Agency problems can occur between administrators in zakat institutions (agents) and *muzakki* (principals). *Muzakki* who entrusts its funds to zakat management to be managed and distributed to the right people in accordance with the provisions of Islamic law. Zakat institutions must work professionally, and have the competence and commitment to manage zakat funds properly to achieve prosperity for the community. The optimal collection, management, and utilization of zakat funds will make zakat as an alternative to poverty alleviation and income distribution. Increased public awareness of the importance of paying zakat can be realized if the community knows that the zakat institution has implemented zakat funds properly and in a trustworthy manner. Therefore, zakat managers are required to work in a transparent, accountable, and responsible manner for the funds that have been deposited by *muzakki*.

2.6 Sharia Enterprise Theory

Sharia Enterprise Theory is an enterprise theory concept that has been combined with Islamic values to produce a transcendental and humanist theory. This theory puts the greatest responsibility on God which is then elaborated in the form of accountability on humans and nature. The sharia enterprise theory is established by the concept that Allah SWT is the Creator and Sole owner who control this world. Meanwhile, the resources owned by the stakeholders are in principle a mandate from God which includes a responsibility to use them in the manner and purpose set by the Trustee. The concept of Shariah Enterprise Theory encourages us to realize the value of justice for humans and the natural environment. The purpose of using these resources is none other than to get God's blessing/permission (*mardhatillah*) by using resources that can bring prosperity to all nature *(rahmatan lil alamin).* Therefore, Shariah Enterprise Theory will bring benefits to *muzakki*, zakat institutions, *mustahik*, society in general, and the environment without leaving the important obligation to pay zakat as a manifestation of worship to Allah.

3 Methodology

This study used a qualitative descriptive approach that focuses on the in-depth identification of microenterprise incubation. A qualitative approach is usually taken to gain a deeper understanding of the nature of a problem and is aimed at making a valid conclusion based on data collected (Sekaran and Bougie 2016). The data collection technique used was a semi-structured interview method which was carried out in-depth with the respondents. This method allows researchers to explore new things related to the research problem. According to Now (2016), interviews were conducted extensively to understand the situation and phenomena that occur. Interviews are used to record a person's ideas, opinions, feelings, emotions, or points of view on a problem. Interviews were conducted face-to-face and by phone with 84 respondents including *muzakki*, *mustahik*, and BAZNAS managers in Central Java.

4 Results

This study proposes a microenterprise incubation model, which is a model that creates jobs and builds prosperity by encouraging the formation of new businesses (Erlewine and Gerl 2004; Marchis et al. 2007) through interest-free financing and business assistance that will improve the performance of micro-businesses assisted by BAZNAS. This model can provide implications for BAZNAS managers in the context of alleviating poverty through micro-business incubation with interest-free financing and business assistance. Religious considerations in sharia loans prohibit the interest charged to the borrower (Shaban et al. 2016). This is expected to provide a solution to the problems of poverty and unemployment and in turn, can absorb a large workforce. Zakat is a tax that distributes wealth from the rich to the poor, meaning that *muzakki* are eligible to distribute their wealth to the poor (Amin 2022). Thus, zakat is a right of the poor according to Islamic ideology (Sohag et al. 2015). Zakat helps improve the welfare of *muzakki* and acts as a bridge to reduce the gap between the rich and the poor (Ahmed et al. 2017). This is very important for the country in the future to be more active in alleviating poverty and unemployment (Fig. 1).

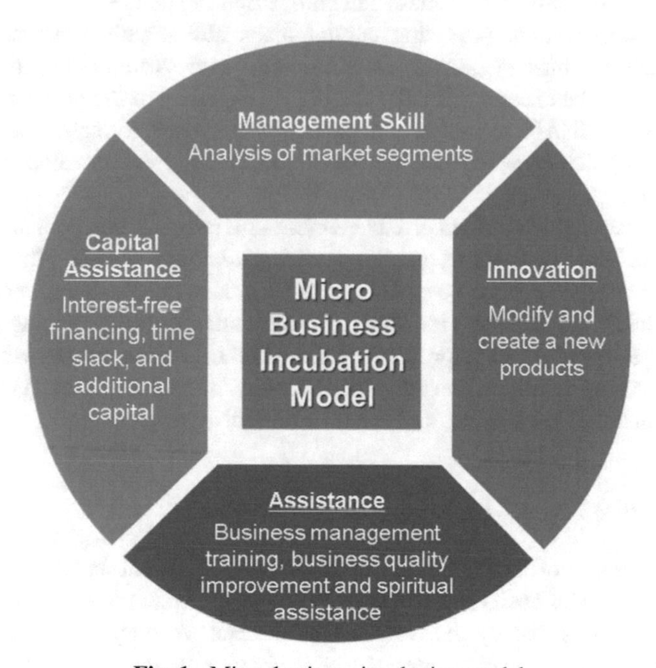

Fig. 1. Micro business incubation model

Based on the interviews, the respondents stated that administrative costs, interest costs, and guarantees do not need to be provided because the source of funding comes from zakat, infaq, and alms. In addition, the existence of other costs will burden the business actor if the initiated business fails because basically, BAZNAS expects a sustainable business. In addition, respondents take longer to repay, because the micro-enterprises they run need time to develop and repay loans. Furthermore, respondents also stated that they wanted additional capital when their business began to grow.

Then, coaching and training to improve the management skill of *mustahik* as business actors is very much needed. Observations to determine market segments carried out by BAZNAS will assist respondents in choosing which segments to take. In addition, in the implementation of the business, it is also necessary to look at perseverance, tenacity, creativity, consumer desires, and the surrounding competitive conditions so that the business carried out can develop more rapidly. Determination of business surplus value and promotion are also things that are considered important so that the efforts carried out can be right on target and in large quantities.

Innovation is also very important for business continuity. In carrying out their business, respondents agree that modifying products, creating new products, reviewing new product development, and observing business products that always vary from one another can help respondents in developing their business. Business actors must actively innovate through observation and assessment of product developments so that product quality and quantity can be increased. Good innovation will be responded well by consumers. Innovation can also make the business can compete in the market.

Based on observation, providing capital loans along with assistance to micro-enterprise managers, business management training, business quality improvement training, and providing business spirituality are highly expected in the assistance of micro-enterprises by BAZNAS. In practice, respondents still need to learn a lot in business development, which is expected that micro business managers can develop independently and sustainably.

With business capital assistance, management skill training, business innovation, and assistance from zakat institutions greatly assist *mustahik* in improving the performance of micro-enterprises. This can be seen from the increase in sales turnover, profits, and working capital. Micro-business incubation through interest-free financing and business assistance by BAZNAS can improve the welfare of micro-businesses which are then expected to be able to run businesses independently. Therefore, they can change their status from *mustahik* to *muzakki*, and can also hire job applicants.

5 Conclusion

Based on the results of this research, the indicators of capital assistance, innovation, skill management, and business assistance are expected to increase the performance of micro-enterprises assisted by the National Zakat Amil Agency or BAZNAZ. Micro-entrepreneurs expect to receive interest-free assistance which is expected to improve the performance of entrepreneurs without incurring debt repayment through installments. In practice, entrepreneurs must always innovate to keep up with market demands and continue to develop management skills and business assistance which is expected to create a sustainable business. Future researchers need to explore how effective the implementation of the incubation model is.

Based on the results of this study, there are several recommendations. Zakat is an effective means for poverty alleviation. Therefore, zakat fund management institutions must maximize their efforts to collect and distribute zakat. This strategy will improve social welfare, security, and social harmony. However, poverty alleviation efforts will be successful if the government, zakat management institutions, social institutions, educational institutions, and the community can collaborate and participate in their respective portions, and also contribute materially and spiritually. The government plays an important role in poverty alleviation because the government has the authority to formulate policies that lead to the creation of equitable economic growth and eradicate the poor who are relatively left behind. Furthermore, educational institutions, especially universities, can provide educational assistance or scholarships to the poor to continue their education until they are ready to work. Meanwhile, lecturers have an obligation to carry out community service that leads to increased social welfare. The community, especially muzakki, can be able to pay their zakat through zakat management institutions which will be distributed with a productive zakat system. If the muzakki pays zakat personally, it will be widely used by mustahik consumptively. Zakat management institutions must be more trustworthy and increase credibility so that muzakki will easily mandate zakat payment. Finally, mustahik who become business actors must have a great commitment and intention to fight poverty. This situation will motivate them to keep running the business.

References

Abdullah, N., Derus, A.M., Al-Malkawi, H.A.N.: The effectiveness of Zakat in alleviating poverty and inequalities a measurement using a newly developed technique. Humanomics **31**(3), 314–329 (2015). https://doi.org/10.1108/H-02-2014-0016

Abdullah, R.B.: Zakat management in Brunei Darussalam: a case study. Abdul Ghafar Ismail Mohd Ezani Mat Hassan Norazman Ismail Shahida Shahimi, pp. 376–407 (2010). https://doi.org/10.1109/incos.2010.100

Abdurraheem, H., Suraju, S.B.: Taming poverty in Nigeria: language, Zakat and national development. QIJIS Qudus Int. J. Islam. Stud. **6**(1), 1–23 (2018). https://doi.org/10.21043/qijis.v1i1.3278

Aderemi, A.M.R., Ishak, M.S.I.: Qard Hasan as a feasible Islamic financial instrument for crowdfunding: its potential and possible application for financing micro-enterprises in Malaysia. Qual. Res. Financ. Mark. (2022). https://doi.org/10.1108/QRFM-08-2021-0145

Adiwijaya, Z.A.: Pengaruh Transparansi Informasi, dan Akuntabilitas BAZIS Terhadap Kepuasan, dan Loyalitas Muzaki Menunaikan Zakat, Infak, dan Shadaqah pada BAZIS DKI Jakarta. Disertasi thesis. Universitas Airlangga, Surabaya (2010)

Adnan, N.I.M., Kashim, M.I.A.M., Hamat, Z., Adnan, H.M., Adnan, N.I.M., Sham, F.M.: The potential for implementing microfinancing from the Zakat fund in Malaysia. Hum. Soc. Sci. Rev. **7**(4), 542–548 (2019). https://doi.org/10.18510/hssr.2019.7473

Ahmad, M.: An empirical study of the challenges facing Zakat and Waqf institutions in Northern Nigeria. ISRA Int. J. Islam. Finance **11**(2), 338–356 (2019). https://doi.org/10.1108/IJIF-04-2018-0044

Ahmed, B.O., Johari, F., Wahab, K.A.: Identifying the poor and the needy among the beneficiaries of Zakat need for a Zakat-based poverty threshold in Nigeria. Int. J. Soc. Econ. **44**(4), 446–458 (2017). https://doi.org/10.1108/IJSE-09-2015-0234

Al Mamun, A., Fazal, S.A., Muniady, R.: Entrepreneurial knowledge, skills, competencies and performance. Asia Pac. J. Innov. Entrep. **13**(1), 29–48 (2019). https://doi.org/10.1108/apjie-11-2018-0067

Alaro, A.A.-M., Alalubosa, A.H.: Potential of Sharī'ah compliant microfinance in alleviating poverty in Nigeria: a lesson from Bangladesh. Int. J. Islam. Middle East. Finance Manag. **12**(1), 115–129 (2019). https://doi.org/10.1108/IMEFM-01-2017-0021

Ali, I., Hatta, Z.A.: Zakat as a poverty reduction mechanism among the Muslim community: case study of Bangladesh, Malaysia, and Indonesia. Asian Soc. Work Policy Rev. **8**(1), 59–70 (2014). https://doi.org/10.1111/aswp.12025

Alim, M.N.: Utilization and accounting of Zakat for productive purposes in Indonesia: a review. Procedia Soc. Behav. Sci. **211**, 232–236 (2015). https://doi.org/10.1016/j.sbspro.2015.11.028

Al-Madhoun, M.I., Analoui, F.: Managerial skills and SMEs' development in palestine. Career Dev. Int. **8**(7), 367–379 (2003). https://doi.org/10.1108/13620430310505322

Amin, H.: Examining new measure of Asnaf muslimpreneur success model: a Maqasid perspective. J. Islam. Account. Bus. Res. **13**(4), 596–622 (2022). https://doi.org/10.1108/JIABR-04-2021-0116

Asah, F., Fatoki, O.O., Rungani, E.: The impact of motivations, personal values and management skills on the performance of SMEs in South Africa. Afr. J. Econ. Manag. Stud. **6**(3), 308–322 (2015). https://doi.org/10.1108/AJEMS-01-2013-0009

Ayuniyyah, Q., Pramanik, A.H., Saad, N.M., Ariffin, M.I.: The impact of Zakat in poverty alleviation and income inequality reduction from the perspective of gender in West Java, Indonesia. Int. J. Islam. Middle East. Finance Manag. (2022). https://doi.org/10.1108/IMEFM-08-2020-0403

Brennan, N.M., Solomon, J.: Corporate governance, accountability and mechanisms of accountability: an overview. Account. Audit. Account. J. **21**(7), 885–906 (2008). https://doi.org/10.1108/09513570810907401

Codini, A.P., Abbate, T., Petruzzelli, A.M.: Business model innovation and exaptation: a new way of innovating in SMEs. Technovation (2022). https://doi.org/10.1016/j.technovation.2022.102548

Cravo, T.A., Piza, C.: The impact of business support services for small and medium enterprises on firm performance in low- and middle-income countries: a meta-analysis. World Bank Policy Research Working Paper No. 7664, May 2016

Djaghballou, C.-E., Djaghballou, M., Larbani, M., Mohamad, A.: Efficiency and productivity performance of Zakat funds in Algeria. Int. J. Islam. Middle East. Finance Manag. **11**(3), 374–394 (2018). https://doi.org/10.1108/IMEFM-07-2017-0185

Erlewine, M., Gerl, E.: A Comprehensive Guide to Business Incubation. National Business Incubation Association NBIA Publications (2004)

Febianto, I., Ashany, A.M.: The impact of Qardhul Hasan financing using Zakah funds on economic empowerment (case study of Dompet Dhuafa, West Java, Indonesia). Asian Bus. Rev. **1**(1), 15–20 (2012). https://doi.org/10.18034/abr.v1i1.332

Fund, S.: From MDGs to SDGs (2021)

Ghoniyah, N., Hartono, S., Sobari, A.: Model Pemberdayaan UMKM Melalui CSR Berbasis Supply Chain Management. Unissula Press (2019)

Gstraunthaler, T.: The business of business incubators. Balt. J. Manag. **5**(3), 397–421 (2010). https://doi.org/10.1108/17465261011079776

Hassan, A.: The challenge in poverty alleviation: role of Islamic microfinance and social capital. Humanomics **30**(1), 76–90 (2014). https://doi.org/10.1108/H-10-2013-0068

Hassan, N.M., Noor, A.H.M.: Do capital assistance programs by Zakat institutions help the poor? Procedia Econ. Finance **31**(15), 551–562 (2015). https://doi.org/10.1016/s2212-5671(15)01201-0

Hayati, K.: Zakat potential as a means to overcome poverty (a study in Lampung). J. Indones. Econ. Bus. **26**(2), 187–200 (2015). https://doi.org/10.22146/jieb.6270

Huda, A.N.: The development of Islamic financing scheme for SMEs in a developing country: the Indonesian case. Procedia Soc. Behav. Sci. **52**, 179–186 (2012). https://doi.org/10.1016/j.sbspro.2012.09.454

Jensen, M.C., Meckling, W.H.: Theory of the firm: Managerial behavior, agency costs and ownership structure. J. Finance Econ. **3**, 305–360 (1976)

Jumaizi, Sudarti, K.: Model Inkubasi Usaha Mikro BAZDA (Badan Amil Zakat Daerah). Ebistek Fakultas Ekonomi Dan Bisnis UNAKI **1**(1), 11–28 (2018)

Karabulut, A.T.: Effects of innovation types on performance of manufacturing firms in Turkey. Procedia Soc. Behav. Sci. **195**, 1355–1364 (2015). https://doi.org/10.1016/j.sbspro.2015.06.322

Manurung, S.: Islamic religiosity and development of Zakat Institution. Qudus Int. J. Islam. Stud. **1**(2), 197–220 (2014). https://doi.org/10.21043/QIJIS.V1I2.186

Marchis, G., Galati, U.D., Structural, E.: Fundamentals of business incubator development. EIRP Proc. **2**(1), 492–496 (2007)

Muhamat, A.A., Jaafar, N., Rosly, H.E., Manan, H.A.: An appraisal on the business success of entrepreneurial Asnaf: an empirical study on the state Zakat organization (the Selangor Zakat board or Lembaga Zakat Selangor) in Malaysia. J. Financ. Report. Account. **11**(1), 51–63 (2013). https://doi.org/10.1108/JFRA-03-2013-0012

Nomran, N.M., Haron, R.: Validity of Zakat ratios as Islamic performance indicators in Islamic banking: a congeneric model and confirmatory factor analysis. ISRA Int. J. Islam. Finance **14**(1), 38–58 (2022). https://doi.org/10.1108/IJIF-08-2018-0088

Nurjanah, F., Kusnendi, Juliana: The impact of economic growth and distribution of Zakat funds on poverty (survey in the third district of West Java Province period 2011–2016. KnE Soc. Sci. **3**(13), 55 (2019). https://doi.org/10.18502/kss.v3i13.4195

Olowu, M.D.Y., Aliyu, I.: Impact of managerial skills on small scale businesses performance and growth in Nigeria. Eur. J. Bus. Manag. **7**(5), 109–114 (2015)

Oyekale, A.S., Adepoju, A.O., Balogun, A.M.: Determinants of poverty among riverine rural households in Ogun State, Nigeria. Stud. Tribes Tribals **10**(2), 99–105 (2012). https://doi.org/10.1080/0972639X.2012.11886647

Razak, S.H.A.: Zakat and Waqf as instrument of Islamic wealth in poverty alleviation and redistribution: case of Malaysia. Int. J. Sociol. Soc. Policy **40**(3/4), 249–266 (2020). https://doi.org/10.1108/IJSSP-11-2018-0208

Rehman, A.U., Aslam, E.: Factors influencing the intention to give Zakat on employment income: evidence from the Kingdom of Saudi Arabia. Islam. Econ. Stud. **29**(1), 33–49 (2020). https://doi.org/10.1108/IES-05-2020-0017

Ridwan, M.: Management of Baitul Maal Wa Tamwil (BMT). UII Press (2005)

Saputro, E.G., Sidiq, S.: The role of Zakat, Infaq and Shadaqah (ZIS) in reducing poverty in Aceh province. J. Islam. Econ. Finance (IJIEF) **3**(2), 63–94 (2020). https://doi.org/10.18196/ijief.3234

Sekaran, U., Bougie, R.: Research Methods for Business: A Skill-Building Approach, 7th edn. Wiley, New York (2016)

Shaban, M., Duygun, M., Fry, J.: SME's Lending and Islamic finance. is it a "win-win" situation? Econ. Model. **55**, 1–5 (2016). https://doi.org/10.1016/j.econmod.2016.01.029

Shleifer, A., Vishny, R.W.: A survey of corporate governance. J. Finance **1**(2), 737–783 (1997). https://doi.org/10.1111/j.1540-6261.1997.tb04820.x

Sinaga, S.R.N., Adilla, N., Sriani, S.: Role of productive zakat funds on mustahik micro business development (case study of Medan City Rumah Zakat). J. Syarikah **6**(2), 130–136 (2020)

Sohag, K., Mahmud, K.T., Alam, M.F., Samargandi, N.: Can Zakat system alleviate rural poverty in Bangladesh? A propensity score matching approach. J. Poverty **19**(3), 261–277 (2015). https://doi.org/10.1080/10875549.2014.999974

Sumiati, S.: Improving small business performance: the role of entrepreneurial intensity and innovation. J. Asian Finance Econ. Bus. **7**(10), 211–218 (2020). https://doi.org/10.13106/jafeb.2020.vol7.n10.211

Widiastuti, T., Auwalin, I., Rani, L.N., Mustofa, M.U.A.: A Mediating Effect of Business Growth on Zakat Empowerment Program and Mustahiq's welfare. Cogent Bus. Manag. **8**(1), 1–18 (2021). https://doi.org/10.1080/23311975.2021.1882039

Yaacob, Y., Azmi, I.A.G.: Entrepreneur's social responsibilities from Islamic perspective: a study of Muslim entrepreneurs in Malaysia. Procedia Soc. Behav. Sci. **58**, 1131–1138 (2012). https://doi.org/10.1016/J.SBSPRO.2012.09.1094

Yumna, A., Clarke, M.: Integrating Zakat and Islamic charities with microfinance initiative in the purpose of poverty alleviation in Indonesia. In: Proceeding 8th International Conference on Islamic Economics and Finance (2011)

Zauro, N.A., Saad, R.A.J., Sawandi, N.: Enhancing socio-economic justice and financial inclusion in Nigeria: the role of Zakat, Sadaqah and Qardhul Hassan. J. Islam. Account. Bus. Res. **11**(3), 555–572 (2020). https://doi.org/10.1108/JIABR-11-2016-0134

Zin, M.L.M., Ibrahim, H.: The influence of entrepreneurial supports on business performance among rural entrepreneurs. Ann. Contemp. Dev. Manag. HR (ACDMHR) **2**(1), 31–41 (2020). https://doi.org/10.33166/ACDMHR.2020.01.004

$\mathbf{T} - \psi$ Schemes for a Transient Eddy-Current Problem on an Unbounded Area

Yiyue Sun(✉)

Engineering University of PAP, Xi'an 710086, China
sunyiyue.cool@163.com

Abstract. In this paper, the problem of the transient eddy-current over an unbounded ares is studied. According to the characteristics of the boundary element and the finite element method, we divide the infinite regions into finite inner region and infinite outer region. Discrete coupled scheme which is based on Euler scheme is discussed fully in this thesis by setting the natural integral operator. In addition, we prove the existence and uniqueness of solution and discuss their energy norm error estimates.

1 Theoretical Analysis of Transient Eddy-Current Problem on an Unbounded Area

1.1 Basic Equation of Eddy-Current Field

First, the Maxwell equations of eddy current problem with displacement current removed are given [1]:

$$\begin{cases} \textbf{curl } \mathbf{H} = \sigma \mathbf{E} + \mathbf{J}, \\ \textbf{curl } \mathbf{E} = \partial(\mu \mathbf{H})/\partial t, \\ \mathrm{div}(\mu \mathbf{H}) = 0. \end{cases} \tag{1}$$

In formula (1), the meaning of the symbols. **E** is the electric field intensity, **H** is the magnetic field intensity, μ is magnetic permeability, σ represent conductivity, and **J** represents the source current density. This model is aimed at low frequency and high conductivity, ignoring the influence of capacitance.

1.2 Field Treatment of Eddy Current Problems on Unbounded Domain

As shown in Fig. 1, Ω_c is a bounded three-dimensional polygonal region with the boundary of Γ_c. The spherical surface is introduced so that the center of the sphere is at the origin and the radius is R and its internal region $\Omega = \overline{\Omega}_c \cup \Omega_e^+$, then Ω_e^+ is the ring region with Γ_c and Γ as the boundary.

Set $\Omega_e = R^3 \backslash \overline{\Omega}_c$, $\Omega_e^- = R^3 \backslash \overline{\Omega}$. For any function u defined in Ω_c or Ω_e, we record the limit of its approximation from Ω_c or Ω_e^+ to Γ_c as u^- or u^+. This definition is also applicable to the approximation from Ω_e^- or Ω_e^+ to Γ.

L. Barolli (Ed.): EIDWT 2023, LNDECT 161, pp. 76–84, 2023.
https://doi.org/10.1007/978-3-031-26281-4_8

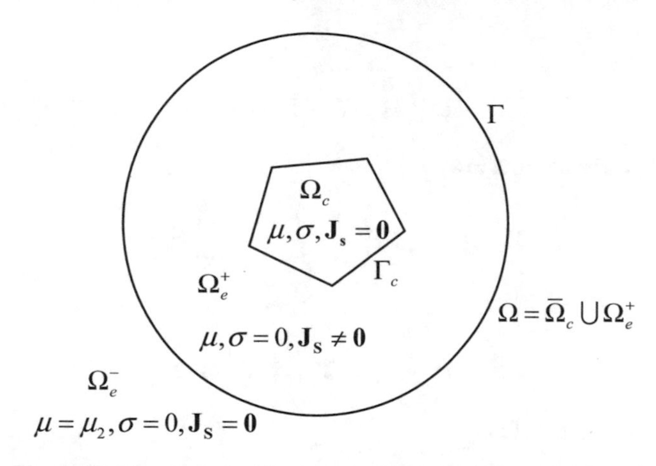

Fig. 1. Field model of eddy current problem in unbounded domain

In addition, we will also give the following assumption: the material parameters are linear and independent of time, that is, $\mu \in L^{\infty}(R^3)$ and $\sigma \in L^{\infty}(\Omega_c)$ are piecewise smooth, bounded real functions. In Ω, $0 \leq \mu_0 \leq \mu(\mathrm{y}) \leq \mu_1$; in $\Omega_e^- \mu(\mathrm{y}) \equiv \mu_2$; in $\Omega_c 0 \leq \sigma_0 \leq \sigma(\mathrm{y}) \leq \sigma_1$; in $\Omega_e \sigma = 0$; Here, μ_0, μ_1, σ_0 and σ_1 are normal numbers. We further assume that $\mathbf{J}_S$ divergence is free and $\mathrm{supp}(\mathbf{J}_S) \subset \Omega_e^+$.

1.3 Potential Based Instantaneous Eddy Current Equation in Unbounded Domain

The mathematical model of $\mathbf{T}-\psi$ method for transient eddy current problems in unbounded regions is given below.

It can be seen from the first equation in (1):

$$\mathrm{div}(\sigma \mathbf{E}) = 0 \quad \mathrm{in}\Omega_c \times (0, T].$$

Here we define

$$\sigma \mathbf{E} = \mathbf{curlT} \tag{2}$$

$$\mathbf{curl\ H} = \sigma \mathbf{E} + \mathrm{J}_s = \mathbf{curlT} + \mathbf{curlH}_s = \mathbf{curl}(\mathbf{T} + \mathbf{H}_s)$$

Thus

$$\mathbf{curl}(\ \mathbf{H} - \mathbf{T} - \mathbf{H}_s) = 0$$

According to Biot Savart law:

$$\mathbf{H}_s(\mathrm{x}) := \mathbf{curl}\left(\frac{1}{4\pi}\int_{R^3}\frac{J\_s(\mathrm{y})}{|\mathrm{x}-\mathrm{y}|}dy\right)$$

Therefore

$$\tilde{\mathbf{H}} = \mathbf{H} - \mathbf{H}_s = \mathbf{T} + \nabla\psi \quad \text{in } \Omega_c \tag{3}$$

For the second equation of (1)

$$\mathbf{curl}\ \mathbf{E} = -\partial(\mu\mathbf{H})/\partial t$$

subtract $\mu\partial\mathbf{H}_s/\partial t$ from both sides simultaneously, then get

$$\mu\frac{\partial(\mathbf{H} - \mathbf{H}_s)}{\partial t} + \mathbf{curl}\left(\frac{1}{\sigma}\mathbf{curlT}\right) = -\mu\frac{\partial\mathbf{H}_s}{\partial t}$$

i.e.

$$\mu\frac{\partial(\mathbf{T} + \nabla\psi)}{\partial t} + \mathbf{curl}\left(\frac{1}{\sigma}\mathbf{curlT}\right) = -\mu\frac{\partial\mathbf{H}_s}{\partial t} \tag{4}$$

For the third expression of (1)

$$\mathrm{div}(\mu\mathbf{H}) = 0$$

Substituting **H**, we get:

$$\mathrm{div}(\mu(\mathbf{H}_S + \mathbf{T} + \nabla\psi))0$$

Therefore

$$\mathrm{div}(\mu(\mathbf{T} + \nabla\psi)) = -\mathrm{div}(\mu\mathbf{H}_s)$$

It can be seen that according to Helmholtz's rule we can ensure the uniqueness of the solution. Therefore, we use a penalty function to limit the free divergence of **T**. Consider adding an entry $-s\nabla\left(\frac{1}{\mu}\mathrm{div}\mathbf{T}\right)$ to Eq. (4) on the left, where s is determined by the strength of the specification conditions. In this paper, $s = 1$.

Considering outside Ω_c, $\sigma = 0$ and ignoring the influence of source current in Ω_e^-, we have:

$$\tilde{\mathbf{H}} = \mathbf{H} - \mathbf{H}_s = \nabla\psi \quad in\ \Omega_e^+ \tag{5}$$

By strengthening the infinite radiation condition, the format $\mathbf{T}-\psi$ of the whole unbounded region problem can be obtained, namely [2]:

$$\begin{cases} \psi = O\left(|\mathrm{x}|^{-1}\right), \quad \nabla\psi = O\left(|\mathrm{x}|^{-1}\right) & \text{for } |\mathrm{x}| \to 0, \\ \mu_2 \Delta\psi = 0 & in\ \Omega_e^- \times (0,T], \\ \psi^- = \psi^+, \quad \mu_2 \nabla\psi^- \cdot \mathbf{n} = \mu\nabla\psi^+ \cdot \mathbf{n} & on\ \Gamma_c \times (0,T], \\ \mathrm{div}(\mu\nabla\psi) = -\mathrm{div}(\mu\mathbf{H}_s) & in\ \Omega_e^+ \times (0,T], \\ \psi^+ = \psi^- & on\ \Gamma_c \times (0,T], \\ \mathbf{T}\times\ \mathbf{n} = \mathbf{0}, \quad \dfrac{1}{\sigma}\mathrm{div}\,\mathbf{T} = 0 & on\ \Gamma_c \times (0,T], \\ \mu\nabla\psi^+ \cdot \mathbf{n} = \mu\left(\mathbf{T} + \nabla\psi^-\right)\cdot\mathbf{n} & on\ \Gamma_c \times (0,T], \\ \mu\dfrac{\partial(\mathbf{T}+\nabla\psi)}{\partial t} + \mathbf{curl}\left(\dfrac{1}{\sigma}\mathbf{curl}\mathbf{T}\right) & \\ \quad -\nabla\left(\dfrac{1}{\sigma}\mathrm{div}\,\mathbf{T}\right) = -\mu\dfrac{\partial\mathbf{H}_s}{\partial t} & in\ \Omega_c \times (0,T], \\ \mathrm{div}\left(\mu(\mathbf{T}+\nabla\psi)\right) = -\mathrm{div}(\mu\mathbf{H}_s) & in\ \Omega_c \times (0,T]. \end{cases} \tag{6}$$

We can assume the initial value problem for this instantaneous problem, $\mathbf{H}(\mathbf{y}, 0) = \mathbf{H}_s(\mathbf{y}, 0)$, let $\mathbf{T}_0(\mathbf{y}, 0) = \mathbf{0}$, $\psi_0(\mathbf{y}) = \mathbf{0}$.

2 Finite Element Analysis of $\mathbf{T}-\psi$ Method for Transient Eddy Current Problems on An Unbounded Area

Let $\mathcal{T}^h$ be a regular triangular mesh of region Ω, and its mesh parameter is h. On Γ_c, it is the matching grid. $\mathcal{T}_c^h$ is the corresponding mesh division on Ω_c. We can introduce the space V_h^0, W_h:

$$V_h^0 = \left(\mathbf{v} \in \hat{H}_0^1(\Omega) \cap C(\Omega)^3; \mathbf{v}|_\kappa \in (\mathcal{P}_1)^3, \forall\kappa \in \mathcal{T}_c^h\right),$$
$$W_h = \left\{\varphi \in H^1(\Omega) \cap C(\Omega); \varphi|_\kappa \in \mathcal{P}_1, \forall\kappa \in \mathcal{T}^h\right\}.$$

Take $\varphi_h \in W_h$, first multiply the fourth equation and the ninth equation in (6) by φ_h, calculate their integrals, and then we can get:

$$(\mu\nabla\psi, \nabla\varphi_h)_{\Omega_e^+} - \int_\Gamma \mu_2 \nabla\psi^- \cdot \mathbf{n}\varphi_h ds = (\mathrm{div}(\mu\mathbf{H}_s), \varphi_h)_{\Omega_e^+} + \int_{\Gamma_c} \mu\nabla\psi^+ \cdot \mathbf{n}\varphi_h ds$$

$$(\mu(\mathbf{T}+\nabla\psi), \nabla\varphi_h)_{\Omega_c} + \int_{\Gamma_c} \mu(\mathbf{T}+\nabla\psi^-)\cdot\mathbf{n}\varphi_h ds = (\mathrm{div}(\mu\mathbf{H}_s), \varphi_h)_{\Omega_c} \tag{7}$$

Next, express $\nabla\psi^{-}\cdot\mathbf{n}=\frac{\partial\psi^{-}}{\partial\mathbf{n}}(p)$ in the appropriate form. Let $G(p,p')$ represent the Green function of Laplace on Ω_e^{-}.

Let $v=G(p,p')$, $u=\psi$, thus

$$\psi(p)=-\int_{\Gamma}\frac{\partial}{\partial\nu'}G(p,p')\cdot\psi^{-}(p')\,ds' \qquad \forall p\in\Omega_e^{-} \tag{8}$$

where, p is a function of x, y, z, and $\nu(\nu')$ is the outer normal vector of Γ.

Now, we introduce the following operators Now, we introduce the following: $\mathcal{K}$[3]

$$\mathcal{K}\psi(p)=-\frac{\partial\psi^{-}}{\partial\mathbf{n}}(p) \qquad \forall p\in\Gamma$$

since $\nabla\psi^{-}\cdot\mathbf{n}=\frac{\partial\psi^{-}}{\partial\mathbf{n}}(p)$, $\forall p\in\Gamma$, $\mathcal{K}\psi$ can be substituted into:

$$(\mu\nabla\psi,\nabla\varphi_h)_{\Omega_e^{+}}+\mu_2\langle\mathcal{K}\psi,\varphi_h\rangle=(\mathrm{div}(\mu\mathbf{H}_s),\varphi_h)_{\Omega_e^{+}}+\int_{\Gamma_c}\mu\nabla\psi^{+}\cdot\mathbf{n}\varphi_h ds \tag{9}$$

Since Γ is a sphere whose center is the origin and radius is R, the Green's function $G(p,p')$ can be written directly, namely:

$$\psi(r,\theta,\varphi)=\frac{R}{4\pi}\int_0^{2\pi}\int_0^{\pi}\frac{(r^2-R^2)\sin\theta'}{(R^2+r^2-2Rr\cos\gamma)^{\frac{3}{2}}}\psi^{-}(\theta',\varphi')d\theta'd\varphi' \quad \forall r>R$$

And

$$\frac{\partial\psi^{-}(\theta,\varphi)}{\partial\mathbf{n}}=-\frac{1}{16\pi R}\int_0^{2\pi}\int_0^{\pi}\frac{\sin\theta'}{\sin^3\frac{\gamma}{2}}\psi^{-}(\theta',\varphi')d\theta'd\varphi' \tag{10}$$

Therefore, taking the derivative of t, we can get

$$\left(\mu\frac{\partial\mathbf{T}}{\partial t},\nabla\varphi_h\right)_{\Omega_c}+\left(\mu\nabla\frac{\partial\psi}{\partial t},\nabla\varphi_h\right)_{\Omega}+\mu_2\left\langle\mathcal{K}\frac{\partial\psi}{\partial t},\varphi_h\right\rangle_{\Gamma}=-\left(\mu\frac{\partial\mathbf{H}_s}{\partial t},\varphi_h\right)_{\Omega} \tag{11}$$

Similarly, if the eighth equation in (6) is multiplied by $\mathbf{Q}_h\in V_h^0$, then:

$$\left(\mu\frac{\partial\mathbf{T}}{\partial t},\mathbf{Q}_h\right)_{\Omega_c}+\left(\mu\nabla\frac{\partial\psi}{\partial t},\mathbf{Q}_h\right)_{\Omega}+\left(\frac{1}{\sigma}\mathbf{curlT},\mathbf{curlQ}_h\right)_{\Omega_c}+\left(\frac{1}{\sigma}\mathrm{div}\mathbf{T},\mathrm{div}\mathbf{Q}_h\right)_{\Omega_c}$$
$$=-\left(\mu\frac{\partial\mathbf{H}_s}{\partial t},\mathbf{Q}_h\right)_{\Omega_c} \tag{12}$$

2.1 Fully Discrete Coupled Scheme of $\mathbf{T} - \psi$ Method

The backward Euler difference format is used at Point $t = t_{i+1}$. So we get: [4]

$$
\begin{aligned}
&(\mu \frac{\mathbf{T}^{i+1} - \mathbf{T}^{i}}{\tau}, \mathbf{Q}_h)_{\Omega_c} + (\mu \nabla \left(\psi^{i+1} - \psi^{i} / \tau \right), \mathbf{Q}_h)_{\Omega_c} + \left(\frac{1}{\sigma} \mathbf{curlT}^{i+1}, \mathbf{curlQ}_h \right)_{\Omega_c} + \\
&\left(\frac{1}{\sigma} \mathrm{div} \mathbf{T}^{i+1}, \mathrm{div} \mathbf{Q}_h \right)_{\Omega_c} = - \left(\mu \left(\frac{\partial \mathbf{H}_s}{\partial t} \right)^{i+1}, \mathbf{Q}_h \right)_{\Omega c} - \left(\mu \mathbf{R}_1^{i+1}, \mathbf{Q}_h \right)_{\Omega_c} - \left(\mu \mathbf{R}_2^{i+1}, \mathbf{Q}_h \right)_{\Omega_c} \\
&\forall \mathbf{Q}_h \in V_h^0
\end{aligned} \tag{13}
$$

$$
\begin{aligned}
&(\mu \frac{\mathbf{T}^{i+1} - \mathbf{T}^{i}}{\tau}, \nabla \varphi_h)_{\Omega_c} + \left(\mu \nabla \left(\psi^{i+1} - \psi^{i} / \tau \right), \nabla \varphi_h \right)_{\Omega} + \mu_2 \left\langle \mathcal{K} \frac{\psi^{i+1} - \psi^{i}}{\tau}, \varphi_h \right\rangle_{\Gamma} \\
&= - \left(\mu \left(\frac{\partial \mathbf{H}_s}{\partial t} \right)^{i+1}, \mathrm{div} \varphi_h \right)_{\Omega} - \left(\mu \mathbf{R}_1^{i+1}, \mathrm{div} \varphi_h \right)_{\Omega_c} \\
&- \left(\mu \mathbf{R}_2^{i+1}, \mathrm{div} \varphi_h \right)_{\Omega} + \mu_2 \left\langle \mathcal{K} \mathcal{R}_3^{i+1}, \varphi_h \right\rangle_{\Gamma} \qquad \forall \varphi_h \in W_h
\end{aligned} \tag{14}
$$

where

$$
\boldsymbol{R}_1^{i+1} = \partial \mathrm{T}(t_{i+1}) / \partial t - \left(\mathrm{T}^{i+1} - \mathrm{T}^{i} \right) / \tau = \frac{1}{\tau} \int_{t_i}^{t_{i+1}} (t - t_i) \frac{\partial^2 \mathrm{T}}{\partial t^2} dt \tag{15}
$$

$$
\mathbf{R}_2^{i+1} = \mathrm{div} \partial \psi(t_{i+1}) / \partial t - \mathrm{div} \psi^{i+1} - \psi^{i} / \tau = \frac{1}{\tau} \int_{t_i}^{t_{i+1}} (t - t_i) \nabla \frac{\partial^2 \psi}{\partial t^2} dt \tag{16}
$$

$$
\mathcal{R}_3^{i+1} = \partial \psi \left(t_{i+1} \right) / \partial t - \left(\psi^{i+1} - \psi^{i} \right) / \tau = \frac{1}{\tau} \int_{t_i}^{t_{i+1}} \left(t - t_i \right) \frac{\partial^2 \psi}{\partial t^2} dt \tag{17}
$$

Substitute approximate solution $(\mathbf{T}_h, \psi_h) \in V_h^0 \times W_h$ for $(\mathbf{T}, \psi)$ in sum and omit its error term. Then let:

$$
T_h^0 = 0, \psi_h^0 = 0 \tag{18}
$$

so the coupling format of the fully discrete $(\mathbf{T}, \psi)$ method can be obtained:

Find $\left(\mathbf{T}_h^{n+1}, \psi_h^{n+1} \right) \in V_h^0 \times W_h$, $n = 0, 1, \cdots, M - 1$, such that:

$$
\begin{aligned}
&(\mu \frac{\mathbf{T}_h^{n+1} - \mathbf{T}_h^{n}}{\tau}, \mathbf{Q}_h)_{\Omega_c} + (\mu \nabla \frac{\psi_h^{n+1} - \psi_h^{n}}{\tau}, \mathbf{Q}_h)_{\Omega_c} + \left(\frac{1}{\sigma} \mathbf{curlT}_h^{n+1}, \mathbf{curlQ}_h \right)_{\Omega_c} \\
&+ \left(\frac{1}{\sigma} \mathrm{div} \mathbf{T}_h^{n+1}, \mathrm{div} \mathbf{Q}_h \right)_{\Omega_c} = - \left(\mu \left(\frac{\partial \mathbf{H}_s}{\partial t} \right)^{n+1}, \mathbf{Q}_h \right)_{\Omega c}
\end{aligned} \tag{19}
$$

$$
(\mu \frac{\mathbf{T}_h^{n+1} - \mathbf{T}_h^{n}}{\tau}, \nabla \varphi_h)_{\Omega_c} + \left(\mu \nabla \frac{\psi_h^{n+1} - \psi_h^{n}}{\tau}, \nabla \varphi_h \right)_{\Omega}
$$

$$+\mu_2\left\langle \mathcal{K}\frac{\psi_h^{n+1}-\psi_h^n}{\tau},\varphi_h\right\rangle_\Gamma=-\left(\mu\left(\frac{\partial \mathbf{H}_s}{\partial t}\right)^{n+1},\nabla\varphi_h\right)_\Omega \tag{20}$$

where $(\mathbf{Q}_h, \varphi_h) \in V_h^0 \times W_h$.

For the fully discrete coupled scheme, the solution $\left(\mathbf{T}_h^{n+1}, \psi_h^{n+1}\right)$ is unique. It can be proved according to Lax-Milgram theorem and the symmetry, continuity and coercive of $a(\cdot\,, \ \cdot)$.

3 Error Estimation Based on Euler Scheme

First, the necessary inequalities are given.

The definitions of $\mathbf{T}$ and ψ are extended from $[0, T]$ to $[-\tau\,, T]$ according to the time variable t.

(1) Let (T, ψ) and (T_h, ψ_h) be the solution and approximate solution of the problem and the coupling scheme sd respectively, assuming that:

$$\mathrm{T}\in H^2\left(0,T;\hat{H}_0^1(\Omega_c)\cap H^2(\Omega_c)^3\right),\psi\in H^2\left(0,T;H^1(\Omega)\cap H^2(\Omega)\right)$$

Then we can get

$$\|(\mathbf{T}_h+\nabla\psi_h)-(\mathbf{T}+\nabla\psi)\|^2_{l^\infty(L^2(\Omega_c)^3)}+\|\nabla\psi_h-\nabla\psi\|^2_{l^\infty\left(L^2(\Omega_e^+)^3\right)}\le C\left(\tau^3+h^2\right) \tag{21}$$

The idea of proof: ① when $n = I$ [5]

$$\begin{aligned}&(\mu\frac{\left(\mathbf{T}_h^{i+1}-\mathbf{T}^{i+1}\right)-\left(\mathbf{T}_h^i-\mathbf{T}^i\right)}{\tau},\mathbf{Q}_h)_{\Omega_c}+(\mu\nabla\frac{\left(\psi_h^{i+1}-\psi^{i+1}\right)-\left(\psi_h^i-\psi^i\right)}{\tau},\mathbf{Q}_h)_{\Omega_c}\\&+\left(\frac{1}{\sigma}\mathrm{div}\left(\mathbf{T}_h^{i+1}-\mathbf{T}^{i+1}\right),\mathrm{div}\mathbf{Q}_h\right)_{\Omega_c}+\left(\frac{1}{\sigma}\mathbf{curl}\left(\mathbf{T}_h^{i+1}-\mathbf{T}^{i+1}\right),\mathbf{curl}\mathbf{Q}_h\right)_{\Omega_c}\\&=\left(\mu\mathbf{R}_1^{i+1},\mathbf{Q}_h\right)_{\Omega_c}+\left(\mu\mathbf{R}_2^{i+1},\mathbf{Q}_h\right)_{\Omega_c}\quad\forall\mathbf{Q}_h\in V_h^0\end{aligned} \tag{22}$$

$$\begin{aligned}&(\mu\frac{\left(\mathbf{T}_h^{i+1}-\mathbf{T}^{i+1}\right)-\left(\mathbf{T}_h^i-\mathbf{T}^i\right)}{\tau},\nabla\varphi_h)_{\Omega_c}+\mu_2\left\langle\mathcal{K}\frac{\left(\psi_h^{i+1}-\psi^{i+1}\right)-\left(\psi_h^i-\psi^i\right)}{\tau},\varphi_h\right\rangle_\Gamma\\&+(\mu\nabla\frac{\left(\psi_h^{i+1}-\psi^{i+1}\right)-\left(\psi_h^i-\psi^i\right)}{\tau},\nabla\varphi_h)_\Omega\\&=\left(\mu\mathbf{R}_1^{i+1},\mathrm{div}\,\varphi_h\right)_{\Omega_c}+\mu_2\left\langle\mathcal{K}\mathcal{R}_3^{i+1},\varphi_h\right\rangle_\Gamma+\left(\mu\mathbf{R}_2^{i+1},\mathrm{div}\,\varphi_h\right)_\Omega\qquad\forall\varphi_h\in W_h\end{aligned} \tag{23}$$

Set
$\mathbf{M}_h^{k+1} = \mathbf{T}_h^{k+1} - P_h\mathbf{T}^{k+1},\ \eta_h^{k+1} = \psi_h^{k+1} - P_h\psi^{k+1}$
$\mathbf{Q}_h = \mathbf{M}_h^{k+1} - \mathbf{M}_h^k,\ \varphi_h = \eta_h^{k+1} - \eta_h^k$
Using the projection definition, the formula (3.4) and (3.5) can be transformed into:

$$\begin{aligned}
&\frac{1}{\tau}\left(\mu\left(\mathbf{M}_h^{i+1} - \mathbf{M}_h^i\right), \mathbf{M}_h^{i+1} - \mathbf{M}_h^i\right)_{\Omega_c} + \frac{1}{\tau}\left(\mu\left(\eta_h^{i+1} - \eta_h^i\right), \mathbf{M}_h^{i+1} - \mathbf{M}_h^i\right)_{\Omega_c} \\
&+ \left(\frac{1}{\sigma}\mathbf{curl}\mathbf{M}_h^{i+1}, \mathbf{curl}\left(\mathbf{M}_h^{i+1} - \mathbf{M}_h^i\right)\right)_{\Omega_c} + \left(\frac{1}{\sigma}\mathrm{div}\mathbf{M}_h^{i+1}, \mathrm{div}\left(\mathbf{M}_h^{i+1} - \mathbf{M}_h^i\right)\right)_{\Omega_c} \\
&= \left(\mu\mathbf{R}^{i+1}, \mathbf{M}_h^{i+1} - \mathbf{M}_h^i\right)_{\Omega_c} + \left(\frac{1}{\sigma}\mathbf{curl}\left(\mathbf{T}^{i+1} - P_h\mathbf{T}^{i+1}\right), \mathbf{curl}\left(\mathbf{M}_h^{i+1} - \mathbf{M}_h^i\right)\right)_{\Omega_c} \\
&+ \left(\frac{1}{\sigma}\mathrm{div}\left(\mathbf{T}^{i+1} - P_h\mathbf{T}^{i+1}\right), \mathrm{div}\left(\mathbf{M}_h^{i+1} - \mathbf{M}_h^i\right)\right)_{\Omega_c} \\
&= \left(\mu\mathbf{R}^{i+1}, \mathbf{M}_h^{i+1} - \mathbf{M}_h^i\right)_{\Omega_c} \\
&- \left(\mu\left(\mathbf{T}^{i+1} + \nabla\psi^{i+1}\right) - \mu\left(P_h\mathbf{T}^{i+1} + \nabla P_h\psi^{i+1}\right), \mathbf{M}_h^{i+1} - \mathbf{M}_h^i\right)_{\Omega_c}
\end{aligned} \tag{24}$$

② Using Cauchy-schwarz's inequality [6].

③ Estimate each type separately, then get it from the Triangle inequality.

In Ω_c and Ω_e^+, wo set $\mathbf{H}_h^{n+1} = \mathbf{T}_h^{n+1} + \nabla\psi_h^{n+1} + \mathbf{H}_s$, $\mathbf{H}_h^{n+1} = \nabla\psi_h^{n+1} + \mathbf{H}_s$ respectively, then it can be seen from the above conclusion

$$\|\mathbf{H}_h - \mathbf{H}\|_{l^\infty(L^2(\Omega)^3)}^2 \le C\left(\tau^3 + h^2\right)$$

(2) According to conclusion (1), we can get

$$\|\mathbf{curl}(\mathbf{T}_h - \mathbf{T})\|_{l^\infty(L^2(\Omega_c)^3)}^2 \le C\left(\tau^2 + h^2\right) \tag{25}$$

where C is constant and independent of time step τ and grid scale h.

In the conductor area of the eddy current field, the eddy current density meets $\mathbf{J} = \sigma\mathbf{E}$. Let $\mathbf{J}_h^{n+1} = \sigma\mathbf{E}_h^{n+1} = \mathbf{curl}\mathbf{T}_h^{n+1}$, according to conclusion (2), we can get

$$\|\mathbf{J}_h - \mathbf{J}\|_{l^\infty(L^2(\Omega_c)^3)}^2 \le C\left(\tau^2 + h^2\right) \tag{26}$$

4 Conclusion

In this paper, Euler difference is used to obtain the fully discrete form, and the central difference can also be used. Of course, different methods lead to different models. We can continue to try in the following two directions in the subsequent work:

i. On the existing basis, further improve the error accuracy;

ii. Numerical experiments of transient eddy current problems in unbounded regions are carried out to compare the correctness and effectiveness of each algorithm under the coupling conditions of finite element and boundary element.

References

1. Wang, B.: Computational Electromagnetics. Science Press, Beijing (2002)
2. Wang, Y., Wang, R., Kang, T.: A-Φ finite element method for a class of nonlinear eddy current problems with ferromagnetic material parameters. Comput. Math., 137–151 (2016)
3. Badics, Z., Matsumoto, Y., Aoki, K., Nakayasu, F., Uesaka, M., Miya, K.: An effective 3-D finite element scheme for computing electromagnetic field distortions due to defects in eddy-current nondestructive evaluation. IEEE Trans. Magn. **33**(2), 1012–1020 (1997)
4. Nie, X., Li, Y.: Electromagnetic field simulation based on a new numerical algorithm. Mech. Strength (2018)
5. Kang, T., Chen, T., Zhang, H., IkKim, K.: Improved T−ψnodal finite element schemes for eddy current problems. Appl. Math. Comput. **218**(2), 287–302 (2011)
6. Wang, Y., Jiang, X.: Calculation of 2D transient eddy current field by ES-FEM-BEM coupling method. J. Harbin Inst. Technol. **53** (2021)

Zakat Management Model Based on ICT

Bedjo Santoso(✉), Provita Wijayanti, and Fenita Austriani

Faculty of Economics, Universitas Islam Sultan Agung, Jalan Raya Kaligawe KM 4, Semarang, Indonesia
{bedjo.s,provita.w}@unissula.ac.id

Abstract. The study of the role of technology to support is very important and attracted many types of research. However, their approach is still partially and less comprehensive which does not involve important variables such as credibility, transparency, accountability, and trust as the main values. Therefore, this study aims to design a model of *Zakat* development backed up by ICT which combines various elements of *Zakat* management in order to make it more comprehensive. This study seeks to design a conceptual model of ICT-based *Zakat* management that involves 350 respondents in three areas of Indonesia which involve *Zakat* Management Units (UPZ), Muslim scholars, academics, activists of *Zakat*, *Muzakki*, and selected *Mustahiq*. By involving these stakeholders, a comprehensive model is expected to achieve. The results of this research showed that of the five important factors in *Zakat* management to boost the *Zakat* volume ICT is the most factor. Followed by trust, credibility, transparency, and accountability. From the five variables, 21 indicators were gained, of which with further IDC (intelligence, Design, and Choice) modeling design method resulted from a conceptual *Zakat* model based on ICT. Based on the interview results Blockchain technology is the most appropriate means to support *Zakat* management in order to meet the *Zakat* requirement, besides that Fintech, IoT, and AI are also important to support *Zakat* Management. Further research is needed to detail the design of the Blockchain model implemented in *Zakat* management.

Keywords: *Zakat* management · Trust · Credibility · Transparency · Accountability · ICT

1 Introduction

Indonesia has a great number of Muslim populations. The strengthening economics for *Ummah* through *Zakat*, therefore, has a promising potential to be developed [1] According to [2] Zakat has been proven to make a large number of contributions in advancing the life of Muslim [2] Alongside with [2, 3] indicated that if Islamic Finance instruments (includes *Waqaf* and *Zakat*) can be managed properly and seriously it is possible for Muslim countries to reduce drastically their foreign debt as *Zakat* funds can be mobilized to build some infrastructures.

Some researchers, however, indicate that the collection of *Zakat* fund in Indonesia has been far from the target [4]. Regarding with this, the president of Indonesia stated

L. Barolli (Ed.): EIDWT 2023, LNDECT 161, pp. 85–97, 2023.
https://doi.org/10.1007/978-3-031-26281-4_9

that the large *Zakat* potential in Indonesia must be optimized, from the potential of IDR. 252 trillion (USD 17,25 Billion), it was just collected in IDR. 8.1 trillion (USD 620 million) by BAZNAS, the Indonesia *Zakat* National Agency (*Badan Amil Zakat Nasional*/BAZNAS). Due to that situation, some significant efforts are needed to upgrade the revenue of *Zakat* in Indonesia [5].

Several studies indicate that the low collection of *Zakat* in Indonesia is caused by one important factor that the *Zakat* management system is not yet modern and the demands for *Zakat* payer have not been accommodated. Several studies have been carried out, especially those related to the importance of increasing the management and professionalism of *Zakat* institutions, for example [6, 7] who claimed that high public trust in *Zakat* institutions can increase the volume of *Zakat*. Whilst, [8] explains that public trust in *Zakat* institutions can be increased by boosting the credibility of *Zakat* institutions through transparent reporting of *Zakat* management. This indicates that there are still flaws in current *Zakat* management as in line with the findings of [9] research where *Zakat* institutions need to develop good *amil* governance so that people consider *Zakat* management institutions to be credible and professional.

The collection of *Zakat* has been regulated by some Acts and Regulations in Indonesia to promote people and agencies to pay, collect, and distribute *Zakat*. It is important because *Zakat* are undertaken by the respective religious authorities according to Shariah requirements. According to several research papers, however, there are still several challenges such as inefficiency, a lack of transparency in how the funds are collected, managed and distributed. According to [10] these factors are the main causes which make *Zakat* cannot work properly and efficiently.

Alongside with that, some study found that the *Zakat* management is consistent with the current technology development. It is believed that the application of the technology on the *Zakat* management include monitoring, reporting, and evaluating. Technology can strengthen the management system, thus provide better management of *Zakat* ([11] Another study conducted by [12] concluded that the detailed and transparent movement of *Zakat*, from the moment of collection to the point of disbursement could soon be tracked via *Zakat tech*. Then it can be concluded that *Zakat tech* based on sophisticated and modern technology would enable the tracking of funds throughout the whole process of *Zakat*. Furthermore, in the disruption Technology Era 4.0, Indonesia *Zakat* National Agency (*Badan Amil Zakat Nasional/BAZNAS*) as Indonesian *Zakat* authorized agency has developed a collecting *Zakat* online in web www.baznas.go.id.

Nevertheless, some weaknesses of the applied technology of *Zakat* are still partial and not yet integrated with inherent factor in the *Zakat* institution agency. From the description above, it can be said that the technological factors that can guarantee the efficiency of collection, distribution and reporting of *Zakat* are very important in the effort to increase the revenue of *Zakat* in Indonesia. In addition, from the demand perspective of *Zakat* payer, *Zakat* management institutions want to improve *Zakat* management to be efficient, open, and accountable. Thus, accountability, transparency, and trust turn out to be very important factors to be built along with the collecting *Zakat* technology factor. Hence, this paper seeks to answer the current issues by proposing an integrated ICT-based *Zakat* management design.

2 Literary Review

2.1 Potential of Zakat

Indonesia has a large potential of *Zakat* because Indonesia is the biggest Islamic country in the world with 250 million people as a Muslim community. The realization of *Zakat*, however, is still little as amounted as 3.5% [4, 13, 14]. The big gap is basically due to various factors such as supply and demand. The supply factors include, for example, supportive government regulations, trust in *Zakat* management institutions, and the role of scholars in teaching the society about contemporary *Zakat* obligations and explaining what types of incomes are obliged to be paid for its *Zakat* is growing [15, 16].

Meanwhile, the demand factors come from trust, willingness, and awareness in the law of Muslim obligations in paying taxes. The awareness of people's preferences in paying *Zakat* is influenced by several factors, including the ease of paying *Zakat*, public trust in *Zakat* management institutions, and other factors such as the credibility and accountability of *Zakat* management institutions. Besides, the factors of the revenue report transparency and the distribution of *Zakat* are also important.

Several previous studies show that these factors have not been optimized. The study of [15] for example, shows that there is still a lack of public understanding in *Zakat* awareness. *Zakat* has not become a special subject in Islamic educational institutions (universities) as an integrated curriculum. This is what according to [2] and [16] is that *Zakat* has not been understood as a tool to develop civilization. However, there are many *Zakat* institutions in every region in Indonesia. The shariah/*da'wah* for *Zakat* is carried out intensively by *Zakat* institutions. The socialization is carried out by understanding the importance of *Zakat* and the urgency of the obligation to pay *Zakat*. Besides, the introduction of *Zakat* institutions and how to calculate and pay *Zakat* are also given. In addition, there are many *Zakat* institutions which have implemented various innovations related to *Zakat* collection and management strategies. Traditional and conventional methods are being upgraded to be more modern, innovative and expansive. The *Zakat* collection strategy carried out by institutions is usually by visiting *Muzakki* door to door, serving *Muzakki* through service offices, or opening outlets or stands in public places.

Furthermore, *Zakat* service institutions have improved online *Zakat* transaction services through mobile applications or net banking. This institution also create collaboration with marketplaces such as bukalapak.com, as well as create online websites for social or crowdfunding fundraising. This strategy is expected to increase *Zakat* donations and bring *Zakat* institutions closer to the community. This model is in accordance with the millennial generation who is projected to be a new market for *Zakat* movement [9].

The strategy for increasing the volume of *Zakat* donations, therefore, is necessary to upgrade some factors from the demand side such as the role of technology which makes *Zakat* efficient and makes it easier for *Zakat* payers to make *Zakat* donation transactions either way through internet technology such as mobile banking or other forms. In addition, it is also important to improve *Zakat* management from the *Zakat* institution perspective, such as trust of *Zakat* institutions, transparency in *Zakat* management, accountability, and credibility of *Zakat* institutions. The combination with governance of *Zakat* institution and technological aspect is very crucial to be developed.

2.2 Information Communication Technology (ICT)

Information and Communication Technology is a term that covers all technical equipment for processing and conveying information. ICT consists of two aspects, namely information technology and communication technology. Information technology includes everything related to the process, use as a tool, manipulation, and management of information. Information technology is the study or use of electronic equipment, especially computers to store, analyze, and distribute any kind of information, including words, numbers, and pictures [17].

Moreover, ICT can be divided into information technology and communication technology. Information technology is used to process information. Communication technology is used to transfer information from source to receiver. According to [18] in general the media has the following advantages: (1) Clarify the presentation of the message so that it is not too formalistic. (2) Overcoming the limitations of space, time, and senses. In addition, [19] argues that ICT has a function: Capturing rare objects and events that can be immortalized with photos, films, and recorded via video or audio, then the events are stored and used when needed.

The ICT systems are premised on the centrality of information technology in everyday socio-economic life to support meanings, tacit understanding, experiences, interconnections, process of change, and positive effects. ICT Focuses either on human interactions and social structure, then the theory tries to resolve in efficiency. The proposition has resulted from the ICT theory as above: ***Proposition***: ICT is a Power to Facilitate Communication in Order to boost any Activities Based on Some Values i.e., Speed, Efficient, Big Capacity, Lower Cost, and effectiveness Significantly.

2.3 The Importance of Transparency, Credibility, and Accountability in Zakat Management

Research related to the above issue have been carried out one of which by [7] who explained that the wider beneficiaries of *Zakat* are, the more the public trust will increase. Thus, it can be said that the success of *Zakat* collection is determined by how widely the community feels the benefits of the *Zakat*. Another research related to the importance of trust was conducted by [6] He stated that trust can increase people's intention to pay *Zakat* through *Zakat* institutions. Furthermore, the trust of this community can be achieved through effective *Zakat* management methods [8]. Whereas [20] found that increasing trust can be done through accountability and transparency. A similar study was conducted by [21] who concluded that transparency can increase *Zakat* payers.

One indicator of good *Zakat* management is transparency. As stated by [22] transparency can maintain community loyalty to the institution. This was confirmed by [23] Transparency is a part of good governance which must be owned by all agencies or institutions. Such transparency can be in the form of report which can also be accessed by the community [8] Related to the issue of credibility as one of the important factors in *Zakat* management is reviewed by [6] They concluded that the credibility of the *Zakat* institution affects the behavior of the people to pay *Zakat*. Furthermore, the transparency of *Zakat* management significantly influences the trust of *Muzakki* in paying *Zakat*. This is reinforced by the previous study that the credibility of *Zakat* institutions influences people's behavior to pay *Zakat*. The research conducted by [24] revealed the importance of service and technology facilities that accommodate the payment of *Zakat*. From the earlier description in relation to the important factors influencing effective *Zakat* management can be described in the figure as follows (Fig. 1):

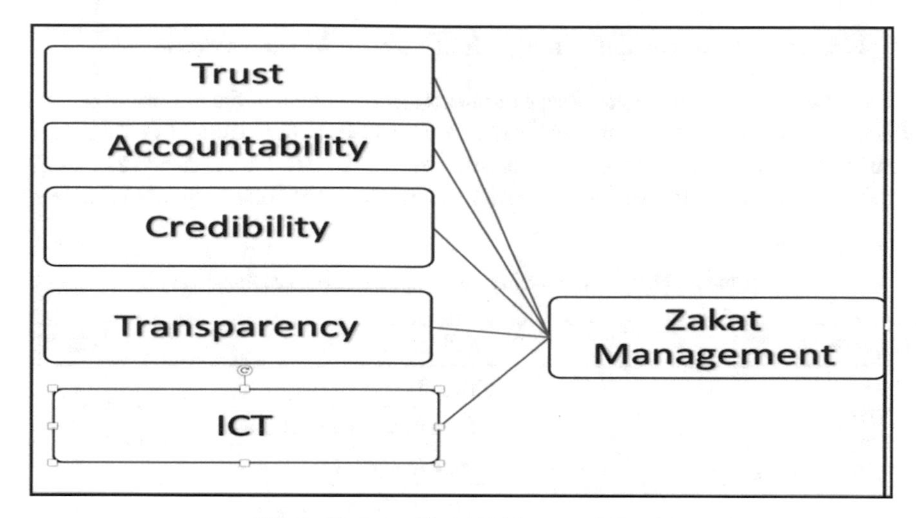

Fig. 1. Factors affected *Zakat* management

From the description above, it can be concluded that accountability, credibility, transparency, and effective management can increase public trust where the increased trust will result in the expansion of the *Zakat* payer segment, loyalty, satisfaction, and community behavior, as well as community preferences towards *Zakat* institutions. Many researchers suggest that realizing the effectiveness of those variables such as accountability, credibility, and transparency is essential to be backed up by an information system and effective management of *Zakat* collection. Information technology in *Zakat* management can be an effective tool in increasing the amount of *Zakat* collection.

The previous researchers were aware that if the *Zakat* institutions wanted to increase their collection, then there should have been obvious evidence that they were quite trustworthy in managing *Zakat* funds. The mandate and responsibility not only merely record financial reports, the total collection and distribution of *Zakat* funds each year to the government and the national *Amil Zakat* Agency, but also must be conveyed to the wider community as a party that has the potential to pay their *Zakat-to-Zakat* institutions. The public expects transparent and credible reporting. The transparent standardization can actually be seen if we intend to look for the financial statements of *Zakat*-related institutions, and also open the websites of *Zakat* institutions to find out the *Zakat* distribution area and what activities are carried out by them (*Zakat* distribution activities). Here, it is visible that there is a gap where people have the convenience to pay *Zakat* anywhere and anytime, yet it will be rather difficult to obtain information related to the distribution of *Zakat* funds.

Moreover, the distribution of *Zakat* funds must be classified according to the area of the beneficiary, based on the agreement for the donation of *Zakat* and charity received, as well as various types of other distribution program specifications such as for mosque construction programs, scholarships, *Da'wah* to prevent the Christianization of poor Muslim communities, venture capital assistance for the poor, and community economic empowerment programs.

2.4 Indicators Development of the Variables Zakat Management

Based on the analysis of the previous sub-chapter, it is found that the important factors in effective *Zakat* management involve trust, accountability, transparency, credibility, and the use of IT. According to the previous researchers, the five factors need to be derived in indicators as a tool to measure the variables shown in the following table (Table 1).

Table 1. Factors influencing the success of *Zakat* management

Variable	Indicators	CODE
Trust [25, 26]	Trust in ICT	T1
	Trust in the future of *Zakat*	T2
	Trust in service quality	T3
	Trust in *Zakat* management	T4
Accountability [25]	Responsible motivation	A1
	Understanding of social obligations	A2
	The Importance of audit knowledge	A3
	The importance of experience	A4
Transparency [25]	Reliable system availability	T1
	Report accessibility	T2
	Report publication	T3
	Report availability	T4
	Availability of *Zakat* information	T5
Credibility	Reliability of the *Zakat* system	C1
	The relevance of the *Zakat* system	C2
	Ease of understanding	C3
Management and information technology [27]	Accuracy	ICT1
	The relevance of the system	ICT2
	Speed of results	ICT3
	Completeness/Features	ICT4
	Convenience	ICT5

3 Research Method

The sampling method used in this research is purposive stratified sampling combined with accidental purposive sampling. The questionnaire method collected 350 data was collected in Indonesia divided into three areas, namely: the east part of Indonesia 50 respondents, the middle part of Indonesia = 100 respondents, and the western part of Indonesia equal to 200 respondents. The Respondents include *Zakat* institutions, *Zakat* collection units, Muslim scholars, scientists, *Zakat* payers, and selected *Zakat* recipients. The respondents were asked to fill out a questionnaire based on the above indicators on a scale of 1–9 by using google Forms and they were asked to assess the important factors which affect the effectiveness of *Zakat* management in Indonesia. Besides that, some selected respondents were interviewed online to assess ICT platforms that deal to increase trust, accountability, transparency, efficiency, and credibility. Based on the respondent's views then will be developed the conceptual model by using the IDC method (Intelligent, Design, and Choice).

4 Results and Discussion

4.1 The Determinant Variable

The collected questionnaire data were processed by using descriptive statistical method consisting of the average for each indicator, the minimum value, the maximum value, the standard deviation, and the average value for each variable. The results of the data processing are presented in the following table.

From the results of the analysis above, the most dominant variable is the use of ICT in *Zakat* management, then followed by public trust, credibility, transparency, and finally, accountability. Thus, *Zakat* institutions need to prioritize the use of information technology, the increase of public trust, the credibility and transparency of *Zakat* management, and the accountability of *Zakat* management report. The order of rank of the most important factors can be seen in the following table (Table 3).

From the table above, it can be seen that in realizing good *Zakat* management, the public requires the existence of a reliable ICT, the level of trust in the *Zakat* institution, the credibility of the *Zakat* management institution, the transparency of the system, the transparency of *Zakat* source, the distribution report, and at last the accountability regarding with how important the *Zakat* institution is in its responsibility.

4.2 The Conceptual Design of ICT-Based *Zakat* Management

Based on the description in the previous sub-chapter and by using the Intelligent Design and Choice (IDC) method from [28] the intelligent Phase includes collecting data and information related to the issue. The phases of design involve experts in the fields to attain the dominant variable into some possible models. Then, the choice is the appropriate selection phases. This design method model is an analytic subjective deductive, which connects some information and data collected so that the conceptual design model for management Zakat based on ICT is obtained as follows (Fig. 2).

Table 2. Descriptive statistics

Variable	Indicator	CODE	Mean	Min	Max	Std.	Average
Trust	Trust in ICT	T1	9,01	1	9	1,80	**8,49**
	Trust in the future of *Zakat*	T2	8,11	1	9	1,91	
	Trust in service quality	T3	8,92	1	9	1,78	
	Trust in *Zakat* management	T4	7,94	1	9	1,77	
Accountability	Responsible motivation	A1	7,23	1	9	1,50	**6,22**
	Understanding of social obligations	A2	6,43	1	9	1,61	
	The importance of audit knowledge	A3	6,31	1	9	1,83	
	The importance of experience	A4	4,92	1	9	0,49	
Transparency	Reliable system availability	T1	8,80	1	9	0,88	**8,42**
	Report accessibility	T2	9,68	1	9	0,97	
	Report publication	T3	8,62	1	9	0,86	
	Report availability	T4	8,44	1	9	0,84	
	Availability of *Zakat* information	T5	6,59	1	9	0,66	
Credibility	Reliability of the *Zakat* system	C1	8,80	1	9	0,88	**8,43**
	The relevance of the *Zakat* system	C2	7,80	1	9	0,78	
	Ease of understanding	C3	8,70	1	9	0,87	
Information and communication Technology	Accuracy	ICT1	8,30	1	9	0,83	**9,18**
	The relevance of the system	ICT2	9,70	1	9	0,97	
	Speed of results	ICT3	8,90	1	9	0,89	
	Completeness/Features	ICT4	9,20	1	9	0,92	
	Convenience	ICT5	9,80	1	9	0,98	

4.3 Respondents' View of the Appropriate ICT Platform to Increase Zakat Management

Based on a study conducted by interviewing 20 respondents which includes ten Academicians including Professors and Doctors who are well informed regarding *Zakat* (AC1 to AC10), three ICT practitioners (ICT1 to ICT3), and four *Zakat* Institutions

Table 3. The rank of the importance variables

Ranking	Variables
1st	Information technology (ICT)
2nd	Trust
3rd	Credibility
4th	Transparency
5th	Accountability

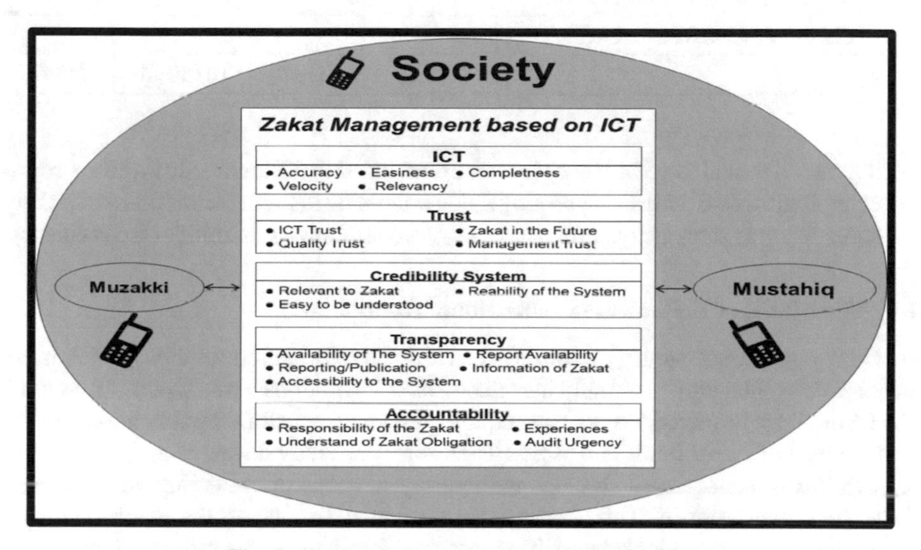

Fig. 2. Design *Zakat* management based on ICT

(ZA1 to ZA4), and three *Zakat* payers (ZAP1 to ZAP3). Questions include what technology can increase *Zakat* collecting through increasing trust, credibility, transparency, and accountability as if these variables can be deals then the *Zakat* payer will satisfy in terms of trust. *Zakat* management will be efficient, transparency, and accountable it does mean everyone can monitor in terms of data openness. Finally, the *Zakat* volume will increase as the target as in Indonesia the volume of *Zakat* is still poor. The respondents were asked to evaluate and choice which ICT technology which urgent in the current time to be implemented in Indonesia in boosting the *Zakat* management, the ICT Technologies offered are: Fintech, Blockchain, Internet of Things (IoT), and Artificial Intelligent (AI) the respondents can choice at least one. The respondents mapping displayed in the Table 4.

From Table 2, it can be concluded that all respondents agree that Blockchain technology can improve *Zakat* management, then Fintech can also increase efficiency and be backed-up by IoT and artificially intelligent. Among respondent's opinions, the highest selected technology is Blockchain (100%), Fintech (70%), IoT (50%), and Artificial

Table 4. Respondents opinion related the appropriate ict technology to deal with the *Zakat* management issues.

No	Respondents	What ICT technology can deal with the issues of trust, transparency, accountability, credibility, and efficiency in terms of *Zakat* management			
		Blockchain	Fintech	IOT	AI
1	Academicians (AC1 to AC10)	10	5	4	2
2	ICT Practitioners (ICT1 to ICT3)	3	3	3	3
3	Zakat Institutions (ZA1 to ZA4)	4	3	2	1
4	Zakat Payers (ZAP1 to ZAP3)	3	3	1	3
		20 (100%)	14 (70%)	10 (50%)	9 (45%)

Intelligent (AI) equal to 45%. Hence it can be concluded that Blockchain is the appropriate technology to deal with the increasing Zakat collecting in Indonesia as this technology has some features such as trust, transparency, accountability, credibility also efficiency.

4.4 The Model of Zakat Based Blockchain Technology

The current collection strategy these days has utilized information technology that it can increase *Zakat* donations, and help introduce *Zakat* institutions to the wider community. The technology, however, has not been equipped with responsibility publications for the management of *Zakat* funds and assets. Therefore, this study develops a more comprehensive *Zakat* management strategy, particularly in terms of collecting and reporting *Zakat* fund management. This strategy can be realized by using the blockchain system (worldwide *Zakat* blockchain). This integrated system can be referred to as worldwide *Zakat* management through blockchain. The Indonesian government also provides support for the use of blockchain technology to improve company performance [29].

Blockchain creates globally and permanently documented records (distributed ledger) [29] The transaction authorization and verification process is faster. Movement of *Zakat* funds and transactions can be traced so easily that transparency increases. The auditing process will be easier to do. Furthermore, a database of the transaction frequency can be processed by the *Zakat* manager to analyze people's behavior towards *Zakat* [30]. There have been many studies on the application of blockchain technology in the field of Islamic Banking products yet blockchain as a comprehensive *Zakat* management system has not been implemented.

Thus, the next research needs to elaborate a *Zakat* management model based on blockchain in detail. This is because the blockchain can deal with such issues as trust, credibility, transparency, and accountability. In addition, blockchain has advantages in security from cybercrime, fraud, and manipulation, and in increasing efficiency. With the implementation of blockchain, *Zakat* management is expected to be optimal. Some of the remarkable advantages of a business when using this technology include the wider access to finance, and the way of running businesses can be executed more efficiently, cheaply and safely [31].

5 Conclusion

From the discussion explained earlier, it can be concluded that to ensure effective zakat management, a design that is supported by ICT is needed. Based on several previous study there are five important variables that can ensure the success of technology-based zakat management. They include trust, accountability, credibility, transparency, and information technology management (ICT).

By using descriptive statistic founded the dominant factors in determining the success of effective zakat management produce five rank sequences; ICT as the first and then followed by Trust, Credibility, Transparency, and at last accountability. Moreover, the ICT-based *Zakat* management is needed to accommodate the variables of trust, credibility, transparency, and accountability where all stakeholders such as *Muzakki, Mustahiq* and society can watch the *zakat* management real time by using mobile phone.

Blockchain is the appropriate technology to be implemented to boosting *Zakat* volume because Blockchain has some features which meet with the *Zakat* requirement such as trust, transparency, accountability, and efficiency. The second rank technology is financial technology (Fintech), IOT and AI. It is therefore, the next research in detail is needed to design a *Zakat* model based on Blockchain technology due to its capability of providing security from the cybercrime, fraud and manipulation. In addition, the blockchain is able to facilitate the access of finance more widely, efficiently and cheaply.

References

1. Center of Strategic Studies: The National Board of Zakat. Indonesia Zakat Outlook 2019. Indonesia Zakat Outlook 2019, Jakarta (2019). https://drive.google.com/file/d/1tNlFdA0UYJmrVV-QHWNi2NcsfIID_9IIET/view. Accessed 08 Dec 2022
2. Kurnia, H.H., Hidayat, H.A.L.: Panduan Pintar Zakat. Qultum Media, Jakarta (2008)
3. Cizakca, M.: A History of Philanthropic Foundations: The Islamic World From the Seventh Century to the Present. Bogazici University Istambul-Turkey, Istambul (2000)
4. Amilahaq, F., Ghoniyah, N.: Compliance behavior model of paying Zakat on income through Zakat management. Share Jurnal Ekonomi dan Keuangan Islam **8**(1), 114–141 (2019). https://doi.org/10.22373/share.v8i1.3655
5. Asmara, C.G.: Potensi Zakat Rp 252 T, Masuk Baznas Cuma Rp 8,1 T, 19 May 2019. https://www.cnbcindonesia.com/syariah/20190516152005-29-72968/potensi-zakat-rp-252-t-masuk-baznas-cuma-rp-81-t. Accessed 06 Dec 2022
6. Azman, F.M.N., Bidin, Z.: Factors influencing Zakat compliance behavior on saving. Int. J. Bus. Soc. Res. **5**(1), 118–128 (2015)
7. Fahrurrozi: Fundraising Berbasis Zis: Strategi Inkonvensional Mendanai Pendidikan Islam. Ta'dib Jurnal Pendidikan Islam **19**(02), 23–42 (2014)
8. Owoyemi, M.Y.: Zakat management. J. Islam. Account. Bus. Res. **11**(2), 498–510 (2020). https://doi.org/10.1108/JIABR-07-2017-0097
9. Amilahaq, F., Ghoniyah, N.: Compliance behavior model of paying Zakat on income through Zakat management organizations. Share Jurnal Ekonomi dan Keuangan Islam **8**(1) (2019). https://doi.org/10.22373/share.v8i1.3655
10. Mahomed, Z.: Zakat modelling for the blockchain age: inspiration from Umar bin Abdul Aziz. International Centre for Education in Islamic Finance, Kuala Lumpur (2019)

11. Nabihah Esrati, S., Abdul-Majid, M., Mohd Nor, S.: Fintech (Blockchain) Dan Pengurusan Zakat di Malaysia/Financial Technology and Zakah Management in Malaysia (2018). https://www.researchgate.net/publication/329389820
12. Reza, Y.: Analisis faktor-faktor sukses sistem Zakat e-payment. Jurnal Riset Sains Manajemen **3**(1), 31–48 (2019)
13. Badan Amil Zakat Nasional: Puzat Baznas (2017). http://pusat.baznas.go.id/lembaga-amil-zakat/daftar-lembaga-amil-zakat/1/3. Accessed 06 Nov 2017
14. Firdaus, M., Beik, I.S., Irawan, T., Juanda, B.: Economic estimation and determinations of Zakat potential in Indonesia. In: IRTI Working Paper Series, vol. WP 1433–07, pp. 1–74, August 2012
15. Hafidhuddin, D.: Zakat dalam Perekonomian Modern, 1st edn. Gema Insani Press, Jakarta (2006)
16. Mufraini, M.A.: Akuntansi dan Manajemen Zakat, 1st edn. Kencana Publisher, Jakarta (2008)
17. Rusman, M.-M.: Berbasis ICT. PT Raja Grafindo Press, Jakarta (2012)
18. Sadiman, A.S.: Media Pendidikan: Pengertian, Pengembangan, dan Pemanfaatannya (2012)
19. Faizi, R., Abdellatif, E.A., Raddouane, C.: Students' perceptions on social media use in language learning based on ICT. In: International Conference, ICT For Language Learning (2013)
20. Kristin, A.: Penerapan Akuntansi Zakat Pada Lembaga Amil Zakat (Studi Pada Laz Dpu Dt Cabang Semarang). Jurnal Akuntansi **7**(2) (2011). https://doi.org/10.26714/vameb.v7i2.698
21. Kashif, M., Faisal Jamal, K., Abdur Rehman, M.: The dynamics of Zakat donation experience among Muslims: a phenomenological inquiry. J. Islam. Account. Bus. Res. **9**(1), 45–58 (2018). https://doi.org/10.1108/JIABR-01-2016-0006
22. Yuliafitri, I., Khoiriyah, A.N.: Pengaruh Kepuasan Muzakki, Transparansi Dan Akuntabilitas Pada Lembaga Amil Zakat Terhadap Loyalitas Muzakki (Studi Persepsi Pada LAZ Rumah Zakat). Islamiconomic: Jurnal Ekonomi Islam **7**(2) (2016). https://doi.org/10.32678/ijei.v7i2.41
23. Jumaizi, J., Wijaya, Z.A.: Good governance Badan Amil Zakat, Infak, dan Sedekah Dan Dampaknya Terhadap Keputusan dan Loyalitas Muzaki. Majalah Ilmiah Informatika (2011). http://www.unaki.ac.id/ejournal/index.php/majalah-ilmiah-informatika/article/view/51/84. Accessed 08 Dec 2022
24. Tajuddin, T.S., Azman, A.S., Shamsuddin, N.: Zakāh compliance behaviour on income among Muslim Youth in Klang Valley. Jurnal Syariah **24**(3), 445–464 (2017). https://doi.org/10.22452/js.vol24no3.5
25. Hasrina, C.D., Yusri, Y., Sy, D.R.A.S.: Pengaruh Akuntabilitas dan Transparansi Lembaga Zakat Terhadap Tingkat Kepercayaan Muzakki Dalam Membayar Zakat Di Baitul Mal Kota Banda Aceh. Jurnal Hum. Jurnal Ilmu Sosial Ekonomi dan Hukum **2**(1), 1–9 (2019). https://doi.org/10.30601/humaniora.v2i1.48
26. Semuel, H., Wijaya, N.: Service quality, perceive value, satisfaction, trust, Dan Loyalty Pada Pt. Kereta Api Indonesia Menurut Penilaian Pelanggan Surabaya. Jurnal Manajemen Pemasaran **4**, 23–37 (2009). https://doi.org/10.9744/pemasaran.4.1.pp.%2023-37
27. Lasmaya, S.M.: Pengaruh Sistem Informasi Sdm, Kompetensi Dan Disiplin Kerja Terhadap Kinerja Karyawan. Jurnal Ekonomi Bisnis Entrepreneurship (e-J.) **10**(1), 25–43 (2021). https://jurnal.stiepas.ac.id/index.php/jebe/article/view/3. Accessed 07 Dec 2022
28. Sprague, R.H.: A framework for the development of decision support systems. MIS Q. **4**(4), 1 (1980). https://doi.org/10.2307/248957
29. Yuliani, A.: Beda blockchain dengan bitcoin. Kementerian Komunikasi dan Informatika (2018). https://www.kominfo.go.id/content/detail/11966/beda-blockchain-dengan-bitcoin/0/sorotan_media. Accessed 07 Dec 2022

30. Muhammad, N.: 5 Kegunaan Teknologi Blockchain Dalam Layanan Keuangan - Coinvestasi. https://coinvestasi.com/blockchain/panduan/pemula/5-kegunaan-teknologi-blockchain. Accessed 07 Dec 2022
31. Safitri, D.: Manfaat Yang Bisa Didapatkan Dari Teknologi Blockchain Dan Mata Uang Virtual - Dunia Fintech (2018). https://duniafintech.com/manfaat-yang-bisa-didapatkan-dari-teknologi-blockchain-dan-mata-uang-virtual/. Accessed 07 Dec 2022

Teaching Method of Advanced Mathematics Combining PAD Classroom with ADDIE Model

Yanyan Zhao(✉), Qiong Li, Xuhui Fan, Lili Su, Jingtao Li, and Xiaokang Liu

Foundation Department, Engineering University of PAP, Xi'an 710086, China
403527705@qq.com

Abstract. Based on the talent training objectives and teaching syllabuses, combined with the characteristics of different professional students and the students' position requirements, this paper proposes a teaching model combining PAD classroom with ADDIE model from the three perspectives of "Professional combination, Project guidance and Task-driven", and establishes a reasonable three-level assessment and evaluation system. Taking the course of Advanced Mathematics as an example, this paper selects a teaching class for pilot teaching, and the experimental effect is remarkable.

Keywords: PAD classroom · ADDIE model · Advanced mathematics · Teaching reform

1 Analysis of the Present Situation of Advanced Mathematics Teaching

Advanced Mathematics is a core cultural basic course offered by universities which is the basis for students to learn the following professional courses and Position development [1]. Learning Advanced Mathematics well is of great value and significance for cultivating students' logical thinking ability and improving their self-learning and sustainable development ability. However, the teaching of Advanced Mathematics is mainly based on traditional teaching. This infusing teaching method makes students become passive receivers of knowledge and it is difficult for them to actively participate in it. In addition, the abstractness, rigor and reasoning of mathematics course content make students not interested in learning it, lack of motivation and fear of difficulties. In particular, the source of students in the same class is significantly different. For example, there are mixed classes for liberal arts and science students, mixed classes for students of different majors, mixed classes for minority students and students with a little higher education, which lead to problems such as different mathematical foundation of students and different understanding ability of knowledge in the same class. The traditional teaching method with single form ignores the individual differences of students and inhibits the cooperation and communication between students. As a result, students with weak foundation cannot keep up with the progress and cannot well meet the personalized development of students, and the cultivation of students' autonomous learning ability

L. Barolli (Ed.): EIDWT 2023, LNDECT 161, pp. 98–107, 2023.
https://doi.org/10.1007/978-3-031-26281-4_10

has not been improved as it should be. Therefore, the teaching reform of Advanced Mathematics course is urgent.

In view of this teaching situation, this paper combines the characteristics and position requirements of students of different majors, and puts forward a teaching model based on PAD classroom and ADDIE mode from the perspectives of "Professional combination, Project guidance and Task-driven". In addition, taking Advanced Mathematics as an example, the class of 2021 was selected to carry out pilot teaching.

2 Advanced Mathematics Teaching Based on PAD Classroom and ADDIE Model

2.1 Introduction to Teaching Model

2.1.1 ADDIE Model

ADDIE teaching model revolves around three aspects: what to learn, how to learn and how to judge whether students have achieved learning results. The systematic method of teaching mainly includes Analysis, Design, Development, Implementation and Evaluation of five stages, which is called ADDIE for short [2]. In the five stages, analysis and design are the premise, development and implementation are the core, and evaluation is the guarantee. The five parts are interrelated and inseparable to ensure efficient course design and implementation.

The analysis focuses on the analysis of the teaching object, teaching content and teaching environment according to the characteristics of students of different majors and position requirements. Design includes teaching objectives, key and difficult points, teaching media and teaching strategies, teaching process and teaching resources and learning evaluation design [3]. Among them, the design of teaching process and teaching resources is based on the three aspects of "Professional combination, Project guidance and Task-driven" after the determination of teaching strategies, centering on learning activities. According to the two stages of analysis and design, development refers to select appropriate offline resources, design and develop various online auxiliary learning resources, pay attention to the combination of online and offline teaching resources, in order to meet the teaching needs. Implementation is to carry out teaching and learning activities. The three-level evaluation mode is established in the evaluation stage. The first-level evaluation is carried out by students filling in questionnaires, the second-level evaluation is the process evaluation, and the third-level evaluation is the final evaluation. Finally, the results of the three-level evaluation can be fed back to the other four stages, and constantly revised.

2.1.2 PAD Classroom

PAD classroom mainly includes three processes: Presentation, Assimilation and Discussion. The core idea of PAD classroom is to allocate half of the class time to the teacher for instruction, and the other half to students for interactive learning in the form of Discussion [4]. The class is divided into two forms: "PAD in-classroom" and "PAD off-classroom". "PAD in-classroom" refers to that the teacher's instruction and the students' discussion is completed in the same class. Between the teacher's instruction and the student's discussion, there should be time for student's self-study and internalization, which is suitable for relatively simple and easy to understand knowledge. "PAD off-classroom" refers to that the teacher's instruction and the students' discussion is completed in the different class. Such as the teacher's instruction is completed in this class and the students' discussion is completed in the next class, self-study internalization and homework is completed after class, which is suitable for more difficult knowledge [5].

2.1.3 Combination of PAD Classroom and ADDIE Model

The teaching is designed and implemented according to the five core steps of "analysis, design, development, implementation and evaluation" of the ADDIE model. The PAD classroom model is integrated into the implementation stage to create a good classroom atmosphere and improve the students' participation and self-learning ability. According to the feedback of the evaluation results, the other four stages can be adjusted at any time to form a systematic teaching closed loop (Fig. 1).

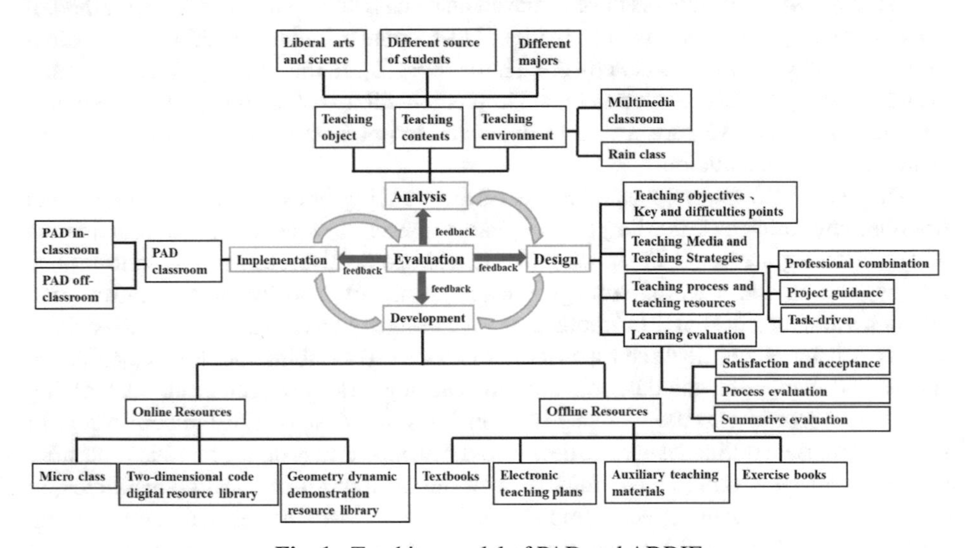

Fig. 1. Teaching model of PAD and ADDIE

2.2 Implementation Steps

2.2.1 Analysis

In the analysis stage, the analysis of teaching objects mainly includes the common characteristics of students, existing knowledge reserves, learning styles and other aspects. The teaching objects of the pilot class are science students from two majors, including local students and military students. The total number of students is 54. The military students have learned some Advanced Mathematics knowledge in local universities, so they have certain knowledge foundation of mathematics, and have a correct learning attitude. The local students are generally younger, more active in thinking, and have strong ability to receive new knowledge.

Combined with the characteristics of Advanced Mathematics, teaching content analysis is to establish a curriculum system suitable for students' different majors according to the different needs of students in liberal arts and science, so as to meet the students' individual needs in the teaching process, make the teaching situation echo with the professional situation, and reflect the application of mathematics to achieve the purpose of learning mathematics meaningfully.

2.2.2 Design

The teaching objectives and the key and difficult points are designed according to the talent training objectives and the teaching syllabus, which is the guiding light of the subsequent stage. The design of teaching media and teaching strategy is to choose the way to organize teaching and learning activities in order to achieve the teaching objectives. Teaching process and teaching resources are designed around learning activities after the determination of teaching strategies.

In the teaching process, the teaching method is PAD classroom. Adhering to the trinity of "Professional combination, Project guidance and Task-driven", the teacher sets tasks of varying difficulty according to the three dimensions of course, major and expansion ability to meet the learning needs of different students. For example, based on the basic situation of the pilot class, taking "Nonhomogeneous linear differential equation with constant coefficients" as an example, the teacher sets the following three tasks:

1. The teacher sets tasks with low difficulty based on the knowledge points of this class. For example, the task is "Summarize the methods to find the general solution of nonhomogeneous linear differential equations with constant coefficients and find the particular solution for two different types". The purpose is to let students consolidate knowledge and enhance their confidence in learning mathematics.
2. For the same knowledge point, different tasks are set for students from different professional backgrounds with moderate difficulty. For example, the task is "Build the corresponding differential equation model about big data classification problem or computer optimization problem". The purpose is to highlight the application of knowledge and enhance the motivation of students to learn mathematics.
3. For students with a good foundation in mathematics, the teacher sets extensive and comprehensive tasks with high difficulty. For example, the task is "How to extend the solution of second order nonhomogeneous linear differential equations with constant coefficients to the case of order n". The purpose is to enable students to extend their knowledge on the basis of mastering it and improve their comprehensive application ability.

Finally, we will realize the teaching model with tasks as the bright line and with the cultivation of students' knowledge and ability as the dark line, which is combined with professional characteristics and led by projects.

Learning evaluation design is to design the evaluation method of students' learning effect and satisfaction after the teaching process. This paper establishes three level evaluation mode, including satisfaction and acceptance evaluation, process evaluation and final evaluation.

2.2.3 Development and Implementation

In the development stage, online and offline teaching resources are excavated. Offline resources include textbooks, auxiliary teaching materials, electronic teaching plans and exercise books. Online resources include micro-class, two-dimensional code digital resource library and geometric dynamic demonstration resource library, so as to generate systematic offline and online teaching resources. According to the requirements of the design stage, the teaching is carried out with PAD classroom. In the teaching link, we should give play to the leading role of teachers, according to the implementation steps of teaching before learning. In the discussion section, the subject status of students should be emphasized, encouraging students to learn independently, relying on the "Bright, Test, Help" to carry out the discussion (Fig. 2).

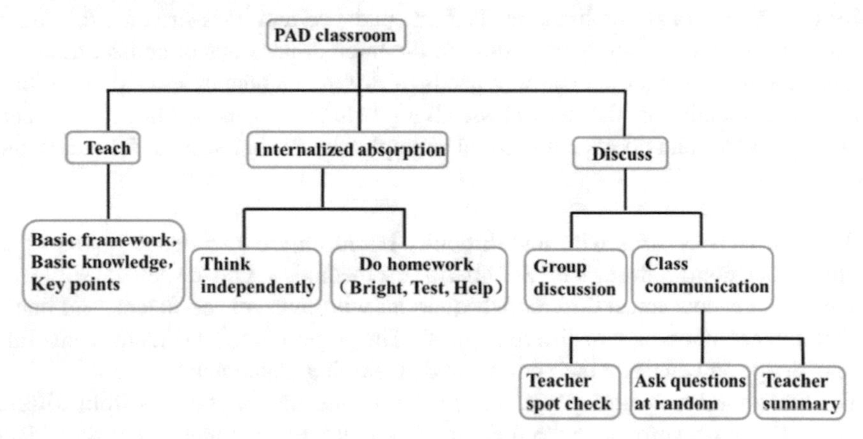

Fig. 2. Flow chart of PAD classroom

The teaching implementation process is as follows:

The first stage is teaching. According to the teaching plan, we arrange the course of Advanced Mathematics three times a week, each time two consecutive classes, each class 45 min. In the first class, the teacher explains the preparatory knowledge, basic framework, basic concepts, key and difficult points and calculation skills of the teaching content, and assigns homework after each class and issues discussed in the next class [6]. Teachers can set some diverging topics for students to learn and discuss independently, so as to arouse their learning enthusiasm. At the same time, according to the design of the previous stage, assign three kinds of tasks of different difficulty for students of different foundations and different specialties to complete. In the second class or the next class, the students internalize and discuss. According to different teaching content, class time does not have to be equal, but can be adjusted appropriately. In the teaching process, according to the needs, we can flexibly select the way of "PAD in-classroom" and "PAD off-classroom", or the combination of them [7].

The second stage is internalized absorption. On the basis of teaching, students make use of the online and offline course resources provided by the teachers, and choose the most suitable way to learn the course content repeatedly according to their own speed of receiving knowledge and professional needs, so as to internalize and absorb the course content independently. According to the content of each lesson, this stage can be completed in class or after class. If it is arranged after class, students will have enough time to prepare, so as to exert their autonomy in learning to a greater extent and achieve the purpose of personalized learning. Through the completion of the homework, the students can in-depth understanding of the teaching content and be able to expand and innovate, to prepare for the next stage of discussion and communication. Compared with traditional teaching, the homework for PAD classroom includes a new section called "Bright, Test, Help". "Bright" is that the students summed up their deepest feelings in the learning process and the most benefited content, light out to share with the class. "Test" is that the students test others in the form of questions what you know but others may be confused. "Help" is to raise questions that I do not understand and ask other students to help me in the discussion [8].

The third stage is discussion, including group discussion, teacher spot check, Ask questions at random and teacher summary, the last three parts collectively referred to as class communication [9]. According to the students' good study, poor study and different majors, the class will be divided into several groups, usually 4–7 people in a group. The teacher selects each group leader, assigns the work of the leader and the members. Then, according to the content taught in the last class and the learning results of the students, the group discussion will be conducted based on the homework and relying on the "Bright Test Help". Such as, discuss "How to extend the solution of second order nonhomogeneous linear differential equations with constant coefficients to the case of order n", for 5–20 min. At the same time, students are required to take class notes. After the group discussion, the teachers randomly check the students to speak on behalf of the group. So the student should share the essence of the group just discussed, make reading notes or homework presentation, and put forward unresolved problems. In view of the common problems and unsolved problems in the group discussion, the teacher organized the whole class to discuss freely. Because the topics of different difficulty were set, students from different backgrounds could share their learning experience with each other and answer questions and doubts. Finally, the teacher will summarize the contents that the students missed and need to improve, and end a class discussion process. In the class discussion, the teacher should guide and demonstrate the link of "Bright Test Help", and guide the students to think and output in a structured way instead of fragmentary accumulation of highlights and questions.

2.2.4 Evaluation

In the evaluation stage, the three-level evaluation system should be established to ensure the better optimization of teaching design (Fig. 3).

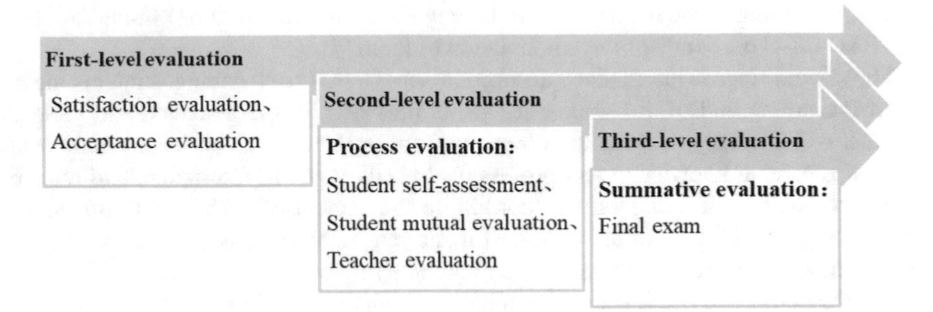

Fig. 3. Three-level evaluation structure diagram

The first-level evaluation refers to the survey of students' satisfaction with the teaching model and the acceptance of the evaluation method, which is convenient for teachers to timely know the students' learning attitude and ideas and further adjust the teaching method.

The Second-level evaluation refers to the process evaluation throughout the semester, including student self-evaluation, student mutual evaluation and teacher evaluation. Among them, student self-evaluation is to let students clearly know how well they have mastered the knowledge; student mutual evaluation can encourage students to learn from each other in the learning process and form healthy competition; teacher evaluation includes class participation, homework, attendance, study notes and unit tests, which are incorporated into regular grades. When evaluating students, teacher should pay attention to the students' attitude toward learning and commitment to the course. Students who actively participate in the discussion of the course and have a high degree of homework completion should be recognized in a timely manner to protect the enthusiasm of students to learn, and at the same time to play an exemplary role for other students. We should be able to grasp the bright spots of the students who are not highly engaged in the discussion, and try to dig out their potential learning ability and stimulate their learning motivation. In order to correct their learning attitude, students who don't write their homework carefully should be punished in time by asking them to do it again or giving them a deduction [10].

The third-level evaluation refers to the final evaluation, that is, the final examination which tests the learning effect of students in the whole semester. The student's final score consists of two parts: the regular score (40%) and the final exam score (60%).

The three-level evaluation method has the function of two-way evaluation and feedback. On the one hand, the student's mastery of the knowledge points can be feedback to the teacher in real time. The teacher can adjust the teaching methods in time according to it, so as to optimize the teaching and improve the teaching level and efficiency. On the other hand, it can enable students to find their shortcomings in learning in time, and then adjust their learning methods flexibly, so as to lay a solid foundation for better learning of follow-up courses and achieve better teaching effect. The following figure shows the results of the three-level evaluation after a semester of pilot teaching (Figs. 4 and 5).

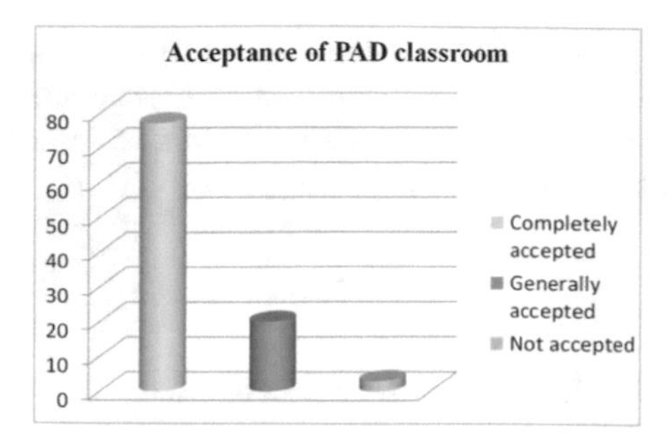

Fig. 4. Acceptance of the PAD classroom

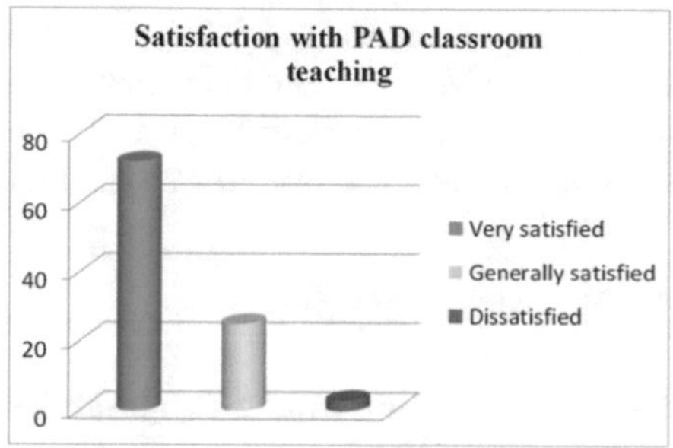

Fig. 5. Satisfaction with PAD classroom teaching

As can be seen from the figures, 77% of the students can fully accept the teaching mode of PAD classroom, 20% of the students generally accept it, 72% of the students are very satisfied with the teaching mode, and only 3% of the students do not accept or not satisfied with the teaching mode.

Because the process evaluation contains many evaluation items, two representative evaluation results are selected here as shown in the figures below (Figs. 6 and 7).

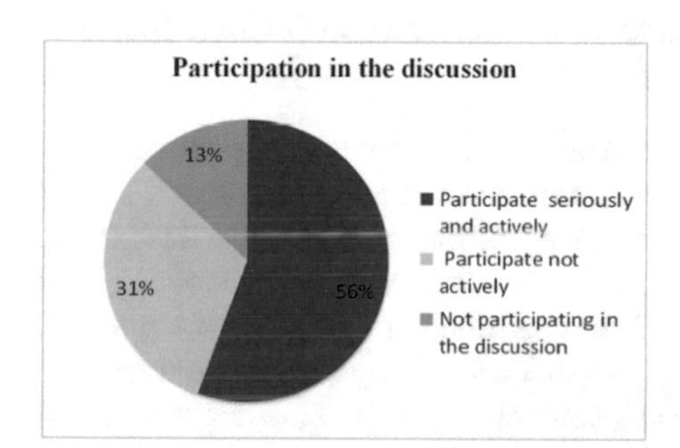

Fig. 6. Participation in the discussion

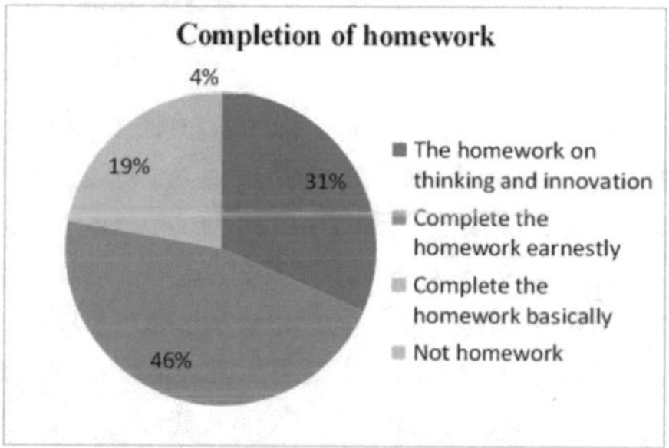

Completion of homework
4%
19%
31%
46%
The homework on thinking and innovation
Complete the homework earnestly
Complete the homework basically
Not homework

Fig. 7. Completion of homework

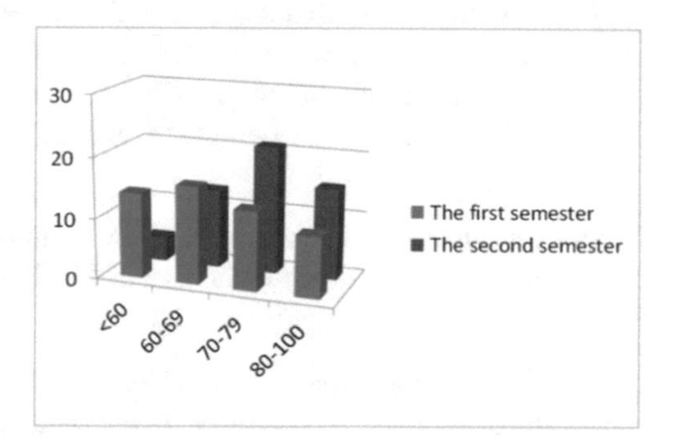

Fig. 8. Comparison of final grades between the first semester and the second semester

As can be seen from Fig. 8, the academic performance of the two semesters was close to the normal distribution. In the first semester, students with scores above 70 accounted for 42.6%, and the failure rate was 25.9%. However, in the second semester, 66.7% of the students scored above 70, and the failure rate was 7.4%.

3 Result Analysis

3.1 The Teaching Model Proposed in This Paper Can Be Implemented and Feasible

According to the teaching model combining PAD classroom and ADDIE, it was able to select a teaching class for pilot teaching of Advanced Mathematics, and the experiment cycle is one semester. At the end of the semester, we conducted a survey on students' satisfaction and acceptance, and organized a final test closely to test the learning effect of students. The evaluation results are obtained. Finally, we can realize the student-centered, combined with professional characteristics and project-oriented, with tasks as the bright line and the cultivation of students' knowledge and ability as the dark line, so as to enhance the students' learning effect and improve the teaching quality.

3.2 The Teaching Model Proposed in This Paper Has Given Full Play to Its Advantages in the Teaching of Advanced Mathematics with Good Teaching Effect

It can be seen from the three-level evaluation results of the pilot teaching that most students like and accept this new teaching method, which can stimulate students' learning enthusiasm and make students actively participate in learning and improve their interest in learning. In the class discussion, students are more involved, forming a strong learning atmosphere of mutual help. This kind of learning mode of independent exploration and cooperation and communication improves students' ability of independent learning, teamwork and problem solving, cultivates students' quality of critical and creative thinking, improves students' interpersonal skills, and realizes a harmonious classroom with students as the main body and effective interaction between teacher and students.

As can be seen from Fig. 8, compared with the scores of the first semester, the scores of the pilot semester were significantly improved, and the failure rate was significantly reduced. It shows that this teaching mode is of great help to improve the teaching performance of Advanced Mathematics course, and the teaching effect is remarkable.

3.3 The Teaching Model Establishes a Scientific and Reasonable Course Evaluation Mechanism to Protect the Enthusiasm of Students to the Greatest Extent

Different from the traditional evaluation standards, the three-level teaching evaluation mechanism sets different evaluation standards according to the students' different majors and different foundations. The teacher doesn't just think about right and wrong when evaluating assignments. The purpose is to protect the students' learning enthusiasm to the maximum extent, and avoid some students entering the vicious circle of "weak—tired of learning" because of the weak foundation, and encourage students to set their own goals.

4 Conclusion

The research in this paper is helpful to enhance the learning effect of students, make students attach importance to daily learning, guide students to improve learning attitude and learning style, change passive learning into autonomous learning and inquiry learning. It is beneficial to exercise the ability of students, train students to have basic theoretical knowledge of Advanced Mathematics and good mathematical literacy, independent thinking, self-innovation and cooperative exploration ability, train students to skillfully use the theoretical knowledge and methods of Advanced Mathematics to analyze and solve various practical problems. It is beneficial to improve the comprehensive quality of students and promote their all-round development. It is beneficial to improve the teaching and research level of teachers, and build a teaching and research team guided by scientific research and integrated with scientific research and teaching. At the same time, it can provide reference value for the teaching reform of other courses.

References

1. Qu, N., Li, Y.: The practice of inquiry-based teaching method in advanced mathematics classroom teaching. Educ. Teach. Forum (43), 197–198 (2018)
2. Qi, H.: The enlightenment of ADDIE model to flipped classroom teaching design. Chin. J. Adult Educ. **17**, 107–109 (2016)
3. Bu, C.: Research on application mode of ADDIE Model in micro-course design. Teach. Manag. (8), 90–93 (2014)
4. Zhang, X.: PAD classroom: the new exploration of college classroom teaching reform. Fudan Educ. Forum **5**, 5–10 (2014)
5. Li, X.: PAD classroom: a new way of university teaching with Chinese characteristics. Learn. Weekly Forum (1) (2016)
6. Liu, S., Qian, P., Wang, M.: PAD classroom teaching model and its success factors. Educ. Teach. Forum **13**, 46–48 (2019)
7. Yang, S.: PAD classroom teaching model and its role analysis of teachers and students. J. Liaoning Norm. Univ. (5) (2015)
8. Jiang, A., Wu, Q.: Discussion teaching and its operating process. J. Sichuan Inst. Educ. (12) (2005)
9. Zheng, L.: The effect comparison between PAD classroom teaching model and traditional teaching model. Educ. Teach. Forum (6) (2017)
10. Mo, J., Yang, M.: The design and development of micro-lecture based on ADDIE model. Educ. Inf. China (14), 81–84 (2020)

A Kind of Online Game Addictive Treatment Model About Young Person

Xiaokang Liu(✉), Jiangtao Li, Yanyan Zhao, Yiyue Sun, and Haibo Zhang

Foundation Department, Engineering University of PAP, Xi'an 710086, China
582342597@qq.com

Abstract. This paper constructs a kind of communication model with forced treatment of online game addicts. First, the regeneration matrix method is used to determine the basic regeneration number of the model, and then the local stability of online game model and online game addiction transmission conditions is judged according to the basic regeneration number. Further, the effect of treatment delay on online game addicts is analyzed. And last, with the help of numerical simulation to verify the stability of the equilibrium point, the author of this paper puts forward effective treatment cycle for online game addicts.

Keywords: Online game addictive equilibrium · Online game non-addictive equilibrium · Time delay · Stability

1 Introduction

Internet addiction refers to an individual can not control, excessive and compulsive to play online games resulting in physical, psychological and social function damage [1]. Teenagers addiction to online games will seriously affect their families and healthy growth, in recent years, many scholars have studied the causes and effects of online game addiction from the perspective of Psychology: online game addiction can bring the sense of spiritual pleasure, improve the subjective well-being [2]; However, there are many violent incidents caused by online game addiction [1]. Long-term online game addiction has a serious impact on social stability and family harmony [3]; Some scholars collect questionnaire data by using ACE model, Davis (2001) online cognitive behavior model, online Satisfaction Compensation Model [4] to Quantitative analysis the influence of various factors, but this kind of method has a high request to the questionnaire design, and it is easy to neglect some unexpected factors, which leads to the validity of the results of the model analysis. Therefore, this paper analyzes the existence and stability of the balance point of online game addiction and non-addiction in the transmission mechanism of online game addiction by means of infectious disease dynamics model, finally, the author puts forward some suggestions and measures for the treatment of online game addiction.

L. Barolli (Ed.): EIDWT 2023, LNDECT 161, pp. 108–117, 2023.
https://doi.org/10.1007/978-3-031-26281-4_11

2 Epdemic Model

SEIR model is a kind of Infectious disease model, which generally divides the population in the epidemic area of Infectious disease into the following categories: Susceptible, Exposed, Infectious and Recovered, it takes a latent period for the infected person to become an addict, so SEIR model is more suitable for the study of the problem of online game addiction. Using SEIR model for reference, the adolescents were divided into 4 groups: Susceptible (S), Exposed (E), Infectious (I), Recovered (R). The total number of teenagers is N(t), $N(t) = S + E + I + R$ (Fig. 1).

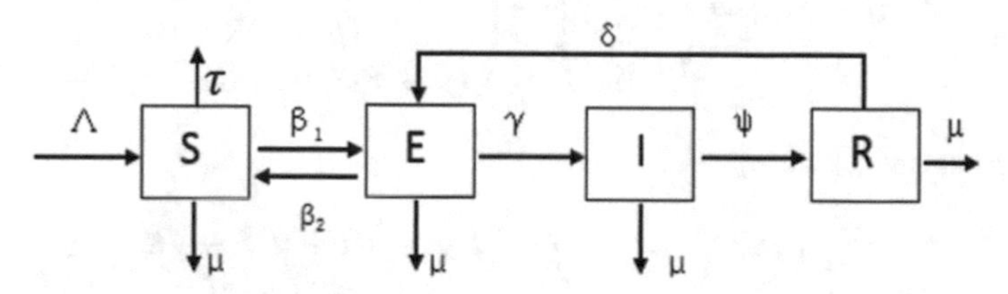

Fig. 1. The diagram of the population flow in each compartment.

SEIR epidemic model:

$$\begin{cases} \dot{S} = \Lambda - \mu S - \beta_1 SE + \beta_2 \mathrm{E} - \tau S \\ \dot{E} = \beta_1 SE - \beta_2 E - \mu E + \delta R - \gamma E \\ \dot{I} = \gamma E - \mu I - \psi I \\ \dot{R} = \psi I - \mu R - \delta R \end{cases} \tag{1}$$

Represents the growth rate of the adolescent population (children aged over 12 growing up to be adolescents), the natural mortality rate of adolescents, the migration rate of adolescents (adolescents aged over 29), the addiction rate of susceptible groups to online games, the rate of self-recovery, the rate of re-addiction, the rate of conversion from mild to severe addiction, and the rate of professional treatment in severe addiction groups. In system (1), there are four parameter variables S, E, I and R, they arc all functions of time, N(t) is the total number of adolescents, and the four equations are the transmission mechanism of adolescents' online game addiction, therefore, all parameters are positive at t = 0, and the feasible solution domain of mathematical model (1) is:

$$\Omega = \{(S, E, I, R) \in R_+^4 | 0 \le S(t) + E(t) + I(t) + R(t) \le \frac{\Lambda}{\mu}\}$$

So

$$\dot{N}(\mathrm{t}) = \dot{S} + \dot{E} + \dot{I} + \dot{R} \le \Lambda - \mu N$$

Then

$$0 \le N(t) \le \frac{\Lambda}{\mu} - N(0)e^{-\mu t}, \lim_{t \to \infty} N(t) = \frac{\Lambda}{\mu}$$

Therefore, all feasible solutions of the model (1) are included and are positive invariant sets of the system (1) [5–7].

3 Model Property Analysis

3.1 Online Game Addiction-Free Balance Point and Basic Reproduction Number

If the right of System (1) is zero, the balance point of online game non-addictive state can be obtained $E_0 = (\frac{\Lambda}{\mu+\tau}, 0, 0, 0)$. The basic regeneration number of system (1) is calculated by the method given by Van and Watmough (2002): if $x = (E, I, R, S)$ then

$$\frac{dx}{dt} = F(x) - V(x) = \begin{pmatrix} \beta_1 SE \\ 0 \\ 0 \\ 0 \end{pmatrix} - \begin{pmatrix} \beta_2 E + (\mu+\gamma)E + \delta R \\ (\mu+\psi)I - \gamma E \\ (\mu+\delta)R - \psi I \\ (\mu+\tau)S + \beta_1 SE - \beta_2 E - \Lambda \end{pmatrix}$$

So $F(x)$ and $V(x)$ were:

$$F(x) = \begin{pmatrix} \beta_1 SE \\ 0 \\ 0 \\ 0 \end{pmatrix} V(x) = \begin{pmatrix} \beta_2 E + (\mu+\gamma)E + \delta R \\ (\mu+\psi)I - \gamma E \\ (\mu+\delta)R - \psi I \\ (\mu+\tau)S + \beta_1 SE - \beta_2 E - \Lambda \end{pmatrix}$$

The Jacobian matrices about the addiction-free equilibrium of online games are:

$$D(F(E_0)) = \begin{pmatrix} F_{2\times 2} & B_1 \\ C_1 & D_1 \end{pmatrix}$$

$$D(V(E_0)) = \begin{pmatrix} \beta_2+\mu+\gamma & 0 & \delta & 0 \\ -\gamma & \mu+\psi & 0 & 0 \\ 0 & -\psi & \mu+\delta & 0 \\ \beta_2 & 0 & 0 & \mu+\tau+\beta_1\frac{\Lambda}{\tau+\mu} \end{pmatrix}$$

$$B_1 = C_1 = D_1 = \begin{pmatrix} 0 & 0 \\ 0 & 0 \end{pmatrix}$$

So

$$F_{2\times 2} = \begin{pmatrix} \beta_1\frac{\Lambda}{u+\tau} & 0 \\ 0 & 0 \end{pmatrix} V_{2\times 2} = \begin{pmatrix} \beta_2+u+\gamma & 0 \\ -\gamma & u+\psi \end{pmatrix}$$

$$R_0 = \rho(F_{2x2}V_{2x2}^{-1}) = \frac{\Lambda\beta_1}{(\mu+\tau)(\gamma+\beta_2+\mu)}$$

3.2 Balance Point of Online Game Addiction

Theorem 1. $R_0 > 1$, There is a unique addiction equilibrium in the system $P^*(E^*, I^*, R^*, S^*)$.

$$E^* = \frac{(\Lambda - S^*(\mu+\tau))\beta_1(\mu+\psi)(\mu+\delta)}{\beta_1(\mu+\gamma)(\mu+\delta)(\mu+\psi) - \delta\psi\gamma}$$

$$I^* = \frac{\gamma}{\mu+\psi} \times \frac{(\Lambda - S^*(\mu+\tau))\beta_1(\mu+\psi)(\mu+\delta)}{\beta_1(\mu+\gamma)(\mu+\delta)(\mu+\psi) - \delta\psi\gamma}$$

$$R^* = \frac{\psi\gamma}{(\mu+\psi)(\mu+\delta)} \times \frac{(\Lambda - S^*(\mu+\tau))\beta_1(\mu+\psi)(\mu+\delta)}{\beta_1(\mu+\gamma)(\mu+\delta)(\mu+\psi) - \delta\psi\gamma}$$

$$S^* = \frac{\beta_2}{\beta_1} + \frac{\mu}{\beta_1} - \frac{\delta\psi\gamma}{(\mu+\psi)(\mu+\delta)\beta_1} + \frac{\gamma}{\beta_1}$$

Proof: we can obtain from system (1)

$$\begin{cases} \Lambda - (\mu+\tau)S - \beta_1 SE + \beta_2 \mathrm{E} = 0 \\ \beta_1 SE - \beta_2 E - \mu E + \delta R - \gamma E = 0 \\ \gamma E - \mu I - \psi I = 0 \\ \psi I - \mu R - \delta R = 0 \end{cases} \tag{2}$$

And then:

$$\begin{cases} I^* = \frac{\gamma E^*}{\mu+\psi} \\ R^* = \frac{\psi I^*}{\mu+\delta} = \frac{\psi\gamma E^*}{(\mu+\psi)(\mu+\delta)} \\ \beta_1 S^* E^* = \beta_2 E^* + \mu E^* - \frac{\delta\psi\gamma E^*}{(\mu+\psi)(\mu+\delta)} + \gamma E^* \end{cases} \tag{3}$$

So

$$S^* = \frac{\beta_2}{\beta_1} + \frac{\mu}{\beta_1} - \frac{\delta\psi\gamma}{(\mu+\psi)(\mu+\delta)\beta_1} + \frac{\gamma}{\beta_1} = W_*$$

$$S^* = \frac{(\Lambda - \mu E^* + \frac{\delta\psi\gamma E^*}{(\mu+\psi)(\mu+\delta)\beta_1} + \gamma E^*)}{(\mu+\tau)}$$

$$E^* = \frac{(\Lambda - W_*(\mu+\tau))\beta_1(\mu+\psi)(\mu+\delta)}{\beta_1(\mu+\gamma)(\mu+\delta)(\mu+\psi) - \delta\psi\gamma}$$

Finally, the conclusion is valid.

3.3 Balance Point of Online Game Non-addiction

Theorem 2. If

$$R_0 < 1, E_0 = (\frac{\Lambda}{\mu+\tau}, 0, 0, 0)$$

is locally asymptotically stable.

Proof: Jacobi Matrix of model (1)

$$J = \begin{pmatrix} -u-\tau-\beta_1 E & -\beta_1 S+\beta_2 & 0 & 0 \\ \beta_1 E & \beta_1 S-\beta_2-u-\gamma & 0 & 0 \\ 0 & \gamma & -u-\psi & 0 \\ 0 & 0 & \psi & -u-\delta \end{pmatrix}$$

Jacobi Matrix of E_0

$$J(E_0) = \begin{pmatrix} -u-\tau & -\beta_1 \frac{\Lambda}{u}+\beta_2 & 0 & 0 \\ 0 & \beta_1 \frac{\Lambda}{u}-\beta_2-u-\gamma & 0 & 0 \\ 0 & \gamma & -u-\psi & 0 \\ 0 & 0 & \psi & -u-\delta \end{pmatrix}$$

$R_0 < 1$, the eigen value of

$$J(E_0): -u-\tau < 0, \beta_1 \frac{\Lambda}{\text{u}} - \beta_2 - u - \gamma < 0, -u-\psi < 0, -u-\delta < 0$$

The real part of all eigenvalues is less than zero, so E_0 is Locally asymptotically stable.

Theorem 3. If

$$R_0 < 1$$

online game addiction-free balance point

$$E_0 = (\frac{\Lambda}{\mu+\tau}, 0, 0, 0)$$

is globally asymptotically stable.

Proof: Construct Lyapunov function

$$V = S - S_0 - S_0 \ln(\frac{S}{S_0}) + E + I$$

$$\begin{aligned} \dot{V} &= \dot{S} - \tfrac{S_0}{S}\dot{S} + \dot{E} + \dot{I} \\ &= (1-\tfrac{S_0}{S})(\Lambda - \mu S - \tau S - \beta_1 SE + \beta_2 \mathrm{E}) + \beta_1 SE - \beta_2 E - \mu E + \delta R - \gamma E + \gamma E - \mu I - \psi I \\ &= (1-\tfrac{S_0}{S})\Lambda - \mu S + \mu S_0 + \tau S_0 + \beta_1 S_0 E - \tfrac{S_0}{S}\beta_2 E - \mu I - \psi I - \mu E + \delta R \\ &\le (1-\tfrac{S_0}{S})\Lambda - \mu S + \mu S_0 + \tau S_0 + \Lambda - \mu S_0 - \tau S_0 - \tfrac{S_0}{S}\beta_2 E - \mu I - \psi I - \mu E + \delta R \\ &\le \Lambda(2 - \tfrac{S_0}{S} - \tfrac{S}{S_0}) - \tfrac{S_0}{S}\beta_2 E - \mu I - \psi I - \mu E + \delta R \end{aligned}$$

$$2 - \frac{S_0}{S} - \frac{S}{S_0} \le 0, \psi I + \delta R < 0$$

if $R_0 < 1$ then $\dot{V} \le 0$; $\dot{V} = 0$ then $\mathrm{A} = \mathrm{A}_0, E = I = 0;$ *when* $t \to \infty, R \to 0$

The Model (1) obtained by LaSalle invariant set principle is globally asymptotically stable when the addiction-free equilibrium of the online game is present [8, 9].

3.4 Time Delay Model Analysis

The time-delay model is studied, a new model is obtained:

$$\begin{cases} \dot{S} = \Lambda - \mu S - \beta_1 SE + \beta_2 \mathrm{E} - \tau S \\ \dot{E} = \beta_1 SE - \beta_2 E - \mu E + \delta R - \gamma E \\ \dot{I} = \gamma E - \mu I - \psi I(t - \tau_0) \\ \dot{R} = \psi I(t - \tau_0) - \mu R - \delta R \end{cases} \tag{4}$$

So, the balance point of online game addiction is $P^*(S^*, E^*, I^*, R^*)$,

Theorem 4: If

$$R_0 > 1, E^* > \frac{\Lambda}{u + \gamma - \beta_2}$$

then the balance point of online game addiction is locally asymptotically stable.

proof: Jacobi Matrix of $P^*(S^*, E^*, I^*, R^*)$.

$$J(P_1^*) = \begin{pmatrix} -u - \tau - \beta_1 E^* & -\beta_1 S^* + \beta_2 & 0 & 0 \\ \beta_1 E^* & \beta_1 S^* - \beta_2 - u - \gamma & 0 & \delta \\ 0 & \gamma & -\psi & 0 \\ 0 & 0 & \psi & -u - \delta \end{pmatrix}$$

$$M = -J(P_1^*) = \begin{pmatrix} u + \tau + \beta_1 E^* & \beta_1 S^* - \beta_2 & 0 & 0 \\ -\beta_1 E^* & -\beta_1 S^* + \beta_2 + u + \gamma & 0 & -\delta \\ 0 & -\gamma & \psi & 0 \\ 0 & 0 & -\psi & u + \delta \end{pmatrix}$$

Δ_i is order principal minor determinant of M, so

$$\Delta_1 = \left| u + \tau + \beta_1 E^* \right| = u + \tau + \beta_1 E^* > 0, \Delta_3 = \Delta_2 \psi, \Delta_4 = (u + \delta)\Delta_3$$

$$\Delta_2 = (u + \gamma)\beta_1 E^* - (u + \tau)\beta_1 \frac{\Lambda + \beta_2 E^*}{u + \beta_1 E^* + \tau} > (u + \gamma)\beta_1 E^* - \beta_1(\Lambda + \beta_2 E^*)$$

If $R_0 > 1$ then $\Delta_2 > (u + \gamma)\beta_1 E^* - \beta_1(\Lambda + \beta_2 E^*) > 0,\ \Delta_3 > 0\ \Delta_4 > 0$

so, M is positive definite matrix, eigen value of M are more than zero. The balance point (P_1^*) of online game addiction is locally asymptotically stable by Routh-hurwitz criterion.

4 Numerical Simulation

The system model is simulated numerically. We fix the parameters

$$u = 0.014, \tau = 0.0109\beta_2 = 0.009, \gamma = 0.52, \delta = 0.04, \Lambda = 0.13944$$

Figure 2 is obtained. S(t-30) and I(t-30) show the population size with time-delay. And same time $R_0 = 0.5156 < 1$, Fig. 2 show that If treated in a timely manner, the number of addicts will gradually decrease and tend to level off; if the treatment of online game addicts will be shelved from the beginning of a sudden surge, with the intervention of treatment of internet addicts fluctuating around stable level, addicts can not be eliminated in the short term.

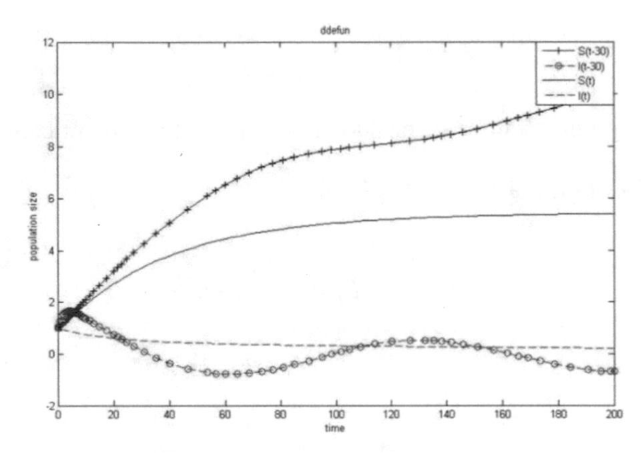

Fig. 2. $R_0 < 1$ Relationship between time delay and online game addiction

We fix the parameters

$$u = 0.014, \tau = 0.0109, \beta_2 = 0.009, \gamma = 0.02, \delta = 0.04, \Lambda = 0.16, \psi = 0.0169$$

then

$$R_0 = 7.4717 \geq 1$$

Observing Fig. 3, we can know the number of online game addicts (I) increased sharply from the beginning, but remained in a large scale group with the decrease of the number of online game addicts treated, and the number of online game addicts did not decrease in a short period; In this case, delayed intervention time, addicts will gradually become more stable state, the size of the number of treatment is smaller than the case directly. The whole ecological environment failed to restrain the growth of online game addiction, and the online game addiction became a endemic disease to be maintained.

If the parameters can be affected artificially, then the value of the basic reproductive number will be less than 1. Now let's study the effect of different parameters on basic reproduction number.

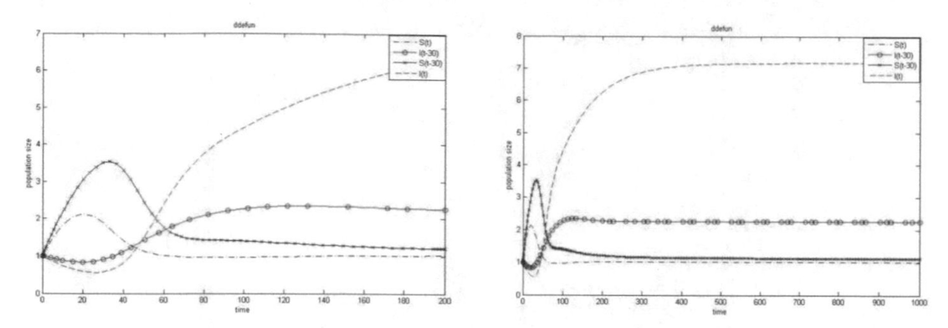

Fig. 3. $R_0 > 1$ Relationship between time delay and online game addiction

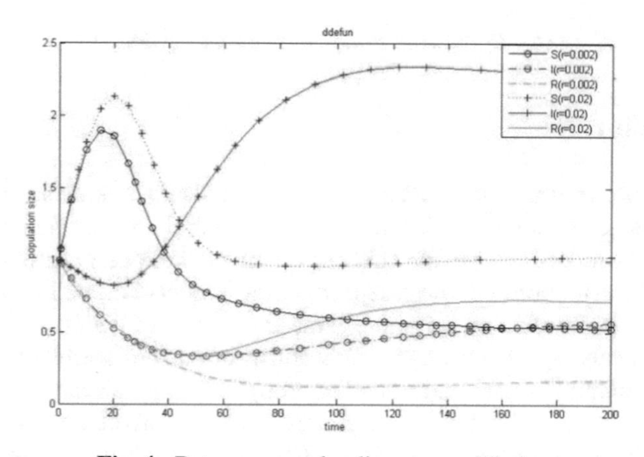

Fig. 4. Between γ and online game addiction

1. **γ fix different values**

Observe the above, γ respectively takes different values of the number of latecomers and infected, when the increase, the number of addicts I, the more likely the system tends to stabilize, if measures are taken to make the parameters, the smaller γ, it can inhibit the transformation speed and quantity of the latent person to the infected person, thus playing a certain inhibitory role to the online game addiction (Fig. 4).

2. **β_1 fix different values**

Observe the above, β_1 takes different values of the number of addicts, the more β_1 the bigger number of addicts, when the value of the basic number of regeneration is greater than 1, the group of addicts to maintain a stable size, become endemic disease to be maintained (Fig. 5).

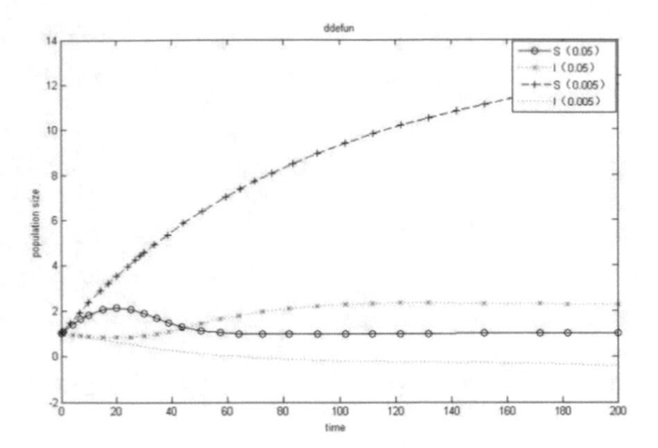

Fig. 5. Between β_1 and online game addiction

5 Conclusion

In this paper, we analyze the SEIR model of compulsory treatment of online game addicts with stage structure, obtain the threshold condition of online game addiction, and further compare and analyze the effect of treatment delay on the group of addiction. According to the definition of the basic reproduction number of the online game addicts, if $R_0 < 1$ the equilibrium point of the online game addiction-free is globally asymptotically stable through analysis, then the online game addiction can be controlled by compulsory treatment, the number of new students infected by online games has been reduced to a very low state, and the spread of online games has been curbed. At that time, $R_0 \geq 1$, the balance point $P^*(E^*, I^*, R^*, S^*)$ was reached. At that time, through the transmission mechanism, online game addiction continued to increase the number of new online game addicts, far outnumbering those who were cured and reaching a stable state, online game addiction will develop into an endemic disease that persists. When the treatment delay is considered, the group of online game addicts fluctuates greatly and can not reach a stable state at the first time. Moreover, $R_0 \geq 1$, the scale of internet addicts is actually smaller than the situation of timely compulsory treatment.

References

1. Wang, J., Yu, C., Li, W.: Peer victimization and online game disorder in adolescents. J. Central China Norm. Univ. (Hum. Soc. Sci.) **59**(04), 184–192 (2020)
2. Zhou, F., Liu, R., Guo, M., Jiang, S.: The influence of teenagers' negative emotion on internet addiction: the moderating effect of happiness tendency. Chin. J. Clin. Psychol. **25**(02), 208–212 (2017)
3. Gao, T.: A Study on the Development of Internet Addiction and its Influencing Factors in Senior High School Students. Jilin University (2020)
4. Zhang, Y., Su, P., Ye, Y., Zhen, S.: Psychosocial factor influencing online game disorder in university students. J. South China Norm. Univ. (Soc. Sci. Ed.), (04), 86–91+191 (2017)
5. Liu, P.: Study of Stability on Three Classes of Nonlinear Dynamical Systems. Northwest A&F University (2018)

6. Ma, Z., Chou, Y.: Mathematical Modeling and Research on Dynamics of Infectious Diseases, pp. 34–38. Science Press, Beijing (2004)
7. Huang, S., Chen, Y., Zhou, Q., Wang, Y.: The relationship between boredom proneness and online game addiction of freshmen: the moderating effect of sense of meaning of life. China J. Health Psychol. **28**(04), 580–585 (2020)
8. Wei, H., Zhou, Z., Tion, Y., Bao, N.: Online game addiction: effects and mechanisms of flow experience. Psychol. Dev. Educ. **28**(06), 651–657 (2012)
9. Li, X., Huang, R., Li, H.: Dynamic analysis of epidemic model with vertical transmission. J. Jilin Univ. (Sci. Ed.) **58**(03), 569–574 (2020)

Research on E-commerce Customer Value Segmentation Model Based on Network Behavior

Jing Zhang(✉) and Juan Li

Department of Information and Communication, Police Officers College of the Chinese People's Armed Force, Chengdu, Sichuan, China
30145002@qq.com

Abstract. The traditional customer segmentation model is based on the value of the customer's consumed data, and the customer's consumption habit is obtained to predict its potential consumption value, and then the marketing strategy and customer retention strategy are determined. According to the characteristics of e-commerce enterprises and the recordability of historical network behavior of e-commerce customers, this paper constructs the AFCS customer segmentation model based on the traditional customer segmentation model customer value matrix which represents the existing value of e-commerce customers, added two potential value factors representing e-commerce customers, one is the total number of clicks of users who represent the activity of e-commerce customers, the other is the total number of user collections and shopping carts representing the potential purchase intention of users. Then the AFCS model is tested with K-Means, SOM and SOM + K-Means, the experimental results prove that the AFCS model based on the SOM + K-Means algorithm is superior to the AFCS model using the SOM or K-Means algorithm alone, and its customer segmentation results are more accurate, which can provide reference for effective customer retention strategies and targeted marketing for e-commerce.

1 Introduction

With the rapid development of network information technology and the impact of the COVID-19 epidemic in recent years, the number of Internet users has increased sharply. According to statistics in 2022, the total global population will be about 7.98 billion, and the number of Internet users will reach about 4.95 billion, accounting for about 62.5% of the total population. In 2021, more than 2.14 billion people worldwide bought goods and services online, and globally, nearly 77% of Internet users aged 16 to 64 said they shopped online every month. With the rapid expansion of the scale of e-commerce users, more and more e-commerce companies have sprung up, and their competition is becoming increasingly fierce. In the context of the "customer economy" era, like traditional retail, how to effectively use existing data to accurately segment customers, and then identify the value of customers is the key for e-commerce companies to implement.

L. Barolli (Ed.): EIDWT 2023, LNDECT 161, pp. 118–128, 2023.
https://doi.org/10.1007/978-3-031-26281-4_12

2 AFCS E-commerce Customer Segmentation Model

2.1 Traditional Customer Segmentation Model

1. RFM Customer Segmentation Model

In 1994, Hughes proposed the RFM customer segmentation method, which uses three behavioral variables to describe and distinguish customers. R (Recency): recent purchase time, refers to the time interval from the last purchase to the present; F (Frequency): purchase frequency, refers to the number of purchases in a certain period of time; M (Monetary): total purchase amount, refers to the total amount purchased during the certain period. RFM analysis scores the variable of each customer, calculates the product of the three variables, and sorts the results. All customers are classified by 20%, 60%, and 20% according to the sorting results, and the top 20% are the best customers, enterprises should keep them as much as possible, the middle 60% of customers need to try to transfer them to the front 20%, and the latter 20% are customers to try to avoid.

RFM model are all behavioral factors, which are easy to obtain and can predict customer purchase behavior. However, the analysis process is complex and time-consuming. Furthermore, there are too many subdivided customer groups. If each variable uses three values, there will be 27 customer groups, and there is multi-collinearity between the variable number of purchases F and the total purchase amount M.

2. Customer Value Matrix Model

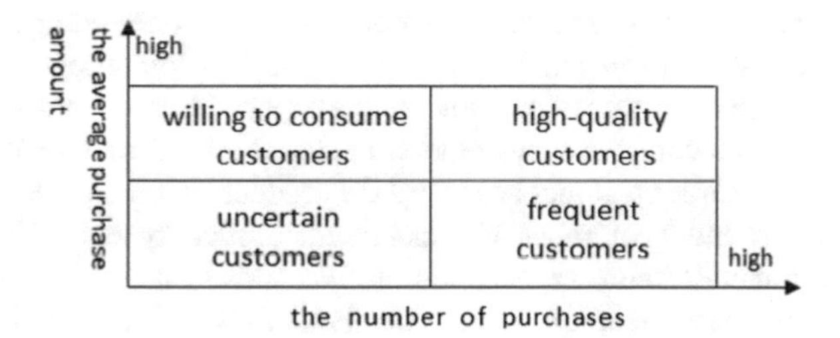

Fig. 1. Customer value matrix

Aiming at the defects of RFM, Marcus put forward the revision scheme of customer value matrix in 1998. This method replaces the total purchase amount M with the average purchase amount A, and constructs a two-dimensional customer value matrix with the number of purchases F. The information in the matrix is obtained from the customer number, purchase date, and daily purchase amount. The number of purchases F is the number of purchases on different dates. The average purchase amount is equal to the quotient of the total daily purchases divided by the number of purchases. The third variable purchase interval R in the RFM analysis is eliminated, and the customer value matrix analysis is more simplified. The customer value matrix divides customers into four types, that is, willing to consume customers, high-quality customers, frequent customers and uncertain customers, as shown in Fig. 1.

In addition to the classic RFM and customer value matrix models, some researchers in the retail field proposed other customer segmentation models based on these two models, trying to solve a specific problem in a certain retail field. For example, in 2018, Sun Jing studied the customer lifetime value model based on complaint behavior; Wang Kefu proposed the AFH model to try to understand the current value and value-added potential of customers, and so on.

2.2 Proposal of AFCS E-commerce Customer Segmentation Model

Traditional customer segmentation models are all aimed at offline sales. Data mining methods are used to obtain the existing value of customers from the past sales records of commodities, and predict the potential value or behavior of customers on the basis of the existing value. There is no direct basis for its potential value to carry out classified management and implement corresponding marketing strategies. Compared with traditional enterprises, e-commerce enterprises use information technology as the carrier, which determines that various online behaviors of consumers, such as a large number of buying, browsing and clicking behaviors, collection and shopping cart behaviors, these behaviors are recorded by the network. In these behaviors the purchased goods are stored in the past sales records, and the traditional customer segmentation model can be used to analyze their existing value, those behaviors such as browsing and clicking, collecting and adding shopping cart directly represent the needs of customers, that is, the potential value of customers. E-commerce companies mine the hidden needs behind behavioral data, and carry out accurate classification management and precise marketing of customers. It is possible to achieve purchases and bring benefits to the company. Therefore, on the basis of the customer value matrix of the traditional customer segmentation model that represents the existing value of customers, this paper adds two factors that represent the potential value of e-commerce customers: Factor C(clicks) on behalf of the user's activity of the e-commerce customer browsing the product which include the number of clicks on the product and the number of clicks on the product type. Factor S (Shopping carts) the total number of items, represented by the user's favorites and shopping carts representing the purchase intention. These two factors C and S represent the potential value of e-commerce, together with A and F of the customer value matrix representing the existing value of e-commerce, build an AFCS model of e-commerce customer

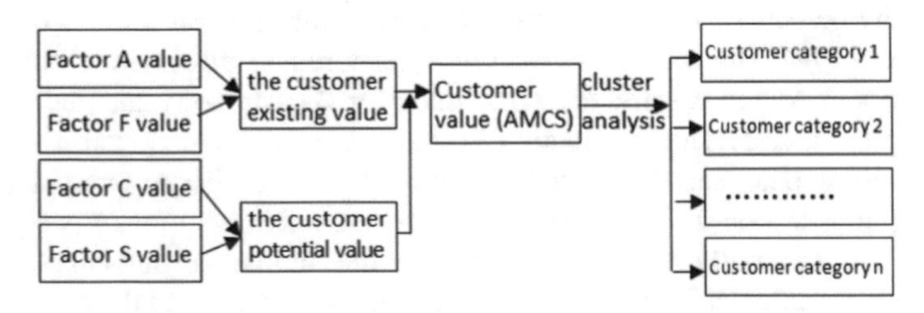

Fig. 2. The AFCS e-commerce customer segmentation model

value, then cluster analysis is carried out on AFCS model to obtain customer category segmentation. AFCS customer segmentation model is shown in Fig. 2.

3 Selection of Clustering Algorithm

The technology used in customer segmentation is mainly the classification and clustering technology in data mining. The more classic ones are distance-based clustering algorithm, hierarchical clustering algorithm, density clustering algorithm and neural network model clustering algorithm. K-Means clustering based on distance is one of the most commonly used classical clustering algorithms.

3.1 K-Means Algorithm

Basic principle of the K-Means algorithm is to use the distance function as the similarity index to divide the sample objects with similar distances into the same cluster (category). The algorithm steps are as follows:

① Randomly select K objects as the initial cluster centroid.
② Divide the remaining sample data to the initial cluster center with the smallest distance according to their distance from the initial cluster center.
③ Recalculate the mean value of all objects in each cluster as the new cluster center;
④ Cycle step ②③ Until the standard measure function starts to converge until. The Euclidean distance formula shown in general formula (1) is used as the sample clustering distance, and the mean square error of formula (2) is used as the standard measurement function.

In this paper, the Euclidean distance is selected, and the sum of squared error (SSE) (sum of squared error) is used as the objective function to measure the quality of clustering, and the centroid is the mean value. The calculation formula of European distance is shown in the following formula (1).

$$d_2(i,j) = \sqrt{|x_{i1} - x_{j1}|^2 + |x_{i2} - x_{j2}|^2 + \cdots + |x_{ip} - x_{jp}|^2} \tag{1}$$

Among them, $i = (x_{i1}, x_{i2}, \ldots, x_{ip})$ and $j = (x_{j1}, x_{j2}, \ldots, x_{jp})$ are two p-dimensional data objects.

The standard measure function is shown in the following formula (2).

$$\mathrm{E} = \sum\nolimits_{i=1}^{k} \sum\nolimits_{p \in C_i} |p - m_i|^2 \tag{2}$$

Here, E is the sum of the squared errors of all objects in the database, p is a point in space, and m_i is the mean value of cluster C_i.

3.2 SOM Neural Network

The SOM (Self-Organizing feature Map) neural network algorithm is shown in Fig. 3. It is an unsupervised learning neural network algorithm composed of an input layer and a competition layer. It can cluster unknown samples. The input layer is mainly responsible for receiving external information and passing the input data to the mapping. The competition layer mainly organizes and trains the data, and divides the data into different classes according to the number of training times and the selection of the field.

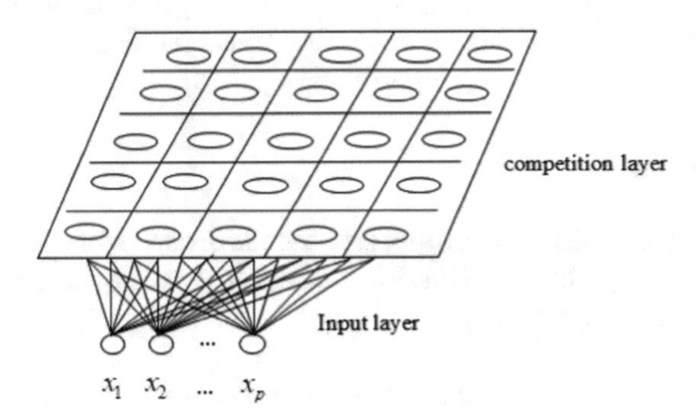

Fig. 3. SOM network structure diagram

The steps of the SOM algorithm are as follows:

① Network initialization: The weight w_j of each node in the competition layer is randomly given a small initial value, and the training end condition is defined.
② Input the vector $x = [x_1, x_2, x_3 ..., x_p]^T$, according to the formula (3) and calculate the Euclidean distance between the input vector and the weight vector of the competition layer.

$$d_j = \sum_{i=1}^{n} (x_i - w_{ji})^2 \tag{3}$$

In the formula (3), w_{ji} is the weight between the neurons of the input layer i and the j neurons of the mapping layer.
③ Determine the winning neuron according to the minimum Euclidean distance: take the smallest d_j^* corresponding neuron in the mapping layer as the winning neuron j^* in the competition. Then according to the formula (4) to correct and update the weight to move closer to x_i.

$$w_{ij}(n+1) = \begin{cases} w_{ij}(n) + \eta(n)h(j, j^*)[x_i - w_{ji}(n)], & j \in h(j, j*) \\ w_{ij}(n) & j \notin h(j, j*) \end{cases} \tag{4}$$

In the formula, n is the number of learning times,η is the learning rate, and $h(j, j^*)$ the neighborhood function. A typical choice is expressed by a Gaussian function as the following formula (5):

$$h(j, j^*) = \exp(-\frac{|j - j^*|^2}{2\sigma^2(n)}) \tag{5}$$

The above formula $\sigma(n)$ is the "effective width" of the neighborhood, which decreases as the learning progresses.

④ Reduce the learning rate η and neighborhood width σ;

⑤ With the increase of learning times n, repeat steps ②③④, and stop training when the training end condition is reached.

⑥ Output specific cluster numbers and cluster centers.

3.3 SOM + K-Means Algorithm

The comparison between the SOM algorithm and the K-Means algorithm is shown in the following Table 1.

Table 1. The comparison between the SOM algorithm and the K-Means algorithm

Algorithm	Advantage	Shortcoming
K-Means	The algorithm is simple and efficient, and can quickly process large amounts of data	The number of clusters and cluster centers must be specified in advance, otherwise the algorithm may not converge or be optimized locally, and it is difficult to select the number of clusters scientifically when the clusters are not clear
SOM	Automatic clustering based on data characteristics, less affected by noise data	The algorithm Can not provide accurate clustering information after classification

The different from the K-Means algorithm, the SOM algorithm can automatically cluster the input data without the need to determine the number of clusters in advance. However, some neurons may not always win when training data, resulting in inaccurate classification results. Therefore, it is difficult to determine the initial cluster number and cluster center of the K-Means algorithm in advance.

In this paper, the SOM and K-Means algorithm are combined. First, the SOM network is used to find out the number of categories of customer groups and the centroids of each cluster, and then the number of categories and each cluster center are used as the initial input of the K-Means clustering algorithm. The combining algorithm proceeds in two phases:

1) The first stage

Use the SOM algorithm to cluster the sample data once to obtain the number K of clusters and the cluster centroid $Z = \{Z_1, Z_2, \ldots Z_c\}$.

2) The second stage

Initialize the K-Means algorithm with the results output by SOM, and then use the K-Means algorithm to cluster sample data, and obtain the final clustering results.

In this way, by combining the SOM neural network with the K-Means algorithm, the K-Means clustering algorithm with a suitable initial value input has a strong local search ability, the convergence speed is improved, and the clustering process has a strong self-adaption performance, and satisfactory clustering results can be obtained after limited iterations.

4 Simulation Experiment

This article conducts an experiment on 20,000 pieces of data from 700 users of an e-commerce platform for one month. After preprocessing the data, such as cleaning, integration transformation, and discretization, the processed data are clustered with K-Means, SOM, and SOM + K-Means, and then the results are compared and analyzed.

4.1 K-Means Algorithm Clustering

Specify the number of clusters K = 4, and the centroids after K-Means clustering are shown in Table 2.

Table 2. K-Means clustering centroids

Cluster	Cluster centroid			
	Average purchase amout	Number of purchases	Activity	Purchase intention
C1	0.17	0.058	0.553	0.251
C2	0.345	0.398	0.53	0.147
C3	0.419	0.081	0.509	0.276
C4	0.307	0.095	0.264	0.52

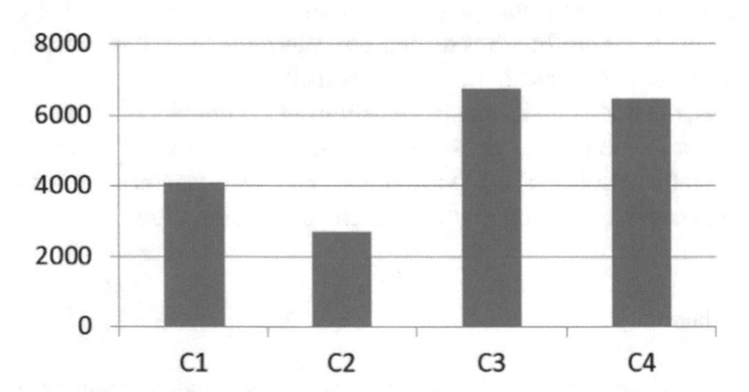

Fig. 4. Distribution of cluster members of K-Means

The distribution of cluster members is shown in Fig. 4: Among them, there are 4090 people in the cluster C1, 2699 people in the cluster C2, 6737 people in the cluster C3, and 6474 people in the cluster C4.

4.2 SOM Algorithm Clustering

In the SOM neural network algorithm in this paper, the neighborhood functions are all Gaussian functions, and when the algorithm is executed the training parameters are set as shown in Table 3:

Table 3. SOM training parameters

Training times	Initial neighborhood width	Network size	Neighborhood type	Initial learning rate
500	1	2 * 3	square	0.4

Centroid of each cluster: After a predetermined number of trainings, customers are automatically divided into 4 categories, and the cluster centroids of each cluster are as shown in the Table 4:

Table 4. SOM clustering centroid

Cluster	Cluster centroid			
	Average purchase amout	Number of purchases	Activity	Purchase intention
C1	0.334	0.113	0.782	0.185
C2	0.241	0.076	0.627	0.413
C3	0.334	0.114	0.524	0.217
C4	0.235	0.08	0.418	0.341

The distribution of cluster members is shown in Fig. 5: 7086 people belonged to the C1 type of purchase behavior, 3725 people belonged to the C2 type of purchase behavior, 2339 people belonged to the C3 type of purchase behavior, and 6850 people belonged to the C4 type of purchase behavior.

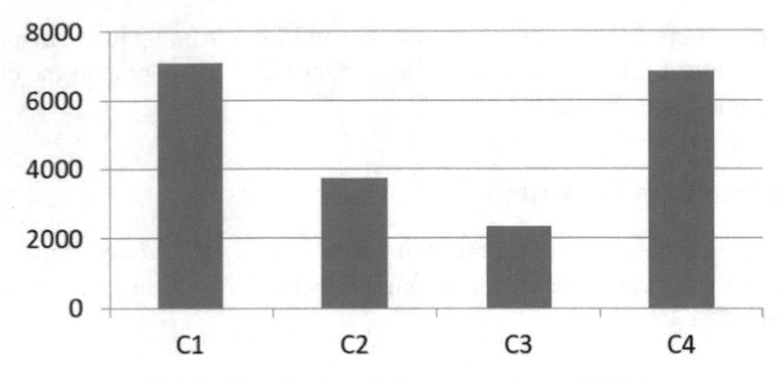

Fig. 5. Distribution of cluster members of SOM

4.3 SOM + K-Means Algorithm Clustering

According to the previous algorithm analysis, this experiment is divided into two stages. The first step is to perform SOM clustering, and output the number of clusters and the centroid of each cluster; the second step is based on the output of the first step as the initial input of the K-means algorithm. Then execute the K-means algorithm for clustering.

1) The first stage

In order to compare the clustering effects of using these two algorithms alone and their combined algorithms, the previous 4.2 experimental results are taken as the first step in the two-step implementation, and the number of clusters is set to K = 4.

2) The second stage

The centroids of each cluster in Table 5, run K-means clustering, and finally get a new set of centroids:

Table 5. Combined algorithm clustering centroid

Cluster	Cluster Centroid			
	Average purchase	Number of purchases	Activity	Purchase intention
C1	0.362	0.078	0.28	0.207
C2	0.413	0.067	0.588	0.145
C3	0.154	0.047	0.445	0.205
C4	0.279	0.066	0.803	0.304

The distribution of cluster members is shown in Fig. 6: there are 7799 people with C1 type purchasing behaviors, 3246 people with C2 type purchasing behaviors, 3339 people with C3 type purchasing behaviors, and 5616 people with C4 type purchasing behaviors.

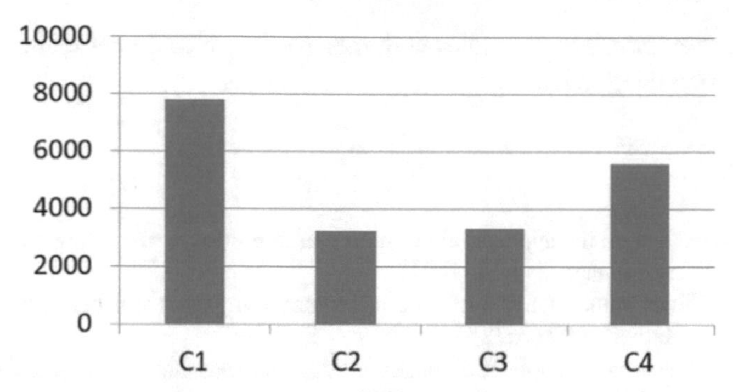

Fig. 6. Distribution of cluster members of SOM + K-Means

5 Conclusion

The intra-cluster standard deviation of the SOM, K-means, and SOM + K-means clustering methods are as shown in the Table 6. From the comparison of the intra-cluster standard variance of the three clustering analysis algorithms, it can be seen that under the same number of categories, the standard deviation of the intra-cluster distance clustering of AMCS customer segments based on the SOM + K-means combination algorithm is basically higher than that of the two. This algorithm is much smaller when used alone. This shows that the distribution of samples in each cluster is relatively uniform, and the effect of clustering is better. Therefore, the clustering effect of the AMCS customer segmentation model of SOM + K-means is better than that of SOM and K-means alone, and

Table 6. Intra-cluster standard deviation of three clustering methods

cluster	Intra-cluster Distance Standard Deviation (SOM)			
	Average purchase amount	Buy frequency	Activity	Purchase intention
C1	0.121	0.118	0.092	0.154
C2	0.111	0.071	0.036	0.099
C3	0.124	0.117	0.032	0.135
C4	0.111	0.072	0.037	0.168
cluster	Intra-cluster Distance Standard Deviation (K-Means)			
	Average purchase amount	Number of purchases	Activity	Purchase intention
C1	0.089	0.045	0.101	0.134
C2	0.065	0.147	0.162	0.117
C3	0.106	0.054	0.088	0.116
C4	0.105	0.074	0.088	0.124
cluster	Intra-cluster Distance Standard Deviation (SOM+K-Means)			
	Average purchase amount	Number of purchases	Activity	Purchase intention
C1	0.095	0.042	0.085	0.097
C2	0.103	0.037	0.086	0.101
C3	0.08	0.033	0.138	0.113
C4	0.11	0.051	0.049	0.126

it can provide e-commerce companies with more accurate customer retention strategies and targeted marketing guidance.

References

1. Zhou, H.: Research on the application of e-commerce precision marketing based on big data technology. Econ. Manage. Abs **11** (2021)
2. Hu, H.: Customer value analysis of enterprise marketing platform based on RFM model improvement. Marketing **08** (2022)
3. Wang, K.F.: Research on application of AFH customer classification based on data mining technology. Tech. Econ. Manage. Res. **11** (2012)
4. Sun, J.: Research on Improvement and Application of Customer Lifetime Value Model Based on Complaint Behavior, vol. 4. South China University of Technology (2018)
5. Gu,Y.R., Chen, Y.Z.: Research on commodity review based on SOM-K-means algorithm. Software Guide **12** (2021)
6. Feng, M.: Research on e-commerce daily sales forecast based on data mining. Comm. Circul. **12** (2021)
7. Zhu, M.: Data Mining. China University of Science and Technology University Press (2002)
8. Shao, F., Yu, Z.: Data Mining Principles and Algorithms. China Water and Power Press (2003)
9. Fu. L.M.: Video recommendation system based on K-means optimized SOM neural network algorithm. Softw. Eng. **10** (2022)

Blockchain Applications for Mobility-as-a-Service Ecosystem: A Survey

Elis Kulla[1(✉)], Leonard Barolli[2], Keita Matsuo[2], and Makoto Ikeda[2]

[1] Department of System Management, Fukuoka Institute of Technology (FIT), 3 -30 -1 Wajiro -higashi, Higashi -ku, Fukuoka 811 -0295, Japan
kulla@fit.ac.jp

[2] Department of Information and Communication Engineering, Fukuoka Institute of Technology (FIT), 3-30-1 Wajiro-Higashi, Higashi-Ku, Fukuoka 811 -0295, Japan
{barolli,kt-matsuo,m-ikeda}@fit.ac.jp

Abstract. The emerging technologies on automatic driving and connected vehicles will eventually enable a vast amount of applications regarding Mobility-as-a-Service (MaaS). However, the design and management of proprietary MaaS operators and integrator will have a great impact in developing this new ecosystem. The complexity of MaaS ecosystem requires decentralized and scalable solutions in order to accommodate a wide range of transportation services. Permissionless blockchain can be leveraged to support MaaS in managing transactions, identity and smart contracts. In this paper, we analyze some of the pioneering works in the field and identify he potential contributions on how can MaaS benefit from blockchain technology.

Keywords: Mobility as a service · Blockchain · Survey

1 Introduction

Recently, increased user mobility needs have created a lot of congestion and pollution in many countries. Vehicle usage is not very efficient in countries where personally owned vehicles are preferred instead of public transportation or green alternatives, such as bicycles and scooters. The reason behind that is that mos transportation services offer predefined routes and using multiple services everyday is complex and time consuming. In order to provide an aggregation of multiple transportation services into one single-ticket journey, MaaS has been introduced. Mobility as a Service (MaaS) integrates various forms of transportation services into a single on-demand mobility service. MaaS aims to provide mobility access to users through a single application and a single payment channel, instead of multiple payments. Recent development in vehicle technology have been spread in different areas, such as autonomous driving, inter-vehicle communication, intra-vehicle communication and so on. These technologies will enable

L. Barolli (Ed.): EIDWT 2023, LNDECT 161, pp. 129–140, 2023.
https://doi.org/10.1007/978-3-031-26281-4_13

and support MaaS in providing useful data from vehicles and passenger. However, many issues are envisioned in the future of MaaS ecosystem. Some of them are:

- Passengers' identity management
- Transportation services' trust and reliability
- Separation of services into multiple MaaS operators
- Data synchronisation
- Payments

All these issues need to be addressed before MaaS systems can be integrated into our lives. In this paper we envision MaaS ecosystems closely connected to blockchain technology, because blockchain provides a flexible, trustworthy, secure and decentralized platform, which can be used to solve the above-mentioned issues and develop novel functionality with practical applications. This paper is organized as follows. In Section II, we present a MaaS overview. Then we discuss some key aspects of blockchain technology in Section III. IN Section IV, we introduce some of the most recent technical approaches to integrating MaaS with blockchain technology. Finally, we conclude our work in Section V.

2 Mobility as a Service (MaaS)

2.1 MaaS Overview

Mobility as a Service (MaaS) integrates various forms of transportation services into a single on-demand mobility service. MaaS aims to provide mobility access to users through a single application and a single payment channel, instead of multiple payments. An important player in implementing MaaS services is the MaaS integrator. A MaaS integrator aims to combine different transportation services, in order to fulfill users' demand, while contributing to society and environment. Transportation services include public transportation, vehicle sharing, bicycle sharing, taxi, car rental, walking and so on. Embracing MaaS services introduces new business models, by offering users an alternative to private vehicles. It supports sustainability and will help reduce congestion and pollution.

The main actors in the MaaS ecosystem include, but not limited to the following:

- Public transportation authorities and operators.
- Local and national regulators.
- Mobility service providers.
- Technology providers.
- MaaS integrator.
- Users' demand.

Obviously, there is a need for these actors to operate on the same ecosystem, by sharing the same data and ensuring transparency.

In [1], the authors propose a topology of MaaS distinguishing 5 levels of integration, based on the complexity and goal of integration service, as shown in Fig. 1.

1. No Integration (**Level 0**): All transportation services operate independently.
2. Integration of Information (**Level 1**): Individual transportation services' routes, timetables and price information can be queried from a single interface and offer unique routes based on users' needs.
3. Integration of booking and payment (**Level 2**): In addition to being able to get specific routes and prices, users will be able to receive the service with a single payment, and seeing it as a single multi-modal trip.
4. Integration of the service offer (**Level 3**): The MaaS integrator will be able to offer monthly subscriptions, travel plans, contracts and so on.
5. Integration of societal goals (**Level 4**): The ultimate goal of integration is the ability to contribute to society and environment, by offering reduced rates for green alternatives, or alternatives that reduce congestion and waiting times.

A typical example of using MaaS is shown in Fig. 1. From the technical point of view, transportation service providers (TSPs), MaaS integrator and users interact continuously and populate the ecosystem respectively with basic data, derived data and demands. TSPs provide the basic data for the ecosystem, such as routes, fares and schedules. Users provide mobility demands, such as start and end destinations, price demands, accessibility demands, and so on. Finally, MaaS integrator, calculates optimal routes based on the basic data provided by TSPs and users' demand patterns.

A TSP should be able to provide basic data regarding vehicles, booking, tickets, coverage and routes, payments and other products. We describe them in the following:

Vehicles : vehicle types, vehicle locations and vehicle availability
Bookings : booking options such as *create*, *read*, *update*, *cancel* and so on
Tickets : Ticket formats and actions such as *create*, *read*, *update*
Coverage and Routes : Routes, stops and coverage areas
Payments : TSPs should accept electronic payments and offer flexibility

MaaS providers should be able to offer mobility services that simplify users' experience. They should be able to support roaming, when users travel through different jurisdictions, countries and continents. Users should be able to access their data, before, during and after the trip. Another service that needs addressing is identity management and privacy. Users should provide minimal personal information and their privacy should be of high priority.

To summarize, MaaS ecosystem requires different actors to provide data and services in a real-time, transparent and global fashion. MaaS providers should also be able to securely manage identities of its users. We believe that using blockchain to enable MaaS can simplify most of MaaS providers' tasks and give users, MaaS providers and TSPs more flexibility.

Users	① Input Start and Destination		③ Choose option				⑦ Start trip and use MaaS ticket	
MaaS Integrator		② Display potential routes, including schedule and cost		④ Request reservation of services		⑥ Merge tickets into a single MaaS ticket		
Transportation Services					⑤ Confirm reservation s and issue tickets			⑧ Verify tickets and provide service

Fig. 1. A typical example of e MaaS ecosystem.

3 Blockchain

3.1 Shared Databases

In most applications different people often need to access the same data. To meet this need, shared databases have emerged, where certain chunks of data can be accessed by more than one person at once. For example, to make it easier to distribute class materials, a teacher uses a web server to share all materials needed for the class, instead of printing one set of materials for each student.

But when databases are shared online, there are some concerns.

- Trust
- Identity management
- Permission
- Synchronization of data
- Ownership

There are many other practical issues with sharing a database, and various solutions are proposed. One way to share data that can help solve these problems and introduce novel applications is through blockchain technology.

3.2 Blockchain Overview

Blockchain is a public database that is updated and shared across many computers around the world. It is rapidly gaining traction in fields such as UAVs [2–4], IoT [5–8], smart cities [9], [10], smart grids [8], supply chain management [11,12], VANETs [13], and so on. It was originally proposed by Satoshi Nakamoto in his whitepaper [14] on Bitcoin. Blockchain is a type of data structure holding records of digital transactions, formally known as a distributed ledger. Identical copies of the database exist across multiple different computing machines, called nodes in blockchain terminology, connected in a peer-to-peer network. Transactions being the fundamental units of blockchain, a definite number of transactions are stored in a block, and blocks are continuously appended in sequence to form a chain. It emphasizes the importance of decentralization where the majority of entities participating in the blockchain are assumed to be genuine and take the decision collectively with the help of the process known as a consensus mechanism [27].

A blockchain is easily accessible from everyone. One can use the public Ethereum blockchain [28], or run a local blockchain by using Hyperledger's open source blockchain solutions [29].

Digital Signatures. Public key cryptography is one of the core concepts of blockchain technology. Each account is assigned a private key and a public key. Anything encrypted using the private key can only be decrypted using the public key, and vice versa. The public key serves as an address for each node, and each digital asset is associated with its owners public key. The piece of data that needs to be transferred is signed using the private key. This can be used to authenticate information; if a piece of data is signed cryptographically using a private key, then the only thing that can decrypt it is the same users public key. Blockchains commonly use elliptic curve digital signature algorithms.

Hashing. Hashing algorithms are arguably the backbone of blockchain technology. The hash function is a type of cryptographic algorithm which takes an input of variable size and returns an output of fixed length, called a hash. SHA family (SHA-1 and SHA-2) are popular hashing algorithms. There are two conditions a good hash algorithm must obey:

1. It must be non-invertible; i.e., it should not be possible to retrieve the input given the output.
2. The chances of two different inputs giving the same output hash must be very small.

The reason this is useful for security is that a small input change will completely change the hash value, and that makes tampering evident.

Transactions. Transactions are cryptographically signed instructions from accounts. An account will initiate a transaction to update the state of the blockchain. The simplest transaction is transferring currency from one account to another.

Blocks. Blocks are batches of transactions with a hash of the previous block in the chain. This links blocks together (in a chain) because hashes are cryptographically derived from the block data. This prevents fraud, because one change in any block in history would invalidate all the following blocks as all subsequent hashes would change and everyone running the blockchain would notice.

Smart Contracts. A "smart contract" is simply a program that runs on the blockchain. It's a collection of code (its functions) and data (its state) that resides at a specific address on the blockchain.

First proposed in 1997, Smart contracts are deployed to the network and run as programmed. User accounts can then interact with a smart contract by submitting transactions that execute a function defined on the smart contract. Smart contracts can define rules, like a regular contract, and automatically enforce them via the code. Smart contracts cannot be deleted by default, and interactions with them are irreversible.

Consensus Algorithm. Nodes in the peer-to-peer network take the responsibility of verifying the transactions and adding them to the blockchain. This process is known as mining and is one of the most important elements of the blockchain network because it is responsible for its decentralized nature. The fundamental idea behind consensus is that nodes must undergo a process that is hard to perform yet easy to validate - discouraging malicious entities from acquiring the necessary conditions required to validate invalid transactions. Putting it all together - suppose Alice desires to send a digital asset to Bob. Then Alice would have to sign the asset using her private key and broadcast a transaction request with the item and Bobs address. A miner, upon receiving the transaction, would bundle that transaction along with several other transactions in the block body. The miner would also create the block header and subsequently, broadcast the header to other blockchain nodes. These blockchain nodes then perform a pre-decided consensus algorithm. If the block is approved, then it is added as the latest block and all the nodes update the ledger to reflect the change. The fundamental role of the miner in all this is to collect, verify, and package transactions into a block, though the specifics of how they would do are dependent on the type of blockchain and consensus mechanisms agreed upon (Fig. 2).

Access Rights. Based on how users can access or manage the blockchain, there are two major categories of blockchains - permissioned and permissionless:

- Permissionless blockchains are public and open access; anyone is capable of joining the blockchain and take part in the consensus mechanism. Interested users having an Internet connection can join become a part of the network, and participants identities are hidden which is a security concern.
- Permissioned blockchains place restrictions on the member nodes in terms of read access or participation in the consensus process, or both. This often helps in computation and network communication overhead, which is a major cause for delay in permissionless networks.

4 Blockchain for MaaS

A few works have been proposed various blockchain-based systems to support MaaS issues, such as privacy and identity management. Some more works are

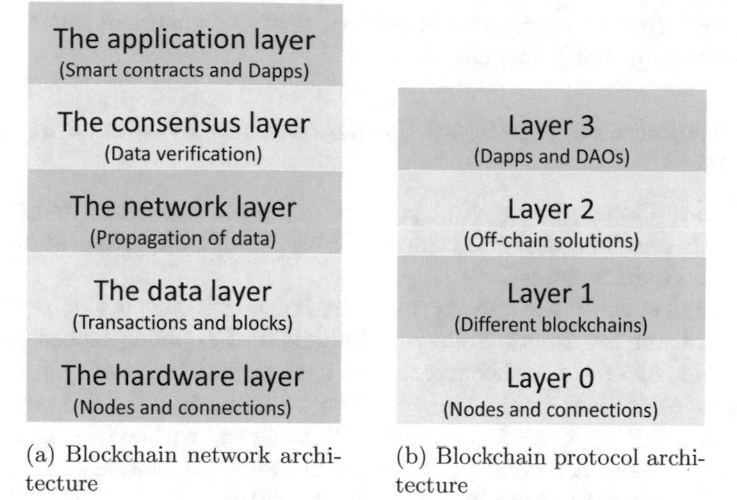

(a) Blockchain network architecture

(b) Blockchain protocol architecture

Fig. 2. Blockchain architecture (a) as a network and (b) as a protocol.

focused on MaaS-enabling technologies, such as Internet of Vehicles and Vehicular adhoc networks (VANET) .

In [15], the authors propose a mechanism to verify user identity and view their performance records, without revealing their identity. The system exploits a blockchain with the proof-of-stake (PoS) consensus. The complexity of the proposed scheme is evaluated in different scenarios.

In [16], blockchain-based Mobility-as-a-Service (MaaS) is envisioned as an application of edge computing. In most MaaS systems, a central MaaS operator plays a crucial role serving an intermediate layer which manages and controls the connections between TSPs and users. In order to eliminate central management, the authors propose a new blockchain-based scheme, in order to improve trust and transparency for all stakeholders as well as eliminates the need to make commercial agreements with separate MaaS agents. From a technical perspective, the power of computing and resources are distributed to different transportation providers at the edge of the network providing trust in a decentralised way.

The authors in [17] present a model of a MaaS ecosystem, enabling easy, quick and trusted transactions taking advantage of artificial intelligence and blockchain-enabled smart contracts. The work also presents a demonstration of this model, TravelToken, which utilizes QR code that stores and uses travel information in smart contract over Ethereum. The benefits are that all travel data can be stored in one ticket, information stays unaltered in blockchain, and value-share as well as compensations in case of delays will be automatic.

In [18], the authors propose a decentralized blockchain-based pseudonym distribution scheme which allows the reuse of existing pseudonyms to different vehicles. The results from simulations demonstrate that the proposed scheme can reuse existing pseudonyms and achieve a better degree of anonymity at a lower cost than existing schemes.

4.1 Blockchain for Intelligent Transportation Systems and Smart City

ITS in Smart Cities plays a vital part in implementing MaaS systems. With efficient ITS and well-desisgned Smart CIties, MaaS implementation is more seamless and quickly spread.

A survey on the latest advancement in blockchain for IoV is presented in [19]. The authors investigate some key challenges of applying blockchain in IoV as an enabler of ITS. Another interesting survey, which focuses on limitations and challenges of blackchain-based IoV, can be found in [20]. The paper in [21] conducts a study of blockchain-based ITS systems, without focusing on any specific application. A seven-layer conceptual model for blockchain is presented and some key research issues in each layer are given.

In [9], a framework that integrates the blockchain technology with smart devices is proposed. Blockchain is mainly used to provide security in a smart city scenario, where different actors access the same data. The authors in [10] propose a framework for smart cities with 3 main perspectives: human, technology and organization. Blockchain is the main focus of technological perspective, and the authors discuss the effect it might have in fundamental factors of a shared economy in smart cities.

In [22] this paper, the authors propose a heterogeneous blockchain-based trust-evaluation strategy for 5G-ITS, which utilizes federated deep learning technology. Extensive experiment results show that the proposed strategy can achieve reasonable and fair trust evaluations.

The authors in [23] propose a risk assessment framework for blockchain applications in smart mobility. As a case study, they analyze the risk associated to a multi-layered Blockchain framework for smart mobility data-markets.

In [24], the authors propose a blockchain-enabled marketplace for Digital Twin as a Service (DTaaS) for ITS. They propose an on-demand DTaaS architecture to fully utilize the sensing capabilities of ITS, a price adjustment algorithm to realize the optimal DT matching for ITS requesters and finally a permissioned blockchain and a novel DT-DPoS consensus mechanism are established to enhance the security and efficiency of DTaaS. Simulation shows that the proposed DTaaS can efficiently stimulate and facilitate DT transactions.

In [25] is designed a new blockchain-enabled certificate-based authentication scheme for vehicle accident detection and notification in ITS. Due to blockchain technology usage, it is shown that the proposed scheme is secure against various potential attacks, maintains transparency and immutability of the information. A comprehensive comparative analysis shows that the proposed scheme achieves better security and has low communication and computational overheads as compared to other competitive authentication schemes in IoV.

In [26] a blockchain-based protocol is proposed to prioritize emergency data in ITS. It uses 2 blockchains: one is to store the authentication information of the vehicle, and another one to store and distribute blockchain services. Experimental analysis revealed that the proposed blockchain-based protocols are better sthan the existing ones in terms of various metrics.

5 Blockchain for Vehicular Communication

Many researchers have proposed and designed blockchain systems to support vehicular communications in air, ground and naval/marine environment. However, most works focus on applications of vehicular communications, rather than communication performance.

The paper in [2] reviews various applications of blockchain in UAV networks such as network security, decentralized storage, inventory management, service coordination, and so on. Blockchain is mainly used to increase security and privacy. An automated inventory system, based in UAV and blockchain, is presented in [3]. The system utilizes blockchain, to make collected inventory data available to the interested parties. In [27], 75 blockchain-based security schemes for vehicular networks are analyzed from an application, security, and blockchain perspective.

In [13], the authors propose a blockchain-based anonymous reputation system (BARS) to preserve privacy, by hiding correlations between real identities and public keys. They design a trust model to improve the trustworthiness of messages relying on the reputation of the sender based on both direct historical interactions and indirect opinions about the sender. Experiments are conducted to evaluate BARS in terms of security and performance and the results show that BARS is able to establish distributed trust management, while protecting the privacy of vehicles.

[30] proposes a blockchain-based framework for enabling secure payment and communication in Intelligent Transportation Systems. The proposed framework uses Ethereum network to facilitating Vehicle-to-Everything (V2X) communications and parking payments.

In [31] a blockchain-based vehicular authentication architecture is proposed. To guarantee the anonymity and traceability, a pseudonym-based privacy-preserving authentication method is also proposed. The efficiency of the proposed scheme is compared to other works, by security analysis and experiments.

In [32], the authors propose a resource management scheme that aims to improve the performance of blockchain systems. A resource control method and a resource monitoring system are developed to cooperate with the system. The proposed method shows better performance than the existing baseline method.

The paper in [33] presents a taxonomy of different architectures while focusing on blockchain technology inclusion in vehicular networks.The authors analyze those architectures, by discussing their limitations, and how can blockchain usage overcome those challenges.

The authors in [34], propose a directed acyclic graph (DAG) enabled knowledge-sharing framework in which vehicular knowledge is encapsulated in the DAG. A tip selection algorithm and a fast authentication scheme for cross-regional vehicles are designed to reduce computation and storage needs. Simulation results show that the proposed DAG framework can achieve a higher knowledge sharing quality and lower authentication latency compared with traditional DAG systems.

6 Conclusions and Future Works

In this paper, we presented some of the most recent approaches in integrating MaaS services with blockchain technology. Most studies are focused on security, trust and identity management and system performance. Based on all the works presented here, we could say that there is still room to implement blockchain solutions in MaaS ecosystem in different fields. MaaS will benefit from blockchain by being able to gain global knowledge of TSPs data as well as users demands. Blockchain can be used to improve connectivity between actors as well. In the future, we plan ti implement our own MaaS integrator based on SUMO mobility simulator and hyperledger blockchain.

References

1. Sochor, J., Arby, H., MariAnne Karlsson, I.C., Sarasini, S.: A topological approach to Mobility as a Service: A proposed tool for understanding requirements and effects, and for aiding the integration of societal goals. Res. Trans. Business Manage. **27**, 3–14 (2018). ISSN 2210-5395, https://doi.org/10.1016/j.rtbm.2018.12.003
2. Alladi, T., Chamola, V., Sahu, N., Guizani, M.: Applications of blockchain in unmanned aerial vehicles: a review. Veh. Commun. **23** 100249 (2020)
3. Fernandez-Carames, T.M., Blanco-Novoa, O., Suarez-Albela, M., Fraga-Lamas, P.: UAV and blockchain-based system for industry 4.0 inventory and traceability applications. Multidiscipl. Digit. Publ. Inst. Proc. **4**(1), 26 (2018)
4. Kapitonov, A., Lonshakov, S., Krupenkin, A., Berman, I.: Blockchain-based protocol of autonomous business activity for multiagent systems consisting of UAVs. In: Proceedings of the Workshop on Research Education Development of Unmanned Aerial System (RED-UAS), pp. 84–89 (2017)
5. Dorri, A., Kanhere, S.S., JurdaK, R.: Towards an optimized blockchain for IoT. In: Proceedings of the IEEE/ACM 2nd International Conference Internet Things Design and Implementation (IoTDI), pp. 173–178 (2017)
6. Novo, O.: Blockchain meets IoT: An architecture for scalable access management in IoT. IEEE Internet Things J., **5**(2), 1184–1195 (2018)
7. Panarello, A., Tapas, N., Merlino, G., Longo, F., Puliafito, A.: Blockchain and IoT integration: a systematic survey. Sensors **18**(8), 2575 (2018)
8. Alladi, T., Chamola, V., Rodrigues, J., Kozlov, S.A.: Blockchain in smart grids: a review on different use cases. Sensors **19**(22), 4862 (2019)
9. Biswas, K., Muthukkumarasamy, V.: Securing smart cities using blockchain technology. In: Proceedings of the IEEE 18th International Conference High Performance Computing and Communication. IEEE 14th International Conference Smart City IEEE 2nd International Confernce Data Science System. (HPCC/SmartCity/DSS), pp. 1392–1393 (2016)

10. Sun, J., Yan, J., Zhang, K.Z.K.: Blockchain-based sharing services: What blockchain technology can contribute to smart cities. Financial Innovation **2**(1), 1–9 (2016). https://doi.org/10.1186/s40854-016-0040-y
11. Tribis, Y., El Bouchti, A., Bouayad, H.: Supply chain management based on blockchain: a systematic mapping study. MATEC Web Conf. **200** 20 (2018)
12. Subramanian, G., Thampy, A.S.: Implementation of hybrid blockchain in a pre-owned electric vehicle supply chain. IEEE Access **9**, 82435–82454 (2021). https://doi.org/10.1109/ACCESS.2021.3084942
13. Lu, Z., Wang, Q., Qu, G., Liu, Z.: BARS: A blockchain-based anonymous reputation system for trust management in VANETs. In: Proceedings of the 17th IEEE International Confernce Trust Security Privacy Computing and Communication/12th IEEE International Conference Big Data Science Engineering. (TrustCom/BigDataSE), pp. 98–103 (2018)
14. Nakamoto, S.: Bitcoin: a peer-to-peer electronic cash system. Decentralized Bus. Rev. 21260 (2008)
15. Kong, Q., Lu, R., Yin, F., Cui, S.: Blockchain-based privacy-preserving driver monitoring for maas in the vehicular iot. IEEE Trans. Veh. Technol. **70**(4), 3788–3799 (2021). https://doi.org/10.1109/TVT.2021.3064834
16. Nguyen, T.H., Partala, J., Pirttikangas, S.: Blockchain-Based Mobility-as-a-Service. In: 2019 28th International Conference on Computer Communication and Networks (ICCCN), pp. 1–6 (2019). https://doi.org/10.1109/ICCCN.2019.8847027
17. Karinsalo, A., Halunen, K.: Smart Contracts for a Mobility-as-a-Service Ecosystem. In: 2018 IEEE International Conference on Software Quality, Reliability and Security Companion (QRS-C), pp. 135–138 (2018). https://doi.org/10.1109/QRS-C.2018.00036.
18. Bao, S., et al.: Pseudonym management through blockchain: cost-efficient privacy preservation on intelligent transportation systems. IEEE Access **7**, 80390–80403 (2019). https://doi.org/10.1109/ACCESS.2019.2921605
19. Mollah, M.B., et al.: Blockchain for the internet of vehicles towards intelligent transportation systems: a survey. IEEE Internet Things J. **8**(6), 4157–4185 (2021). https://doi.org/10.1109/JIOT.2020.3028368
20. Jabbar, R., et al.: Blockchain technology for intelligent transportation systems: a systematic literature review. IEEE Access **10**, 20995–21031 (2022). https://doi.org/10.1109/ACCESS.2022.3149958
21. Yuan,Y., Wang, F.-Y.: Towards blockchain-based intelligent transportation systems. In: 2016 IEEE 19th International Conference on Intelligent Transportation Systems (ITSC), pp. 2663–2668 (2016). https://doi.org/10.1109/ITSC.2016.7795984.
22. Wang, X., Garg, S., Lin, H., Kaddoum, G., Hu , J., Hassan, M.M.: Heterogeneous Blockchain and AI-Driven Hierarchical Trust Evaluation for 5G-Enabled Intelligent Transportation Systems. IEEE Trans. Intell. Transp. Syst. **99** 1–10 (2021) https://doi.org/10.1109/TITS.2021.3129417
23. Mallah, R.A., Lopez, D., Farooq, B.: Cyber-security risk assessment framework for blockchains in smart mobility. IEEE Open J. Intell. Transp. Syst. vol. **2** 294–311 (2021) https://doi.org/10.1109/OJITS.2021.3106863
24. Liao, S., Wu, J., Bashir, A.K., Yang, W., Li, J., Tariq, U.: Digital twin consensus for blockchain-enabled intelligent transportation systems in smart cities. IEEE Trans. Intell. Transp. Syst. **23**(11), 22619–22629 (2022). https://doi.org/10.1109/TITS.2021.3134002

25. Vangala, A., Bera, B., Saha, S., Das, A.K., Kumar, N., Park, Y.: Blockchain-enabled certificate-based authentication for vehicle accident detection and notification in intelligent transportation systems. IEEE Sensors J. **21**(4), 15824–15838 (2021). https://doi.org/10.1109/JSEN.2020.3009382
26. Ahmed, M., et al.: A blockchain-based emergency message transmission protocol for cooperative VANET. IEEE Trans. Intell. Transp. Syst. **23**(10), 19624–19633 (2022). https://doi.org/10.1109/TITS.2021.3115245
27. Alladi, T., Chamola, V., Sahu, N., Venkatesh, V., Goyal, A., Guizani, M.: A comprehensive survey on the applications of blockchain for securing vehicular networks. IEEE Commun. Surv. Tutorials **24**(2), 1212–1239 (2022)
28. Ethereum Blockchain. https://ethereum.org/
29. Hyperledger Foundation. https://www.hyperledger.org/
30. Jabbar, R., Fetais, N., Kharbeche, M., Krichen, M., Barkaoui, K., Shinoy, M.: Blockchain for the Internet of Vehicles: how to use blockchain to secure vehicle-to-everything (v2x) communication and payment?. IEEE Sensors J. **21**(14), 15807–15823 (2021). https://doi.org/10.1109/JSEN.2021.3062219
31. Yang, Y., Wei, L., Wu, J., Long, C., Li, B.: A blockchain-based multidomain authentication scheme for conditional privacy preserving in vehicular ad-hoc network. IEEE Internet of Things J. **9**(11), 8078–8090 (2022). https://doi.org/10.1109/JIOT.2021.3107443
32. Gao, L., et al.: Resource management for blockchain-enabled internet of vehicles. In: 2021 International Conference on Information and Communication Technologies for Disaster Management (ICT-DM), pp. 164–170 (2021). https://doi.org/10.1109/ICT-DM52643.2021.9664129
33. Diallo, E.H., Dib, O., Agha, K.A.: The journey of blockchain inclusion in vehicular networks: a taxonomy. In: Third International Conference on Blockchain Computing and Applications (BCCA) pp. 135–142 (2021). https://doi.org/10.1109/BCCA53669.2021.9657050
34. Chai, H., Leng, S., Wu, F.: Secure Knowledge Sharing in Internet of Vehicles: A DAG-Enabled Blockchain Framework. In: ICC 2021 - IEEE International Conference on Communications, pp. 1–6 (2021). https://doi.org/10.1109/ICC42927.2021.9500503

Construction of a Fully Homomorphic Encryption Scheme with Shorter Ciphertext and Its Implementation on the CUDA Platform

Dong Chen[1], Tanping Zhou[1,2,3](✉), Wenchao Liu[1,2], Zichen Zhou[1], Yujie Ding[1], and Xiaoyuan Yang[1,2]

[1] College of Cryptography Engineering, Engineering University of People's Armed Police, Xi'an 710086, China
tanping2020@iscas.ac.cn, liuwch3@mail3.sysu.edu.cn

[2] Key Laboratory of Network and Information Security Under the People's Armed Police, Xi'an 710086, China

[3] Institute of Software, Chinese Academy of Sciences, Beijing 100080, China

Abstract. Homomorphic encryption supports meaningful operations on ciphertext and is widely used in outsourcing computing, secure multi-party computing and other scenarios. The TFHE-type fully homomorphic encryption scheme has an efficient bootstrapping process, so it supports efficient homomorphic logic gate circuits. However, its single-bit encryption feature results in a large ciphertext expansion (the ciphertext expansion rate of the CGGI16, CGGI17, and ZYL + 18 scheme is 16032). Therefore, how to reduce the scale of ciphertext without affecting the efficiency is a practical problem. TFHE-type fully homomorphic encryption scheme with shorter ciphertexts is constructed. By applying the round function to the encryption process, the ciphertext scale is reduced by 62%; Using the GPU's ability to implement large-scale matrix operations efficiently, a fully homomorphic encryption scheme with shorter ciphertexts is implemented based on the cuFHE. The experimental results show that, compared with the CGGI17 scheme, the running time of a single gate circuit (including the bootstrapping process) in this scheme does not exceed 1 ms on the CUDA platform, and the ciphertext scale of this algorithm is reduced by 62%.

1 Introduction

Cryptography researchers believe that "public key encryption has opened up a new direction of cryptography, and a practical fully homomorphic encryption scheme will give birth to a new distributed computing model" [1]. After more than ten years of development, the efficiency of fully homomorphic encryption has been greatly improved, and it has even gradually moved closer to practical applications. In 1995, Benaloh proposed the first public key cryptography scheme focusing on homomorphism [2], which can perform an addition or multiplication homomorphic operation. Such schemes are also known as semi-homomorphic encryption [3]. In 2005, Boneh et al. proposed the first somewhat homomorphic encryption scheme (SHE) that supports homomorphic addition

L. Barolli (Ed.): EIDWT 2023, LNDECT 161, pp. 141–151, 2023.
https://doi.org/10.1007/978-3-031-26281-4_14

and multiplication, and can run a lower number of polynomial circuits [4]. In 2009, Gentry constructed the first fully homomorphic encryption scheme (FHE) that supports any number of additions and any number of multiplications based on the difficult problem on ideal lattices and the sparse subset sum problem [3]. In 2011, Brakerski and Vaikuntanathan constructed the BV11b scheme [5], which utilizes the LWE assumption, introduces relinearization and dimensionality reduction and modulo reduction techniques. In 2012, Brakerski, Gentry and Vaikuntanathan et al. optimized the mode reduction technology based on the BV11b scheme to reduce the noise growth from exponential growth to linear growth, and constructed the BGV12 scheme [6], which is one of the most efficient fully homomorphic encryption schemes at present. Subsequently, Halevi and Shoup implemented the BGV12 scheme [6] and the corresponding ciphertext packing technology and the optimization technology [8, 9] in GHS12b [7] using the C++ language and NTL math function library, which is called HElib. In 2018, Halevi and Shoup re-wrote the Helib code [10] and optimized the linear transformations used in the bootstrapping process and other processes. The new algorithm is 30–75 times faster than the original HElib algorithm, and the size of the calculation key is reduced by 33%–50%.

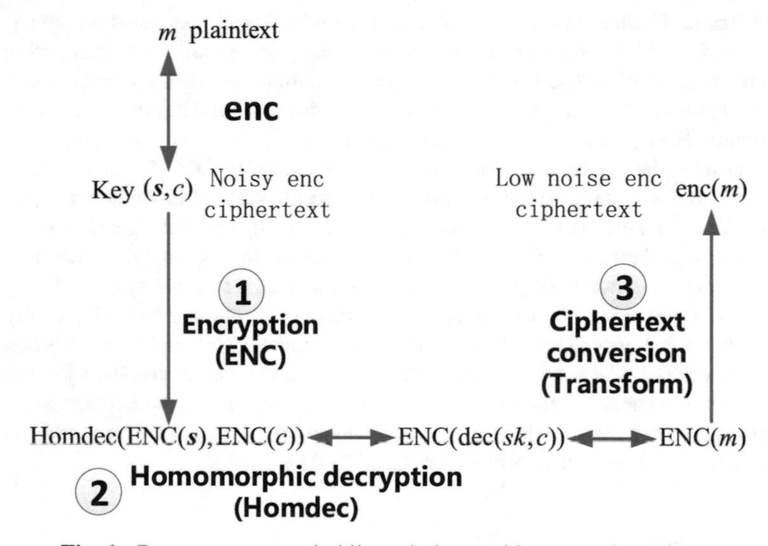

Fig. 1. Bootstrap process in bilayer holomorphic encryption scheme

The speed of the bootstrapping process affects the speed of the fully homomorphic encryption scheme. The construction and optimization of the bootstrapping process is a hot and difficult issue in the research of fully homomorphic encryption. In CRYPTO'2014, Alperin and Peikert constructed the first double-layer fully homomorphic scheme AP14 [11], which used a specially designed outer scheme to run the decryption circuit of the original scheme (inner scheme). The advantage of the double-layer fully homomorphic encryption scheme is that different outer-layer schemes can be designed according to the characteristics of the decryption circuit of the inner-layer scheme, so that the decryption circuit of the inner-layer scheme can be run efficiently. The bootstrapping process of this scheme is less noisy than BV14 [12]. However, as shown in Fig. 1, compared with the traditional bootstrapping process, using the bootstrapping process in the double-layer fully homomorphic encryption scheme needs to increase the third step of ciphertext conversion, that is, it is required to convert the outer ciphertext after running the decryption circuit into the inner ciphertext. The existence of this transformation step largely limits the form of the outer ciphertext. In EUROCRYPT'2015, Ducas and Micciancio constructed a more efficient bilayer fully homomorphic scheme FHEW [13]. In ASIACRYPT '2016, Chillotti et al. [10] constructed an efficient two-layer fully homomorphic scheme TFHE [14] with the structure of $T = (0,1]$, in which the bootstrap operation time is less than 0.1 s and the bootstrap key is reduced from 1GB bytes to 23MB. In ASIACRYPT'2017, Chillotti et al. further optimized the accumulation process in the TFHE scheme, which reduced the computation time of the bootstrapping process to 13 ms [15]. The cuFHE library [16] is an open-source GPU-based fully homomorphic encryption software library. Compared with the CGGI17 implementation library TFHE on the CPU side, it has achieved about 20 times speed improvement on the NVIDIA Titan Xp graphics card.

Currently, efficient fully homomorphic encryption schemes include BGV type and TFHE type. The BGV scheme and its optimization scheme are typical high-efficiency hierarchical fully homomorphic schemes. The depth of the homomorphic computing circuit is related to the security parameters, and the suitable scenario is multi-bit parallel computing [17]. TFHE scheme is a highly efficient pure full homomorphism, which is more suitable for serial operation and logical operation. Pure fully homomorphic encryption can efficiently construct arbitrary logic circuits (operations) and does not require a preset number of multiplication operations. The disadvantage is that the plain ciphertext expansion of the scheme is relatively large, reaching 16032. Therefore, how to reduce the size of the ciphertext without affecting the efficiency of the scheme, thereby reducing the amount of communication in the actual operation process is a practical problem that needs to be solved urgently.

The idea of this paper: In the CGGI17 scheme, the single-bit ciphertext is 501-dimensional 32-bit data (the plain ciphertext expansion rate is 16032), and the ciphertext is directly operated during the homomorphic calculation. However, when the scheme runs the bootstrapping process on the ciphertext, it needs to convert the component $\boldsymbol{c}$ $[i]$ of the ciphertext to $X^{c[i]}$ in the ring $\mathbb{Z}[X]/(X^N + 1)$. This results in a collision between the 32 bits of the ciphertext component and the X^i index in $\mathbb{Z}[X]/(X^N + 1)$. In order to improve the efficiency, the CGGI17 scheme implements the *round* function on the ciphertext before the bootstrapping process, which reduces the 501-dimensional 32-bit ciphertext to 501-dimensional 11-bit ciphertext. Therefore, most of the redundant information of the ciphertext will be discarded before the core steps of the bootstrapping process run. In this paper, we consider discarding redundant information when generating ciphertext, that is, running the *round* function during the ciphertext generation process. The analysis shows that by setting the *round* parameters reasonably, the size of the ciphertext can be effectively reduced, the noise can be reduced by 62%, and the efficiency of the scheme can be further improved. The comparasion of our scheme and CGGI17 is showing in Fig. 2, where c_i is ciphertext, P is the parameters in round, **Homdec** is homomorphic decrypt process. The main difference lines in that the ciphertexts of our scheme is rounded, which deduce to an shorter ciphertexts.

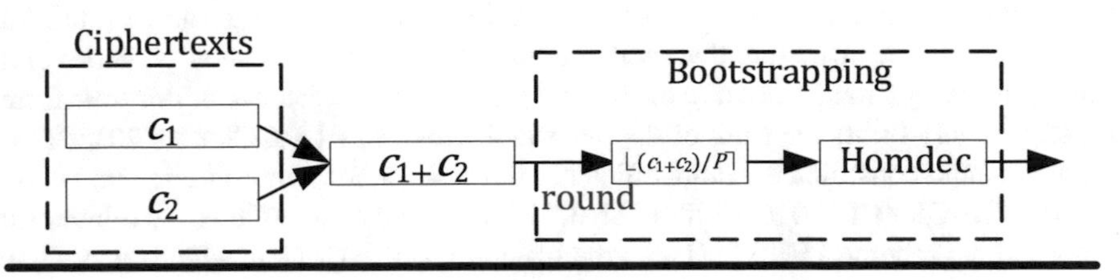

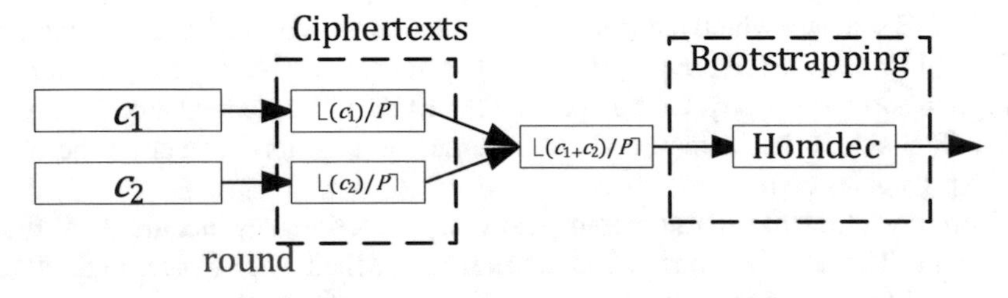

Fig. 2. The comparasion of our scheme and CGGI17

2 Related Knowledge and Key Technologies

2.1 Related Knowledge

This section introduces the relevant knowledge and key technologies of scheme design. λ represents the security parameter. Define $\mathbb{B} = \{0, 1\}$, $\mathbb{T} = \mathbb{R}/\mathbb{Z} \in [0, 1)$ is a real ring of real numbers modulo 1, $\mathcal{R} = \mathbb{Z}[X]/(X^N + 1)$ and $\mathbb{T}_N[X] = \mathbb{R}[X]/(X^N + 1)$ mod 1 are two polynomial rings, where $X^N + 1$ is a (2N)-th cyclotomic polynomial. $\mathcal{M}_{\mathrm{p,q}}(\mathrm{E})$ is represented as a set of matrices with $p \times q$ elements belonging to E. Define $\|x\|_p = \min_{u \in x+\mathbb{Z}^k}(\|\boldsymbol{u}\|_p)$, where $\boldsymbol{x} \in \mathbb{T}^{\mathrm{k}}$. $\|\boldsymbol{x}\|_p$ represents the p-norm of the vector of coefficients in $(-1/2, 1/2]$. Extending these concepts to polynomials, $\|P(X)\|_p$ represents the norm of a vector of coefficients of a polynomial with real or integer coefficients. If the polynomial is modulo $\left(X^N + 1\right)$, then the polynomial is uniquely represented as a polynomial of degree less than $N - 1$. Define the noise $Err(\boldsymbol{a}, b) := b - \boldsymbol{a} \cdot \boldsymbol{s}$ of the TLWE instance for $\boldsymbol{c} = (\boldsymbol{a}, b) \in \mathbb{T}_{\mathrm{N}}[\mathrm{X}]^{\mathrm{k}} \times \mathbb{T}_{\mathrm{N}}[\mathrm{X}]$ and $\boldsymbol{s} \in \mathbb{B}_{\mathrm{N}}[\mathrm{X}]^{\mathrm{k}}$.

Distribution χ on a ring $\mathbb{T}$ is defined to be concentrated if the distribution is concentrated in a sphere of radius 1/4 in $\mathbb{T}$. The variance and expectation of distribution are defined as $\mathrm{Var}(\chi)$ and $\mathbb{E}(\chi)$, respectively, where $\mathrm{Var}(\chi) = \min_{\bar{x} \in \mathbb{T}} \sum p(x)|x - \bar{x}|^2$ and $\mathbb{E}(\chi)$ is equal to $\bar{x} \in \mathbb{T}$ that minimizes the variance. Extending these concepts to polynomials, the distribution χ' over $\mathbb{T}^{\mathrm{n}}$ or $\mathbb{T}_{\mathrm{N}}[\mathrm{X}]^{\mathrm{k}}$ is centralized if and only if the distribution of each coefficient of the vector is centralized. Vector expectation $\mathbb{E}(\chi')$ is defined as a vector of expectations for each component. The variance $\mathrm{Var}(\chi')$ of a vector is defined as the maximum value of the variance of all components.

Definition 1 [14] (TLWE): Let $n \geq 1$ be an integer, N be a power of 2, the noise parameter $\alpha \geq 0$, and the private key $\boldsymbol{s} \in \mathbb{B}_{\mathrm{N}}[\mathrm{X}]^{\mathrm{k}}$ is randomly and uniformly selected.

The message $\mu \in \mathbb{T}_N[X]$ is encrypted to $\boldsymbol{c} = (\boldsymbol{a}, b) \in \mathbb{T}_N[X]^k \times \mathbb{T}_N[X]$ and $b \in \mathbb{T}_N[X]$ follows a Gaussian distribution $D_{\mathbb{T}_N[X], \alpha, \boldsymbol{sa}+\mu}$. The instance $\boldsymbol{c} = (\boldsymbol{a}, b)$ is random when the vector $\boldsymbol{a}$ is taken from a uniform distribution; The instance is called a trivial instance when $\boldsymbol{a} = 0$; It is called a noise-free instance when $\alpha = 0$; It is called a homogeneous instance when $\mu = 0$.

Definition 2 [14] (TGSW): Let l and $k \geq 1$ be two integers, $\alpha \geq 0$ is the noise parameter, and $\boldsymbol{h}$ is the decomposition tool defined in Eq. 1. Let $\boldsymbol{s} \in \mathbb{B}_N[X]^k$ be a randomly chosen key, and define $\boldsymbol{C} \in \mathcal{M}_{(k+1)l,k+1}(\mathbb{T}_N[X])$ to be a fresh TGSW instance of $\mu \in \Re/\boldsymbol{h}^{\perp}$ with noise parameter if and only if $\boldsymbol{C} = \boldsymbol{Z} + \mu \cdot \boldsymbol{h}$, where each row of the matrix $\boldsymbol{Z} \in \mathcal{M}_{(k+1)l,k+1}(\mathbb{T}_N[X])$ is a homogeneous TLWE instance with Gaussian noise parameter α. If there is a unique polynomial $\mu \in \Re/\boldsymbol{h}^{\perp}$ and a unique key $\boldsymbol{s}$ such that each row of $\boldsymbol{C} - \mu \cdot \boldsymbol{h}$ is a legal TLWE instance with plaintext 0 and key $\boldsymbol{s}$, then $\boldsymbol{C} \in \mathcal{M}_{(k+1)l,k+1}(\mathbb{T}_N[X])$ is said to be a legal TGSW instance, and polynomial μ is called the plaintext of $\boldsymbol{C}$, denoted as $msg(\boldsymbol{C})$.

$$\boldsymbol{h} = \left\{ \begin{matrix} \frac{1}{B_g} & 0 \\ \vdots & 0 \\ \frac{1}{B_g^l} & 0 \\ 0 \;\vdots\vdots & 0 \\ 0 & \frac{1}{B_g} \\ 0 & \vdots \\ 0 & \frac{1}{B_g^l} \end{matrix} \right\} \in \mathcal{M}_{(k+1)l,k+1}(\mathbb{T}_N[X]) \tag{1}$$

2.2 External Product

Definition 3 [14] (External Product): A is a legal TGSW instance whose plaintext is μ_A, and $\boldsymbol{b}$ is a legal TLWE instance whose plaintext is μ_b. The external product is defined as:$\boxdot$: TGSW $\times$ TLWE $\rightarrow$ TLWE, i.e.A, b $\rightarrow$ $A \boxdot \boldsymbol{b}$ $=\text{Dec}_{\boldsymbol{h},\beta,\varepsilon}(\boldsymbol{b})$A. The ciphertext decomposition function $\text{Dec}_{\boldsymbol{h},\beta,\varepsilon}$ can decompose a random matrix into a large dimensional matrix whose component coefficients are taken from {0,1}, which is formally defined in TFHE scheme [14]. For external products, it has the following properties.

Theorem 4 [14] (Correctness pf External Product): A is a legal TGSW instance whose plaintext is μ_A, $\boldsymbol{b}$ is a legal TLWE instance whose plaintext is μ_b, $\beta = \frac{B_g}{2}$ and $\varepsilon = \frac{1}{2B_g^l}$ are the quality and precision used in the decomposition process $\text{Dec}_{\boldsymbol{h},\beta,\varepsilon}(\boldsymbol{b})$. Then the external product $A \boxdot \boldsymbol{b}$ outputs a TLWE instance whose plaintext is $\mu_A \mu_B$, satisfying $\|Err(A \boxdot \boldsymbol{b})\|_{\infty} \leq (\text{k}+1)\text{lN}\beta \|\text{Err(A)}\|_{\infty} + \|\mu_A\|_1 (1+kN)\varepsilon + \|\mu_A\|_1 \|Err(\boldsymbol{b})\|_{\infty}$ and $Var(Err(A \boxdot \boldsymbol{b})) \leq (\text{k}+1)\text{lN}\beta^2 Var(Err(A)) + (1+kN)^2 \|\mu_A\|_2^2 \varepsilon^2 + \|\mu_A\|_2^2 Var(Err(\boldsymbol{b}))$.

2.3 Ciphertext Conversion

In the bootstrapping process, the RLWE ciphertext corresponding to the key $s'' \in R^k$ will be output after the homomorphic operation of the inner product, to ensure that the output can be decrypted by the initial key $s \in \mathbb{B}^n$. It is necessary to use the TLWE extraction technology to homomorphically truncate the constant term of the polynomial to obtain the LWE ciphertext corresponding to the key $s' \in \mathbb{Z}^{kN}$; Then use the keyswitch technology to convert the ciphertext of the private key s' into the ciphertext of the private key s without changing the plaintext. The properties of TLWE extraction and keyswitch are as follows:

Definition 5 [14] (TLWE SampleExtract): The input is $TLWE_{s''}(\mu)$ instance (a'', b''), key $s'' \in \mathcal{R}^k$. Output key $s' := (coefs(s''_1(X), ..., coefs(s''_k(X))) \in \mathbb{Z}^{kN}$ and corresponding LWE instance $(a', b') := ((coefs(a''_1(\frac{1}{X}), ..., coefs(a''_k(\frac{1}{X}))), b''_0) \in \mathbb{T}^{kN+1}$, b''_0 is the constant term of the polynomial b''. Then there is $\varphi_{s'}(a', b') = \varphi_{s''}(a'', b'')_0$ (a constant term equal to $\mu = msg(a'', b'')_0$), $\|Err(a', b')\|_\infty \le \|Err(a'', b'')\|_\infty$ and $Var(Err(a', b')) \le Var(Err(a'', b''))$.

Lemma 6 [14] (KeySwitch): Let $(a', b') \in LWE_{s'}(\mu)$ be an LWE instance with key $s' \in \{0, 1\}^{n'}$, noise $\eta' \in ||Err(a', b')||_\infty$, noise variance $\eta'' \in Var(Err(a', b'))$, and transformation key $KS_{s' \to s, \gamma, t} = \{k_{i,j,v}\}$, where $k_{i,j,v} \in \text{LWE}_s^{q/q}(v \cdot s'_i B_{ks}^j)$, $s \in \{0, 1\}^n$, $v \in \mathbb{Z}_{B_{ks}}$. The keyswitch process KeySwitch((a, b), $KS_{s' \to s, \gamma, t}$) $= (0, b) - \sum_{i=1}^{N} \sum_{j=0}^{d_{ks}-1} k_{i,j,a_{i,j}}$ will output an LWE instance $(a, b) \in LWE_s(\mu)$ that satisfies $\|Err(a, b)\|_\infty \le \eta' + n't\gamma + n'2^{-(t+1)}$, $Var(Err(a, b)) \le \eta'' + n't\gamma^2 + n'2^{-2(t+1)}$.

3 Construction of a Fully Homomorphic Encryption Scheme with Shorter Ciphertext

The TFHE-type fully homomorphic encryption scheme is a typical double-layer fully homomorphic encryption scheme. The inner layer scheme is a typical lattice encryption scheme based on the LWE problem, and the outer layer scheme is a hybrid fully homomorphic encryption scheme based on the RLWE problem. The difference between our scheme and the CGGI17 scheme is that the *round* function in the bootstrapping process of the original scheme is advanced to the encryption process, and the effect of reducing the size of the ciphertext is achieved by setting parameters reasonably. The specific algorithm of the fully homomorphic encryption scheme with shorter ciphertext is as follows:

Initialization $Setup(1^\lambda)$**:** Input security parameter λ, define LWE dimension n, key distribution χ, Gaussian distribution related parameter α, decomposition base B_{ks}, decomposition order d_{ks}, $g' = (B_{ks}^{-1}, ..., B_{ks}^{-d_{ks}})$, and output system parameter $pp^{\text{LWE}} = (n, \chi, \alpha, B_{ks}, d_{ks})$.

Key generation KeyGen(pp^{LWE})**:** The LWE key $s \leftarrow \chi^n$ and the GSW key $s'' \in \mathbb{B}_N[X]^k$ are randomly selected. Generate bootstrap key s, transform key $KS_{s' \to s, \gamma, t} = \{k_{i,j,v}\}$, where $k_{i,j,v} \in \text{LWE}_s^{q/q}(v \cdot s'_i B_{ks}^j)$, $i \in [1, n']$, $j \in [1, d_{ks}]$, $v \in \mathbb{Z}_{B_{ks}}$.

Encryption $Enc(m, \boldsymbol{s})$: Input plaintext $m \in \{0.1\}$, private key $\boldsymbol{s}$, uniformly select $\boldsymbol{a}' \leftarrow \mathbb{T}^n$, $e \leftarrow \chi$, calculate $b' = -\langle \boldsymbol{a}', \boldsymbol{s}\rangle + m/4 + e(\text{mod}1)$, and output ciphertext $(b, \boldsymbol{a}) = round_{p,q}(b', \boldsymbol{a}') \in \mathbb{Z}_{2N}^{n+1}$. The *round* function used in this paper: $round_{p,q}(\boldsymbol{c}) = \lfloor (p/q)\boldsymbol{c} \rceil$, in the actual algorithm, this paper takes $p/q = 4N$, and the *round* function can also be expressed as $round_{p,q}(\boldsymbol{c}) = \lfloor 4N\boldsymbol{c} \rceil$.

Decryption $Dec(\boldsymbol{c}, \boldsymbol{s})$: Input ciphertext $\boldsymbol{c}$, private key $\boldsymbol{s}$, output m', such that $b + \langle \boldsymbol{a}, \boldsymbol{s}\rangle \approx m'/4(\text{mod}2N)$.

Homomorphic NAND gate (including bootstrapping process) HomNAND($\boldsymbol{c}_1$, $\boldsymbol{c}_2$): Input the ciphertext $\boldsymbol{c}_1$ corresponding to μ_1, the ciphertext $\boldsymbol{c}_2$ corresponding to μ_2, and output the ciphertext $\boldsymbol{c}$ corresponding to $NAND(\mu_1, \mu_2)$. The specific process is shown in Algorithm 1.

Algorithm 1: Homomorphic NAND gate process (HomNAND)

Input: $(\mathrm{a}_1, b_1) \in TLWE_{s,\eta}(\mu_1)$,

$(\mathrm{a}_2, b_2) \in TLWE_{s,\eta}(\mu_2), BK_{s \to s'',\alpha}$,

$KS_{s' \to s,\gamma}, s' = KeyExtract(s'')$.

Output: $\mathrm{c} = (\mathrm{a}, b) \in TLWE_{s,\eta}(1 - \mu_1\mu_2)$

1.$\bar{\mathrm{c}}_1 = \lfloor 4N\mathrm{c}_1 \rceil, \bar{\mathrm{c}}_2 = \lfloor 4N\mathrm{c}_2 \rceil$

2.$\bar{\mathrm{c}} = \lfloor 2N(0,5/8) \rceil - \lfloor (\bar{\mathrm{c}}_1 - \bar{\mathrm{c}}_2)/2 \rceil, i \in [1, n]$

3.$v := (1 + X + ... + X^{\bar{N}-1}) \cdot X^{\frac{\bar{N}}{2}} \cdot \frac{1}{4} \in \mathbb{T}[X]$

4. $\begin{cases} (1).ACC \leftarrow X^{-\bar{b}} \cdot (0, v) \in \mathbb{T}[X] \\ (2).\text{for } i \in [1, \underline{n}] \\ (3).\ ACC \leftarrow BK_i \boxdot (X^{\bar{a}_i} \cdot ACC - ACC) + \\ \quad ACC \in \mathbb{T}[X]; \end{cases}$

Return ACC

$\mathrm{c}' = (0, \mu) + \text{SampleExtract}(ACC)$

$\mathrm{c} = \text{KeySwitch}_{KS_{s' \to s,\gamma}}(\mathrm{c}')$

Theorem 7 (Bootstrapping Theorem): $\boldsymbol{h} \in \mathcal{M}_{(k+1)l,k+1}$ is the decomposition tool, $\text{Dec}_{\boldsymbol{h},\beta,\varepsilon}$ is the corresponding decomposition function, the keys are $\boldsymbol{s} \in \mathbb{B}^n$, $\boldsymbol{s}'' \in \mathbb{B}_N[X]^k$, the noise parameters α, γ, the bootstrapping key $BK = BK_{s \to s'',\alpha}$, the key $\boldsymbol{s}' = \text{Sample}Extract(\boldsymbol{s}'') \in \mathbb{B}^{\text{kN}}$, the transformation key $KS = KS_{s' \to s,\gamma,t}$, the two plaintexts μ_0,μ_1 corresponding to the given ciphertext $(\boldsymbol{a}_1, b_1) \in TLWE_{s,\eta}(\mu_1)$,$(\boldsymbol{a}_2, b_2) \in TLWE_{s,\eta}(\mu_2)$. Algorithm 1 will output an instance $\boldsymbol{c} = (\boldsymbol{a}, b) \in TLWE_{s,\eta}(1 - \mu_1\mu_2)$ where the rounding error δ is equal to $(n + 1)/2N$ in the worst case, then Algorithm 1 outputs a vector $\boldsymbol{c}$ that satisfies the noise ceiling $\|Err(\boldsymbol{v})\|_\infty \leq 2n(k + 1)l\beta N\alpha + kNt\gamma + n(1 + kN)\varepsilon + kN2^{-(t+1)}$.

The bootstrap theorem is obtained from the above analysis. Except for the range of the rounding error δ, other conclusions are the same as CGGI, and the specific analysis process is similar to the CGGI16 scheme.

Correctness analysis: The overall structure of the scheme is similar to the CGGI17 scheme. The correctness of our scheme depends on the noise scale of the ciphertext output during the bootstrapping process. Except for steps 1–2 in this algorithm, the rest of the process is the same as the CGGI17 scheme, so focus on analyzing the noise growth in steps 1–2.

Steps 1–2 in our scheme can be expressed as $\bar{c} = \lfloor 4N(0,5/8) \rceil - \lfloor 4Nc_1 \rceil - \lfloor 4Nc_2 \rceil$. Let $\boldsymbol{c}'' = \boldsymbol{c}_1 + \boldsymbol{c}_2$, for a random input ciphertext $\boldsymbol{c}_i$, the rounding noise generated by our scheme is analyzed as follows:

$$\begin{aligned} \bar{\varphi} &\triangleq \bar{b} - \sum_{i=1}^{n} \bar{a}_i s_i \bmod 2N \\ &= (\lfloor 2N \cdot 5/8 \rceil - \lfloor (\lfloor 4Nb_1 \rceil + \lfloor 4Nb_2 \rceil)/2 \rceil) - \\ &\quad \sum_{i=1}^{n} (\lfloor (\lfloor 4Na_1 \rceil + \lfloor 4Nb_2 \rceil)/2 \rceil) s_i \bmod 2N \end{aligned} \tag{2}$$

The upper bound analysis of rounding noise is as follows:

$$\begin{aligned} \left|\varphi - \tfrac{\bar{\varphi}}{2N}\right| &= |(5/8 - (b_1 + b_2) + \sum_{i=1}^{n} (a_{1,i} + a_{2,i}) \cdot s_i \bmod 2N)) \\ &\quad - \frac{(\lfloor 2N \cdot 5/8 \rceil - \lfloor (\lfloor 4Nb_1 \rceil + \lfloor 4Nb_2 \rceil)/2 \rceil)}{2N} \\ &\quad - \frac{\sum_{i=1}^{n} (\lfloor (\lfloor 4Na_{1,i} \rceil + \lfloor 4Na_{2,i} \rceil)/2 \rceil) s_i}{2N} | \\ &= |(5/8 - \frac{\lfloor 2N \cdot 5/8 \rceil}{2N}) - (b'' - \frac{\lfloor (\lfloor 4Nb_1 \rceil + \lfloor 4Nb_2 \rceil)/2 \rceil}{2N}) \\ &\quad + \sum_{i=1}^{n} (a_i'' - \frac{\lfloor (\lfloor 4Na_1 \rceil + \lfloor 4Nb_2 \rceil)/2 \rceil}{2N}) \cdot s_i \bmod 2N)) | \\ &\leq | \tfrac{1}{2N} + \sum_{i=1}^{n} \tfrac{1}{2N} | \leq \tfrac{n+1}{2N} < \delta \end{aligned} \tag{3}$$

The above analysis shows that, compared with the rounding noise $\frac{n+1}{4N}$ of CGGI17, the rounding noise $\frac{n+1}{2N}$ of our scheme increases slightly. But in the actual selection of parameters, $n = 500$, $N = 1024$, the rounding noise is much smaller than other noises; and this paper is the worst case analysis, in most cases the rounding noise will be smaller.

Security analysis: In terms of the security of the underlying TLWE instance, our scheme and TFHE [14] use the same TLWE parameters and similar encryption process. Our scheme is based on the TFHE scheme, and performs the round function on the ciphertext. Since this function only performs public operations on the ciphertext and does not involve private key and plaintext information, the security of the scheme will not be reduced. Specifically, for the parameter $l = 2$, $B_g = 512$, $\beta = 256$, $n = 500$, $N = 1024$, $k = 1$, $\varepsilon = 2^{-31}$, $\alpha = 9.0 \times 10^{-9}$, $\gamma = 3.05 \times 10^{-5}$, $t = 15$. The proposed scheme is the same as CGGI17 scheme, and can achieve 136bit security under the CPA security model. Ciphertext scale analysis: In this paper, the round function in the boot-up process of CGGI17 scheme is prepositioned to the encryption process, and the ciphertext $\boldsymbol{c}$ of 501-dimension 32 bits is reduced to 501-dimension 12 bits, so that the plaintext can better match the structure of the polynomial ring, the ciphertext size can be reduced by 62%, and the expansion rate of plain ciphertext can be reduced to 6012.

4 Experimental Test of the Scheme on the CUDA Platform

In this section, aiming at the low efficiency of the bootstrapping process of the fully homomorphic encryption scheme, the algorithm calculation is accelerated by CUDA parallel computing, which further improves the implementation efficiency of the scheme. In the practical application scenarios of bootstrapping fully homomorphic encryption schemes, millions of homomorphic gate circuits (bootstrapping process) may be required, so efficiency improvement is crucial. In the early stage, cryptographers have done a good job in the implementation of software acceleration, and the cuFHE library [16] is a typical representative. Under the CUDA platform, based on the cuFHE library [16], this paper tests the CGGI17 scheme and our scheme, and the test samples are the method of calculating the average time within the group for multiple tests. The test environment is: Inter i7-9750H CPU, 16G memory, NVIDIA Quadro P3200, using Ubuntu 20.04 LTS 64-bit operating system and CUDA 11 version.

The test items are: the calculation time of the homomorphic gate circuit, the time of encryption and decryption, etc., to realize the encryption of two sets of 896-bit data, and to run the homomorphic basic gate circuit NAND (including the bootstrap process). The experimental results show that the ciphertext expansion rate of the scheme is reduced from 16032 to 6012, the average encryption time of a single bit is 0.0711633 ms, the average decryption time is 0.0008012 ms, and the average time of the basic gate circuit (including the bootstrap process) is 0.785347 ms. The comparison experimental data of the scheme is shown in Table 1. Therefore, our scheme effectively reduces the size of the ciphertext when other performances are close.

Table 1. Comparison between CGGI17 scheme and our scheme.

Efficiency	cuFHE	Our scheme
Homomorphic gate calculation time (including bootstrappping process)	0.804462 ms/bit	0.785347 ms/bit
Homomorphic gate calculation time (including bootstrappping process)	0.0819882 ms/bit	0.0711633 ms/bit
Decryption time	0.00090945 ms/bit	0.00080125 ms/bit
Ciphertext size	**2004 Byte**	**751 Byte**
Plaintext size	1bit	1bit
Plaintext extension ratio	**16032**	**6012**
Private key size	500bit	500bit
Evaluation key size	32763904 Byte	32763904 Byte

5 Conclusion

In this paper, a single-bit bootstrapping fully homomorphic encryption scheme with shorter ciphertext is designed by prepending the round function to the encryption process. Compared with the same type of scheme, the ciphertext size of the proposed scheme is reduced by 62%. Based on the feature of GPU supporting parallel computing, the scheme is implemented on CUDA platform, and the high efficiency and shorter ciphertext properties of the scheme are verified.

Acknowledgments. This work was supported by Innovative Research Team in Engineering University of PAP (KYTD201805), National Natural Science Foundation of China (Grant Nos. 62172436, 62102452).

References

1. Zhou, Y.B.: Research progress of homomorphic cryptography. China Cryptography Development Report 2010, pp. 34–40. Publishing House of Electronics Industry, Beijing (2011)
2. Benaloh, J.: Dense probabilistic encryption [EB/OL]. In: Proceedings of the Workshop on Selected Areas of Cryptography, pp. 120–128 (1994). http://citeseerx.ist.psu.edu/viewdoc/summary?doi=10.1.1.33.3710
3. Gentry, C.: Fully homomorphic encryption using ideal lattices. In: Proceedings of the 41st Annual ACM Symposium on Theory of Computing, STOC, pp. 169–178. ACM, New York (2009)
4. Boneh, D., Goh, E.-J., Nissim, K.: Evaluating 2-DNF formulas on ciphertexts. In: Kilian, J. (ed.) TCC 2005. LNCS, vol. 3378, pp. 325–341. Springer, Heidelberg (2005). https://doi.org/10.1007/978-3-540-30576-7_18
5. Brakerski, Z., Vaikuntanathan, V.: Efficient fully homomorphic encryption from (standard) LWE. In: Proceedings of 52nd Annual Symp on Foundations of Computer Science, pp. 97–106. IEEE Computer Society, Los Alamitos,CA (2011)
6. Brakerski, Z., Gentry, C., Vaikuntanathan, V.: (Leveled) fully homomorphic encryption without bootstrapping. Proceedings of the 3rd Innovations in Theoretical Computer Science Conference, pp. 309–325. ACM (2012)
7. Gentry, C., Halevi, S., Smart, N.P.: Fully homomorphic encryption with polylog overhead. In: Pointcheval, D., Johansson, T. (eds.) EUROCRYPT 2012. LNCS, vol. 7237, pp. 465–482. Springer, Heidelberg (2012). https://doi.org/10.1007/978-3-642-29011-4_28
8. Halevi, S., Shoup, V.: Design and Implementation of a Homomorphic-Encryption Library [EB/OL] (2012). http://eprint.iacr.org/2012/181
9. Halevi, S., Shoup, V.: Bootstrapping for HElib. In: Oswald, E., Fischlin, M. (eds.) EUROCRYPT 2015. LNCS, vol. 9056, pp. 641–670. Springer, Heidelberg (2015). https://doi.org/10.1007/978-3-662-46800-5_25
10. Halevi, S., Shoup, V.: Faster homomorphic linear transformations in HElib [EB/OL] (2018). https://eprint.iacr.org/2018/244
11. Alperin-Sheriff, J., Peikert, C.: Faster bootstrapping with polynomial error. In: Garay, J.A., Gennaro, R. (eds.) CRYPTO 2014. LNCS, vol. 8616, pp. 297–314. Springer, Heidelberg (2014). https://doi.org/10.1007/978-3-662-44371-2_17
12. Brakerski, Z., Vaikuntanathan, V.: Lattice-based FHE as secure as PKE. In: Proceedings of the 5th Conference on Innovations in Theoretical Computer Science, pp. 1–12. ACM (2014)

13. Ducas, L., Micciancio, D.: FHEW: Bootstrapping homomorphic encryption in less than a second. In: Oswald, E., Fischlin, M. (eds.) EUROCRYPT 2015. LNCS, vol. 9056, pp. 617–640. Springer, Heidelberg (2015). https://doi.org/10.1007/978-3-662-46800-5_24
14. Chillotti, I., Gama, N., Georgieva, M., Izabachène, M.: Faster fully homomorphic encryption: bootstrapping in less than 0.1 seconds. In: Cheon, J.H., Takagi, T. (eds.) ASIACRYPT 2016. LNCS, vol. 10031, pp. 3–33. Springer, Heidelberg (2016). https://doi.org/10.1007/978-3-662-53887-6_1
15. Chillotti, I., Gama, N., Georgieva, M., Izabachène, M.: Faster packed homomorphic operations and efficient circuit bootstrapping for TFHE. In: Takagi, T., Peyrin, T. (eds.) ASIACRYPT 2017. LNCS, vol. 10624, pp. 377–408. Springer, Cham (2017). https://doi.org/10.1007/978-3-319-70694-8_14
16. Dai, W., Sunar, B.: CUDA-accelerated Fully Homomorphic Encryption Library [EB/OL] (2018). https://github.com/vernamlab/cuFHE
17. Che, X.L., Zhou, H.N., Yang, X.Y., et al.: Efficient multi-key fully homomorphic encryption scheme from RLWE. J. Xidian Univ. **48**(1), 9 (2021)

Traffic-Oriented Shellcode Detection Based on VSM

Pengju Liu(✉), Baojiang Cui, and Can Cui

School of Cyberspace Security, Beijing University of Posts and Telecommunications, Haidian District Xitucheng Road No.10, Beijing, China
{liupengju,cuibj}@bupt.edu.cn

Abstract. Shellcode is the core part of an attacker exploiting a vulnerability in a binary program, and it is an essential piece of binary bytes to gain control of the target machine. Therefore, the detection of Shellcode is an important part of binary program security protection. However, the currently common static analysis and simulation execution methods for Shellcode detection have problems of low accuracy and low efficiency, resulting in limited actual role. Machine learning models have strong learning and generalization capabilities, and can extract hidden features that are difficult to find manually. This paper proposes a system for detecting Shellcode in network traffic based on the VSM machine learning model. Through the VSM model, the payload data in the network traffic is matched with the Shellcode library to achieve the effect of detecting unknown Shellcode. The experimental results show that the Shellcode detection system based on the VSM model proposed in this paper can effectively detect the known Shellcode, and still has a certain ability to detect the unknown Shellcode.

1 Introduction

With the rapid development of the Internet, the computer network has provided more and more convenience and benefits to people and has been closely linked with our lives, but at the same time, various attacks against the computer network have also become more and more. Especially with the large-scale deployment of 5G, more and more smart devices such as cameras and smart door locks are connected to the Internet, providing attackers with a large attack surface. At the same time, with the country's vigorous development of infrastructure and industrial upgrading, the production and monitoring systems of important industries related to the national economy and people's livelihood, such as electric, energy, aerospace and other industries, have begun to be gradually upgraded to more intelligent industrial control systems. The safety of these systems is extremely important. Once the attacked person leaks state secrets, the serious person may lead to large-scale social problems. Therefore, in recent years, hackers with a national background have also attacked industrial control systems more and more. The impact of the CVE-2018–1303 vulnerability alone has reached an astonishing 4.59 million devices, and the network security situation has become more and more severe.

Shellcode is the payload that the attacker actually executes malicious behavior in the vulnerability attack. Its essence is a piece of executable binary code. In the early

L. Barolli (Ed.): EIDWT 2023, LNDECT 161, pp. 152–162, 2023.
https://doi.org/10.1007/978-3-031-26281-4_15

stage of the development of the attack, this code is usually used to obtain the shell of a target machine and that's why we call it Shellcode. With the development of Shellcode, attackers can not only use Shellcode to gain control rights of the system or perform other operations, but also many worms and viruses can even use Shellcode to infiltrate and spread laterally. Therefore, Shellcode detection is an important part of binary security protection. Shellcode can not only be embedded in malicious files, such as PDF, Word and other formats, and executed after the user opens the file, it can also be directly sent to the target host through network traffic for execution. For the first shellcode, detecting malicious files on the host side can achieve very good results, but this requires the installation of corresponding detection tools on each host. But generally speaking, when attackers use malicious files to execute Shellcode, they usually first send the malicious files to the target host through the network, then this will leave traces in the network traffic. For uncompressed malicious files, We can directly detect Shellcode through network traffic.

Therefore, this paper proposes a shellcode detection scheme based on the VSM machine learning model. The VSM machine learning model can be used for generalized feature extraction and comparison, and the similarity between the payload of the network traffic packet and the existing Shellcode library is compared., so as to detect whether the current input network traffic packet contains Shellcode. The experimental data show that this scheme has high detection accuracy for Shellcode, and has a certain detection ability for unknown Shellcode.

This paper is organized according to the following structure: the second part discusses the method of shellcode detection and the contributions of other scholars in this area; the third part introduces the overall architecture of the model proposed in this paper; the fourth part introduces the construction of the dataset and the Data preprocessing; the fifth part presents the influence of some parameters in the model on the performance of the model; the sixth part summarizes this paper and looks forward to the future development.

2 Related Word

The current research on Shellcode detection technology mainly exists in two directions, dynamic detection and static detection.

The dynamic detection technology is to monitor and track the system calls, memory read and write, and memory changes during the simulation execution process by simulating the execution of the payload in the network traffic packet. This method can detect obfuscated and encrypted shellcode very well. Polychronakis [1] et al. detected the decryption sequence in Shellcode by relying on the embedded CPU emulator of NIDS, but this method was mainly designed to detect encrypted Shellcode. Zhang [2] et al. determined the starting position of the encrypted or obfuscated Shellcode by using the bidirectional data lookup method to find GetPC in the given data, thus solving the problem of low efficiency of dynamic detection technology to a certain extent. D. Lukan [3] and others released Libemu in 2014, which can be used for x86 simulation and Shellcode detection. It has a complete simulation function and has been used in many open source projects as an open source detection tool. Detect Shellcode based on GetPC heuristic rules. However, although the dynamic detection technology has a high detection rate, it has a big disadvantage that it consumes a lot of resources.

Static analysis technology mainly focuses on byte analysis and disassembly analysis in the payload, and detects through feature extraction and signature generation. Zhao [4] et al. proposed a new method of shellcode detection and classification based on disassembled instruction sequences, which proved the possibility of analyzing disassembled instruction sequences, and proved that the encoded shellcode will also have certain characteristics. Verma [5] found that there is a high degree of similarity between different Shellcode samples by comparing multiple shellcodes of pure numeric letter type, and proposed that this method can be used to detect Shellcode in the future. Qi Yudong [6] and others adopted the idea of dimensionality reduction, and used the Hash calculation method to map the high-order input data into a four-byte signature feature, and calculated the Hamming distance and cosine similarity between the malicious code and its variants. Enables quick classification of malicious code. The proposed method can effectively detect variants of malicious code. However, one feature of this method is that feature extraction is complex and requires manual extraction.

With the continuous development of machine learning technology, machine learning models can find deep but universal features that are difficult for humans to find. Compared with static analysis technology, the features that need to be manually extracted are more in-depth, generalized and comprehensive. At the same time, machine learning The performance overhead of the model is much smaller than that of the dynamic detection technology. Therefore, this paper proposes a Shellcode detection technology based on the VSM model.

3 Shellcode Detection Model Based on VSM

The overall architecture of the model is shown in Fig. 1.

In order not to affect the normal function of the gateway, the solution firstly mirrors the traffic routed through the gateway to the traffic storage server. In the mirrored traffic storage server, the traffic distribution component sends the data packets to the parallel VSM model detection module for detection. The VSM model sequentially reads the Shellcode from the Signature Library. After word segmentation, thesaurus construction, TF-IDF calculation, and similarity comparison, the similarity between the currently input traffic data Payload and the Shellcode bytecode is finally calculated, and the the similarity is compared with a given threshold, and if it is greater than the threshold, alarm processing is performed, otherwise, it is considered that the currently input traffic data packet does not contain Shellcode.

3.1 VSM

The Vector Space Model (VSM) was proposed by G.salton [7]. The basic idea of the vector space model is that the position and arrangement order of the feature items in the text will not affect the classification of the text, so the position and arrangement order of the feature items in the text can be removed here, so that the text can be regarded as composed of composed of multiple features. Feature items are generally formed by selecting letter, words and phrases in the text, but because the length of the letter is too

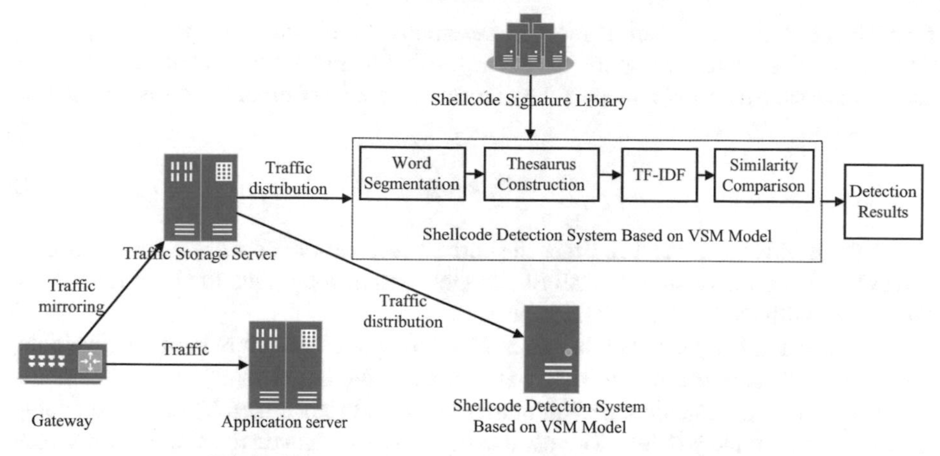

Fig. 1. The shellcode detection system based on VSM

short to represent the features of the text, the feature items of general texts are represented by words or phrases.

In fact, Shellcode can be regarded as a special string sequence, that is, specific text. Combined with the vector space model, we can express a piece of Shellcode as formula 1.

$$S = (\mathrm{t}_1, \mathrm{w}_1, \ldots, \mathrm{t}_\mathrm{n}, \mathrm{w}_\mathrm{n}) \quad (1)$$

n represents the total number of feature items in the current Shellcode string, $\mathrm{t_i}$ represents the $\mathrm{i_{th}}$ feature item, and $\mathrm{w_i}$ represents the weight of the $\mathrm{i_{th}}$ feature item, which is used to indicate the importance of the $\mathrm{t_i}$ feature item in the current Shellcode.

The commonly used weight calculation methods are mainly composed of two.

1. TF

   Term Frequency Weighting (TF). The core idea of TF is that the more times the feature item appears in the current text, the more important it is, that is, the higher the weight value, so the value of the TF weight of the current feature item is the number of its value occurrences in the current text, the calculation formula is expressed as formula 2.

$$W_i = C_i \quad (2)$$

   where C_i represents the number of times the current feature item appears in the Shellcode. But in fact, it is not that the more frequent the feature item, the more important it is.

2. TF-IDF

Term Frequency-Inverse Document Frequency (TF-IDF). The core idea of TF-IDF is that if a feature item appears more frequently in a text, but rarely appears in other texts, then The current feature item is more representative of the current text. Among them, TF is used to measure the importance of the current feature item in the text, and IDF is used to measure the importance of the current feature item in all texts. Its calculation formula is as formula 3.

$$W_i = C_i * log\left(\frac{n}{n_i}\right) \tag{3}$$

C_i represents the number of times the current feature item appears in the Shellcode, n represents the total number of all Shellcodes, and n_i represents the total number of Shellcodes with the i_{th} feature item appearing.

By representing Shellcode as a space vector composed of feature items and weights, the VSM model converts the similarity between texts, which originally required complex calculation, into the calculation of similarity between vectors, thereby greatly reducing the complexity of the problem. There are many ways to calculate the similarity between vectors, such as the common angle cosine similarity calculation.

4 Dataset and Data Preprocessing

The core content of this solution is the construction of the Shellcode Signature Library, which will directly determine the overall performance and performance of the model, so it is very important to build the Shellcode Signature Library here. At present, there are a large number of Shellcode samples for security engineers to learn and test on the two websites exploit-db.com and shell-storm.org. These Shellcodes have been tested and verified by professional security engineers. Therefore the shellcode samples on these two sites are very credible. And the most popular penetration testing software Metasploit also has built-in Shellcodes with multiple platforms, multiple architectures, and multiple functions. These Shellcodes can be directly used in the penetration testing process, and Metasploit also provides a variety of different encoder, to obfuscate the Shellcode. Therefore, the construction of the Shellcode Signature Library used for testing in this article is carried out through the above three methods.

4.1 Shellcode Signature Library

The sources of Shellcode in the Shellcode Signature Library are mainly divided into three parts, exploit-db.com, shell-storm.org, and Metasploit. Through automated crawling and extraction of Shellcodes in the two public websites, exploit-db.com and shell-storm.org, the Shellcodes of the corresponding architectures in the above two websites are extracted and stored in the Shellcode Signature Library. At the same time, the msfvenom command line tool provided by Metasploit is used, and the msfvenom command is executed in batches in combination with Python, so as to use Metasploit to batch generate shellcodes that specify the architecture and different functions of the platform and store them in the signature database. At the same time, the built-in encoders of different architectures and different platforms in msfvenom are used to obfuscate the Shellcode of the specified function, and the obfuscated Shellcode samples are generated and stored in the Shellcode Signature Library. In the end, a total of 282 plaintext Shellcode sample data were generated, including 182 number of x64 architecture Shellcodes, 29 number of x86 architecture Shellcodes, and 71 number of Mips and Arm architecture Shellcodes, as shown in Table 1.

Table 1. Shellcode signature library composition table

Type	Number
x64	182
X86	29
Mips	15
Arm	56
Total	282

4.2 Normal Dataset

In order to test the accuracy and performance of the scheme proposed in this paper, the experiment also needs corresponding normal traffic data samples. The number of normal data samples should be similar to the number of Shellcodes in the Shellcode signature database. Here we deploy the traffic capture module in the school gateway to capture the normal traffic data routed through the school gateway, and randomly select n traffic data packets as the normal data sample set for testing.

5 Experiment Results and Evaluation

The experimental configuration in the process of evaluating the model parameters is shown in Table 2.

Table 2. Experimental configuration

OS	CentOS Linux release 7.5.1804
CPU	Intel(R) Xeon(R) CPU E5–2620 v3 @ 2.40 GHz
RAM	126 GB
Anaconda	4.5.11
Keras	2.2.2
Python	3.6.5

5.1 Metrics

In this paper, three commonly used parameters for evaluating intrusion detection systems are adopted: Accuracy (ACC, Precision), True Positive Rate (TPR, Recall) and F1-Score. Among them, ACC represents the overall performance of the model, TPR represents the proportion of all positive samples occupied by real positive samples in the current positive samples, and F1-Score is the combination of Precision and Recall. The higher the F1 score, the better the model. Steady. Its defined formula is as formula 4–6.

$$ACC = \frac{TP+TN}{TP+FP+FN+TN} \tag{4}$$

$$TPR = \frac{TP}{TP+FN} \tag{5}$$

$$F1 = \frac{2*\mathrm{ACC}*\mathrm{TPR}}{\mathrm{ACC}+\mathrm{TPR}} \tag{6}$$

The definitions of TP, TN, FP, and FN are shown in Table 3.

Table 3. The definition of TP, TN, FP and FN

	Relevant	NonRelevant
Retrieved	TP	FP
Not retrieved	FN	TN

5.2 Known Shellcode Detection

In this part of experiment, the Shellcode signature library uses all Shellcode, and the test data is randomly selected from 20% of the normal traffic sample data and 20% of the Shellcode feature data. The purpose of setting the known shellcode test is to evaluate the overall performance of the model by detecting the existing shellcode in the shellcode signature library. Influence of Vector Similarity Calculation Method on Model Performance.

Influence of Vector Similarity Calculation Method on Model Performance. This paper tests the effects of cosine similarity calculation and vector product similarity calculation on the performance of the model. The experimental results are shown in Table 4.

From the experimental results, the accuracy of the model using the vector product similarity method can always be kept above 98%, while the model using the cosine similarity calculation method has an accuracy rate of no more than 70%, and from the value of TPR it can be seen that the detection ability of the model using the cosine similarity calculation method for positive cases, that is, normal traffic, is much higher than that of the model using the vector product similarity calculation method. From this, we can draw the conclusion: the detection accuracy of normal samples using the cosine similarity calculation method is higher, but the detection accuracy of abnormal samples is much lower than that of the model using the vector product similarity calculation method. On the whole, the model adopts the calculation method of vector product similarity to be more reasonable.

Table 4. The experiment results of the different vector similarity calculation method and different word segmentation bytes of the model in known shellcode detection

Vector Similarity Calculation Method	Cosine			Vector product		
Word Segmentation Bytes \ Metrics	ACC	TPR	F1	ACC	TPR	F1
1	0.6792	1.0000	0.8082	0.9924	0.8467	0.9130
2	0.6903	1.0000	0.8156	0.9882	0.8478	0.9117
3	0.6646	1.0000	0.7977	0.9948	0.8394	0.9100
4	0.6789	1.0000	0.8074	0.9909	0.8512	0.9149
5	0.6701	1.0000	0.8018	0.9900	0.8453	0.9113
6	0.6814	1.0000	0.8087	0.9864	0.8605	0.9183
7	0.6887	1.0000	0.8149	0.9916	0.8415	0.9095
8	0.6813	1.0000	0.8087	0.9881	0.8575	0.9176

Influence of Shellcode Word Segmentation Bytes on Model Performance. Since the word segmentation in this model uses a fixed byte method, the number of Shellcode word segmentation bytes here may have a greater impact on the model. Therefore, this paper designs a set of experiments to determine the most suitable Shellcode word segmentation bytes. The experimental results are shown in Table 4.

From the experimental results, with the increase of the number of word segmentation, the overall accuracy rate, recall rate and F1 value of the model using the vector product similarity calculation method all show a trend of increasing first and then decreasing. The 3 bytes selected by the model are used as the number of word segmentations of Shellcode.

5.3 UnKnown Shellcode Detection

The purpose of setting the unknown shellcode experiment is to detect the detection ability of the model for unknown shellcode.

Influence of Vector Similarity Calculation Method on Model Performance. Here, the Shellcode signature library uses 80% of the Shellcode, and the test data is randomly selected from 20% of the normal traffic sample data and 20% of the Shellcode feature data. This paper tests the effects of cosine similarity calculation and vector product similarity calculation on the performance of the two vector similarity calculation methods. The experimental results are shown in the Table 5.

From the experimental results, the accuracy of the model using the vector product similarity method can always be kept above 97%, while the model using the cosine similarity calculation method has an accuracy rate of no more than 70%. From the value of TPR, we can see that the model using the cosine similarity calculation method has a much higher detection ability for positive cases, that is, normal traffic, than the model using the vector product similarity calculation method. From this, we can draw a conclusion: in the detection process of unknown traffic among them, the cosine similarity calculation method has a higher detection accuracy for normal samples, but its detection accuracy for abnormal samples is much lower than that of the model using the vector product similarity calculation method. On the whole, the model adopts the calculation method of vector product similarity to be more reasonable.

Table 5. The experiment results of the different vector similarity calculation method and different word segmentation bytes of the model in unknown shellcode detection

Vector Similarity Calculation Method	Cosine			Vector product		
Word Segmentation Bytes \ Metrics	ACC	TPR	F1	ACC	TPR	F1
1	0.6774	1.0000	0.8073	0.9925	0.9392	0.9647
2	0.6814	1.0000	0.8104	0.9961	0.9571	0.9761
3	0.6880	1.0000	0.8147	0.9814	0.9428	0.9616
4	0.6900	1.0000	0.8164	0.9782	0.9571	0.9673
5	0.6864	1.0000	0.8140	0.9927	0.9464	0.9688
6	0.6844	1.0000	0.8122	0.9888	0.9500	0.9688
7	0.6722	1.0000	0.8037	0.9890	0.9714	0.9801
8	0.6870	1.0000	0.8142	1.0000	0.9571	0.9780

Influence of Shellcode Word Segmentation Bytes on Model Performance. From the experimental results, with the increase of the number of word segmentation, the overall accuracy rate, recall rate and F1 value of the model using the vector product similarity calculation method all show a trend of increasing first and then decreasing. The 3 bytes selected by the model are used as the number of word segmentations of Shellcode.

Influence of Unknown Shellcode to Shellcode Signature Database Ratio on Model Performance. In this experiment, the collected Shellcode is added to the signature library in the proportion of n%, and the proportion of 1-n% is used as unknown Shellcode to participate in the detection to test the performance of the model under the condition of imperfect signature library. The calculation method of similarity is vector product calculation, and the number of Shellcode word segmentation is 3 bytes. The experimental results are shown in Table 6.

From the experimental results, as the proportion of unknown Shellcodes in the Shellcode signature library gradually increases, the accuracy of the model shows a gradual

Table 6. The experiment results of the different Shellcode to Shellcode signature database ratio of the model in unknown shellcode detection

Ratio \ Metrics	ACC	TPR	F1
10%	1.0000	0.9428	0.9699
20%	0.9814	0.9428	0.9616
30%	0.9949	0.9547	0.9744
40%	0.9928	0.9696	0.9809
50%	0.9827	0.9687	0.9756
60%	0.9831	0.9680	0.9755
70%	0.9797	0.9715	0.9755
80%	0.9626	0.9795	0.9709
90%	0.9351	0.9833	0.9584

downward trend, but when the proportion of unknown Shellcodes reaches 90%, the model can still maintain a good detection rate. The model has a certain ability to detect unknown Shellcode.

6 Conclusion and Future Work

In order to detect potential shellcode attacks in traffic, this paper proposes a shellcode detection system based on the VSM machine learning model. It collects the shellcodes that have been disclosed to form a signature library, and converts them into special vector features in the same way as network traffic data, and calculate the similarity between network traffic and Shellcode by the vector product similarity calculation method, so as to realize the detection of network traffic containing Shellcode. In order to evaluate the model, this paper evaluates the performance of the model detection shellcode through the known shellcode detection experiment, and sets up the unknown shellcode detection experiment to evaluate the model's detection ability for the unknown shellcode. The experimental results show that the detection model proposed in this paper has a good detection effect on both known and unknown Shellcodes.

However, the model in this paper still has certain deficiencies. For example, attackers currently use obfuscation and encryption to bypass detection. Through analysis, we find that the obfuscated and encrypted Shellcode will still have decrypted or de-obfuscated Shellcode code, but the length is much shorter than general shellcode, so in order to detect encrypted and obfuscated shellcode, the model still needs to be optimized.

References

1. Detection of Intrusions and Malware, and Vulnerability Assessment. Springer International Publishing (2018)

2. Zhang, Q., Reeves, D.S., Ning, P., et al.: Analyzing network traffic to detect self-decrypting exploit code. In: Proceedings of the 2nd ACM Symposium on Information, Computer and Communications Security, pp. 4–12 (2007)
3. Lukan, D.: Shellcode detection and emulation with libemu (2014)
4. Zhao, Z., Ahn, G.J.: Using instruction sequence abstraction for shellcode detection and attribution. In: 2013 IEEE Conference on Communications and Network Security (CNS). IEEE, pp. 323–331 (2013)
5. Verma, N., Mishra, V., Singh, V.P.: Detection of alphanumeric shellcodes using similarity index. In: 2014 International Conference on Advances in Computing, Communications and Informatics (ICACCI), pp. 1573–1577. IEEE (2014)
6. Qi, D.Y.: jiyu tezheng pipei dee yidai mabian zhongjian ce [Malicious code variant detection based on feature matching]. Jisuan jiyu shuzi gongcheng **47**(5), 1179–1183 (2019)
7. Chowdhury, G.G.: Introduction to Modern Information Retrieval. Facet Publishing (2010)

Supply Chain Finance Mediates the Effect of Trust and Commitment on Supply Chain Effectiveness

Lisa Kartikasari(✉) and Muhammad Ali Ridho

Department of Accounting, Universitas Islam Sultan Agung, Jalan Raya Kaligawe KM 4, Semarang, Indonesia
lisakartika@unissula.ac.id

Abstract. Supply chain financc aims for building trust, commitment to the timely delivery of goods, negotiation about terms of payment, sharing information about customer needs, and supply-related matters. It creates a money-related win-win situation for buyers, suppliers, and financial institutions. This research examines the relationship between trust, relationship commitment, supply chain finance, and supply chain effectiveness for small and medium manufacturing enterprises (MSMEs). Supply chain finance is also essential for MSMEs to maintain business operations smoothly. The target sample was owners of small and medium manufacturing enterprises (MSMEs) in Semarang, Demak, and Kendal city, Indonesia. This research is expected to contribute to identifying supply chain finance adoption and how MSMEs can implement supply chain financc to increase company liquidity in achieving supply chain effectiveness.

1 Introduction

The contribution of Micro, Small, and Medium Enterprises (MSMEs) to improving people's welfare in developing countries becomes an important part of the country's economic life. Their contribution to gross domestic product increased from 57,84% to 60,34% in the last five years. Labor absorption in this sector also increased from 96,99% to 97,22% in the same period. In many countries, MSMEs contribute as much as in Indonesia. South Africa is one of the countries where 95% of the business sector is MSMEs. This sector annually contributes an average of 35% to the gross domestic product and is able to reduce as much as 50% of the unemployment rate in the country [27].

Micro, Small, and Medium Enterprises (MSMEs) are the main force to promote national social economic development. In reality, difficulties in financing and expensive financing problems always limit the development of MSMEs. Since 2020, there has been a COVID-19 pandemic that has affected the world economy and MSMEs. In general, the world economy is facing severe cash flow pressures which exacerbates the difficulty of surviving. Overcoming financial difficulties in MSMEs is an essential matter. MSMEs without financial support are unable to perform sufficient productivity and have difficulties in the development of information technology. Therefore, they will experience

L. Barolli (Ed.): EIDWT 2023, LNDECT 161, pp. 163–170, 2023.
https://doi.org/10.1007/978-3-031-26281-4_16

slow growth [10]. Financing constraints lead to the underdevelopment of MSMEs in the development of efficient productivity [6].

To overcome the financial constraints for SMEs, Supply Chain Finance (SCF) is expected to solve these obstacles. According to [4], the definition of SCF is based on two perspectives. Firstly, SCF is financially oriented, which focuses on short-term solutions and involves financial institutions. Secondly, SCF focuses on short-term solutions and does not involve financial institutions. In its development, SCF has been widely used in the manufacturing industry, including computer communication equipment, electricity, automobiles, chemicals, steel, medicine, building materials, and furniture. After more than ten years of development, SCF has become an important financing channel for the manufacturing industry in accordance with the increased cash flows from MSMEs as well as to expand market share [13].

Several things that have an important influence on SCF are trust and commitment. In SCF, trust is an important concept that explains transaction behaviour between suppliers and the leading behaviour between banks and MSMEs. SCF provides a financial system to effectively support the real economy and establish a new business model that involves multiple entities' participation [21]. Research from [13], stated that there are shortcomings in the form of transparency and information sharing among parties involved in SCF. To resolve this drawback, this research raises the factors from SCF in the form of trust and commitment. Antecedent factors such as trust and commitment can help MSMEs to adopt SCF [1]. Research from [9], stated that the SCF framework is able to build trust, commitment to the on-time delivery of goods, negotiation regarding payments, and collaboration to share information. SCF often increases the level of commitment to trust and profitability for Supply Chain (SC) partners [19]. In implementing SCF, trust and commitment are important factors that support the success of the flow of information [24].

Research on supply chain finance in Micro, Small, and Medium Enterprises (MSMEs) mostly focuses on how MSMEs integrate the relationship between MSME owners and financial institutions providing capital credit. As conducted by [4, 13] and [1], research that focuses on supply chain finance on MSMEs with a total cost approach has not been widely conducted. This research examined the variables of trust, commitment, and supply chain finance on supply chain effectiveness with a total cost approach to Micro, Small, and Medium Enterprises (MSMEs). Supply chain finance can mitigate total costs and offer a risk-free financial solution for MSMEs in optimizing liquidity and increasing supply chain effectiveness.

2 Literature Review

2.1 Supply Chain Finance

Nowadays, many works of literature consider the use of a supply chain as a set of inter-company products between companies, information, and financial flow [17]. Many efforts have been made toward aligning product flow and information flow concerning product integration. The combination of product flow, information flow, and financing is integrated with supply chain finance. Research from [4] suggested supply chain finance which is a set of financial solutions where the bank is the party that holds a key role in financing the parties involved in product flow and information. Another important aspect of financial institutions' role is as a financing solution focusing on debt and receivables. Research from [7], states that SCF as a solution trigger in trading is a process of product flow such as receipt of goods, shipments, invoices, and payment due dates. According to [2], SCF is divided into three categories: pre-shipment financing, in-transit financing, and post-shipment financing. The role of SCF as a solution in trading is often referred to as a working capital-oriented financial solution [20].

2.2 Trust has a Positive Effect on Supply Chain Finance

Trust is a core concept that explains the business behavior of the organization [13]. Research from [18] stated that trust arises from expectations for future sustainability, past experiences, and interactions between humans. In SCF, trust is an essential concept that explains transaction behavior between suppliers and the leading behavior between banks and MSMEs. When upstream suppliers make transactions with downstream suppliers, commercial credit reflects the trust relationship between buyers and sellers. Commercial credit means the seller follows the buyer to postpone payment after the payment period after receiving the goods. According to [3], commercial credit eases the financing limit for buyers. Trust in SCF is the subject's expectation for MSMEs, suppliers, and financial institutions. It is in line with the statement of [8] that Trust is the subject's expectation of the subjective possibility for object-specific behavior.

H1. Trust between MSMEs and financial institutions has a positive effect on supply chain finance

2.3 Commitment has a Positive Effect on Supply Chain Finance

According to [25], commitments that are part of collaboration in the supply chain structure have a positive effect on collaboration. It has a function to develop long-term partnerships in micro-business as assets that are used constructively by companies. Commitment is the willingness of the parties to invest resources into the relationship. According to [16], commitment is seen as the result of a relationship in the supply chain. The relationship between the parties involved in the SCF must have a strong bond to support the flow of products and information to produce a smooth flow. The research by [14] showed that the commitment between buyers and sellers in SC increases SC performance.

H2. Commitment between MSMEs and financial institutions has a positive effect on supply chain finance

2.4 Supply Chain Finance has a Positive Effect on Supply Chain Effectiveness

The advantage of companies adopting SCF is a benefits strategy because it is generally associated with financial risk management in the supply chain [10]. This can be achieved by redistributing net operative working capital and other activities to increase suppliers' and distributors' access to financing [12]. SCF performs as an adaptation approach strategy that is capable of helping organizations such as MSMEs to reduce risk [10, 4] stated that the supply chain perspective can be identified as innovative financing solutions such as seller-based invoices as part of the SCF solution. [1] stated that SC effectiveness shows how successful the company is in achieving financial and operational goals in SC.

Operational and financial goals achieved by the company during SC operations can be balanced between efficiency and cost [15]. SC effectiveness minimizes shipping and handling costs and reduces distribution costs. Therefore, it lowering lowers product and logistics costs. The low cost of debt can reduce the capital cost of the supply chain. Thus, it increases the effectiveness of the supply chain [4]. The results of research by [1, 15] and [24] conclude that the adoption of SCF factors can achieve the output of SC. Therefore, it increases the effectiveness of SC.

H3. Supply Chain Finance has a positive effect on Supply Chain Effectiveness

3 Research Methods

3.1 Population and Sample

The population in this research were all micro, small and medium enterprises (MSMEs) in Semarang, Demak, and Kendal. Sampling was done by purposive sampling method. The purposive sampling method was applied because this research required intensive interaction with the research subjects. Therefore, the research subjects were selected based on the researcher's judgment about the subject's location and the subject's willingness to be involved in this study.

3.2 Research Settings

This research is causality research because it tested the hypothesis about the causality relationship between one or several variables with one or several other variables. Based on the research model developed, it is expected to further explain the causal relationship between the variables analyzed and be able to make useful research implications for the development of science as well as a method and technique for solving problems in the field. This research is focused on empirical testing of the structured model that developed from the theoretical framework. Integration of supply chain finance determinant variables in micro, small and medium enterprises (MSMEs) was conducted to achieve supply chain effectiveness through empirical research models.

3.3 Measurement Method

Variable measurements were taken from previous research to design the questionnaire's structure and ensure the items' content validity. The trust variable was measured by five items adapted from [23], emphasising information sharing among supply chain actors. The commitment variable was measured by five items adapted from [1]. The SCF variable was adopted from [26] by looking at how strongly MSME actors view SCF based on the risk prevention system, increasing capital flow coordination, and preventing high-risk capability. The SCE variable was adopted from [8] using six items, including those related to the cost of transportation, warehousing, inventory, logistics administration, product, and delivery orders.

3.4 Data Collection Plan

The type of data used in this research was quantitative data, namely, data stated by numbers that indicate the size of the value of the variable studied. Meanwhile, the data source in this research was primary data. It is information obtained directly by the researcher is in accordance with the factors or variables needed in the research. The data collecting method was by distributing questionnaires to respondents online via google Forms or direct distribution.

4 Data Analysis Method

4.1 Data Analysis

This research used Partial Least Square (PLS) to determine and analyze the relationship between the variables of trust, relationship commitment, supply chain finance, and supply chain effectiveness. The use of Partial Least Square was with consideration due to relationship complexity. Another consideration was the use of the PLS model because this model is more appropriate for prediction, as an implication of the research results on theory [5].

4.2 Parameter Estimation

Parameter estimation in PLS is the least squares method. The parameter estimation category consists of three things; First, the weight estimate has a function to create a score for the latent variable. Second, path estimates are useful for connecting latent variables and estimate loading, as well as latent variables and their indicators. Third, means and regression constant value for latent variables and their indicators.

4.3 Path Analysis Model

The path analysis model in this research has two equations as follows:

1. The effect of trust and commitment to supply chain finance $SCE = \alpha_1 + \beta_1 T + \beta_2 K + e1$
2. The effect of trust, commitment, and supply chain finance on supply chain effectiveness.

$$SCE = \alpha_1 + \beta_1 T + \beta_2 K + \beta_3 SCF + e2$$

Information:
T = Trust
K = Commitment
SCF = Supply Chain Finance
SCE = Supply Chain Effectiveness

4.4 Hypothesis Test

Hypothesis testing in this research used the t-test to determine the effect of each independent variable dependent on the dependent variable. The t-test was used to determine the effect of the independent variable on the dependent variable by using the following method:

The formula is:

$$t = \frac{b1}{Sb1}$$

Information:

t = count

b_1 = regression coefficient

Sb_1 = standard regression coefficient

It was conducted by comparing the value of the t_{count} with the t_{table}.

If $t_{count} < t_{table}$ value or p-value > α (0.05), then the decision accepts Ho, meaning that the independent variable has no significant effect on the dependent variable.

If $t_{count} > t_{table}$ value or p-value < α (0.05), then the decision rejects Ho, meaning that the independent variable has a significant effect on the dependent variable.

4.5 Goodness of Fit

Testing the model's ability to explain the variation of the dependent variable was carried out with the difference between the observed correlation and the model implied in the correlation matrix. It was indicated by the SRMR value < 0.08. It shows that the model is fit. Standardized Root Mean Square Residual (SRMR) is a goodness of fit measure for PLS-SEM that can be used to avoid model misspecification.

To see the goodness of fit model, the Normed Fit Index (NFI) can also be used as an alternative test. The Normed Fit Index is an index of an additional fit measure that calculates the Chi-square value of the proposed model and compares it with a meaningful standard (Bentler and Bonett 1980). The acceptable NFI match value is above 0.9.

Besides using SRMR and NFI, RMS Theta measurements can be used to test the goodness of fit model. RMS Theta assesses the extent to which the residuals of the outer model are correlated. The fit model is indicated by the RMS Theta value <0.12.

4.6 Coefficient of Determination

To find out how large the model's ability predicts the variation of the independent variables used in the coefficient of determination. The coefficient of determination is used to measure how far the model's ability explains the variation of the dependent variable. The calculation of the coefficient of determination uses the following formula:

$$R^2 = (r)^2 x\, 100\%$$

R^2: coefficient of determination

r: correlation coefficient

The coefficient of determination value is between zero and one. The small value of R^2 means that the ability of the independent variables in explaining the variation of the dependent variable is very limited. A value close to one means that the independent variables provide almost all the information needed to predict the dependent variables.

5 Conclusion

The necessity for supply chain finance for SMEs, suppliers, distributors and customers is quite high. This can be seen from the high demand for communication, information sharing and knowledge among parties involved in supply chain finance. However, SMEs still need to gain more awareness of being involved in supply chain finance. The limitation of this research is that a few SMEs still participate in strengthening and being involved in supply chain finance with their suppliers, distributors and customers. For further research, it is possible to add more variables that affect supply chain finance and effectiveness, such as financial alignment variables, and intelligent fraud.

References

1. Ali, Z., Gongbing, B., Mehreen, A.: Predicting supply chain effectiveness through supply chain finance: evidence from small and medium enterprises. Int. J. Logist. Manag. **30**(2), 488–505 (2019). https://doi.org/10.1108/IJLM-05-2018-0118
2. Basu, P., Nair, S.K.: Supply chain finance enabled early pay: unlocking trapped value in B2B logistics. Int. J. Logist. Syst. Manag. **12**, 334–353 (2012)
3. Burkart, M., Ellingsen, T.: In-kind finance: a theory of trade credit. Am. Econ. Revol. **94**(3), 569–590 (2004)
4. Caniato, F., Gelsomino, L.M., Perego, A., Ronchi, S.: Does finance solve the supply chain financing problem?. Supply Chain Manag. **21**(5), 534–549 (2016). https://doi.org/10.1108/SCM-11-2015-0436
5. Chin, W.W.: The partial Least Square Approach to Structural Equation Model. Publiser, London, Lawrence Elbaum Associates (1998)
6. Duan, Y., Mu, C., Yang, M., Chin, T.: Study on early warning of strategis risk during the process of firm sustainable innovation based on an optimized genetic bp neural network model: evidence from Chinese manufactures industries. Int. J. Product. Econ. **233**, 108293 (2021)
7. Lamoureux, J.F., Evans, T.A.: Supply chain finance: a new means to support the competitiveness and resilience of global value chains. Available at SSRN 2179944 (2011)
8. Fugate, B.S., Stank, T.P., Mentzer, J.T.: Linking improved knowledge management to operational and organizational performance. J. Oper. Manag. **27**(3), 247–264 (2009)
9. Gelsomino, L.M., Mangiaracina, R., Perego, A., Tumino, A.: Supply chain finance: a literature review. Int. J. Phys. Distrib. Logist. Manag. **46**(4), 1–19 (2016)
10. Gomm, M.L.: Supply chain finance: applying finance theory to supply chain management to enhance finance in supply chains. Int. J. Log. Res. Appl. **13**(2), 133–142 (2010)
11. Gorodnichenko, Y.: Financial Constraints and Innovation: why poor countries don't catch up. J. Europe Econ. Assos. **11**(5), 1115–1152 (2013)
12. Hofmann, E.: Supply chain finance: some conceptual insights, Logistik Management Innovation Logistikkonzepte, pp. 203–214. Dtsch. Univ, Wiesbaden (2005)

13. Jiang, R., Kang, Y., Liu, Y., Liang, Z., Duan, Y., Sun, Y.: A trust transitivity model of small and medium-sized manufacturing enterprises under blockchain-based supply chain finance. Int. J. Product. Econ. **247**, 108469 (2022)
14. Kanwal, A., Rajput, A.: A transaction cost framework in supply chain relationships: a social capital perspective. J. Relat. Mark. **15**(1/2), 92–107 (2016)
15. Nguema, J.N.B.B., Bi, G., Ali, Z., Mehreen, A., Rukundo, C., Ke, Y.: Exploring the factors influencing the adoption of supply chain finance in supply chain effectiveness: evidence from manufacturing firms. J. Bus. Ind. Mark. **36**(5), 706–716 (2021). https://doi.org/10.1108/JBIM-01-2020-0047
16. Paluri, R.A., Mishal, A.: Trust and commitment in supply chain management: a systematic review of literature. Benchmark. Int. J. **27**(10), 2831–2862 (2020). https://doi.org/10.1108/BIJ-11-2019-0517
17. Pfohl, H.-C., Gomm, M.: Supply chain finance: optimizing financial flows in supply chains. Logist. Res. **1**(3–4), 149–161 (2009). https://doi.org/10.1007/s12159-009-0020-y
18. Poppo, L., Zheu, K.Z., Li, J.J.: When can you trust? calculative trust, relational trust, and supplier performance. Strateg. Manag. J. **37**(4), 724–741 (2016)
19. Randall, W.S., Farris, M.T.: Supply chain financing: using cash-to-cash variables to strengthen the supply chain. Int. J. Phys. Distrib. Logist. Manag. **39**(8), 669–689 (2009)
20. Seifert, D., Seifert, D.: Financing the chain. Int. Comerce Revol. **1**, 33–34 (2011)
21. Song, H., Lu, Q., Yu, K., Qian, C.: How do knowledge spillover and access in supply chain network enhance SMEs' credit quality? Ind. Manag. Data Syst. **119**(2), 274–291 (2019). https://doi.org/10.1108/IMDS-01-2018-0049
22. Vieira, J.G.V., Yoshizaki, H., Ho, L.: The effects of collaboration on logistical performance and transaction costs. Int. J. Bus. Sci. Appl. Manage. **10**(1), 1–14 (2015)
23. Waheed, K.: Measuring trust in supply chain partners relationships. J. Measur. Bus. Excellence **14**(3), 53–69 (2010)
24. Eko, W., Purwanto, E.: The Mediating Role of the Supply Chain Financing on the Relationship between Negotiation, Collaboration and Digitalization with Supply Chain Effectiveness Technologie Report of Kansai University, **63**(05) (2021)
25. Yunus, E.N.: Leveraging supply chain collaboration in pursuing radical innovation. Int. J. Innov. Sci. **10**(3), 350–370 (2018). https://doi.org/10.1108/IJIS-05-2017-0039
26. Zhang, R.: The research on influence facts of supply chain finance operation. In: Proceedings of International Conference on Management Engineering and Management Innovation, Atlantis Press, Changsha, January, pp. 88–92 (2015)
27. Zimele, A.: The SMME Business Toolkit. New York: SBDA(Pty) Ltd (2009)

Blockchain Technology and Financing Risk in Profit Loss Sharing Financing of Indonesian Islamic Bank

Mutamimah Mutamimah[1(✉)] and Indri Kartika[2]

[1] Department of Management, Faculty of Economics, Universitas Islam Sultan Agung, Semarang, Indonesia
mutamimah@unissula.ac.id

[2] Department of Accounting, Faculty of Economics, Universitas Islam Sultan Agung, Semarang, Indonesia
indri@unissula.ac.id

Abstract. Profit loss-sharing contracts (*mudharabah* and *musyarakah*) have higher financing risk than debt financing contract (*murabahah*). The purpose of this study is to develop a blockchain technology model as a mechanism to reduce financing risk in profit loss-sharing contracts in Indonesian Islamic banks. This is a conceptual paper with an integrative literature review related to the financing feasibility evaluation mechanism, financing risks, smart contracts, and blockchain technology. The results show that blockchain technology can reduce asymmetric information and financing risk in Islamic bank profit loss-sharing contracts because, in blockchain technology, there are smart contracts that can omit asymmetric information, and all stakeholders involved in the blockchain can access and monitor data and none of them can change the data.

Keywords: Profit loss sharing contract · Financing risk · Smart contract · Blockchain technology

1 Introduction

Islamic banks in Indonesia are experiencing good development. This is shown by Islamic banking data in 2021 that the market share is 6.52%; asset value of 646.2 trillion; total financing 413.3 trillion and third party funds 503.8 trillion [1]. Islamic banks as intermediary institutions that carry out fundraising and financing, where in financing there are *murabahah* contracts, *mudharabah* contracts, and *musyarakah* contracts. Based on data from [2], it shows that the amount of *murabahah* financing is 46.22%, *musyarakah* is 45.69%, and *mudharabah* is 2.65%. This means that *murabahah* financing dominates both *musyarakah* financing and *mudharabah* financing. The reason is that *musyarakah* and *mudharabah* financing has a higher risk than *murabahah* financing [3, 4]. In fact, the two financing contracts are actually in accordance with the essential objectives of Islamic banks, namely empowering the real sector and improving the economic welfare of community. In addition, based on the results of research [5] found that there are 4

L. Barolli (Ed.): EIDWT 2023, LNDECT 161, pp. 171–179, 2023.
https://doi.org/10.1007/978-3-031-26281-4_17

obstacles in the financing of Profit Loss Sharing, namely: risk, difficulty in choosing the right partner; requests come from customers with low creditworthiness; and lack of capital security. Thus, it is necessary to manage the risk of Islamic bank financing, especially for profit loss sharing contracts, namely: *musyarakah* and *mudharabah* so that risk can be reduced. The low risk indicates that Islamic bank managers are able to manage risk professionally so as to improve the reputation and sustainability of Islamic banking. Moreover, the sources of Islamic banking fund collection in Indonesia are dominated by Third Party Funds which must be managed professionally [1]. If Islamic bank managers are able to manage risk well, they will be able to improve financial performance [6].

Financing risk is the risk that occurs if the debtor does not return the loan according to the initial agreement [7, 8]. One of the causes of financing risk in Islamic banks is due to the asymmetric information between Islamic banks as principals and customers as agents. One of the mechanisms to reduce financing risk is the implementation of corporate governance. Corporate governance is a mechanism, system and structure to monitor and control the behavior of managers so that they carry out business activities in accordance with the objectives of stakeholders [9]. One form of corporate governance implementation is the application of an assessment and evaluation mechanism for prospective debtors by Islamic banks as a basis for determining the feasibility of obtaining financing, with the aim that risk can be reduced. However, the implementation of corporate governance is not effective in reducing financing risk in *musyarakah* financing [6].

Therefore, the existence of these problems encourages researchers to use blockchain technology in the profit loss sharing (*mudharabah* and *musyarakah*) financing mechanism, so that the financing risk can be reduced. The use of blockchain technology in Islamic banks in Indonesia is still very limited. In fact, in the current technological era, banks are already using digital in their transactions where based on data from the [1] shows the value of digital transactions at Islamic banks in Indonesia reaches Rp 39,841 trillion. Blockchain technology can reduce asymmetric information because blockchain technology is a form of ecosystem and smart contract that connects stakeholders through blocks [10, 11]. Blockchain can facilitate recording, financial reporting, and storage of business transactions by all stakeholders on a digital block network that encourages stakeholder behavior to always be honest, transparent, resilient and trusting between stakeholders, all stakeholders can monitor all business processes properly, thereby reducing credit risk [12]. Therefore, the purpose of this study is to develop a conceptual blockchain technology model to reduce the risk of profit loss sharing financing at Indonesia Islamic banks.

This paper is divided into five parts: (1) introduction, (2) literature review, (3) research methods, (4) finding and discussion, and (5) conclusion, limitations, and future research.

2 Literature Review

2.1 Financing Based Profit Loss Sharing in Islamic Bank

Islamic banking has 2 types of contracts, namely debt financing and Profit Loss Sharing (PLS) financing. Debt financing consist of *murabahah* financing which means a sales and purchase contract wherein the Islamic bank as seller buys goods and then sells to the customer with payment is made based on agree two party, Islamic bank and customer [6]. Financing based on PLS is divided into 2, namely: *mudharabah* and *musyarakah* contracts [5]. *Mudharabah* financing means where the bank and the customer agree to work together on a business project and where the bank acts as a provider of capital and the customer provides the knowledge and skills to run the project [13]. *Mudharabah* financing has potential risks, because the profits obtained by the *mudarib* are uncertain, and if a loss occurs, the bank must be prepared to bear all the losses of the project. *Musyarakah* financing is profit-sharing financing, where Islamic banks and customers both collect funds and work together to fund projects, and the results are shared by both parties according to the agreement [13].

2.2 Risk Management in Islamic Banking

Potential risk always occurs in all Islamic bank activities. Islamic banks as intermediary institutions that collect funds and channel funds to customers and carry out social functions, must be able to manage risk professionally. Risk mitigation is a part of risk management [14]. The whole series of activities are integrated each other as an ecosystem that must be managed professionally. Islamic banks as intermediary institutions are required to implement risk management, because one indicator of the performance and reputation of Islamic banks is the ability of managers to manage risk. According to Bank Indonesia Regulation Number 13/23/PBI/2011, Islamic banking risk is divided into 10 risks, namely: Credit Risk, Market Risk, Liquidity Risk, Operational Risk, Legal Risk, Reputational Risk, Strategic Risk, Compliance Risk, Return Risk, and Investment Risk. Risks are interrelated each other. For example, if an Islamic bank is not able to manage financing risk, it will have an impact on liquidity risk and reputation risk. One of the reasons for the high risk of financing is the asymmetric information between Islamic banking as creditors and MSMEs as debtors, thus encouraging high deviations in the use of loan capital, which in turn MSMEs cannot repay loans on time, which is called bad credit. Various efforts have been made so that asymmetric information can be reduced, one of which is the implementation of corporate governance so that there are no irregularities in the use of loan capital, and the risk of financing decreases. However, the implementation of corporate governance is only effective on *Mudharabah* financing, but is not effective when applied to *Murabahah* and *Musyarakah* financing schemes [6].

2.3 Blockchain Technology and Islamic Banking Profit Loss Sharing Contract

Blockchain technology is a technological innovation that is currently developing. According to [15], blockchain is a new, decentralized technology with a ledger system, capable of storing information, and recording all transactions made by stakeholders without third parties. All information from stakeholders that is stored on a computer, can be monitored by all parties in real time and no one party can change the data so that the validity of the data is well maintained. Blockchain can facilitate the recording, financial reporting, and storage of business transactions by all stakeholders on the digital block network so that they are valid, transparent, and robust so that all stakeholders in this block can monitor all business processes properly, so as to reduce credit risk [12]. Blockchain technology can make it easier to connect between stakeholders with smart contracts, so that all stakeholders involved in financing profit loss sharing can enter data, monitor and access digitally recorded data and those owned by other stakeholders in a fast and cost-efficient manner. This means that the existence of blockchain makes it easier to make profit loss sharing financing decisions so that financing risk can be reduced.

3 Research Method

This study aims to develop a blockchain technology model in reducing the risk of financing in profit loss sharing contracts for Islamic banks in Indonesia. These contracts consist of two parties (Islamic Bank and Debtor) in cooperation in capital and skill. Furthermore, the costs of searching, screening, and contracting will increase [22]. This model does not only apply in Indonesia but applies to all Islamic banks around the world, because all Islamic banks have profit loss sharing contracts (*mudharabah* and *musyarakah*).

This is a conceptual paper that uses an integrative literature review approach [16] through reviewing and critiquing previous literature related to Islamic bank financing risks, corporate governance, financing worthiness and profit loss sharing contracts and the adoption of blockchain technology as the basis for developing a conceptual model of blockchain technology in reducing financing risk in profit-loss sharing contracts for Indonesia Islamic bank. This blockchain technology model has a smart contract, all stakeholders involved in the block, namely Islamic banks, customers/investors, governments, and debtors can find out, analyze, and evaluate all data and documents of other stakeholders through the block network (see Fig. 2).

4 Finding and Discussion

Finding and discussion explain subchapters as follows: corporate governance, financing worthiness and Profit Loss Sharing (PLS) contract, smart contract and blockchain technology in decreasing financing risk at profit loss sharing in Islamic bank, and implementation of blockchain technology in financing Profit Loss Sharing (PLS).

4.1 Corporate Governance, Financing Worthiness and Profit Loss Sharing Contract

In carrying out the financing function, there are 2 contracts, namely: debt financing contract and Profit Loss Sharing (PLS) contract. To reduce the risk of financing, Islamic banking implements a mechanism to evaluate the feasibility of obtaining financing as a form of corporate governance. Based on the results of research by [17], it shows that the distribution of Islamic bank funds is predominantly to MSMEs with greater risk than non-MSME financing. The Islamic banking party evaluates the feasibility of customers to obtain financing using a 5C analysis consisting of: Character, Capacity, Capital, Collateral, and Condition [18]. In addition, Islamic banking also analyzes the clarity of the debtor's business that meets sharia principles and DIS (Debtor Information System). However, based on this mechanism, it turns out that there are still unresolved financing risks until today.

Figure 1 shows several weaknesses of the mechanism for analyzing the feasibility of financing as a form of corporate governance, including: a). In submitting financing requirements to Islamic banks, prospective debtors still use proof of identity that must be photocopied as well as administrative documents in the form of papers that are easy to forge and change the data, so that it has the potential for financing risks to arise. b). In this mechanism, there is still a third party; namely Islamic banking, where when the prospective debtor does not have financial data and does not have the necessary documents, the feasibility of getting financing is only based on the perception of the bank manager [19]. This is certainly a potential risk of financing. c). The process of assessing the feasibility of obtaining financing approval takes a very long time, with a gradual and inefficient process. d). The implementation of corporate governance is only between Islamic banking as principal and debtor as agent, without involving other stakeholders in monitoring business transactions and other activities carried out by debtors. With this mechanism, of course there is still risk of Islamic bank financing.

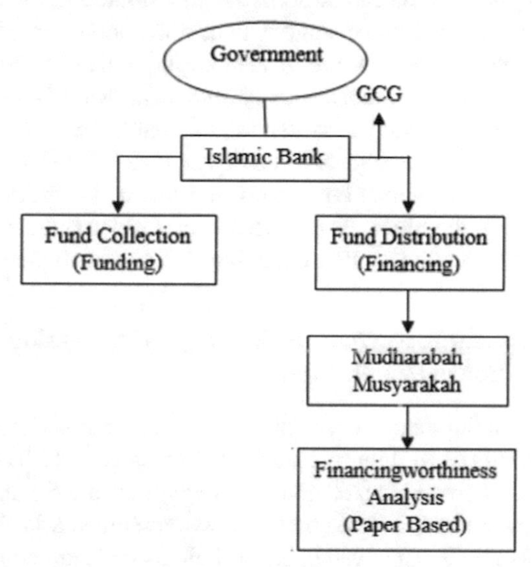

Fig. 1. Corporate governance, financing worthiness and profit loss sharing contract (existing model)

Figure 1 explains the function of an Islamic bank as an intermediary institution that collects funds from customers/investors, then these funds are channeled to debtors under *murabahah, mudharabah* and *musyarakah* contracts. The government acts as a regulator and monitors the operations of Islamic banks. In this Fig. 1, the stakeholders are not connected in blocks, so there is still asymmetric information between the Islamic bank and the debtor, where the information held by the debtor is more than the information held by the Islamic bank. This encourages the emergence of moral hazard and irregularities in the use of funds by debtors so that financing risk increases.

While Fig. 2 explains that all stakeholders, namely: Islamic banks, consumers/investors, government, debtors are connected in blocks. Through this block, all

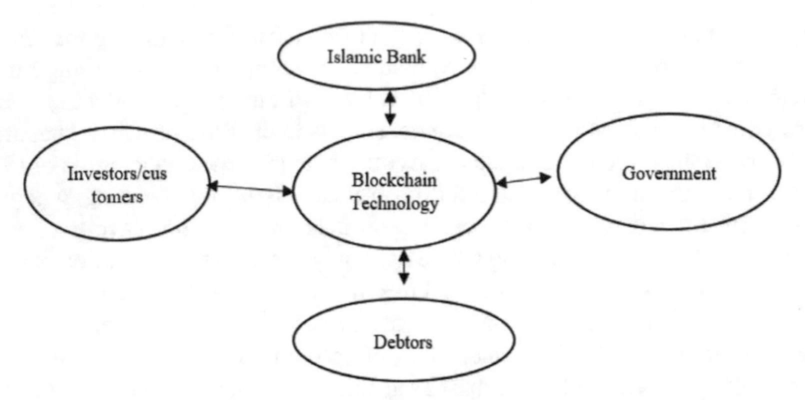

Fig. 2. Blockchain technology and profit loss sharing financing (new model)

stakeholder data and information can be accessed and monitored by all stakeholders and no third party, in fact, no party can change data or documents except with the approval of all parties in the block. Besides, Fig. 2 also explains that in the financing mechanism with profit-loss sharing contracts, namely *mudharabah* and *musyarakah* contracts, Islamic banks can use smart contracts so that Islamic banks can know, analyze, evaluate the feasibility of obtaining financing based on debtor data that is connected in the block. All business transactions, financial conditions and debtor prospects are known by all stakeholders involved in the block. Thus, there is no hidden debtor data or documents and no asymmetric information, so that financing risk can be reduced.

4.2 Smart Contract and Blockchain Technology in Decreasing Financing Risk Profit Loss Sharing in Islamic Bank

Islamic banks carry out the financing function with 2 contracts, namely: debt financing and Profit Loss Sharing (PLS). Figure 2 shows the existence of blockchain technology in profit loss sharing, where in blockchain technology there are 5 stakeholders involved, namely: investors/ customers, Islamic banks, government, Sharia Supervisory Board, and Debtors. All stakeholders involved in blockchain using smart contracts can monitor, and evaluate the data of other stakeholders, there is no third party and no party can change the data unless there is a mutual consensus. Asymmetric information can be eliminated, because all stakeholders involved in the block can provide information, monitor, track information quickly as a basis for decision making. The existence of blockchain technology will be able to avoid wrong perceptions in making creditworthiness decisions that are only based on perceptions, especially for MSMEs as debtors who do not have data and documents systematically. Thus, blockchain technology can increase transparency and transactions according to sharia, there is no asymmetric information, and the risk of financing can be reduced. This is reinforced by [20], that smart contracts in accordance with the Qur'an Surah al-Baqarah 282–283 emphasize the need to record agreements with full accuracy, fairness and accountability.

Blockchain technology can provide benefits through smart contracts which are part of the transactions that are used and executed on the blockchain system. Asymmetric information as the cause of financing risk can be eliminated by the presence of smart contracts [11]. Thus, it is very appropriate if blockchain Technology is used to deal with asymmetric information problems, so that it can reduce financing risk. Smart contracts in Islamic banks are different from traditional contracts because they are in the form of actual code on a computer, while traditional contracts are in written form and in a language that is easily understood by the contracting party [20]. But in blockchain, computer code, counterparts can rely on consistent execution (automated trust) between the stakeholders involved. The code cannot be changed without the coordination of all parties involved. [21] states that *mudharabah* can be developed with smart contracts. If a smart contract is implemented, it will be efficient, secure, and transparent. [20] states that through blockchain technology, all information and transactions are recorded systematically through a cryptographic process in a public database that allows all stakeholders in this public network to participate and contribute in validating all information and no one party can change or even delete it. Data. Thus, the existence of data is very helpful for stakeholders in every decision making. Through this smart contract, it will be easier to track transactions quickly and validly. The absence of a third party will actually speed up the process of evaluating the feasibility of financing at Islamic banks and save the cost of applying for financing [20]. The existence of blockchain technology will be able to avoid wrong perceptions in making creditworthiness decisions that are only based on perceptions, especially for MSMEs as debtors who do not have data and documents systematically.

4.3 Implementation of Blockchain Technology in Financing Profit Loss Sharing

Implementation of blockchain technology in Islamic banking is not easy, and there are several challenges, including: a). The quality of human resources for all stakeholders must be improved, especially information technology. b). There must be adequate infrastructure to implement blockchain technology. c). In Islamic banking there are sharia principles that must be obeyed. Thus, a strategy is needed so that sharia principles can be encoded computationally. d). There are no regulations and policies for implementing blockchain technology in Islamic banking involving all stakeholders, and there is no fatwa from the MUI (Indonesian Ulema Council) related to blockchain technology [20].

5 Conclusion, Limitations, and Future Research

Islamic banks carry out the financing function with 2 contracts, namely: debt financing contracts and Profit Loss Sharing (PLS) contracts. However, the issue of financing risk has not been resolved until now. Profit Loss Sharing (PLS) financing is higher risk than debt financing. One of the efforts to reduce the risk of Profit Loss Sharing (PLS) financing is by implementing corporate governance. However, this mechanism is not effective in reducing the financing risk of Islamic bank. Therefore, blockchain technology is needed as a mechanism to reduce financing risk on profit loss sharing in Indonesian Islamic banks. With smart contracts on blockchain technology, asymmetric information can be eliminated, because all stakeholders are connected to blocks that can convey information, monitor and evaluate other stakeholder data as a basis for decision making. Even the blockchain technology has a ledger record that contains all transactions that can be monitored by all parties according to the agreement in the smart contract, so that the risk of financing can be lowered. This article only analyzes risk financing on profit loss sharing financing at Islamic banks, which can still be developed for other contracts, for example in Islamic insurance, Islamic Micro Finance. In addition, this article is in the form of a conceptual model, so it still needs to be tested empirically in the future, especially for Islamic banks in evaluating and making decisions on the feasibility to get financing, so that financing risk can be reduced.

References

1. Financial Services Authority, "Islamic banking snapshot september (2021). www.ojk.go.id
2. Financial Services Authority, Islamic banking snapshot february 2022. (2022) www.ojk.go.id
3. Yaya, R., Saud, I.M., Hassan, M.K., Rashid, M.: Governance of profit and loss sharing financing in achieving socio-economic justice. J. Islam. Account. Bus. Res. **12**(6), 814–830 (2021). https://doi.org/10.1108/JIABR-11-2017-0161
4. Ishak, M.S.I., Rahman, M.H.: Equity-based Islamic crowdfunding in Malaysia: a potential application for mudharabah. Qual. Res. Financ. Mark. **13**(2), 183–198 (2021). https://doi.org/10.1108/QRFM-03-2020-0024
5. Abdul-rahman, A., Nor, S.M.: Challenges of profit-and-loss sharing financing in Malaysian islamic banking. Geogr. Malays. J. Soc. Sp. **12**(2), 39–46 (2017)
6. Mutamimah, M., Saputri, P.L.: Corporate governance and financing risk in Islamic banks in Indonesia. J. Islam. Account. Bus. Res. (2022). https://doi.org/10.1108/JIABR-09-2021-0268
7. Haryono, Y., Ariffin, N.M., Hamat, M.: Factors affecting credit risk in Indonesian islamic banks. J. Islam. Financ. **5**(1), 12–25 (2016). https://doi.org/10.12816/0027649
8. Mustafa, O.A.O.: Why do islamic banks concentrating finance in murabaha mode? performance and risk analysis (Sudan: 1997–2018). Int. Bus. Res. **13**(7), 208 (2020). https://doi.org/10.5539/ibr.v13n7p208
9. Mutamimah, M., Tholib, M., Robiyanto, R.: Corporate governance, credit risk, and financial literacy for small medium enterprise in Indonesia. Bus. Theory Pract. **22**(2), 406–413 (2021). https://doi.org/10.3846/btp.2021.13063
10. Schinckus, C.: The good, the bad and the ugly: an overview of the sustainability of blockchain technology. Energy Res. Soc. Sci. **69**, 101614 (2020). https://doi.org/10.1016/j.erss.2020.101614
11. Lacasse, R., Lambert, B., Khan, N.: Islamic banking - towards a blockchain monitoring process. In: Conference: 5th International Conference on Entrepreneurial Finabce, CIFEMA 2017, Journal of Business and Economics, vol. 6, pp. 33–46 (2017)
12. Osmani, M., El-Haddadeh, R., Hindi, N., Janssen, M., Weerakkody, V.: Blockchain for next generation services in banking and finance: cost, benefit, risk and opportunity analysis. J. Enterp. Inf. Manag. **34**(3), 884–899 (2021). https://doi.org/10.1108/JEIM-02-2020-0044
13. Warninda, T.D., Ekaputra, I.A., Rokhim, R.: Do mudarabah and musyarakah financing impact islamic bank credit risk differently?. Res. Int. Bus. Financ. **49**, 166–175 (2019). https://doi.org/10.1016/j.ribaf.2019.03.002
14. Mutamimah, M., Zaenudin, Z., Cokrohadisumarto, W.B.M.: Risk management practices of Islamic microfinance institutions to improve their financial performance and sustainability: a study on Baitut Tamwil Muhammadiyah, Indonesia. Qual. Res. Financ. Mark. (2022). https://doi.org/10.1108/QRFM-06-2021-0099

15. Singh, H., Jain, G., Munjal, A., Rakesh, S.: Blockchain technology in corporate governance: disrupting chain reaction or not? Corp. Gov. **20**(1), 67–86 (2020). https://doi.org/10.1108/CG-07-2018-0261
16. Snyder, H.: Literature review as a research methodology: an overview and guidelines. J. Bus. Res. **104**, 333–339 (2019). https://doi.org/10.1016/j.jbusres.2019.07.039
17. Mutamimah, H.: Islamic financial inclusion: supply side approach. In: 5th ASEAN International University Conference on Islamic Finance (5th AICIF), pp. 1–9 (2017)
18. Wasiuzzaman, S., Nurdin, N., Abdullah, A.H., Vinayan, G.: Creditworthiness and access to finance: a study of SMEs in the Malaysian manufacturing industry. Manag. Res. Rev. **43**(3), 293–310 (2020). https://doi.org/10.1108/MRR-05-2019-0221
19. Karlan, D., Bryan, G., Jakiela, P., Keniston, D.: Direct and indirect impacts of credit for SMEs, Res. Note, no. 670, (2015). http://pedl.cepr.org/sites/default/files/ResearchNote_670_KarlanBryanJakielaKeniston.pdf
20. Chong, F.H.L.: Enhancing trust through digital islamic finance and blockchain technology. Qual. Res. Financ. Mark. **13**(3), 328–341 (2021). https://doi.org/10.1108/QRFM-05-2020-0076
21. Rejeb, D.: Smart contract's contributions to mudaraba. Tazkia Islam. Financ. Bus. Rev. **15**(1), 1–18 (2021). https://doi.org/10.30993/tifbr.v15i1.236
22. Badaj, F., Radi, B.: Empirical investigation of SMEs' perceptions towards PLS financing in Morocco. Int. J. Islamic Middle East. Financ. Manag. (2017). https://doi.org/10.1108/IMEFM-05-2017-0133

Privacy-Preserving Scheme for Nearest Life Services Search Based on Dummy Locations and Homomorphic Encryption Algorithm

TieSen Zhao[1] and LiPing Shi[1,2](✉)

[1] College of Computer Science, Sichuan University, Chengdu, China
413234498@qq.com

[2] Police Officers College of the Chinese People's Armed Force, Chengdu, China

Abstract. With the rapid development of emerging internet technologies, Location Based Services (LBS) have brought great convenience to the people's lives. However the locations will be exposed to service providers posing serious threats to users' information security. No trusted entities except the user himself as a basic assumption, This paper proposes a privacy-preserving the nearest life services search scheme with cloud server instead of LBS provider, which the user and the life service providers both do not need to provide their actual locations to the cloud server. First, the user constructs a K-anonymous area composed of all dummy locations. The relevant life service providers respond according to the distance to the dummy address. Then, with the computing and communication assist of the cloud server which is based on homomorphic properties of Paillier cryptosystem, the user finds and connects to the nearest life service provider. Our scheme not only reduces the cost of communication between users and servers but also protects the location and track information, which achieves trade-off of privacy and availability in mobile web applications, it can also be used as a business model applicable to any location-based service industry.

1 Introduction

Location Based Service (LBS) is the most widely used service in mobile intelligent terminals. By providing their real location and life service needs to the server, users can search the nearest life service providers returned by the server, such as online car booking, takeout delivery, querying the nearest gas station, cinema, hotel, etc. However, direct submission of such query requests can bring serious problems of privacy leakage of user's location and activity track. Attackers can infer sensitive information such as home address, workplace, habits and hobbies from users' location information. In traditional LBS, location privacy leakage is caused by attacks on the platform by external illegal users due to inadequate security management of location-based service providers, or LBS platform can obtain the current location of users for tracking without obtaining special permission from users. At the same time, the LBS platform can also learn the user's trajectory and points of interest (POIs) through the life service provider that the user eventually chooses, so the location information of the life service provider also

L. Barolli (Ed.): EIDWT 2023, LNDECT 161, pp. 180–189, 2023.
https://doi.org/10.1007/978-3-031-26281-4_18

needs to be protected. To avoid unauthorized access to location information by LBS platform and hackers, users and life service providers can encrypt location information locally before uploading it to the service platform. However, once the ciphertext data is stored in the service platform, most of the traditional encryption methods do not support the computation of ciphertext when the service platform needs to perform distance calculation on the ciphertext data. How to guarantee both the privacy and availability of location data in the process of distance computation remains a major challenge, homomorphic encryption technique is a key means to solve this problem. In addition, there are a large number of life service providers providing certain types of services, for example, cabs in a city are distributed in various areas of the city, and cabs that are far away from the user are meaningless to the user, so it is not necessary to calculate the distance between the user and all life service providers, but we can delineate the acceptable distance range by the current location of the user, and the life service providers within this range are used as candidate.

Based on the above reasons, we propose a nearest living service query scheme based on K-anonymous dummy location region construction algorithm and Paillier homomorphic encryption algorithm. It protects the user's location and trip information effectively in an untrustworthy environment, and completes the user's query for the nearest life service based on his own location. In this scheme, the traditional location-based service provider is removed and replaced by a cloud server that assists in computing and communication. The cloud server acts as a bridge for communication between the user and the life service provider, and does not acquire and track the real-time locations of the user and the life service provider. First, the user constructs a K-anonymous dummy locations area based on its own location and broadcasts them to life servers through the cloud server. The life servers calculate the distance to the dummy locations, and if the distance is within the acceptable range, they respond to the user through the cloud server and become candidates. Then, the user and the candidate encrypt their real location data using the Paillier encryption algorithm, and the cloud server calculates the ciphertext of the distance according to the homomorphic encryption feature, and the ciphertext is sent to the user, who decrypts the distance and selects the life service provider corresponding to the minimum distance, and establishes a communication connection with the nearest life service provider through the cloud server, Finally they communicate encrypted through the traditional public key encryption system RSA.

2 Related Research and Theoretical Basis

To solve the problem of location privacy leakage, there are two main ideas, one is fuzzy generalization [1] and the other is encryption. Fuzzy generalization protects the user's real location by constructing a cloaking region (CR) [1, 2] and generating dummy locations [3, 4]. Encryption method is a way to prevent location information leakage by encrypting location data with cryptography techniques. Homomorphic encryption algorithm combined with secure multi-party computation [5] makes it possible to provide location-based services while protecting location privacy.

2.1 Location K-Anonymous Method

The location k-anonymization method serves the purpose of protecting the privacy of the user's location by mixing the user's location with at least k-1 candidates to form an anonymization region of the user [6]. The k-anonymization dummy locations method [7] is that the user generates k-1 dummy locations based on his own address and sends them to the server along with the user's real location information to form an anonymous set, and the server will perform query operations on all candidate locations and return the query results to the user, who selects the desired real information from the candidate results. The dummy locations is used to confuse the service providers so that they cannot identify the real user's location information from the anonymized set to achieve the purpose of location privacy protection.

In this paper, we use the hidden area based on the all dummy locations to replace the user's real location, which on the one hand hides the user's location in a certain area size to reduce the attacker's ability to distinguish the user for the purpose of location privacy protection, and on the other hand the hidden area can be used as the basis for subsequent distance judgment.

2.2 Homomorphic Encryption Algorithm

Homomorphic encryption (HEE) is a major breakthrough in the field of cryptography [8]. The output obtained by processing the ciphertext of the original data encrypted using homomorphic encryption algorithm and decrypting this output is the same as the output obtained by processing the original unencrypted data by the same method. The development and application of homomorphic encryption technology provides researchers with new research ideas [9–12], where users send their private data to a service provider in ciphertext, and the service provider does not need to decrypt the data, but can directly process the ciphertext data, and the user can get the processed plaintext data by decrypting it, so that the privacy data is protected and the corresponding computing service is available.

Paillier encryption algorithm [13] is a partial homomorphic encryption (PHE) technique.

(1) Key generation process

Choose two large prime numbers p and q randomly as well as g $\in Z^*_{N^2}$, let $N = pq, \lambda = lcm(p-1)(q-1)$, set the function $L(u) = \frac{u-1}{N}$, Choose g to satisfy $gcd\left(L\left(g^{\lambda} modN^2\right), N\right) = 1$, then public key $Pk = (N, g)$, the private key $Sk = \lambda$.

(2) Encryption process

Select a random number $r \in Z^*_N$, for plaintext $M \in Z_n$, Use the public key to encrypt to obtain the ciphertext C:

$$C = Enc(m) = [m] = g^M \cdot r^N modN^2 \quad (1)$$

(3) Decryption process

According to the ciphertext C, use the private key Sk to decrypt the plaintext M:

$$M = Dec(C) = \frac{L\left(C^{\lambda} modN^{2}\right)}{L\left(g^{\lambda} modN^{2}\right)} modn \tag{2}$$

Since

$$\begin{aligned} \text{Enc}(\text{M}_1) \cdot \text{Enc}(\text{M}_2) &= \left(g^{M_1} \cdot \text{r}_1^N\right)\text{mod}N^2 \cdot \left(g^{M_2} \cdot \text{r}_2^N\right)\text{mod}N^2 \\ &= g^{M1+M2}(r_1 r_2)^N \\ &= Enc(M_1 + M_2) modN^2, \end{aligned}$$

So the plaintext addition operation corresponds to the ciphertext multiplication operation, so the Paillier encryption algorithm satisfies the additive homomorphism, which is simplified as

$$[m_1] \cdot [m_2] = [m_1 + m_2] \tag{3}$$

In this paper, we use the Paillier encryption algorithm to encrypt the user's real location data, and the Cloud computing servicer uses the homomorphism to calculate the distance ciphertext between the user and the life service provider based on their location ciphertext without knowing their real locations.

3 Scheme Overview

The scheme has three types of entities, i.e., user client U, life service provider client P, and cloud server S, where both user client and life service provider client are mobile intelligent terminals with GPS or other positioning function modules. The interaction of the entities in the scheme is shown in Fig. 1. The search process is as follows:

Step1. The user client generates K dummy locations based on its own real location, and sends these dummy locations and requests for using life service type t to the cloud server.

Step2. The cloud server filters out online life service providers that can provide life service of type t in the list of life service providers, and sends dummy locations to the relevant life service providers.

Step 3. The life service provider responds to the life service provision request to the cloud server if the distance is within the acceptable range based on the Euclidean distance between its own geographic location and each of the K fake locations. The cloud server numbers the candidate life service providers as P_i ($\text{i} \in [1, \text{m}]$).

Step4. The user and the candidate life service providers encrypt their own locations and send the location ciphertext to the cloud server.

Step5. The cloud server calculates the received location ciphertext and obtains the distance ciphertext between the user's location and each candidate life service provider and sends them to the user.

Step6. The user decrypts the distance ciphertext, calculates the distance between himself and each life service provider, and selects the nearest candidate life service provider P_w.

Step7. The user and the nearest life service provider P_w communicate confidentially using RSA public key encryption system.

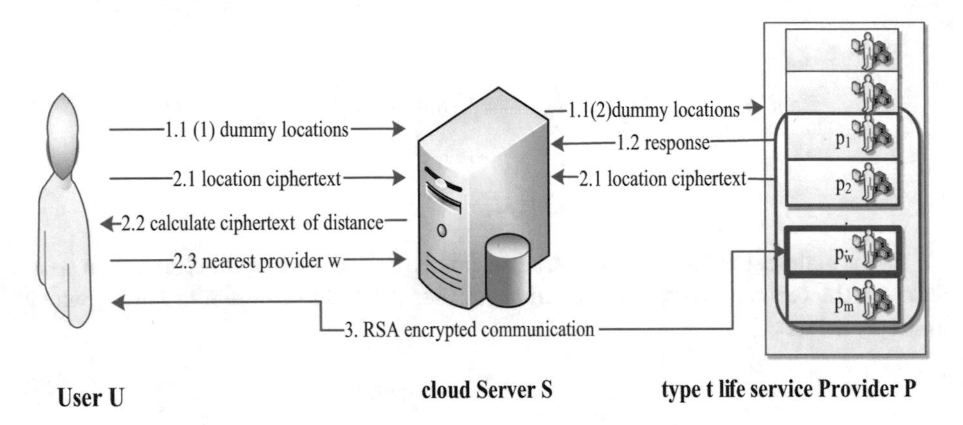

Fig. 1. Workflow diagram of nearest life service search

4 Nearest Life Services Search Algorithm

The process of querying the nearest living service initiated by users involves two main algorithms: one is the k-anonymity region construction based on dummy locations generation algorithm; the other is the distance calculation based on the Paillier encryption algorithm.

4.1 Dummy Locations Area Construction

The dummy locations generation algorithm is to generate a series of dummy locations that satisfy the user's anonymity needs within the user's acceptable anonymity region. The anonymous area is a circular area formed by the user's real location as the center of the circle. The user sets the number of dummy locations K according to his needs, and the maximum and minimum values of the acceptable anonymous area, which are represented by the shortest radius r_{min}, the longest radius r_{max}, and the coordinates of the real location (pos_x, pos_y) are obtained by the positioning system. Since the randomly generated dummy locations may differ significantly from the user's real address and is easily rejected by the attacker, the distance $d_{(1,pos)}$ between the first dummy location and the real location is required to satisfy $r_{min} < d_{(1,pos)} < r_{max}$. The next dummy location is generated randomly in turn on a clockwise rotated chord with angle $\theta(\theta = \frac{2\pi}{K-1})$ from the previous dummy location. If the dummy locations are all within the range$r_{min} < d_{(i,pos)} < r_{max}$, then the central part of the anonymized region is too blank and the attacker can easily narrow the anonymized region to that blank area. Therefore, the randomly generated dummy locations are cross-distributed in the range of $r_{min} < d_{(i,pos)} < r_{max}$ and0 $< d_{(i,pos)} < r_{min}$, so that the dummy locations are distributed evenly as a whole. Assuming k = 9, the schematic diagram of generating dummy locations is shown in Fig. 2.

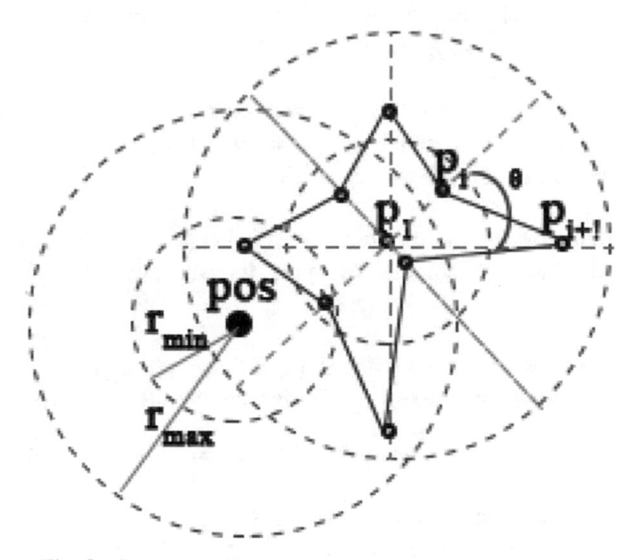

Fig. 2. Schematic diagram of generating dummy locations

In summary, the dummy address generation algorithm is shown below.

All Dummy generation algorithms

Input:pos,rmin,rmax,K

Output:dummy location set P

$P \leftarrow \emptyset, \theta = \frac{2\pi}{k-1}$;

$pos_{1,x} = ((rand(1) \geq 0.5) * 2 - 1) * rand(r_{min}, r_{max}) + pos_x$;

$pos_{1,y} = ((rand(1) \geq 0.5) * 2 - 1) * rand(r_{min}, r_{max}) + pos_y$;

for i from 2 to K do

  if (i%2==0)

    $r_i = rand(0, r_{min})$;

  else

    $r_i = rand(r_{min}, r_{max})$;

  $pos_{i,x} = r_i \cos(i * \theta) + pos_{1,x}$;

  $pos_{i,y} = r_i \sin(i * \theta) + pos_{1,y}$;

  Append posi to P;

Return P;

4.2 Distance Calculation

The distance between the user and the candidate life service providers are calculated as follows:

(1) The user generates a public key PK and a private key SK, publishes the public key PK to the public, and stores the private key SK in secret. The specific operations for generating Paillier encryption algorithm keys:

<1>. Randomly choose two large prime numbers p and q of equal length.
<2>. Compute $N = p \cdot q$ and $\lambda = \varphi(N) = (p-1)\cdot(q-1)$.
<3>. Public key PK:N; private key SK:λ.

(2) The user client obtains the real-time coordinate positions(XU,YU) through the satellite positioning system and encrypts them using the Paillier encryption algorithm:

<1>. Generate two random numbers:$r_{\mathrm{UX}} \in Z_N^*$,$r_{\mathrm{UY}} \in Z_N^*$.
<2>. Compute the ciphertext($C_{X_{\mathrm{U}}}, C_{Y_{\mathrm{U}}}$)of the real location ($X_{\mathrm{U}}$,$Y_{\mathrm{U}}$):

$$C_{X_{\mathrm{U}}} = (N+1)^{X_{\mathrm{U}}} \cdot r_{\mathrm{UX}}^{N} modN^2, C_{Y_U} = (N+1)^{Y_{\mathrm{U}}} \cdot r_{\mathrm{UY}}^{N} modN^2 \tag{4}$$

(C_{X_U}, C_{Y_U}) will be send to the cloud server S.

(3) The life service provider $\mathrm{P_i}$(i = 1, 2, …, m) encrypts its own location with Paillier encryption algorithm using the public key of user U to form a ciphertext ($C_{X_{\mathrm{P}_i}}, C_{Y_{\mathrm{P}_i}}$)of the real location coordinates($X_{\mathrm{P}_i}, Y_{\mathrm{P}_i}$), and sends it to the cloud server S.

(4) The cloud server calculates the ciphertext of the coordinate difference $\{X_U - X_{\mathrm{P}i}, Y_{\mathrm{U}} - Y_{\mathrm{P}i}\}$ between user U and $\mathrm{P_i}$ according to the Paillier encryption homomorphic property:

$$C_{(X_{\mathrm{U}}-X_{\mathrm{P}i})} = C_{X_{\mathrm{U}}} \cdot {C_{X_{\mathrm{P}_i}}}^{-1}, C_{(Y_{\mathrm{U}}-Y_{\mathrm{P}i})} = C_{Y_{\mathrm{U}}} \cdot {C_{Y_{\mathrm{P}_i}}}^{-1} \tag{5}$$

Generate a coordinate difference ciphertext pair $\{C_{(X_{\mathrm{U}}-X_{\mathrm{P}i})}, C_{(Y_{\mathrm{U}}-Y_{\mathrm{P}i})}\}_{i\in[1,m]}$, and send it to user U.

(5) The user decrypts $\{C_{(X_{\mathrm{U}}-X_{\mathrm{P}i})}, C_{(Y_{\mathrm{U}}-Y_{\mathrm{P}i})}\}_{i\in[1,m]}$ using his private key to decrypt $\{X_{\mathrm{U}} - X_{\mathrm{P}i}, Y_{\mathrm{U}} - Y_{\mathrm{P}i}\}_{i\in[1,m]}$.

(6) Euclid distance calculation formula can calculate the distance $d_{\mathrm{U},\mathrm{P}_i}$ between user U and life service provider P_i:

$$d_{\mathrm{U},\mathrm{P}_i} = \sqrt{(X_{\mathrm{U}} - X_{\mathrm{P}i})^2 + (Y_{\mathrm{U}} - Y_{\mathrm{P}i})^2}, i \in [1, m] \tag{6}$$

Since this paper only needs to find the minimum value, and does not care about the specific value, the above distance formula can be simplified into:

$$d_{\mathrm{U},\mathrm{P}_i} = |X_{\mathrm{U}} - X_{\mathrm{P}i}| + |Y_{\mathrm{U}} - Y_{\mathrm{P}i}|, i \in [1, m] \tag{7}$$

Find the minimum value in the set $\{d_{U,P_i}\} i \in [1, m]$, and let the number of life service provider corresponding to the minimum value be w.

(7) User U successfully matches with the recent life service provider P_W, and uses RSA public key encryption system for confidential communication. Set the RSA key pair of the user as (RPK_U, RSK_U) and the key pair of the nearest life server Pw as (RPK_{Pw}, RSK_{Pw}). The user sends the RPK_{Pw} encrypted message to P_W, and P_W decrypts it with RSK_{Pw}, and vice versa.

5 Security and Performance Analysis

From the above scheme algorithm, the scheme performance and security are influenced by the following two factors.

(1) Dummy locations. The construction of dummy locations not only ensures the security of users' real location, but also narrows the candidate range of life providers, reduces the communication volume and computation overhead, and improves the query efficiency. A larger number of dummy locations means a larger anonymization time and anonymization area needed to achieve anonymity, which of course has a higher security level, but an increase in computation and a decrease in performance. The performance can be evaluated in terms of both the time spent to generate dummy locations and the uniformity of the dummy locations distributed in the anonymization area. Experiments prove that as the number of dummy location K increases, the time increases, but the overall time required increases less and does not bring too much computational pressure on the user client. As the number K increases, the randomness is stronger and the uniformity of the location distribution tends to decrease, but it is always at a high level, and it is difficult for the attacker to narrow down the anonymous area and determine the real location of the user.
(2) Homomorphic encryption algorithm. In this scheme, there is no traditional LBS, but only a cloud server that helps with computing and communication. The cloud server obtains ciphertext from each party's location. It uses the characteristics of homomorphic encryption algorithm to help calculate the ciphertext of distance. It is impossible to obtain and save the user's real location, nor to obtain the location of life service providers, so it is impossible to infer the user's location and track, so as to ensure a high degree of privacy protection. Experimentally, Paillier homomorphism is relatively stable and consumes relatively less time in ciphertext operation, which is an advantage. In this scheme, the cloud server involves more ciphertext addition operations when calculating distance, which precisely takes the advantage of Paillier in ciphertext operations. However, since the work of encryption and decryption using homomorphic encryption algorithm is done by the user client, the current computation of homomorphic encryption is large, which is a test for the computing power of mobile terminals, and this is also the direction for further improvement.

6 Conclusion

Focusing on the features of LBS and the demands of user location privacy protest, we propose a nearest living service provider search scheme The scheme uses a K-anonymous dummy locations region construction algorithm to filter out life service providers within a certain distance, and then uses a homomorphic encryption algorithm to design a nearest life service provider search method for encrypted locations. Finally, we analyze the security and performance of the proposed scheme, and combine the experimental results to show that the scheme has good security performance and query efficiency. This scheme has extensive application significance and can be used as a business model applicable to any location-based service industry. However, the encryption and decryption computation overhead of the mobile client is a bottleneck for the application of this scheme and the establishment of a new business model. How to improve the encryption and decryption computation efficiency of the mobile client or design a tightly secure computation protocol while ensuring the existing security remains unchanged is the future research direction.

References

1. Zhang, X.J., Gui, X.L., Wu, Z.D.: Privacy preservation for location based services: a survey. Ruan Jian Xue Bao/J. Softw. **26**(9), 2373–2395 (2015). https://doi.org/10.13328/j.cnki.jos.004857. (in Chinese with English abstract). http://www.jos.org.cn/1000.9825/4857.htm
2. Gao, S., Ma, J.F., Yao, Q.S., Sun, C.: Towards cooperation location privacy-preserving group nearest neighbor queries in LBS. J. Commun. **36**(3), 146–154 (2015). (in Chinese with English abstract)
3. Niu, B., Zhang, Z., Li, X., et al.: Privacy-area aware dummy generation algorithms for location-based services. In: Proceedings of the 2014 IEEE International Conference on Communications (ICC), pp. 957–962. IEEE (2014)
4. Gong, Z., Sun, G.Z., Xie, X.: Protecting privacy in location-based services using k-anonymity without cloaked region. In: Proceedings of the 2010 11th International Conference on Mobile Data Management (MDM), pp. 366–371. IEEE (2010)
5. Bendlin, R., Damgård, I., Orlandi, C., Zakarias, S.: Semi-homomorphic encryption and multiparty computation. In: Paterson, K.G. (ed.) EUROCRYPT 2011. LNCS, vol. 6632, pp. 169–188. Springer, Heidelberg (2011). https://doi.org/10.1007/978-3-642-20465-4_11
6. Gruteser, M., Grunwald, D.: Anonymous usage of location-based services through spatial and temporal cloaking. In: Proceedings of the 1st International Conference on Mobile Systems, Applications and Services, San Francisco, pp. 31–42. ACM (2003)
7. Zhao, H., Yi, X., Wan, J.: Privacy-area aware all-dummy-based location privacy algorithms for location-based services. In: The International Conference on Artificial Intelligence and Computer Engineering (2016)
8. Gentry, C.: A fully homomorphic encryption scheme. Ph.D. thesis, Stanford University (2009)
9. Lin, H.-Y., Tzeng, W.-G.: An efficient solution to the millionaires' problem based on homomorphic encryption. In: Ioannidis, J., Keromytis, A., Yung, M. (eds.) ACNS 2005. LNCS, vol. 3531, pp. 456–466. Springer, Heidelberg (2005). https://doi.org/10.1007/11496137_31
10. Li, J., Wang, Q., Wang, C., et al.: Fuzzy keyword search over encrypted data in cloud computing. In: Proceedings of the Conference on Information Communications, pp. 441–445. IEEE Press (2010)

11. Wang, C.N., Li, C., Ren, M., Lou, W.K.: Privacy preserving multi-keyword ranked search over encrypted cloud data. IEEE Trans. Parallel Distrib. Syst. **25**(1), 222–233 (2014)
12. Yi, X., Paulet, R., Bertino, E., et al.: Practical approximate k nearest neighbor queries with location and query privacy. IEEE Trans. Knowl. Data Eng. **28**(6), 1546–1559 (2016)
13. Paillier, P.: Public-key cryptosystems based on composite degree residuosity classes. In: Stern, J. (ed.) EUROCRYPT 1999. LNCS, vol. 1592, pp. 223–238. Springer, Heidelberg (1999). https://doi.org/10.1007/3-540-48910-X_16

Terminology Extraction of New Energy Vehicle Patent Texts Based on BERT-BILSTM-CRF

Cheng Zheng, Na Deng(✉), Ruiyi Cui, and Hanhui Lin

School of Computer Science, Hubei University of Technology, Wuhan 430068, China
iamdengna@163.com

Abstract. Automatic extraction of domain terminology plays an important role in constructing domain knowledge graphs, translating domain documents and learning domain core knowledge. To improve the accuracy of terminology extraction in the field of New Energy Vehicles (NEV) and learn the core content of the NEV patent texts. Combining the BERT pre-training model with the BILSTM-CRF deep learning model, we propose a BERT-BILSTM-CRF based terminology extraction method for NEV patents. Experimenting on a self-built text dataset of 885 NEV patent abstracts, the proposed model achieves an accuracy of 88.62%, which is better than other deep learning terminology extraction models.

1 Introduction

New energy vehicles (NEV) are vehicles with advanced technical principles, modern technologies and new structures that use unconventional automotive fuels as power sources, combined with advanced technologies in automotive power control and driving. In recent years, people have been more willing to opt for NEV as a means of travel due to the volatile international situation affecting the normal functioning of the oil market, which has resulted in a continuous increase in fuel prices for fuel vehicles. In addition, coupled with the strong support of government policies and the increasing development of battery technology, the production of NEV in China has increased significantly. Based on Chinese NEV production data and statistics from 2014 to 2021, domestic NEV production grew from 83,900 units in 2014 to 3.545 million units in 2021.

While the production of NEV has increased greatly, the patents related to this field have also grown explosively. A search on China National Knowledge Internet (CNKI) with the keyword of "new energy vehicles" revealed that the number of NEV patent applications in 2014 was only 176, while the number of applications in 2021 is 8,142. Patents carry the results of various human inventions, are of great scientific value and industry core knowledge, and are one of the most crucial ways for ordinary researchers and companies to gain knowledge when inventing and creating. Patent documents contain all relevant documents in the process of patent application, examination and approval. As the number of patents in the field of NEV is increasing year by year, a timely understanding of the terminology in the NEV patent literature is helpful to learn the core knowledge in the field and to grasp the latest research hotspots and predict the trends in the NEV field.

L. Barolli (Ed.): EIDWT 2023, LNDECT 161, pp. 190–202, 2023.
https://doi.org/10.1007/978-3-031-26281-4_19

Patent terminology are binding symbols that express or qualify specialized concepts in words, and entities include terminology but are not limited to them [1]. The dense terminology, extensive domain, and frequent use of non-canonical names in patent documents are common difficulties in the English translation of patent documents [2], and the traditional translation of patent documents requires specialized domain translators, which is extremely demanding for translators and requires the use of domain glossaries to accurately translate the domain terminology in patent documents.

Knowledge graph is a hot topic of current research in natural language processing, which uses graphs to represent the relationships between knowledge, and entity extraction is the first step in the construction of knowledge graph. Compared to ordinary knowledge graph, domain knowledge graph is extremely domain specific, which makes ordinary entity extraction unable to extract certain specialized terminology with long sequences, in which case terminology extraction is particularly crucial.

Using deep learning methods to automatically extract domain terminology from patent documents plays a crucial role in patent retrieval, patent text translation and domain knowledge graph construction. With the progress of human society and the continuous development of the NEV industry, the number of patents in the field of NEV is increasing year by year, and it is difficult to meet the demand for terminology extraction in the field of NEV by traditional machine learning methods alone, and using deep learning methods to achieve domain terminology extraction has become the mainstream development trend.

For the task of terminology extraction in the NEV domain, the deep learning model constructed in this paper achieves the task with great accuracy. The main contributions of this paper are as follows.

(1) A corpus of patent texts in the field of NEV is constructed, and a method for processing annotated documents is designed.
(2) We constructed the BERT-BILSTM-CRF model for terminology extraction in the NEV patent domain.

2 Related Work

Terminology extraction is an essential and fundamental task in natural language processing, and numerous domestic and international scholars have used traditional methods to extract terms, such as rule- and lexicon-based methods, statistics-based methods, and traditional machine learning algorithms.

In 2012, Pan [3] proposed to construct external and internal rules and perform Chinese entity recognition by probabilistic statistics, and experimented entity recognition on the People's Daily corpus, and its average F-value reached 80%. In 2014, Yan et al. [4] developed 152 entity recognition rules on Vietnamese language and experimented on Vietnamese political and economic corpus, and its accuracy rate reached more than 90%. In 2015, Xu et al. [5] first extracted the entities by terminology table to subdivide the text and lexical annotation, and then identified medical entities according to the customized language rules. Medical entities such as diseases, symptoms, and causative agents can be well identified in drug instructions using this method. In 2016, Fan et al.

[6] fused language rules and statistical strategies, firstly, counted domain text word frequencies, took out words with frequencies greater than a set threshold, extracted candidate terminology lists based on left-right information entropy expansion with the words as the center, and finally calculated the TF-IDF value of each terminology, and remove the terminology and generic words that are smaller than the set threshold to get the final terminology list. This method achieves 84.33% accuracy in terminology extraction in computer domain literature. In 2016, Jiang et al. [7] first extracted the terminology lexical collocation template through the terminology database to extract the candidate terminology, and then used the samples that satisfy the extracted features to train the SVR model, and used the model to calculate each candidate terminology becomes terminology and set a threshold to extract the terminology, this method treats terminology extraction as a problem of predicting the probability of a word becoming terminology, and achieves great results in terminology extraction in the library intelligence domain.

Domain terminology with fewer words can be cut accurately in sentences, while domain terminology with large word lengths are easily cut incorrectly. In 2017, Liu et al. [8] integrated terminology length and grammatical features and proposed a word length ratio concept to improve the weight of domain terminology with large word length by first extracting the set of candidate terminology using a machine learning approach combined with constraint rules and language rules, and then this method achieves an accuracy of 89.5% in the extraction of terminology in the domain of "computer virus".

In 2018, Xu et al. [9] automatically annotated the text of food safety events based on the deep learning model BILSTM-CRF, which is important for improving the supervision of food safety events. In 2022, Chen et al. [10] obtained word vectors and sentence vectors with contextual semantics through the training of the BERT model, took the average of the word vectors to obtain the entity vectors, and after attention mechanism combines sentence vectors and entity vectors, and finally the optimal result is obtained by sequence annotation from conditional random fields, which effectively improves the efficiency of entity extraction in the process of knowledge graph construction.

3 Model Building

3.1 BERT Pre-training Model

Pre-training divides the training task into two steps, that is, common learning and feature learning. Common learning is the training of models with common features through huge low-cost training data. In feature learning, only a small amount of special domain data is required to fine-tune the model, thus reducing the learning burden for specific tasks such as entity recognition, text classification, and others. In 2018 Jacob Devlin et al. [11] proposed the BERT pre-training model. Due to the strong feature extraction capability of transformer, BERT uses the encoder layer of Transformer model for feature extraction, uses a pre-training and fine-tuning training model with masked language model (MLM) and next sentence prediction (NSP) to learn word-level and sentence-level features, is trained and tested by fine-tuning in different downstream tasks, and it achieves excellent results in 11 natural language processing tasks.

The powerful feature extraction capability of Transformer is based on its unique multi-headed attention mechanism. The main effect of the attention mechanism is that more attention can be allocated for keywords and less attention for other parts [12]. The computational flow of attention mechanism is shown in Fig. 1.

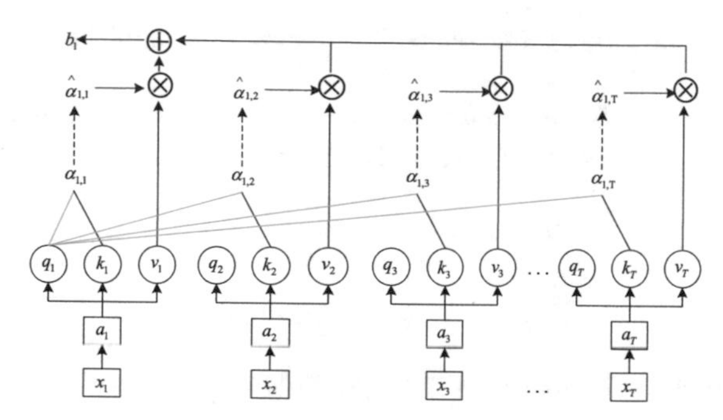

Fig. 1. Transformer's attention mechanism.

The input sequence data is $(x_1, x_2, x_3, \ldots, x_T)$ represents the vector corresponding to the first word of a sentence. After processing this vector sequence through the embedding layer, we get $a_1, a_2, a_3, \ldots, a_T$ and then multiply the three trainable parameter matrices w^q w^k w^v with each of them to get $q_i, k_i, v_i, i \in (1, 2, 3, \ldots, T)$, compute the vector dot product with q_1 in $k_1, k_2, k_3, \ldots, k_T$ respectively to get $\alpha_{1,1}, \alpha_{1,2}, \alpha_{1,3}, \ldots, \alpha_{1,T}$, then processed by the SoftMax function to obtain the values of attention weights $\widehat{\alpha}_{1,1}, \widehat{\alpha}_{1,2}, \widehat{\alpha}_{1,3}, \ldots, \widehat{\alpha}_{1,T}$ with the value domain between [0, 1], Multiply this set of attention weights with the corresponding values of v_1, v_2, v_3, ..., v_T. Finally, sum up to get the final output b_1. x_2 corresponding to b_2 is the same as the above processing flow. "Multi-head" refers to the multiplication of multiple parameter matrices w^q w^k w^v with x_i to obtain multiple $q_i, k_i, v_i, i \in (1, 2, 3, \ldots, T)$ and b^i_{head}, in which i is the number of "Multi-head". The formula for calculating attention is shown in Eq. (1). d_k is the dimension of vector K [13].

$$Attention(Q, K, V) = softmax\left(\frac{QK^T}{\sqrt{d_k}}\right)V. \quad (1)$$

3.2 BILSTM Layer

In natural language processing tasks, especially entity extraction and terminology extraction, the Bi-directional Long Short-Term Memory (BILSTM) is widely used, which consists of a combination of forward and backward Long Short-Term Memory (LSTM). LSTM is a temporal recurrent neural network that addresses the long-term dependency problem that exists in general recurrent neural networks. LSTM is composed of the forget gate F_t, input gate I_t, and output gate O_t. The LSTM cell structure is shown in Fig. 2.

Firstly, the forget gate F_t decides which information to discard from the cell state C_{t-1} based on the previous output h_{t-1} and the current input X_t via the sigmoid function, as shown in Eq. (2).

$$F_t = \sigma\left(X_t w_{xf} + h_{t-1} w_{hf} + b_f\right) \tag{2}$$

σ is a sigmoid function. w_f is the connection weight matrix of the forget gate F_t, b_f is the offset value of the forget gate F_t.

Next, the input gate I_t and $\tan h$ function will determine what new information is stored and update the cell status C_{t-1}, as shown in Eq. (3) and Eq. (4).

$$I_t = \sigma(X_t w_{xi} + h_{t-1} w_{hi} + b_i) \tag{3}$$

$$C_t^{\sim} = \tan(X_t w_{xc} + h_{t-1} w_{hc} + b_c) \tag{4}$$

w_i is the connection weight matrix of the input gate I_t. b_i is the offset value of the input gate I_t.

Later on, the output gate determines which information will be output, as shown in Eq. (5).

$$O_t = \sigma(X_t w_{xo} + h_{t-1} w_{ho} + b_o) \tag{5}$$

Finally, the cell state C_t is processed by the $\tan h$ function to control its value between $[-1, 1]$, and then multiplied by O_t to obtain the final output h_t, as shown in Eq. (6).

$$h_t = O_t \tan(C_t) \tag{6}$$

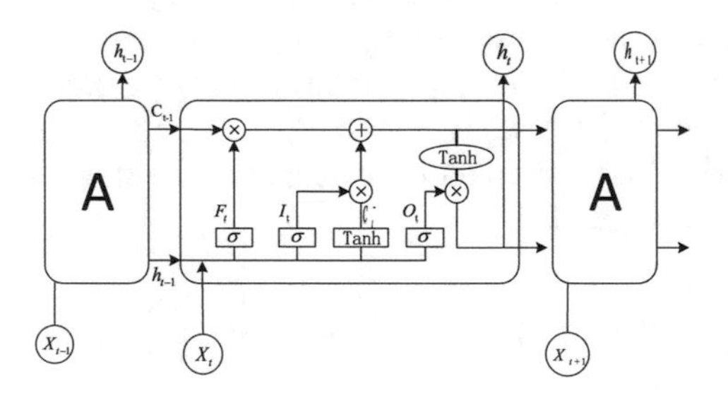

Fig. 2. LSTM cell structure [14].

However, unidirectional LSTM cannot reverse the extraction of sentence features, which makes it unable to better extract semantic information. By constructing forward and backward LSTM to perform forward and backward feature extraction on sentences, the BILSTM can better capture semantic dependencies in both directions.

3.3 Conditional Random Field

The Conditional Random Field (CRF) was proposed by Lafferty et al. in 2001 as an undirected graph model for sequence labeling in machine learning [15]. In recent years, it has been widely used in the fields of word separation, lexical annotation, and named entity recognition. CRF is a special case of Markov random field. Let X and Y be random variables. $P(Y|X)$ is the conditional probability distribution of the random variable Y given the random variable X. If the random variable Y constitutes a Markov random field represented by an undirected graph $G = (V, E)$. Satisfying Eq. (7) holds for any node v. Then the conditional probability distribution $P(Y|X)$ is said to be the conditional random field where $w \sim v$ denotes all nodes w that have connections to v nodes in graph $G = (V, E)$.

$$P(Y_v|X, Y_w, w \neq v) = P(Y_v|X, Y_w, w \sim v) \tag{7}$$

The CRF model can take into account the relationship between the tags, and can avoid errors such as using $I - TERM$ as the starting tag or $E - TERM$ as the intermediate tag. By defining the set of feature functions, let each feature function score a set of annotated sequences and then combine the scores of all feature functions on the same set of annotated sequences. The selected best score is the final score of this set of annotated sequences. As shown in Fig. 3, the annotation sequence $B-TERMI-TERMI-TERME-TERM$ with the highest score of "电动汽车" in CRF level is its final output.

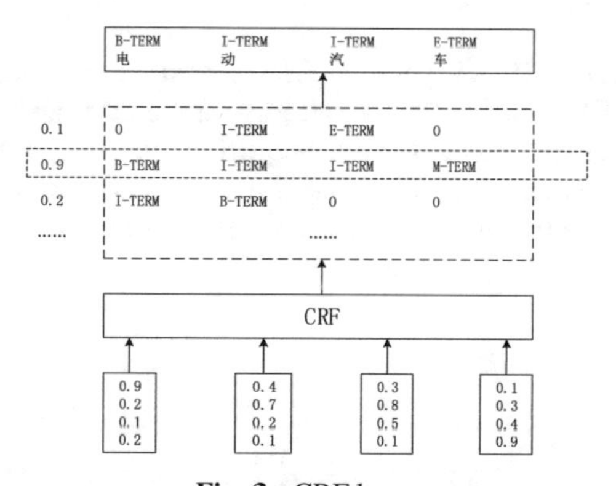

Fig. 3. CRF layer.

3.4 The Proposed Model

The overall framework of the terminology extraction model in this paper is shown in Fig. 4. The model is divided into three layers. The first layer is the BERT pre-trained model with the input of the annotated character-level corpus. The BERT learns syntactic and semantic information of the sentences. Transform each labeled character into a low-dimensional word vector, which is then fed into the second layer BILSTM. The sentence features are automatically extracted by BILSTM, which outputs a score for each character corresponding to each tag. Finally, the optimal sequence is output by the third layer of CRF, which takes into account the dependencies between tags.

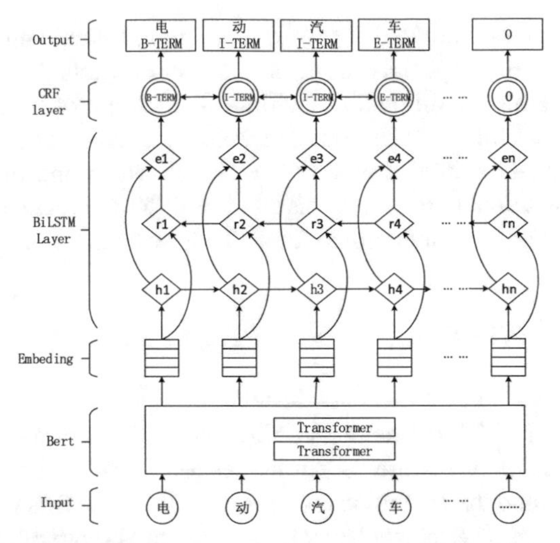

Fig. 4. BERT-BILSTM-CRF new energy vehicle patent terminology extraction model.

4 Experiment

In this paper, we apply the BERT-BILSTM-CRF model for terminology extraction from patent texts in the NEV domain. The overall flow of the experiment is shown in Fig. 5. The main steps of the experiment include acquiring text data, cleaning and manually annotating the acquired text data using the Baidu EasyData platform, processing the JSON files, dividing the training set into test set and validation set, and finally training and evaluating the terminology extraction model.

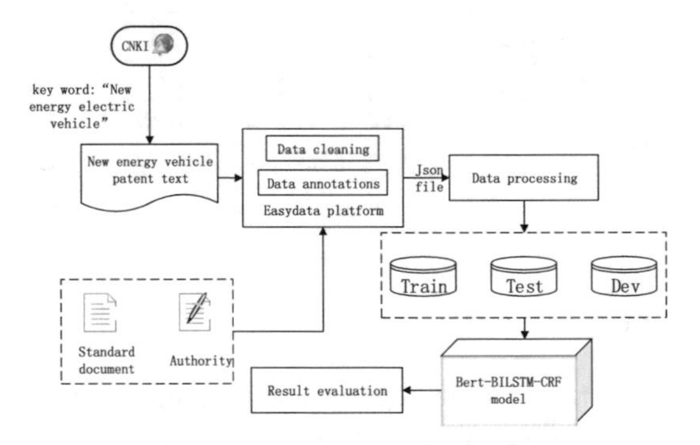

Fig. 5. The overall flow chart of the experiment.

4.1 New Energy Vehicle Patent Text Corpus Construction

The data used in the experiment were downloaded from the CNKI, and the NEV patents were searched with "new energy vehicle" as the keyword, and the retrieved NEV patent texts were downloaded and pre-processed to remove some irrelevant patent texts and extract the abstract texts from the patent texts. Due to the mixed English and Chinese abbreviations in the abstract, various symbols in the abstract were standardized before annotation. After the above processing, we obtained 885 abstract texts in the NEV domain and imported them into the Baidu EasyData platform for manual classification and terminology annotation. Tagged terminology should be domain specific, unique, and concise. The rules of labeling refer to the authoritative documents "GB/T 19596-2017 Electric Vehicle Terminology" [16] and "GB/T 24548-2009 Fuel Cell Electric Vehicle Terminology" [17] issued by the Chinese National Standardization Administration Committee. A sample label display is shown in Fig. 6.

本发明公开了一种**新能源电动汽车电池快速连接器**,涉及**新能源电动汽车**相关领域,为解决现有技术中的**电动汽车**的**电池**与**电池仓**通过导电体的连接接触实现电路的连接,而插头插入插座后在外力作用下容易出现连接位置的松动,影响**电池**与**电池仓**的电性连接,而采用螺钉等传统固定方式的连接耗时长,不能快速连接的问题。

The invention discloses a **new energy electric vehicle battery quick connector**, which relates to the related field of **new energy electric vehicles**. In order to solve the problem that in the prior art, **the battery of the electric vehicle** and **the battery compartment** can realize the connection of the circuit through the connection contact of the conductive body, and after the plug is inserted into the socket, it is easy to loose the connection position under the effect of external force, which affects the electrical connection between **the battery** and **the battery compartment**. However, the connection using traditional fixing methods such as screws takes a long time, Can't connect quickly.

Fig. 6. Sample terminology labeling display.

After sorting and labeling the above information, we obtain 16434 terms in the NEV domain. The dataset is divided into training set, validation set and test set according to 8:1:1, where the training set contains 13251 terms, the validation set contains 1597 terms, and the test set contains 1555 terms. The details of the dataset are shown in Table 1.

Table 1. Dataset information.

Category	Number of sentences	Number of terms
Training set	1400	13251
Validation set	167	1597
Test set	184	1555

4.2 Annotation Text Processing Flow

To better process NEV patent texts and annotate them through Baidu EasyData platform, this paper designs a method to process annotated documents. First, read the JSON file exported from EasyData. The internal structure of the JSON file is shown in Fig. 7. Writing the "tag", "offset" and "span" corresponding to each abstract text in the JSON file into the annotation_file. "content" is written to the text_file file, and the abstract text is automatically annotated by Algorithm 1 to get the final annotated text. In this paper, we use the BIEO quadratic labeling method. B-TERM indicates the first word of the terminology. I-TERM indicates the non-first non-last word of the terminology. E-TERM indicates the last word of the terminology. O indicates the non-terminology words and various symbols. The labels are shown in Table 2 with a space separating the word from the label and a line break separating the sentences.

[{"content": "本发明涉及电动汽车技术领域。目的是提供一种简单可靠的新能源电动汽车剩余里程估算方法。采用的技术方案是：该估算方法包括剩余里程S'预设、计算单位SOC可行驶里程K值和计算剩余里程S'。所述的剩余里程S'预设是根据前次存储的单位SOC可行驶里程K值，读取和记录车辆初始的总里程信息S0及SOC0，计算出剩余里程S'的预设值。所述的计算单位SOC可行驶距离K值，是通过间隔读取里程信息S值和SOC值，将累计记录十次的有效变化值相加再相除得到的。本发明简单可靠，能够在占用较少计算资源的前提下满足基本使用需求，并警示驾驶员及时充电，有效降低了因电量耗尽而半路抛锚的概率。", "records": [{"span": "电动汽车技术领域", "offset": [5, 12], "tag": "TERM"}, {"span": "新能源电动汽车", "offset": [26, 32], "tag": "TERM"}]}, {"content": "本实用新型公开了一种新能源电动汽车空调压缩机安装支架，涉及空调压缩机用安装支架领域，包括第一支架，第一支架的顶部通过安装螺栓安装有第二支架。本实用新型通过设置有导热垫、减震弹簧、散热片和散热槽，当汽车行驶并发生颠簸时，空调压缩机会产生振动，进而使得导热垫受力形变，导热垫受力后会挤压减震弹簧，并使减震弹簧受力压缩，从而可对空调压缩机产生的振动力进行缓冲，避免振动力过大而造成空调压缩机内部电子元件的损坏，同时提高了对空调压缩机的保护效果。同时空调压缩机工作所产生的热量会经导热垫传导至散热片出，并经散热片传导至散热槽处，从而可经散热槽散出，并提高了空调压缩机的散热效率。", "records": [{"span": "新能源电动汽车空调压缩机安装支架", "offset": [10, 25], "tag": "TERM"}, {"span": "空调压缩机用安装支架领域", "offset": [29, 40], "tag": "TERM"}, {"span": "支架", "offset": [46, 47], "tag": "TERM"}, {"span": "支架", "offset": [51, 52], "tag": "TERM"}, {"span": "支架", "offset": [67, 68], "tag": "TERM"}, {"span": "导热垫", "offset": [80, 82], "tag": "TERM"}, {"span": "减震弹簧", "offset": [84, 87], "tag": "TERM"}, {"span": "散热片", "offset": [89, 91], "tag": "TERM"}, {"span": "散热槽", "offset": [93, 95], "tag": "TERM"}, {"span": "空调压缩机", "offset": [109, 113], "tag": "TERM"}, {"span": "导热垫", "offset": [124, 126], "tag": "TERM"}, {"span": "导热垫", "offset": [132, 134], "tag": "TERM"}, {"span": "减震弹簧", "offset": [141, 144], "tag": "TERM"}, {"span": "减震弹簧", "offset": [148, 151], "tag": "TERM"}, {"span": "空调压缩机", "offset": [161, 165], "tag": "TERM"}, {"span": "空调压缩机", "offset": [187, 191], "tag": "TERM"}, {"span": "空调压缩机", "offset": [208, 212], "tag": "TERM"}, {"span": "空调压缩机", "offset": [221, 225], "tag": "TERM"}, {"span": "导热垫", "offset": [236, 238], "tag": "TERM"}, {"span": "散热片", "offset": [242, 244], "tag": "TERM"}, {"span": "散热片", "offset": [249, 251], "tag": "TERM"}, {"span": "散热槽", "offset": [255, 257], "tag": "TERM"}, {"span": "散热槽", "offset": [264, 266], "tag": "TERM"}, {"span": "空调压缩机", "offset": [274, 278], "tag": "TERM"}]}, {"content": "本发明公开了一种新能源电动汽车蓄电池保护装置，包括铁丝笼、导电片、弹簧安装孔、通气孔、调节螺杆和

Fig. 7. JSON file internal structure.

Algorithm 1.

```
Input: annotation_file, text_file
Output: BIEO-tagged format file
1.for each line in text_file do
2.   word_list ← split line with "\n"
3.end for
4.for each line in annotation_file do
5.   for word in word_list do
6.          ann_list ← split line with "\r\n"
7.          start = ann_list[1]
8.          end = ann_list[2]
9.          outfile ← word[start] + B-TERM
10.         for i in range((start +1), end) do
11.                outfile ← word[i] + I-TERM
12.         end for
13.         outfile ← word[end] + E-TERM
14.         outfile ← "\n"
15.  end for
16.end for
```

Table 2. New energy vehicle patent text labeling sample.

本 O	池 I-TERM	· O	种 O	汽 I-TERM
新 O	包 I-TERM	且 O	新 B-TERM	车 I-TERM
型 O	技 I-TERM	公 O	能 I-TERM	电 I-TERM
涉 O	术 I-TERM	开 O	源 I-TERM	池 I-TERM
及 O	领 I-TERM	了 O	电 I-TERM	包 E-TERM
电 B-TERM	域 E-TERM	一 O	动 I-TERM	。 O

4.3 Experimental Parameter Setting

The specific settings of the hyperparameters for this experiment are given in Table 3.

Table 3. Parameter settings.

Parameters	Value
BERT model	Bert-base-Chinese
Bert hidden layer dimension	768
Transformer number of layers	12
Attention-head	12
BERT learning rate	3e−5
CRF learning rate	2e−2
LSTM hidden layer dimension	128
Dropout	0.5
Optimizer	Adam

4.4 Experimental Results and Analysis

Three metrics are used to measure the terminology extraction task: Precision (P), Recall (R), and F1 value. Their calculation formulae are as follows.

$$P = \frac{Identify\ the\ correct\ number\ of\ terms}{Total\ number\ of\ terms\ identified} \times 100\% \tag{8}$$

$$R = \frac{Identify\ the\ correct\ number\ of\ terms}{Total\ number\ of\ terms\ labeled} \times 100\% \tag{9}$$

$$F1 = \frac{2 \times P \times R}{P + R} \times 100\% \tag{10}$$

4.4.1 Experimental Results

To validate the effectiveness of the BERT-BILSTM-CRF model for terminology extraction from patent texts in the NEV domain, the following eight groups of models are designed for comparison in this experiment: BERT, CRF, BILSTM-CRF, BERT-CRF, BERT-BILSTM, BERT-BILSTM-CRF, IDCNN-CRF, BERT-IDCNN-CRF. The experimental results are shown in Table 4.

Table 4. Experimental results.

No.	Model	Evaluation indicators		
		Precision (P)	Recall (R)	F1 value
1	BERT	0.8482	0.7794	0.8124
2	CRF	0.7791	0.6534	0.7107
3	BILSTM-CRF	0.8328	0.7300	0.7780
4	BERT-CRF	0.8583	0.7784	0.8164
5	BERT-BILSTM	0.8530	0.7796	0.8146
6	BERT-BILSTM-CRF	**0.8862**	**0.7920**	**0.8364**
7	IDCNN-CRF	0.8233	0.7494	0.7846
8	BERT-IDCNN-CRF	0.8509	0.7759	0.8117

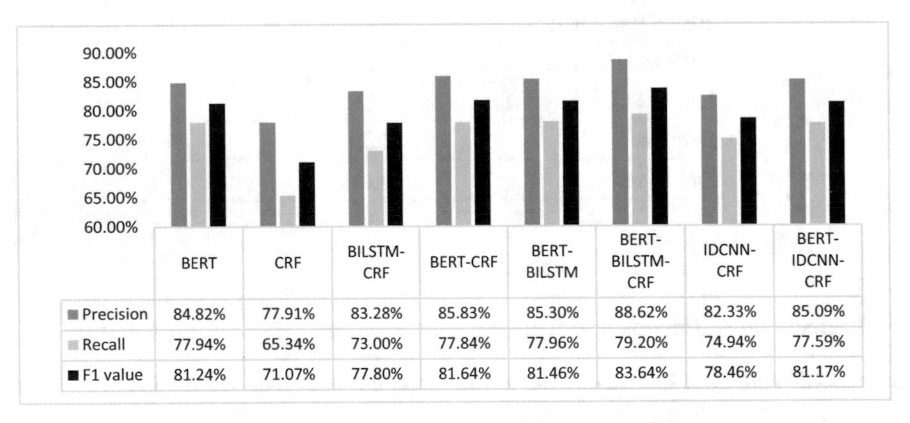

Fig. 8. Experimental results of different models.

4.4.2 Experimental Analysis

As shown in Fig. 8, the Precision (P) of the BERT-BILSTM-CRF model trained in this paper reached 88.62% for the extraction of patent text terminology in the NEV domain. Compared with the BILSTM-CRF model, the Precision (P) is improved by 3.32% and the F1 value is improved by 2.18%. It shows that using the BERT pre-training model can better learn the semantic information in the text and more accurately learn the relationship between terminology and non-terminology words. Compared with the BERT-BILSTM model, the Precision (P) is improved by 3.32% and the F1 value is improved by 2.18%. It shows that the constraint between tags after CRF layer can effectively reduce the number of incorrect tags and improve the Precision (P). The IDCNN-CRF model is able to achieve parallel computation compared to the BILSTM-CRF model, which can improve the model training speed but decrease the accuracy rate, while BILSTM-CRF performs better regardless of the training time.

5 Conclusion

To automatically extract domain terminology from NEV patent texts, we construct a corpus of NEV patent texts and propose a BERT-BILSTM-CRF model. The annotated characters are transformed into low-latitude word vectors using BERT, and then BILSTM-CRF is used to obtain sentence semantic features, identify and extract domain terminology. The next work focuses on expanding the patent text dataset, developing a thesaurus for the NEV domain, and continuing to optimize the model presented in this paper to make its extraction results more accurate and generalizable.

References

1. Li, J.Q., Li, B.A.: Computing similarity of patent terms based on knowledge graph. Data Anal. Knowl. Disc.. **4**(10), 104–112 (2020)
2. Xu, P.W., Leng, B.B.: Common difficulties and practical strategies for English translation of patent literature terms. Chin. Sci. Technol. Transl. J. **32**(04), 28–31 (2019)
3. Pan, Z.G.: Research on the recognition of Chinese named entity based on rules and statistics. Inf. Sci. **30**(05), 708–712 (2012)
4. Yan, D.H., Bi, Y.D.: Rule-based recognition of Vietnamese named entities. J. Chin. Inf. Process. **28**(05), 198–205+214 (2014)
5. Xu, H., Liu, M.F.: Disease and bacteria entity extraction based on linguistic rule. J. Wuhan Univ. (Nat. Sci. Ed.) **61**(02), 151–155 (2015)
6. Fan, M.J., Duan, D.S.: Domain-specific terms extraction algorithm based on combination of statistics and rules. Appl. Res. Comput. **33**(08), 2282–2285+2306 (2016)
7. Jiang, T., Sun, J.J.: Research on automatic Chinese domain terminology extraction based on SVR model - for library intelligence domain. Inf. Stud. Theory Appl. **39**(01), 24–31+15 (2016)
8. Liu, L., Xiao, Y.Y.: A statistical domain terminology extraction method based on word length and grammatical feature. J. Harbin Eng. Univ. **38**(09), 1437–1443 (2017)
9. Xu, F., Ye, W.H.: Part-of-speech automated annotation of food safety events based on BiLSTM-CRF. J. China Soc. Sci. Tech. Inf. **37**(12), 1204–1211 (2018)
10. Chen, W., Zhang, R., Yin, Z.: Knowledge graph entity extraction method based on BERT model and entity vector. J. Chin. Comput. Syst. **43**(08), 1577–1582 (2022)
11. Devlin, J., et al.: BERT: pre-training of deep bidirectional transformers for language understanding. arXiv preprint arXiv:1810.04805 (2018)
12. Ma, J.H., Zhang, Y.M., Yao, S.: Terminology extraction for new energy vehicle based on BLSTM_attention_CRF model. Appl. Res. Comput. **36**(05), 1385–1389+1395 (2019)

13. Vaswani, A., et al.: Attention is all you need. In: Advances in Neural Information Processing Systems 30 (2017)
14. Shi, M., Huang, J., Li, C.: Entity relationship extraction based on BLSTM model. In: 2019 IEEE/ACIS 18th International Conference on Computer and Information Science (ICIS). IEEE (2019)
15. Li, Z.W., Ding, D., Li, C.W.: Chinese word segmentation method for short Chinese text based on conditional random fields. J. Tsinghua Univ. (Sci. Technol.) **55**(08), 906–910+915 (2015)
16. GB/T 19596-2017, Electric Vehicle Terminology. General Administration of Quality Supervision, Inspection and Quarantine of the People's Republic of China, Beijing. Standardization Administration of the People's Republic of China (2017)
17. GB/T 24548-2009, Fuel Cell Electric Vehicle Terminology. General Administration of Quality Supervision, Inspection and Quarantine of the People's Republic of China, Beijing. Standardization Administration of the People's Republic of China (2009)

Conceptual Paper of Environmental Disclosure and Financial Performance: The Role of Environmental Performance

Luluk Muhimatul Ifada[1(✉)], Naila Najihah[1], Farikha Amilahaq[1], and Azizah Azmi Khatamy[2]

[1] Faculty of Economics, Universitas Islam Sultan Agung, Semarang, Indonesia
{luluk.ifada,naila.najihah,farikha}@unissula.ac.id
[2] Student of Master of Accounting, Faculty of Economics, Universitas Islam Sultan Agung, Semarang, Indonesia
azizahazmi05@std.unissula.ac.id

Abstract. This research examines the company's environmental responsibility in responding to current environmental issues. The regulations from stakeholders and the increasing public awareness of pollution are new challenges for companies. Therefore, companies must increase their environmental responsibility and pay more attention to their business activities. This research aims to identify the role of environmental disclosure on the company's financial and environmental performance. In addition, this study also observes the role of environmental performance in affecting environmental disclosure and financial performance. This research used an exploratory quantitative approach. The population was large companies listed on the Indonesia Stock Exchange (IDX) from 2006–2020. The method used purposive sampling with secondary data. This research is still in the form of a conceptual paper. Therefore, further research is needed to increase the company's knowledge and competence to respond to environmental issues.

Keywords: Environmental disclosure · Environmental performance · Financial performance

1 Introduction

Global warming has been one of the central issues of the world since the end of the twentieth century. The United Nations Framework Convention on Climate Change (United Nations Conference Declaration on the Human Environment, 1992, p. 3) defines global warming as "a climate change which is attributed directly or indirectly to human activity that alters the composition of the global atmosphere [1]. Another issue related to the environment is the preservation of the environment and natural resources. Environmental issues are growing rapidly. The initial concerns were pollution, wilderness conservation, population growth, and depletion of natural resources. These concerns have merged with energy supply, biodiversity, species extinction, climate change, and other disturbances to the Earth's systems [2]. United Nations Framework Convention on Climate Change

L. Barolli (Ed.): EIDWT 2023, LNDECT 161, pp. 203–214, 2023.
https://doi.org/10.1007/978-3-031-26281-4_20

(1972) states that "Earth's natural resources, including water, water, soil, flora and fauna and especially representative samples of natural ecosystems, must be protected appropriately for the benefit of present and future generations through careful planning or management. Environmental issues continue to be in the spotlight and current discussion. In the business world, there is an increase in business in terms of environmental management and preservation [3].

To overcome climate change, stakeholders such as governments, international associations, and other related stakeholders immediately require companies' involvement in preserving the environment through rules and regulations [4]. These issues will affect the company in conducting its daily operations, at the same time, society will also become more sensitive to the pollution caused by the company [5]. As a result of getting much pressure from its stakeholders, the private industry needs to be responsible for the impact of its business activities on the community. Along with environmental damage and more serious environmental problems, stakeholders are paying attention to the company's environmental responsibility [6, 7].

Environmental disclosures should cover key environmental issues and their impact on the company's future performance and position, risks and uncertainties, and policies on significant environmental issues such as emissions trade [8]. Organizations must report emissions trading schemes and include greenhouse gas emission reports for direct emission calculations, such as combustion of fuel in boilers, and indirect emissions such as waste disposal, disclosure of energy expenditures, and direct energy reporting, e.g., oil and coal. Moreover, the indirect energies are electricity purchases, reporting the amount of water taken, reporting the amount of waste, reporting on water use policies, and more. They should also explain how environmental damage can affect tangible and intangible assets [9].

Environmental performance is an essential topic to discuss due to climate change, global warming, and environmental damage caused by the production process. It creates many changes in manufacturing technology to understand environmental care or environmental awareness [10]. Businesses depend on natural and human resources. When managers struggle to compete in a global economy, they must perform within social constraints characterized by increasing environmental accountability. This accountability includes high public scrutiny of a company's environmental performance and its public performance disclosure. The elements of corporate environmental accountability affect the company's profitability [5]. Research on environmental disclosure, environmental performance, and financial performance by [5] found a positive relationship between environmental performance and financial performance and a relationship between environmental disclosure and environmental performance.

Legitimacy also certainly affects the company in a way that makes them strive to optimize its environmental performance and disclose its environment. Therefore, companies are required to voluntarily report activities if management believes that these activities are expected by the communities where they operate [11]. Research from [12] also emphasised that corporate social disclosure is motivated by the company's necessity to legitimize its activities. In other words, environmental performance and environmental disclosure are very important for companies to gain, maintain, and enhance their legitimacy status in front of stakeholders. Based on legitimacy theory, companies apply environmental practices related to environmental performance, financial performance, and environmental disclosure. The higher the financial performance of the company, the more environmental disclosure will be released if good environmental performance strengthens it.

Research from [13] stated that environmental disclosure affects how the environmental performance and its impact will be a company risk. According to [14], companies can gain legitimacy by providing environmental disclosures. In addition, participating in external environmental performance assessments is another way for companies to gain legitimacy. The premise is that companies with an adequate level of environmental disclosure have more opportunities and may have a tendency to raise higher environmental performance. Several studies found a positive correlation between environmental performance and the environmental disclosure [1]. While others show a negative correlation [4].

Previous studies have shown many results on the relationship between a company's financial performance and the level of environmental disclosure. Larger companies tend to provide more comprehensive information about their environmental activities and are more visible to their external audiences and stakeholders [15]. Therefore, large companies can improve their reputation by communicating their environmental disclosures to the public [14].

Research from [16] explained that companies with high environmental performance are determined to maintain investors and other stakeholders through more voluntary environmental disclosures compared to companies with lower environmental performance. From these findings, financial and environmental performance are the key factors that determine the extent of environmental disclosure.

This research provides an integrated analysis of the overall management strategy that affects (1) environmental disclosure, (2) environmental performance, and (3) financial performance. Environmental accounting has an important role in the financial performance of a company. Financial performance is presented in the form of financial variables that will be associated with environmental performance variables and environmental disclosures. Meanwhile, environmental performance is presented in the form of environmental variables which will be associated with environmental disclosure variables and financial performance. The environmental disclosure variables will be associated with the financial performance variables and intervening variables in environmental influences of financial performance.

This research differs from previous research in several dimensions. The difference is located in the data and variable measurement, which examine the relationship between the variables used. This difference strengthens previous research, which only tested the relationship between variables because it tested the effect and added that the intervening variable used in this research was environmental performance. It is expected to add insight to report users. The contribution of this research is expected to be helpful for the company and to increase its awareness of environmental management, which has only focused on the company's short-term profits, regardless of the environmental damage that will occur. The findings of this research are expected to assist in the decision-making process related to environmental disclosure initiated by companies, investors, and regulators. Furthermore, our results are expected to enrich knowledge related to environmental disclosure. This research also attempts to answer the following research questions:

RQ1. How does environmental disclosure affect financial performance?
RQ2. How does environmental disclosure affect environmental performance?
RQ3. How does environmental performance affect the relationship between environmental disclosure and financial performance?

Based on the description above, the researchers are interested in conducting this research because of the differences in the results from previous studies. The researcher also measured the effect of environmental disclosure on financial performance and wanted to identify how environmental performance can mediate the impact of environmental disclosure on financial performance.

2 Literature Review

Legitimacy Theory
Legitimacy theory is based on "the general perception or assumption that the actions of an entity are desirable, appropriate, or conforms in some socially constructed system of norms, values, beliefs, and definitions" [12]. Based on the legitimacy theory perspective, companies prefer to disclose more environmental information to gain legitimacy [17]. Investors are more likely to buy company stock that discloses more environmental information due to the low legitimacy risk and high transparency of environmental information. According to the legitimacy theory of Dowling and Pfeffer (1975), corporate social disclosure is one of the ways companies respond to political and public pressures and corporate activities must be fully in line with the goals of the general public. Legitimacy is essential because it affects not only how stakeholders perceive and understand a company but also how they react to the organization of the company. If a company's corporate activities deviate from socially recognized values, social legitimacy will be threatened. On the other hand, the legitimacy theory [2] proposed that corporate environmental disclosure and social information are a function of social and political pressures. Companies are becoming more concerned about disclosing information as they face more social and political pressures. Therefore, the theory explains that companies with poor environmental performance face more public pressure. In such a scenario, actors tend to make greater and more positive disclosures of environmental information to avoid the threat of bad legitimacy. This means that environmental disclosure and performance are very important for companies to gain, maintain and enhance their legitimacy status in front of stakeholders. Based on legitimacy theory, companies apply environmental practices related to environmental performance, financial performance, and environmental disclosure. Lindblom (1994) in [11] shows that companies adopt one of three disclosure strategies to legitimize their actions. Companies may try to: (1) inform the public; (2) change their perception; or (3) distract stakeholders from their corporate actions.

3 Hypothesis Development

Environmental Disclosure and Financial Performance

Companies require to do activities that can be used to demonstrate their responsibilities to stakeholders, one of them is by providing environmental disclosures [13]. Building a relationship between environmental disclosure and environmental performance is important from a social responsibility perspective in a positive relationship [5]. Environmental disclosure leads to an asymmetry reduction of information between the company and external stakeholders, which enhances the reputation and brand firm value, enabling it to obtain more financing opportunities, reducing costs financing and increasing its value [15]. The company must disclose environmental information in accordance with guidelines that can limit behaviour-selective corporate disclosure (greenwashing) and ensure objective and fair disclosure of information [5].

Companies are forced to increase their environmental spending and invest limited resources in environmental protection, though they can improve environmental performance rather than growth in the financial performance [18]. Companies with environmentally responsible practices are more likely to generate positive perceptions from stakeholders, leading to superior financial performance [19]. Environmental disclosure is a company's medium to describe its environmental performance, so the existence of this disclosure is an important factor in ensuring the sustainability of the company's environmental performance program [7]. Companies that disclose more information have better stock liquidity, reduce transaction and capital costs, and improve financial performance [20]. Companies that disclose their environmental policies signify transparency and reduce the risk of uncertainty and competitive advantage. Meanwhile, companies that disclose fewer items show various risks, such as the risk of litigation, penalties for pollution, future environmental costs and low future cash flows [9]. [21] explained that companies will benefit economically from preparing the expanded social and environmental disclosures in the form of higher stock prices.

Companies with high levels of profitability tend to present high environmental disclosures because profitable companies tend to have more resources to make environmental disclosures. Ownership of large resources can be used to show the company's contribution to the environment to reduce social pressure from the community and give a positive impression to stakeholders [22]. When companies realize that environmental information disclosure is positively related to financial performance, they will be more likely to increase the level of environmental information disclosure. Companies with high profits are able to allocate their expenses to many aspects, including involvement in social activities. A high level of profitability leads to more social disclosure [1]. Measurement of a company's environmental aspects, such as the amount of waste or greenhouse emissions, tends to significantly increase a company's spending. Environmental disclosure also requires high tangible costs, including the costs of establishing systems, and identifying, measuring, and reporting information. Therefore, only profitable companies can afford the costs [21].

H1: Environmental Disclosure has a positive effect on Financial Performance.

Environmental Disclosure and Environmental Performance

The voluntary disclosure literature shows that companies tend to report good news, while they are discouraged from disclosing bad news [9]. Indeed, these companies use earnings management or income smoothing to reduce the adverse effects of bad news. Voluntary environmental disclosures include reporting information about financial capabilities, environmentally sensitive operations, shareholder ownership, previous environmental law involvement, media exposure, environmental concerns and risks, and previous involvement with environmental groups.

Companies with high levels of environmental performance due to a proactive environmental strategy have incentives to disclose more environmental information voluntarily to investors and other stakeholders [23]. Companies that are environmentally sensitive and adopt healthy environmental policies will be motivated to provide voluntary environmental disclosures to inform investors about their global environmental strategy [24]. According to [5], companies with good environmental performance should disclose more environmental information (in quantity and quality) than companies with poor environmental performance.

Companies that provide voluntary environmental disclosures tend to use practices that are less harmful to the environment. Empirical studies from [11] found a positive and statistically significant relationship between the disclosures of environmental accounting information and environmental performance. They stated that the better the environmental performance, the higher the quality of environmental disclosure. The authors found that disclosure of environmental information had a positive and statistically significant impact on environmental performance. Legitimacy theory [25] proposes that the disclosure of corporate environmental and social information is a function of social and political pressures. As companies face more social and political pressures, they become more concerned about disclosing information. The theory explains that companies with poor environmental performance face more public pressure. In such a scenario, bad actors will tend to make greater and more positive disclosures of environmental information to offset the threat to legitimacy.

H2: Environmental disclosure has a positive effect on environmental performance.

Environmental Performance and Financial Performance

Many companies are implementing environmental activities to improve environmental performance. Companies should engage in environmental activities to generate capability development, which can have a positive impact on reducing costs and improving reputation [26]. Proactive environmental companies are no longer asking how much environmental activities cost but how much benefit they will provide [27].

Implementing environmental activities to achieve better environmental performance depends on a favourable cost-benefit relationship. Whilst, companies should have incentives to implement environmental activities proactively only if the associated (expected) benefits outweigh the costs. Research according to [28] stated that more companies have voluntarily published environmental information because top managers, generally believe that positive environmental concerns can increase stakeholder tendencies in investment decisions. It also proves that environmental trends can improve financial performance. In research on environmental and financial performance, [19] found a positive effect of environmental performance on financial performance, because companies with excellent environmental performance tend to get positive responses from their stakeholders, resulting in sustainable profit growth.

According to [29], stated that the application of environmental management practices to improve environmental performance not only generates business opportunities, but also reduces environmental pollution, environmental conflicts, organisational risks, and production costs as well as improvements in product quality and production efficiency, which will improve the organizational image and financial performance. Environmentally responsible behaviour can encourage innovation that helps improve business efficiency and commercial competitiveness [30]. There are various findings in the literature considering the nexus.

Research from [22] proved that companies that are actively involved in initiatives to improve environmental performance show positive economic benefits. In general, employees, customers, and government, which are the main stakeholders, can react positively to the green image created by the company and therefore develop a positive attitude. Research from [31] analysed the effect of disclosure of greenhouse gas emissions on Tobin's Q and found a negative effect. It indicates that stakeholders respond negatively to activities that damage the environment such as greenhouse gas emissions. Companies involved in environmental business operations can create affirmative stakeholder perceptions that resulted in increased financial performance. Research by [11] found that firms with a high environmental performance report more information. It most likely emerged due to the economic benefits derived from announcing positive news.

H3: Environmental performance has a positive effect on financial performance.

Research Framework
Hypotheses:

H1: Environmental performance has a positive effect on environmental disclosure (Fig. 1).

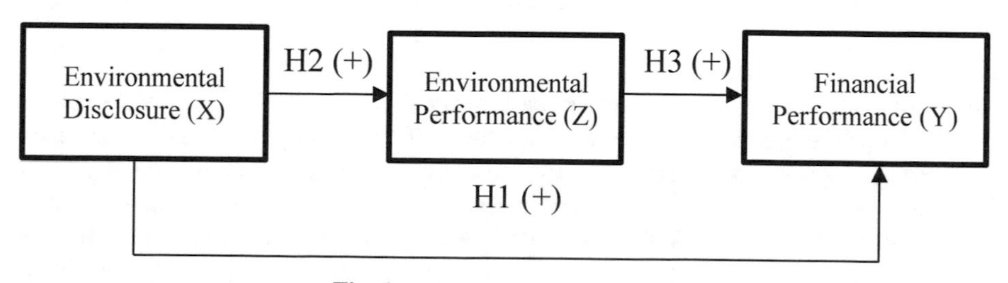

Fig. 1. Empirical research model

H2: Environmental performance has a positive effect on financial performance
H3: Environmental disclosure has a positive effect on financial performance.

4 Research Methods

The population in this research were large companies listed on the Indonesia Stock Exchange (IDX) from 2006–2020. The sampling technique was done by purposive sampling, and the data type was secondary. The selected companies were 1.) Companies that have completed information for research, such as Return on Assets, Return on Equity environmental disclosure (EDS), and environmental performance (EP) for the period 2006–2020, 2.) Companies that participated in the PROPER program 2006–2020, 3.) Companies listed on the Indonesia Stock Exchange (IDX) in 2006–2020.

Variable Description and Empirical Model
This research consisted of three variables: environmental disclosure as the dependent variable, financial performance as the independent variable, and environmental performance as the intervening variable.

Financial Performance
Financial performance is a measurement that can be used to describe a company's performance in the financial sector [13]. The financial performance of a company is required as a tool to measure a company's financial health [5]. Research from [32] found that financial performance is from all activities in utilizing the financial resources owned. In other words, the company's financial performance is the impact of activities carried out by management on an ongoing basis or management as a whole. In the era of a market economy, it is necessary to realize superior and efficient financial performance. Furthermore, it needs to be accompanied by ethical financial performance behaviour, namely realising good corporate social responsibility. The financial performance of the company was measured by the profitability of the company which is recognized through ROA and ROE [33].

Return on Equity (ROE) measures the company's net profit divided by the average. It is the company's efficiency measurement in generating profits from each unit of the shareholder equity [10]. Return On Assets (ROA analyses how successfully the company can utilize its assets to generate revenue and shows the position or level of competitiveness of the company against competitors [27].

Environmental Performance
Environmental performance is a measurement of the assessment carried out on the results of environmental conservation activities as a form of concern for the company as a user of existing environmental resources [16]. According to [3], the author defined environmental performance as "the measurable results of an organization's management of its environmental aspects." According to [27], minimizing the environmental impact of economic activities measure the environmental performance of a company. The environmental performance of companies from environmentally sensitive industries can attract the attention of non-financial stakeholders, including community members concerned

with the company's environmental performance, such as neighbouring communities, environmentalists, or regulatory bodies [34]. The researchers collected GHG emissions data from the Bloomberg database to measure the company's environmental performance, referring to research by [35]. In particular, the main measure for emissions and Bloomberg Total GHG Emissions show how far companies have applied their performance to reduce greenhouse gas emissions from their activities. The value of "Total GHG Emissions" is collected by Bloomberg directly from company filings, reports, and other publicly obtained information. Bloomberg has documentation related to published company reports where data is taken to ensure the validity and traceability of all such GHG emission data.

Environmental Disclosure

Environmental disclosure is an activity carried out by the company voluntarily or as a fulfilment of regulatory requirements to provide information about the company's environmental practices and activities. Research by [5], explained that the limitations of the environmental disclosure definition focus on disclosing future environmental costs and cost drivers which are disclosed in the annual report. Environmental disclosure in this research used the Bloomberg score, which has been used in several related academic studies recently [21]. The Bloomberg environmental disclosure score provides environmental information covering 60 different environmental data points, such as energy consumption and emissions, waste data, environmental initiatives, and environmental policies [21]. This data helps objective assessment in measuring environmental disclosure. The data sources include the company's annual report, press release sustainability report, and third-party research. Three factors measure the sustainability and impact of a company's investments on society, including environmental, social, and governance measures. This variable is an indicator of environmental transparency. Bloomberg summarizes these environmental disclosure scores, with higher scores indicating more clarity on environmental issues.

Analysis Techniques

This research used panel data by combining cross-sectional time series observations. The output of this regression was the coefficient of determination, F-test, and T-test. The panel data format will provide valuable results with less collinearity between many variables. The regression method on panel data has fixed effect models. Fixed effects are a way of considering the individuality of each firm or each cross-sectional unit for intercepting various but still estimating the slope coefficient is constant across firms.

5 Limitations and Future Research

This study has several limitations that create opportunities for future research. The limitation of this study is that many companies listed on the Indonesia Stock Exchange do not publish complete financial report data. Furthermore, there are still many shortcomings of researchers in conducting research and data collection which is still limited. Future research should be able to add measurement tools that are proxied by dependent variables other than those used in this study. It is better to understand which financial performance ratios can be affected by environmental performance and environmental disclosures, such as liquidity ratios and different profitability ratios. Additional research may extend the observation period or increase the number of companies observed to improve data distribution.

References

1. Deswanto, R.B., Siregar, S.V.: The associations between environmental disclosures with financial performance, environmental performance, and firm value. Soc. Responsib. J. **14**(1), 180–193 (2018). https://doi.org/10.1108/SRJ-01-2017-0005
2. Mahmood, Z., Ahmad, Z., Ali, W., Ejaz, A.: Does Environmental disclosure relate to environmental performance? Reconciling legitimacy theory and voluntary disclosure theory. Pak. J. Commer. Soc. Sci. **11**(3), 1134–1152 (2017)
3. Agyemang, A.O., Yusheng, K., Twum, A.K., Ayamba, E.C., Kongkuah, M., Musah, M.: Trend and relationship between environmental accounting disclosure and environmental performance for mining companies listed in China. Environ. Dev. Sustain. **23**(8), 12192–12216 (2021). https://doi.org/10.1007/s10668-020-01164-4
4. Patten Dennis, M.: The relation between environmental performance and environmental disclosure: a research note. Account. Organ. Soc. **27**(8), 763–773 (2002). http://www.sciencedirect.com/science/article/pii/S0361368202000284
5. Al-Tuwaijri, S.A., Christensen, T.E., Hughes, K.E.: The relations among environmental disclosure, environmental performance, and economic performance: a simultaneous equations approach. Account. Organ. Soc. **29**(5–6), 447–471 (2004). https://doi.org/10.1016/S0361-3682(03)00032-1
6. Wang, S., Wang, H., Wang, J., Yang, F.: Does environmental information disclosure contribute to improve firm financial performance? An examination of the underlying mechanism. Sci. Total Environ. **714**(96), 136855 (2020). https://doi.org/10.1016/j.scitotenv.2020.136855
7. Ifada, L.M., Ghozali, I., Faisal: Corporate social responsibility, normative pressure and firm value: evidence from companies listed on Indonesia stock exchange. In: Barolli, L., Poniszewska-Maranda, A., Enokido, T. (eds.) CISIS 2020. AISC, vol. 1194, pp. 390–397. Springer, Cham (2021). https://doi.org/10.1007/978-3-030-50454-0_38
8. Ifada, L.M., Indriastuti, M.: Government ownership, international operations, board independence and environmental disclosure: evidence from Asia–Pacific. J. Din. Akunt. **13**(2), 131–147 (2021). https://doi.org/10.15294/jda.v13i2.30268
9. Iatridis, G.E.: Environmental disclosure quality: evidence on environmental performance, corporate governance and value relevance. Emerg. Mark. Rev. **14**(1), 55–75 (2013). https://doi.org/10.1016/j.ememar.2012.11.003
10. Haninun, H., Lampung, U.B., Lindrianasari, L.: The effect of environmental performance and disclosure on financial performance. Haninun Haninun * Lindrianasari Lindrianasari Angrita Denziana, no. January 2018. https://doi.org/10.1504/IJTGM.2018.092471
11. Tadros, H., Magnan, M.: How does environmental performance map into environmental disclosure?: A look at underlying economic incentives and legitimacy aims. Sustain. Account. Manag. Policy J. **10**(1), 62–96 (2019). https://doi.org/10.1108/SAMPJ-05-2018-0125
12. Garcia, E.A. da R., de Carvalho, G.M., Boaventura, J.M.G., de Souza Filho, J.M.: Determinants of corporate social performance disclosure: a literature review. Soc. Responsib. J. **17**(4), 445–468 (2020). https://doi.org/10.1108/SRJ-12-2016-0224
13. Wahyuningrum, I.F.S., Budihardjo, M.A., Muhammad, F.I., Djajadikerta, H.G., Trireksani, T.: Do environmental and financial performances affect environmental disclosures? Evidence from listed companies in Indonesia. Entrep. Sustain. Issues **8**(2), 1047–1061 (2020). https://doi.org/10.9770/jesi.2020.8.2(63)

14. Xie, J., Nozawa, W., Yagi, M., Fujii, H., Managi, S.: Do environmental, social, and governance activities improve corporate financial performance? Bus. Strateg. Environ. **28**(2), 286–300 (2019). https://doi.org/10.1002/bse.2224
15. Hongjun, W., Qiren, L., Shinong, W.: Corporate environmental disclosure and financing constraints. J. World Econ. **40**(05), 124–147 (2017)
16. Clarkson, P.M., Li, Y., Richardson, G.D., Vasvari, F.P.: Revisiting the relation between environmental performance and environmental disclosure: An empirical analysis. Account. Organ. Soc. **33**(4–5), 303–327 (2008). https://doi.org/10.1016/j.aos.2007.05.003
17. Aragón-Correa, J.A., Marcus, A., Hurtado-Torres, N.: The natural environmental strategies of international firms: old controversies and new evidence on performance and disclosure. Acad. Manag. Perspect. **30**(1) (2016). https://doi.org/10.5465/amp.2014.0043
18. Ren, S., Wei, W., Sun, H., Xu, Q., Hu, Y., Chen, X.: Can mandatory environmental information disclosure achieve a win-win for a firm's environmental and economic performance? J. Clean. Prod. **250**, 119530 (2020). https://doi.org/10.1016/j.jclepro.2019.119530
19. Huynh, Q.L., Lan, T.T.N.: Importance of environmentally managerial accounting to environmental and economic performance. Int. J. Energy Econ. Policy **11**(5), 381–388 (2021). https://doi.org/10.32479/ijeep.11511
20. Ifada, L.M., Saleh, N.M.: Environmental performance and environmental disclosure relationship: the moderating effects of environmental cost disclosure in emerging Asian countries. Manag. Environ. Qual. An Int. J. **33**(6), 1553–1571 (2022). https://doi.org/10.1108/MEQ-09-2021-0233
21. Qiu, Y., Shaukat, A., Tharyan, R.: Environmental and social disclosures: link with corporate financial performance. Br. Account. Rev. **48**(1), 102–116 (2016). https://doi.org/10.1016/j.bar.2014.10.007
22. Sila, I., Cek, K.: The impact of environmental, social and governance dimensions of corporate social responsibility: Australian evidence. Procedia Comput. Sci. **120**, 797–804 (2017). https://doi.org/10.1016/j.procs.2017.11.310
23. Muhimatul Ifada, L., Munawaroh, Kartika, I., Fuad, K.: Environmental performance announcement and shareholder value: the role of environmental disclosure. In: Barolli, L., Yim, K., Enokido, T. (eds.) CISIS 2021. LNNS, vol. 278, pp. 426–434. Springer, Cham (2021). https://doi.org/10.1007/978-3-030-79725-6_42
24. Ahmadi, A., Bouri, A.: The relationship between financial attributes, environmental performance and environmental disclosure: empirical investigation on French firms listed on CAC 40. Manag. Environ. Qual. An Int. J. **28**(4), 490–506 (2017). https://doi.org/10.1108/MEQ-07-2015-0132
25. Deegan, C., Rankin, M., Tobin, J.: An examination of the corporate social and environmental disclosures of BHP from 1983–1997: a test of legitimacy theory. Account. Audit. Account. J. **15**(3), 312–343 (2002). https://doi.org/10.1108/09513570210435861
26. Zhang, K.Q., Tang, L.Z., Chen, H.H.: The impacts of environmental performance and development of financing decisions on economic sustainable performance: from the view of renewable and clean energy industry. Clean Technol. Environ. Policy **23**(6), 1807–1819 (2021). https://doi.org/10.1007/s10098-021-02068-1
27. Nishitani, K., Kokubu, K.: Can firms enhance economic performance by contributing to sustainable consumption and production? Analyzing the patterns of influence of environmental performance in Japanese manufacturing firms. Sustain. Prod. Consum. **21**, 156–169 (2020). https://doi.org/10.1016/j.spc.2019.12.002
28. Liang, D., Liu, T.: Does environmental management capability of Chinese industrial firms improve the contribution of corporate environmental performance to economic performance? Evidence from 2010 to 2015. J. Clean. Prod. **142**(2017), 2985–2998 (2017). https://doi.org/10.1016/j.jclepro.2016.10.169

29. Chuang, S.-P., Huang, S.-J.: The effect of environmental corporate social responsibility on environmental performance and business competitiveness: the mediation of green information technology capital. J. Bus. Ethics **150**(4), 991–1009 (2016). https://doi.org/10.1007/s10551-016-3167-x
30. Singh, S.K., Del Giudice, M., Chierici, R., Graziano, D.: Green innovation and environmental performance: the role of green transformational leadership and green human resource management. Technol. Forecast. Soc. Change **150**, 119762 (2020). https://doi.org/10.1016/j.techfore.2019.119762
31. Putri, V.R., Rachmawati, A.: The effect of profitability, dividend policy, debt policy, and firm age on firm value in the non-bank financial industry. J. Ilmu Manaj. Ekon. **10**(1), 14–21 (2017)
32. Gitman, L.J., Zutter, C.J.: Principles of Managerial Finance, 14th edn. (2015)
33. Nguyen, L.S., Tran, M.D.: Disclosure levels of environmental accounting information and financial performance: the case of Vietnam. Manag. Sci. Lett. **9**(4), 557–570 (2019). https://doi.org/10.5267/j.msl.2019.1.007
34. Adhikary, A., Sharma, A., Diatha, K.S., Jayaram, J.: Impact of buyer-supplier network complexity on firms' greenhouse gas (GHG) emissions: an empirical investigation. Int. J. Prod. Econ. **230**, 107864 (2020). https://doi.org/10.1016/j.ijpe.2020.107864
35. Alipour, M., Ghanbari, M., Jamshidinavid, B., Taherabadi, A.: Does board independence moderate the relationship between environmental disclosure quality and performance? Evidence from static and dynamic panel data. Corp. Gov. Int. J. Bus. Soc. **19**(3), 580–610 (2019)

Applying BERT on the Classification of Chinese Legal Documents

Qiong Zhang[1] and Xu Chen[2](✉)

[1] School of Management Information and System, Zhongnan University of Economics and Law, Wuhan 430073, China
zqiong@stu.zuel.edu.cn
[2] School of Information and Safety Engineering, Zhongnan University of Economics and Law, Wuhan 430073, China
chenxu@zuel.edu.cn

Abstract. Chinese Legal documents contain complex underlying facts, controversies and legal application issues that make most domestic legal document retrieval platforms perform poorly in terms of relevance and accuracy. In this paper, we try to evaluate the performance of BERT on Chinese legal document classification. The data set for this paper is obtained from the legal judgment documents of a single charge on China Judicial Documents Network, with a total of 8 accusations and 2454 legal cases. The experimental result shows that the BERT performs much better than FastText, TextCNN, and RNN on our data set, obtaining a classification accuracy of 0.89.

1 Introduction

Based on an incomplete statistic from China Judicial Documents Network, various types of legal cases in China, such as criminal cases, civil cases and administrative cases, have shown a significant growing in number year by year from 2012 to 2020. Among them, criminal cases have exceeded 970,000 in 2017 and civil cases have exceeded one million. With the increasing imbalance in the distribution of legal cases and legal practitioners, it is advisable to adopt new technologies to retrieve relevant legal documents efficiently and accurately.

Pre-trained language models have achieved impressive results and have become important tools for various natural language processing tasks [1]. In particular, the transformer-based BERT model, has obtained state-of-the-art results in many baseline tests [2]. Currently, BERT and its derivative models have been widely used in tasks such as medical texts processing [3], news texts classification [4], and sentiment analysis [5]. BERT has verified its validity and strength in these tasks.

This paper tries to apply BERT to the classification task of Chinese legal documents to evaluate its classification performance. Firstly, we crawl 2545 single-crime criminal judgment documents from China Judicial Documents Network. The criminal judgment documents in China are the decision documents made by the people's court in accordance with the law to determine the defendant's guilt or innocence, and in the case of guilt, to

L. Barolli (Ed.): EIDWT 2023, LNDECT 161, pp. 215–222, 2023.
https://doi.org/10.1007/978-3-031-26281-4_21

confirm what crime was committed, what penalty is applicable and the range of penalty after the conclusion of the trial of the case brought by the public prosecution or the case brought by the private prosecutor in accordance with the corresponding criminal procedure [6]. Due to the standardization of the format and content of the verdicts, we use regular expressions to extract the case fact part from each criminal verdict. The original criminal verdict contains a lot of noisy information thus making the verdict too long, and the key information is extracted by obtaining the legal case facts to reduce the training burden of the model. Finally, the case fact texts are used as the inputs of the BERT model. The contributions of this paper mainly lie in the following two points:

- Evaluate the performance of the BERT model on a single charge criminal case data set compared to existing text classification models such as FastText and TextCNN.
- Validate the stability of the BERT model by training on our small data set.

The remainder of this paper contains four sections. In Sect. 2, previous works related to legal document classification are presented. In Sect. 3, the process of implementing this paper's methodology is detailed elaborated. The Sect. 4 contains the experimental result and the conclusion of the comparative analysis. The final section summarizes the shortcomings of the thesis approach and future work based on the experimental result in Sect. 5.

2 Related Works

Neha Bansal et al. [7] combined CNN with BiLSTM to develop a deep learning framework for automatic classification of legal documents for online judgment documents from the University of Washington School of Law Supreme Court data set. The optimal feature vector representations of the verdict texts are trained by a genetic algorithm with several iterations of CNN, and these feature vectors are used as the inputs of BiLSTM. Finally, the classification is performed by a soft-max activation function. The experimental result shows that the automatic classification algorithm respectively achieves 0.89 and 0.82 accuracy on the training and test sets, and the comprehensive performance is significantly better than the baseline models such as CRF and CNN.

Chen Haihua et al. [8] propose a machine learning algorithm featuring domain concept and random forest as classifier to study the performance of deep learning for automatic classification of U.S. legal texts containing 50 categories with a total of 30,000 complete U.S. legal cases, and the experimental results show that the random forest model incorporating domain concept has the strongest performance compared with the baseline models such as TextCNN and BiLSTM, obtaining an accuracy of 0.8449, which proves the effectiveness of the domain concept in the classification of U.S. legal texts.

Faced with the problems of presenting a large base and complex content of Chinese legal documents, Ma Yinglong et al. [9] proposed an ontology-driven knowledge block summarizing method to calculate document similarity for the classification of Chinese legal documents. The method is proven to effectively address the inefficiency of traditional machine learning classification models due to the lack of incorporating the overall structure of legal documents and the lack of incorporating additional domain-specific knowledge.

This paper investigates the effectiveness of BERT on Chinese legal document classification by comparing and analyzing the performance effect of BERT model with current baseline models such as FastText, TextCNN and RNN. The modeling approach we use in our paper does not make additional innovations, but we propose to apply the BERT model to the classification of Chinese legal documents to explore the applicability of BERT. Our experiments consist of two main parts, one is data preprocessing and the other is model training. We will elaborate the above two steps in the next section.

3 Proposed Method

In this section, the implementation process of our proposed method in the paper is detailed elaborated. As shown in Fig. 1 [10]. Our experimental process consists of two main parts, one is data pre-processing and the other is model training. We first use regular expression to extract the case facts from the legal judgments as the texts to be trained, and then input them into the BERT model to participate in the training. Finally, we can obtain the type of accusation based on the case facts by the classifier.

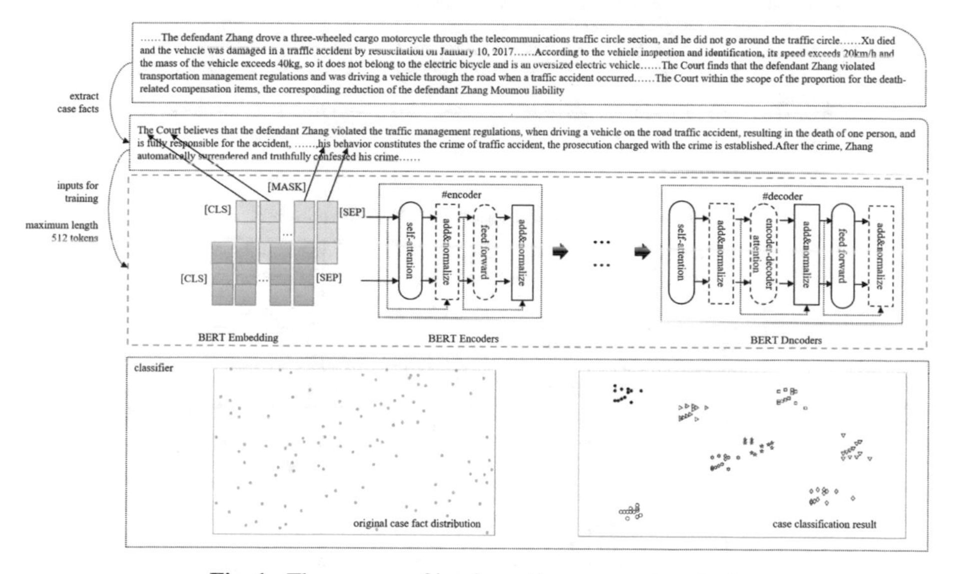

Fig. 1. The process of implementing proposed method.

3.1 Data Preprocessing

Since the original legal judgments are too lengthy in content and there is a large amount of noisy information in the judgments. If the key information is not extracted from the content of the judgment, it will not only cost a great training cost, but also produce a great bias to the results of subsequent model training. Additionally, the maximum position embedding for pre-training in BERT is usually 512, and this maximum position embedding also takes the [CLS] and [SEP] character representation information into account [4, 11].

Therefore, for long texts such as verdicts, the key information is not extracted before training, and after the long texts are forced to be truncated in the training process, a large amount of information in the backward position is directly discarded and fails to participate in training. In this case, the key information in the verdicts may be discarded, so we need to extract the case facts from the verdict to form the dataset to be trained. The process of extracting case facts from Chinese legal documents by using regular expressions is shown in Algorithm 1. The information related to the case fact text dataset formed after processing is shown in Table 1.

Algorithm 1: Pseudo Codes for Key Steps in Data Pre -processing

```
1. input X,
//input Chinese legal judgement documents
2. x=re.findall(pattern,X),
//extract case facts using re
3. len(x)=re.split(r'[.!;?]',x.strip()),
//number of sentences in the statistical case fact
4. segs(x)=jieba.cut(x,cut_all=False),
//use jieba to participle
5. words(x)=segs(x) not in stopwords,
//remove stopwords and obtain the number of words in case fact
6. end
```

Table 1. Case facts' categories and their distribution

Category ID	Doc count	Sentences average	Words average
0	203	31	378
1	242	14	143
2	124	30	339
3	376	32	368
4	398	26	273
5	473	29	318
6	494	21	223
7	144	29	342

3.2 Model Training

The BERT model is a pre-trained model that can fully express the semantic features of the text, based on a huge model and consuming massive computing power, trained from a very large corpus data [11]. BERT uses transformer's encoder structures as feature extractors and uses the accompanying MLM training method to achieve bidirectional encoding of the input sequence texts. Compared with one-way encoders that only use preorder textual information to extract semantics, BERT has a stronger ability to extract semantic information.

BERT trains the semantic comprehension of words by MLM (Masked Language Model) method and the comprehension between sentences by NSP (Next Sentence Prediction) method. The process of the BERT in fully connected layer calculating the loss function is shown in Fig. 2. BERT combines the losses of both MLM and NSP. In the fully connected layer, the smallest loss function after training convergence is used as the optimal case category output of the softmax function.

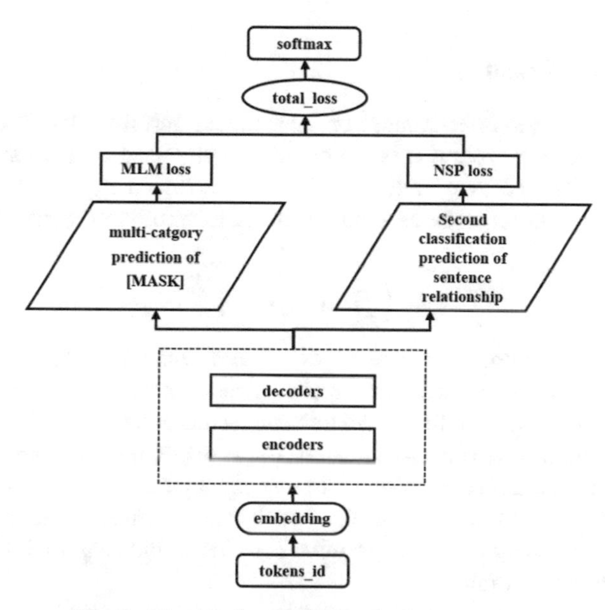

Fig. 2. The details of BERT calculating the loss.

4 Experimental Setup and Result

4.1 Experimental Setup

We use python language to complete the experiment in a computer environment with 2.10 GHz processor and 16 GB RAM. For the use of BERT model, the experiments are fine-tuned based on the Bert-base-Chinese in the BERT model released by google, and the model parameters are set as shown in Table 2.

Table 2. Model hyperparameters and experimental environment

Environment	Configuration parameters
python version	3.7
deep learning framework	1.14.0
learning_rate	2e−5
max_seq_length	256
num_train_epoch	10
train_batch_size	16

4.2 Experimental Result

This paper measures the effectiveness of each model performance by comparing the accuracy rate on each model, as it is a more commonly used evaluation metric in text classification [8]. The accuracy rate is defined as shown in Eq. (1): where TP is the number of correctly classified cases and FP is the number of incorrectly classified cases.

$$accuracy = \frac{TP + TN}{TP + FP + TN + FN} \quad (1)$$

We adopts a 5-fold cross-validation method and uses the average of the accuracy rates under the 5 validation results as the classification accuracy on the BERT model, aiming to reduce the chance of the model training results and improve the generalization ability of the model through cross-validation. In the paper, the legal documents of each type of crime in the data set are divided into five equal parts for 5-fold cross-validation, and in each validation process, one-fifth of each type of crime is selected to form the test set, and the remaining part is the training set. The 5-fold cross-validation results of the model are shown in Table 3.

Table 3. The results of five fold cross validation on BERT model based on accuracy

BERT classification	Accuracy
TRAIN 1	0.9566116
TRAIN 2	0.9013513
TRAIN 3	0.8648649
TRAIN 4	0.8491736
TRAIN 5	0.8976834
Average score	0.8939370

Table 4. Model comparison and experimental results

Model	Accuracy
FastText	0.4848732
TextCNN	0.7896879
LSTM	0.8548168
BERT	0.8939370

We train the data set of this paper on FastText, TextCNN, LSTM, and BERT. As shown in Table 4, FastText obtains an accuracy of 0.48. TextCNN obtains an accuracy of 0.79. LSTM obtains an accuracy of 0.85. BERT obtained an accuracy of 0.89. Obviously, BERT has the best classification results on the Chinese legal document classification task compared with those above models.

5 Conclusion and Future Work

The BERT model obtains an accuracy of 0.89 on the task of Chinese legal document classification, which is better than some baseline models such as: FastText, TextCNN and RNN. This not only illustrates that the BERT model can show relatively significant classification performance on a small data set, but also shows the effectiveness of BERT in the task of classifying Chinese legal documents.

However, there are still two limitations in our proposed method: the first one: due to the limitation of the experimental environment, the experimental data set of this paper is not large enough, and the classification accuracy of BERT in large scale Chinese single-crime legal documents is still unknown in the task of classifying Chinese legal document. The second one, the paper has only trained on single-count Chinese legal cases, while in practice, there are not many cases involving multiple counts. In particular, how should Chinese law cases with mixed charges be resolved? These two points are the problems that need to be considered and solved in depth in the future.

Acknowledgement. This research was funded by National Funds of Social Science (21BXW076), National Natural Science Foundation of China (61602518), Philosophy and Social Science Research Project of Hubei Provincial Department of Education (20G026), Innovation Research of Young Teachers of Central Universities in 2021 (2722021BZ040), The Key Social Science Projects in Wuhan in 2021 (2021010), Prof. Liu Yaqi's Outstanding Youth Innovation team Construction Project (Big Data Intelligent Information Processing and Application Technology Innovation Team) and School-level reform project of Zhongnan University of Economics and Law (YB202158).

References

1. Kong, J., Wang, J., Zhang, X.: Hierarchical BERT with an adaptive fine-tuning strategy for document classification. Knowl. Based Syst. **238**, 107872 (2022)

2. Anna, R., Olga, K., Anna, R.: A primer in BERTology: what we know about how BERT works. Trans. Assoc. Comput. Linguist. **8**, 842–866 (2021)
3. Patricoski, J., et al.: An evaluation of pretrained BERT models for comparing semantic similarity across unstructured clinical trial texts. Stud. Health Technol. Inform. **289**, 18–21 (2022)
4. Ding, M., Zhou, C., Yang, H., et al.: CogLTX: applying BERT to long texts. In: Advances in Neural Information Processing Systems 33, pp. 12792–12804 (2020)
5. Barros, L., Trifan, A., Oliveira, J.L.: VADER meets BERT: Sentiment analysis for early detection of signs of self-harm through social mining. In: CEUR Workshop Proceedings, vol. 2936 (2021)
6. Zhang, W.: Research on criminal judgment. Renmin University of China (2005)
7. Bansal, N., Sharma, A., Singh, R.K.: An evolving hybrid deep learning framework for legal document classification. Ingénierie des Systèmes d'Information **24**(4), 425–431 (2019)
8. Chen, H., Wu, L., Chen, J., Lu, W., Ding, J.: A comparative study of automated legal text classification using random forests and deep learning. Inf. Process. Manag. **59**(2), 102798 (2022)
9. Ma, Y., Zhang, P., Ma, J.: An ontology driven knowledge block summarization approach for Chinese judgment document classification. IEEE Access **6**, 71327–71338 (2018)
10. Devlin, J., Chang, M.W., Lee, K., et al.: BERT: pre-training of deep bidirectional transformers for language understanding. arXiv preprint arXiv:1810.04805 (2018)
11. Jing, Y., Wang, M.Y., Zhou, W.Y.: Research on text summary extraction algorithm based on BBCM-TextRank. J. Northeast Normal Univ. (Nat. Sci. Ed.) **54**(03), 67–75 (2022). https://doi.org/10.16163/j.cnki.dslkxb202107310001

Technology and Efficacy Extraction of Mechanical Patents Based on BiLSTM-CRF

Ruiyi Cui, Na Deng(✉), and Cheng Zheng

School of Computer Science, Hubei University of Technology, Wuhan 430068, China
iamdengna@163.com

Abstract. With the development of industry and technology innovation in China, the number of patent applications in the field of machinery has been increasing year by year. Mechanical patents contain rich and latest technical and legal information. The mining of mechanical patents play an important practical role. In knowledge graph construction and infringement judgment for mechanical patents, extracting the technology and efficacy is an important step. In this paper, we propose a model combining bidirectional long and short-term memory neural network and conditional random field to achieve the technology and efficacy extraction from mechanical patents abstract texts. The model in this paper achieves an accuracy rate of 94.31%, a recall rate of 94.34%, and an F1 value of 94.31%. The results show that the model in this paper can accurately extract the technology and efficacy of patents in the mechanical field.

1 Introduction

As China places more and more emphasis on the development of technology innovation, more attention is being paid to the protection of intellectual property rights. Patents represent the most advanced technology. Countries and enterprises can usually obtain the latest research results in a field from the latest patents and use them to ensure that an enterprise is always in a leading position or to achieve curve overtaking.

The development of any technology is inseparable from the development of the real industry, and the development of the real industry is inseparable from the development of machinery. The development of machinery can drive the development of industry, and the development of industry can promote the development of science and technology. Recently, the application of patents in China increases year by year, and the number of invention patents in the field of machinery disclosed in 2020 is 536,520, while the number of patents in the field of machinery disclosed in 2021 is 694,874. It can be seen that the number of mechanical patents is increasing, and the text analysis and mining for mechanical patents becomes more and more meaningful.

Before analysis and mining for mechanical patents, the extraction of technology and efficacy entities is a key step. You can extract the mechanical technologies or efficacy from patent abstract texts. For example: from the text "the utility model discloses a corrosion-resistant buffer hydraulic cylinder", we can extract the mechanical technology "corrosion-resistant buffer hydraulic cylinder". The text "realizes the effect of continuous cooling during use, and the cooling temperature is relatively controllable", we can

L. Barolli (Ed.): EIDWT 2023, LNDECT 161, pp. 223–234, 2023.
https://doi.org/10.1007/978-3-031-26281-4_22

extract the mechanical efficacy "continuous cooling" and "temperature is relatively controllable". The extracted mechanical entity of patent can be used to determine whether the patent will constitute an infringement, or they can be used to recommend a patent to a relevant enterprise, or can be used as the semantic expansion object to increase the accuracy of patent search. Therefore, the study of extracting technology and efficacy entities of mechanical patents becomes very meaningful.

At present, the patent technology and efficacy identification faces some difficulties: Firstly, for mechanical patent, the most difficult problem is to identify a mechanical technology since most of the technology descriptions of the mechanical patent contain the role of the machinery and mechanical construction composition, identification is relatively more difficult. Secondly, the mechanical patent efficacy entity description is not clear, some efficacy description is too long, such as "fundamentally solve the problem of stress concentration of splitting rod caused by irregular rock holes". Such efficacy description will make the mechanical patent efficacy extraction more difficult and less accurate. Thirdly, there are many proprietary terms in the mechanical field and a combination of terms that express a specific meaning for the first time in the patent. Fourthly, there is no standard dataset for mechanical patents, and it is necessary to construct the dataset manually in the technical efficacy entity extraction task.

The paper focuses on the work related to technology and efficacy entity extraction, the relevant models to implement the extraction and the analysis of the experimental findings.

2 Related Work

Named entity recognition (NER) works as an important task in natural language processing, and the purpose of NER is to be able to extract keywords from a text or a sentence. Named entity recognition's can be used in various aspects such as retrieval, building knowledge graphs, computing text similarity, question and answer systems, etc. Early methods are rule-based and dictionary-based methods, which mainly construct corresponding entity dictionaries and realize entity extraction by comparing dictionary libraries, the problem of this method is that the accuracy depends on the accuracy and completeness of the constructed dictionaries. Currently, the most used methods are machine learning methods based on HMM, MEMM, CRF, or deep learning methods such as RNN-CRF, which support autonomous learning of entity features from the training set.

Jianhong Ma et al. [1] proposed a multi-channel neural network-based entity recognition model for new energy vehicles in order to solve the problem of blurred boundaries of named entities and little extant marker data for new energy vehicles. The conclusion shows that the model can significantly reduce manual labeling while can improve the accuracy and recall rate.

Dandan Qu et al. [2] proposed a method combining a bidirectional recurrent neural network with Markov random fields, namely BiGRU-CRF, using the BERT model to pre-train the custom data to obtain word vectors containing contextual semantics, and then inputting the word vectors to the BiGRU-CRF model. The results show that this model can achieve entity extraction better. Meanwhile, Zhantang Shi et al. [3] combined convolutional neural network with Transformer of multi-headed attention mechanism well solved the problem that RNN cannot make full use of GPU parallel computing, i.e., long-range text dependency. Combining RNN with Transformer improves the accuracy of Chinese named entity recognition. As a special neural network RNN in deep learning, LSTM avoids the problems of RNN gradient disappearance and gradient explosion. Compared with the ordinary RNN, LSTM can selectively remember more important information and ignore the unimportant information, and can perform better on longer text.

Farag Saad et al. [4] used BiLSTM model to identify entities in biological documents and patents with good results. Hengjian Yu et al. [5] used CRF conditional random field and BiLSTM model for entity extraction of government documents in order to construct a knowledge graph of government documents. The results show that BiLSTM-CRF model can effectively extract the entities in the government official documents. Yanling Yang et al. [6] used BiLSTM-CRF model for entity identification of named entities in Chinese medical cases, which effectively improved the accuracy of entity identification in Chinese medical bills. In the authors' previous study [7], we used BiLSTM-CRF model for Chinese medicine patent naming entity recognition to identify entities such as Chinese medicine names, and medicine effects and achieved better results.

3 Conditional Random Field

Lafferty et al. [8] proposed the conditional random field in 2001. The conditional random field (CRF), is an undirected graph model, and the conditional random field combines the features of the maximum entropy model and the hidden Markov model. The conditional random field is a discriminative model suitable for prediction tasks and has been used in recent years in the fields of named entity recognition, lexical annotation and so on with good results. CRF is defined as an undirected graph G = (V, E) given random variables X and Y if the random variable Y satisfies the construction Markovian conditional random field, and the judgment formula is shown in Eq. (1).

$$P(Y_v|X, Y_w, w \neq v) = P(Y_v|X, Y_w, w \sim v) \tag{1}$$

where $P(Y|X)$ is the Y conditional probability distribution for a given X. If Eq. (1) is satisfied for any node v, the conditional probability distribution $P(Y|X)$ is the conditional random field. In Eq. (1) $w \sim v$ denotes all nodes in the undirected graph G that are connected to node v, and $w \neq v$ denotes all nodes other than node v. Y_v and Y_w denotes the nodes v with w of random variables.

The CRF used in the task of named entity identification is a linear chain conditional random field, and the linear chain conditional random field is determined by the Eq. (2).

$$P\big(Y_i|X, Y_1, Y_2, \ldots, Y_{i-1}, Y_{i+1}, \ldots, Y_n\big) = P(Y_i|X, Y_{i-1}, Y_{i+1}) \tag{2}$$

Among them $X = (X_1, X_2, \ldots, X_n)$, $Y = (Y_1, Y_2, \ldots, Y_n)$, X is the observed sequences in the named entity recognition task, Y is the corresponding label sequences. When Y satisfies the judgment formula, the conditional probability distribution $P(Y|X)$ is the conditional random field.

The linear chain condition random field is calculated by Eq. (3).

$$P(y|x) = \frac{1}{Z(X)} \exp(\sum\nolimits_{i,k} \lambda_k t_k(y_{i-1}, y_i, x, i) + \sum\nolimits_{i,l} u_l s_l(y_i, x, i)) \tag{3}$$

$$z(x) = \sum\nolimits_Y \exp(\sum\nolimits_{i,k} \lambda_k t_k(y_{i-1}, y_i, x, i) + \sum\nolimits_{i,l} u_l s_l(y_i, x, i)) \tag{4}$$

When $P(Y|X)$ is the linear chain condition random field, in the condition that X takes the value of x, the conditional probability of Y taking the value of y is shown in Eq. (3). Where t_k and s_l are binary eigenfunctions with values of 0 and 1 defined on the edges and nodes, respectively. The characteristic function of the linear chain conditional random field is also shown in Eq. (3). CRF model can calculate the probability that a named entity is correctly labeled for identification, which can increase the correctness of the whole model.

4 Long and Short-Term Memory Neural Networks

Hochreiter and Schmidhuber [9] proposed the long short-term memory (LSTM) model, which is an improved recurrent neural network. The significance of LSTM is to solve the dependency problem that RNNs cannot handle long distance. Compared to the hidden layer of RNN which has only one state h, LSTM has one more unit state c that preserves the long-term state.

As shown in Fig. 1, at a certain moment t, the LSTM model has three inputs and two outputs. The three inputs of the LSTM model are the input at the current moment x_t, the output of the previous moment h_{t-1} and the cell state at the previous moment c_{t-1}. The two outputs of the LSTM model are the output value at the current moment h_t and the current state c_t The two outputs of the LSTM model are the output value at the current moment and the current state.

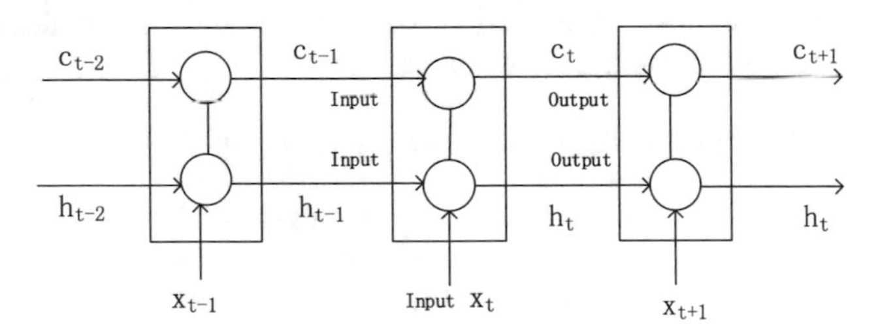

Fig. 1. LSTM input and output diagram.

In order to implement the control of the long-term state c, LSTM model sets three control switches inside it. The three switches control the continuation of the long-term state c, to transform the immediate state to the long-term state c and to control whether to output the current long-term state c. To implement the control switches, the LSTM model sets up a gate [10], which is a fully connected layer with a vector as input and a real vector between 0 and 1 as output, calculated as Eq. (5):

$$g(x) = \sigma(Wx + b) \tag{5}$$

The three gates of the LSTM are the forgetting gate, the input gate and the output gate. The role of the forgetting gate is to determine how much of the state of the cell at the previous moment c_{t-1} is saved to the current state c_t. The calculation formula for the forgetting gate is shown in Eq. (6).

$$f_t = \sigma\left(W_f\left[h_{t-1}, x_t\right] + b_f\right) \tag{6}$$

where W_f is the weight matrix of the forgetting gate, σ denotes the sigmoid function, $\left[h_{t-1}, x_t\right]$ denotes that the input data x_t and the output of the previous moment h_{t-1} are combined into a new vector, and b_f is the bias term of the forgetting gate.

The role of the input gate is to determine how much of the current moment of input x_t is saved to the cell state c_t, that is, the input gate determines the importance of the current input and generates a new memory cell. The calculation formula is shown in Eq. (7).

$$i_t = \sigma\left(W_i\left[h_{t-1}, x_t\right] + b_i\right) \tag{7}$$

The input gate determines the current input state based on the last output and the current input, that is, it determines whether the current input becomes a new candidate memory unit. The calculation formula is shown in Eq. (8).

$$\widetilde{c_t} = \tan\mathrm{h}\left(W_c\left[h_{t-1}, x_t\right] + b_c\right) \tag{8}$$

After that, according to the last cell status c_{t-1} multiplied by the forgetting gate f_t plus the current state c_t multiplied by the input gate i_t. According to this calculation, the current memory is combined with the previous memory to generate a new memory unit. The calculation formula is shown in Eq. (9).

$$c_t = f_t \times c_{t-1} + i_t \times c_t \tag{9}$$

The role of the output gate is to determine the output of the model, and the output gate controls how much of the cell state c_t is output to h_t to control the output of the model. The calculation formula is shown in Eq. (10).

$$o_t = \sigma\left(W_o\left[h_{t-1}, x_t\right] + b_o\right) \tag{10}$$

The final output of the LSTM model is given by the output gate o_t multiplied by $tanh(c_t)$.

The characteristic of LSTM model is that it can remember the influencing factors of past time of input sequence but ignore the influencing factors from backward to forward. For Chinese patents, sometimes the modifier of the entity is after the entity, so if only the semantic information from front to back is considered and the semantic information from back to front is ignored, the accuracy of entity recognition will be reduced. In order to solve this problem, a bidirectional LSTM model is introduced to obtain the semantic relationship between two directions, which can improve the accuracy of entity recognition.

5 Our Model

In mechanical patents, machinery technology usually in the first sentence of the patent abstract, for example, in a sentence described as "the utility model discloses a self-circulating cooling hydraulic cylinder structure", "self-circulating cooling hydraulic cylinder structure" is the mechanical technology. Machinery technology is generally followed by the word "a". Therefore, in identifying the machinery technology, the starting word of the technology depends on the context. However, in the abstract text, the word "a" may be followed by an adjective used to describe the machinery, such as "The present invention relates to a heat dissipating electro-hydraulic rotator with a long service life" where "heat dissipating electro-hydraulic rotator " is the machinery technology and "with a long service life" is the adjective.

The CNN is good at acquiring text sequences, but when it receives the word "a", it incorrectly uses the subsequent adjective words as the patent technology. The unidirectional long and short-term memory neural network LSTM can acquire the medium and long term information in the text sequence, and can effectively eliminate the adjectives before the mechanical technology. However, the disadvantage of LSTM is that it is not able to recognize some descriptions after the entity. For example, "The present invention discloses a passive articulated hydraulic control system and control method for shield machine", where "and control method" is not a machinery technology, but is connected after "a". The LSTM will incorrectly identify it as the machinery technology.

In summary, CNN can not effectively determine the noise words before the entity, and LSTM can eliminate the noise words before the entity, but can not determine the noise words after the entity. Therefore, in this paper, we use BiLSTM model, a bi-directional LSTM model not only records the information of a text sequence from front to back,

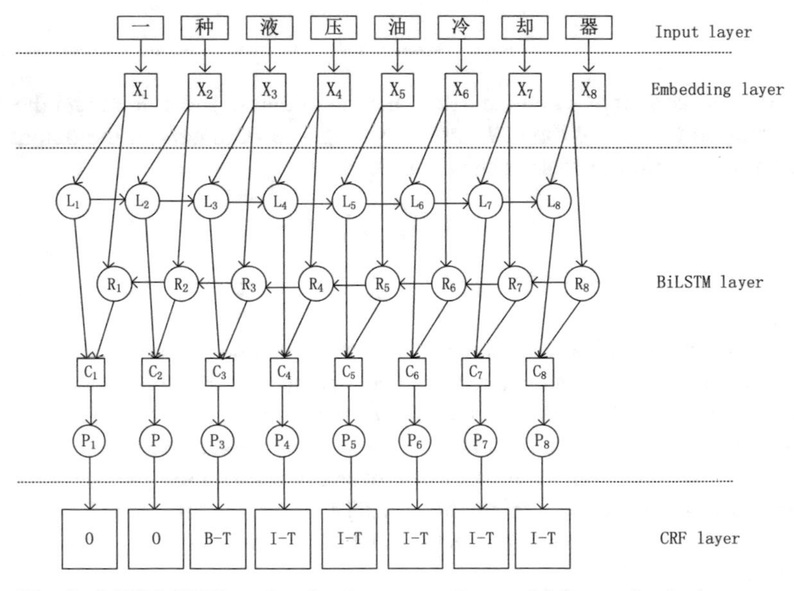

Fig. 2. LSTM-CRF based technology extraction model for mechanical patents.

but also records the information of a text sequence from back to front, which can better combine the contextual information and improve the accuracy of entity recognition. In addition, CRF model can improve the accuracy of the whole named entity extraction. BiLSTM-CRF model is shown in Fig. 2.

As in Fig. 2, the BiLSTM model is divided into four layers [11]: the first layer is the input layer, the second layer is the embedding layer, the third layer is the BiLSTM layer, and the fourth layer is the CRF layer. The input layer inputs a mech patent abstract text, and the encoding layer converts the input patent abstract encoding into a vector. The role of BiLSTM is to extract the abstract features of the text, and the role of CRF is to process the output of BiLSTM and exclude the tag sequences that do not conform to the sentence syntax rules. The output layer outputs the data after processing by CRF.

The input layer is the entrance of mechanical domain patents for technology and efficacy extraction, and the role of the input layer is to input the manually annotated completed mechanical patents into the BiLSTM model. Unlike most other named entity recognition models, the input of BiLSTM model is character-based, while other models are word-based, and the accuracy of other models in recognizing entities depends on the word separation accuracy. In contrast, BiLSTM with character as unit can largely avoid the problem of low recognition rate due to inaccurate word separation.

The embedding layer is to transform the mechanical patents from the input layer into a low-dimensional vector that the computer can recognize and compute. The embedding layer mainly matches each character in the mechanical patent x_i with a pre-trained word embedding matrix to obtain a word vector. e_i for each character. e_i is calculated using Eq. (11).

$$e_i = W^{d|V|} v_i \tag{11}$$

Here, |V| is the length of the pre-trained word list and d denotes the dimensionality of the word embedding matrix.

BiLSTM layer is used to extract sentence features. The BiLSTM layer consists of a forward LSTM and a reverse LSTM. Each word vector $X_i = (x_1, x_2, \ldots, x_n)$ is input to the LSTM model from the forward direction, and then the hidden state of the ith character $L_i = (L_1, L_2, \ldots, L_n)$ is output is the predicted label score of a character, which will be used as input to the CRF layer.

The purpose of the CRF layer is to correct the label prediction scores for characters that are inaccurate in the BiLSTM layer, for example, the label B-T will not follow the label I-T. CRF can label the entire sentence, which solves the dependency problem between adjacent labels, then CRF can correct this error and improve the accuracy of the model. CRF corrects the incorrect labels by using the transfer matrix $A_{y-1,y}$ to calculate the transfer probability of the (y-1)th label to the yth label. For example, a sentence input is $X = (x_1, x_2, \ldots, x_n)$, the output sequence obtained is $Y = (y_1, y_2, \ldots, y_n)$, the output sequence Y is calculated by Eq. (12):

$$S(X, Y) = \sum\nolimits_{t=1}^{n+1} (A_{y_{t-1},y_t} + P_{t,y_t}) \tag{12}$$

where P_{t,y_t} is the probability of label transfer, P_{t,y_t} can be used to measure the probability of x_t labeled as y_t. To compute the score of each training sample X with a possible labeled sequence y, a softmax function is then added to normalize all the scores. The sequence y is calculated by Eq. (13).

$$P(Y|X) = \frac{\prod_{t=1}^{n} e^{s(X,Y)}}{\sum_{\tilde{Y}\in Y_X} e^{s(\tilde{Y},X)}} \tag{13}$$

where $\tilde{Y}$ is all possible state sequences of the input sequence X. The logarithm of the maximum state sequence probability is usually used as the calculation result (14).

$$\log(P(Y|x)) = s(X, Y) - \log(\sum_{\tilde{Y}\in Y_X} e^{s(\tilde{Y},X)}) \tag{14}$$

Finally, the state sequence with the highest score Y^* as the final labeling result. Y^* The calculation method of (15) is used.

$$Y^* = argmax_{\tilde{Y}\in Y_X}(s(X, \tilde{Y})) \tag{15}$$

In this paper, BiLSTM-CRF model is used to extract the technology and efficacy of patents in the mechanical domain. BiLSTM can record the contextual information of a character in both directions for a long period of time, which is beneficial for understanding the context. At the same time, CRF can optimize the label score of a character through sentence level to select the best label, so as to improve the accuracy of the overall identification.

6 Experiments

6.1 Experimental Data

Experimental data used in this paper were 1077 patents in the field of machinery crawled from the State Intellectual Property Office. Mechanical patent entities are labeled through the Baidu AI platform. Three entity categories were designed for the dataset, namely: mechanical technology, mechanical components, and patent efficacy. The dataset is annotated using BIO, where B represents the beginning of an entity, I represents the middle of an entity, and O represents a non-entity. The specific annotations are shown in Table 1. The dataset is divided into training set, test set and validation set at 8:1:1 scale.

Table 1. Experimental BIO tag set.

Entity	Start tag	Middle or end tag
Mechanical technology	B-T	I-T
Mechanical components	B-C	I-C
Efficacy	B-E	I-E
Non-entity	O	O

6.2 Evaluation Criteria

In this paper, we use three evaluation metrics commonly used for named entity recognition: precision (P), F1 and recall (R). The specific formula is as follows:

$$P = \frac{T_p}{T_p + F_p} \times 100\% \tag{16}$$

$$R = \frac{T_p}{T_p + F_n} \times 100\% \tag{17}$$

$$F_1 = \frac{2PR}{P + R} \times 100\% \tag{18}$$

where T_p is the number of entities correctly identified by the model, and F_p is the number of entities that are incorrectly identified by the model, and F_n is the number of entities not identified by the model.

6.3 Experimental Environment and Parameters

This paper experiments with a machine with Windows operating system, Intel Core i7 CPU model, RTX 2060 graphics card model, and 16G video memory. Python version 3.8 and Torch 1.12.0 frameworks were built on the machine. The experimental parameters are listed in Table 2.

Table 2. Experimental parameters.

Parameters	Value
Batch_size	64
Learning rate	0.001
Epoch	100
Hidden size	128

6.4 Experimental Results

Table 3 show the identification scores of the optimal model for technology (T), efficacy (E), and components (C). The accuracy is 94.00%, the recall is 94.03%, and the F1 value is 93.99%. The efficacy of mechanical patent has the lowest recognition accuracy because not every mechanical patent abstract contains efficacy, the number of efficacy entities is relatively small, which affects the training of the model. In addition, the efficacy description is always vague, the labeling of which may be inaccurate.

Mechanical components have the best recognition result because they have the largest number of labels, and they have fixed term expressions, which makes their recognition easier.

However, it is relatively difficult to identify the machinery technology of the patent, which is often the name of the machinery patent. The technology of mechanical patent generally appears after the word "one" or "a/an". For example, in the sentence "one kind of self-lubricating device for oil cylinder ball head", "self-lubricating device" should be a technology, but whether "oil cylinder ball head" should also be marked as a part of the technical entity is ambiguous, making the recognition accuracy is naturally low.

Table 3. Experimental results.

Tag	Accuracy	Recall rate	F1
I-T	79.36	86.13	82.61
B-T	81.52	64.10	71.77
I-C	84.77	85.02	84.89
B-C	84.25	81.64	82.92
I-E	75.76	67.16	71.20
B-E	82.81	60.23	69.74
O	96.58	96.80	96.69
Avg	94.00	94.03	93.99

In order to verify the validity of the models described in this paper, several related models were compared in the same experimental setting, and the results of the comparison experiments are shown in Table 4. It can be seen that the deep learning models BiLSTM and BiLSTM-CRF work better than the traditional machine learning models. Also, we can see that not all deep learning models plus CRF achieve better results. The reason is that the recognition of technology and efficacy relies heavily on contextual relationships, which BiLSTM is able to obtain better than other models, and hence the results will be better.

Table 4. Relevant comparison experiments.

Models	Accuracy	Recall Rate	F1
HMM	88.87	86.94	87.70
BERT-CRF	81.39	78.35	81.26
BiLSTM	93.72	93.80	93.74
BiLSTM-CRF	94.00	94.03	93.99

In order to find the optimal parameters of the proposed model, Table 5 shows the experimental results of the proposed model for different parameter settings. Table 5 shows that the best results are obtained when the number of training rounds is 100, the learning rate is 0.01, and the LSTM hidden vector dimension is set to 128. Table 5 shows that overfitting occurs when the number of training rounds is too large, which leads to unsatisfactory training results. When the learning rate is too small, convergence is slow, resulting in lower recognition accuracy.

Table 5. Comparison experiments of different parameters of BiLSTM-CRF.

Parameter	Value	Accuracy	Recall rate	F1
Epoch	50	93.91	93.98	93.92
Epoch	100	94.00	94.03	93.99
Epoch	200	93.86	94.01	93.89
Learning rate	0.1	94.31	94.34	94.31
Learning rate	0.001	94.00	94.03	93.99
Learning rate	0.0001	93.06	93.14	93.03
Hidden size	128	94.00	94.03	93.99
Hidden size	256	93.99	94.03	94.00

7 Conclusion

Design experiments verify that the proposed model BiLSTM-CRF performs well in terms of technology and efficacy extraction in the mechanical patents. At the same time, the model in this paper suffers from several problems, such as the identification accuracy of the model is relatively low. The reason is that there is no standard dataset for patents in the mechanical domain, and our own constructed data set has many noises. At the same time, the description of mechanical patents is relatively confusing, which increases the difficulty of model learning. To effectively improve the accuracy, we will establish our own dataset annotation criteria to improve model identification accuracy.

References

1. Ma, J., Zhang, B., Zhang, S., Liu, S.: Named entity recognition of new energy vehicles based on active MCNN-SCRF. Comput. Eng. Appl. **55**(07), 23–29 (2019)
2. Qu, D., Yang, T., Zhu, Y., Hu, K.: Research on entity extraction of four diagnosis information of BiGRU-CRF lung cancer medical case based on word. Mod. Tradit. Chin. Med. Materia Medica World Sci. Technol. **23**(09), 3118–3125 (2021)
3. Shi, Z., Ma, Y., Zhao, F., Ma, B.: Chinese entity recognition based on CNN head transformer encoder. Comput. Eng. **48**(10), 1–10 (2022)

4. Saad, F., Aras, H., Hackl-Sommer, R.: Improving named entity recognition for biomedical and patent data using bi-LSTM deep neural network models. In: Métais, E., Meziane, F., Horacek, H., Cimiano, P. (eds.) NLDB 2020. LNCS, vol. 12089, pp. 25–36. Springer, Cham (2020). https://doi.org/10.1007/978-3-030-51310-8_3
5. Yu, H., Huang, H., Yu, Z., Tan, X., Lu, S.: Entity recognition of government official documents based on BiLSTM-CRF. Program. Skills Maint., 119–121 (2022)
6. Yang, Y., Li, Y., Zhong, X., Xu, L.: Named entity recognition of TCM medical records based on BiLSTM-CRF. Inf. Tradit. Chin. Med. **38**(11), 15–21 (2021)
7. Deng, N., Fu, H., Chen, X.: Named entity recognition of traditional Chinese medicine patents based on BiLSTM-CRF. Wirel. Commun. Mob. Comput. **2021**, 1–12 (2021)
8. Lafferty, J., Mccallum, A., Pereira, F.: Conditional random fields: probabilistic models for segmenting and labeling sequence data. In: Proceedings of the 18th International Conference on Machine Learning, pp. 282–289 (2001)
9. Hochreiter, S., Schmidhuber, J.: Long short-term memory. Neural Comput. **9**(8), 1735–1780 (1997)
10. Yang, P., Xie, Z.: Research on entity identification in judicial field based on BiLSTM-CRF. Mod. Comput. **38**(11), 3–8 (2020)
11. Zhao, L.: Research on a method of named entity recognition based on BiLSTM-CRF. Comput. Inf. Technol. **29**(02), 8–11+19 (2021)

Talent Incubator System: A Conceptual Framework of Employee Recruitment Strategy in Digital Era

Olivia Fachrunnisa(✉), Nurhidayati, and Ardian Adhiatma

Department of Management, Faculty of Economics, UNISSULA, Semarang, Indonesia
{olivia.fachrunnisa,nurhidayati,ardian}@unissula.ac.id

Abstract. Recruitment methods in business organization adopt many new approaches. Several approaches have attempted to adapt the benefits of digital technology, including the internet, data, and the web. However, based on existing literature, improvements are still needed to be applied to business intelligence, including combining the recruitment method with the onboarding method. This combination could improve the quality of the workforce by helping to close the gap between the competencies desired by the business and the competencies possessed by the workforce candidate. This study aims to propose a framework of talent incubator system to overcome the competence gaps between the need of business and workforce candidate. We establish a talent incubator prototype to enhance the quality of workforce competency in business organizations. As consequence, all parties, including the company, society, job providers, job seekers, and talent institutions, will benefit from matching the need for workforce competence.

Keywords: Competency · Workforce · Recruitment · Talent management

1 Introduction

Recruitment is one of the important aspects of Human Resources Management (HRM) practices in meeting the availability of a workforce according to qualifications to achieve a competitive advantage in a business organization. A business organization, large, small, or micro, faces competition in recruiting the qualified workforce candidate from a society as a job seeker. Business intelligence can take advantage of the response from job seekers during selection process in recruitment to attract and retain intangible, rate, and unique competencies. It is essential to investigate the embedded talent to manage the competency of the workforce in the recent digital age, as the benchmark of business's success is determined by the contribution of the workforce [1]. Meanwhile, globally, the workforce landscape will experience endurable changes related to the latest technology such as demands for automation, virtual cooperation, IoT, AI, Blockchain, and so forth.

Recent research on Human Resources (HR) digital recruitment shows the relevance of social networks as a determinant of the success of sustainable HR activities [2]. Several studies related to the use of digital technology in HR activities, especially recruitment in small and medium scale business have also been attempted [3–5].

L. Barolli (Ed.): EIDWT 2023, LNDECT 161, pp. 235–242, 2023.
https://doi.org/10.1007/978-3-031-26281-4_23

More specifically on HR activities in the terms of training and recruitment, recent trends in the use of digital technology have been attempted by Hassan Onik et al. [6]; Jeong and Choi [7], Dhanala and Radha [8]; Serranito et al. [9]; Al Hamrani and Al Hamrani, [10] by focusing on recruitment systems in design and application. However, from several studies related to the new technology usage for recruitment, nothing has led to how to match competence availability from job seekers to match the competence qualification of the job provider to the work formation required by the business organization. A few studies also state that HRM practices frequently encounter problems related to workforce recruitment qualification, and are not always able to employ competence workforce in accordance with the expected performance [11, 12]. As a consequence, finding a match workforce with high competence in accordance with the standards required to fill the work formation position provided by the job supplier remains a challenge. One of them is that the workforce that has managed to pass the selection procedure in the recruitment process to take up the available work formation has not been able to meet the minimum competency [4, 13–16].

The talent incubation in recruitment process transformation is expected to enhance workforce competence in accordance with the competency needs of small business in creative industries. Utilization of technology in the form of existing training to meet workforce competency requirement. Recent studies related to the use of digital technology have been proposed for mitigating competency gaps in the workforce. For example, Fachrunnisa and Hussain [17] have proposed the development of a blockchain-based HR framework to match company needs and workforce competencies by involving a corporate training center for skills gap mitigation. Hence, the needs of the workforce with the qualifications required by the business in such industries will always fit the current situation in the long run. As a result, Blockchain assist in analyzing the information and data effectively. However, it has not led to the talent training building serving as an incubator for workforce candidates to learn relevant knowledge, thereby providing an outcome on the attainment of the competencies required to fill various work formations. To address the aforementioned gaps, we propose a new recruitment strategy in the form of a Talent Incubator design. The prototype could serve as a meeting place for job providers to create a workforce with standard competencies that match the work formation available from job providers based on the talent to be transformed into the competencies required.

2 Literature Review

2.1 Recent Technology Application for Recruitment System

Few research has been provided new technological application for the recruitment management system [3, 17]. The results of case study revealed that the proposed system outperformed the current HRM system Hassan Onik et al. [6]; developed a digital certificate management platform to be used in recruitment processes; and used the Blockcerts platform to manage blockchain services and digital certificates. On this framework, applicants receive certificates based on their previous performance, the validity of the certificates is checked, all certificates are stored, and changes to saved record keeping are impossible. Jeong and Choi [7] emphasized that blockchain technology is an effective tool for verifying and securing data, and developed a recruitment system supported by blockchain technology in their study. In their study, they explained the process of this system supported by blockchain technology. They described the system's operation as follows: First and foremost, the recruitment firm submits the candidate list to the system, which automatically checks the candidate information from databases such as schools and law enforcement agencies, and the verified data is stored in the blockchain. Furthermore, the company evaluates the approved candidates on the blockchain and decides whether or not to hire.

This system, which has been developed through smart contract tools on Ethereum platform, has been tested via Rinkeby and Ganache Software. Dhanala and Radha [8]; Serranito et al. 2020 [9] designed an ecosystem for certificate validation on a global scale. This ecosystem is based on the blockchain running smart contracts on the Ethereum platform. While higher education institutions in the ecosystem automatically save their education certificates to the blockchain, recruitment companies can question the certificate accuracy of the candidate by examining the registered information. The system developed in this study was applied to recruitment process in the public sector through a pilot study; by using the science of design research method, presented a model suggestion in which blockchain technology is used for the recruitment processes of people with disabilities in the United Arab Emirates. By means of this model proposal, they aimed to provide the authorities with decision support by accessing the education, health, course and promotion information of disabled people accurately and at low cost [10].

Moreover, [3] pointed out that the candidate information obtained in the traditional recruitment process has many shortcomings such as the risk of being inaccurate, the risk of data loss, being costly and requiring a long time in the data validation process. Moreover, they stated that blockchain technology can minimize the disadvantages of traditional recruitment methods. Several stages of the blockchain-based recruitment and selection process as follows: First, the candidates register to the systems. Second, the candidate information is verified via the blockchain system. Third, the managers make decision and employees are hired in right positions.

Of all the studies that have been carried out related to the use of new technology in the recruitment process, there are still several obstacles, for example, related to security and privacy, a potential risk in employee data, loss of privacy due to data transparency, problems of scalability and cost-effectiveness (system ability to maintain its performance), organizational business model mismatch, lack of knowledge, skills, and abilities of practitioners, and lack of cooperation between stakeholders.

Of the several technology application in the field of recruitment that already exist, among others; Jobstreet, LinkedIn, Karir.com, Glints, Urbanhire, and Kalibrr Disnakertrans [18], have not included the availability of training to produce workforce competence according to the company's needs in the recruitment process. Karir.com and Glints have tried to include training activities in the recruitment process, but they have not been optimal in how to producing training graduates into a workforce that is truly competent to match the work formation required by the company. Furthermore, efforts to match the standard competence needed by the company and the availability of competence from job seekers in 'E-Makaryo' have also been carried out by [19]. The 'E-Makaryo' system has attempted to bring together job providers and BLK (Job Training Centers) throughout Central Java, involving the provincial and district/city governments as managers. This system offers an advantage such as it has tried to reach the entire society from the provincial level to the district/city level. However, job matching efforts are still limited based on the terms of job seeker location (province, district, or city), education, and keywords (for example, desired field of work).

2.2 Talent Management

The essence of talent refers to humans as a workforce who would have a major impact on business performance [20]. This could be through their abilities or, in the long term, by achieving their full potential. Talent management is the method of understanding what talent exist within an business, what talent populations are needed, and investigate workforce/candidates who are particularly valuable to the business by using data from the workforce, performance management, and preparedness tools [21]. It is critical to be able to identify and recruit the most effective methods for developing and retaining talent. It may necessitate specific business intervention, such as development programs, but it is also about the capability to develop custom-designed program to suit employee-employer specific needs. In essence, talent management focuses on how to attract, investigate, involve, retain, and place workforce/candidates who are a real benefit to a business organization. This could be due to their potency or even because they fulfill a crucial role on job formation.

Job will display numerous novel problems in the near future, for example, as automation and AI replace more routine roles and the employment market for highly qualified workers thickens [22]. Business actors will have to concentrate more on advancing than on recruiting new employees. A customizable, institution talent management strategy focuses on human capital investment and elevates talent acquisition to the top of business agenda. To achieve business performance, talent management can contribute to strategic goals such as meaningful work, increased productivity, a long-term learning culture, and high-quality workforce competence.

The talent survey report states several recommendations regarding future talent management [23]. Those recommendation include: remaining focused on skill development, even in difficult times such as pandemic, protecting the strategic learning and development budget; continuing to invest in ways for workforce candidates to access opportunities such as apprenticeships, training, industry placements, and post-A-level routes; ensuring skills training investigate 'essential' traceable competence gap such as problem-solving, teamwork, and communication; and ensuring competence building address 'vital' relevant skills. The emergence technology offers customized solutions based on personalization to support business and workforce matching based on talent in recruitment strategies.

Based on several previous studies and various forms of recruitment strategy to close the gap competences, it can be concluded that there is a need for new model that systematically engaged job provider and job seeker integrated into digital infrastructure in the form of a talent incubator.

3 Proposed Framework

This research proposes a conceptual framework of incubator system by which job seeker and job provider can communicate automatically with the help of AI. In this scenario, talent incubator is a place by which for workforce candidate will have better preparation to attend the recruitment process. Likewise, from job provider interest, it can be easier to find candidate that meet the selection criteria. The framework can be visualized in Fig. 1.

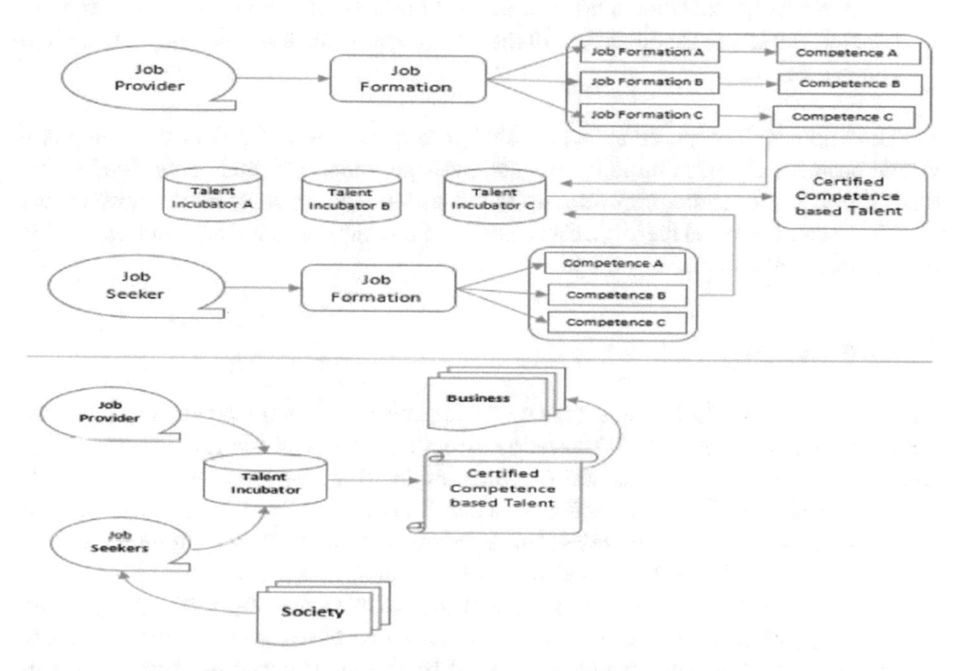

Fig. 1. Conceptual framework of talent incubator system

The detailed working of our framework is as follows:

1. Society
   Society consists of people with different educational background, competencies, work experience, geographical locations, etc.
2. Job Seeker
   Job seekers come from a society that requires information services to find the desired job according to their competence.
3. Job Provider
   A job provider is a provider of work formation data and information needed by the competency standard qualifications required by the company.
4. Talent Incubator
   A talent incubator is a learning and virtual business organization that provides talent investigation and management to train, develop and retain the required competencies based on work formation information from job providers. This incubator

will train job seekers based on the talent that is embedded in them to be trained and developed so that it can be transformed into several competencies needed to fill several available work formations. Furthermore, the talent incubator processes information related to several competencies-based talent needed in curriculum materials to train job seekers. The outcome of incubator talent in the form of workforce competence matches creative industry business competence needs in the form of certified competence-based talent.

5. Business Organization

   Business organization consist of several business in specific sector that require talented workforce candidates to fill the workforce formation following the required competencies.

The framework proposed aims to provide a platform in the form of a learning and virtual business organization to investigate, manage, train, and retain embedded workforce talent. Furthermore, transforming talent into matching competence between the workforce candidate and the job seekers as part of a society with various work formation and competencies required by the job provider.

4 Conclusion

The urgency of the talent incubator as a new approach to match the competence of the workforce of Job Seekers (candidates/early career workers) against the minimum to high competence required by the business organization to fill several job formations available from Job Providers. The talent incubator is a link in the input-to-output process. In this study, input means the talent embedded as resource of competencies required by the small business creative sector provided by job providers as a part of business organization. Process means approval from the talent incubator as investigation, management, and training provider. Output means the competence- based talent of the workforce that is by the qualification standards required by the company to fill several job formations. Meanwhile, to get data and information about the competencies required by job providers, we use technology applications. A technology website can be used as an incubator to transform the tasks of human resources in the field of recruitment strategy. Future research is needed to validate the proposed framework by using artificial intelligence and blockchain.

References

1. Caracol, C., Dias, I.: Business intelligence: an essential tool in the identification of organizational internal talent. In: Rocha, A., Correia, A.M., Costanzo, S., Reis, L.P. (eds.) New Contributions in Information Systems and Technologies. Advances in Intelligent Systems and Computing, vol. 354, pp. 93–104. Springer, Cham (2015)
2. Oncioiu, I., Anton, E., Ifrim, A.M., Mândricel, D.A.: The Influence of social networks on the digital recruitment of human resources: an empirical study in the tourism sector. Sustainability **14**, 1–14 (2022)

3. Rhemananda, H., Simbolon, D.R., Fachrunnisa, O.: Blockchain technology to support employee recruitment and selection in industrial revolution 4.0. In: Pattnaik, P.K., Sain, M., Al-Absi, A.A., Kumar, P. (eds.) Proceedings of International Conference on Smart Computing and Cyber Security: Strategic Foresight, Security Challenges and Innovation (SMARTCYBER 2020), pp. 305–311. Springer Singapore, Singapore (2021). https://doi.org/10.1007/978-981-15-7990-5_30
4. Turcu, C.E., Turcu, C.O.: Digital transformation of human resource processes in small and medium sized enterprises using robotic process automation. Int. J. Adv. Comput. Sci. Appl. **12**, 70–75 (2021)
5. Trașca, D.L., Ștefan, G.M., Sahlian, D.N., Hoinaru, R., Șerban-Oprescu, G.L.: Digitalization and business activity. The struggle to catch up in CEE countries. Sustainability **11**(8), 2204 (2019).https://doi.org/10.3390/su11082204
6. Hassan Onik, M.M., Miraz, M.H., Kim, C.S.: A recruitment and human resource management technique using blockchain technology for industry 4.0. In: IET Conference Publications 'Smart Cities Symposium 2018, pp. 1–16 (2018)
7. Jeong, W.Y., Choi, M.: Design of recruitment management platform using digital certificate on blockchain. J. Inf. Process. Syst. **15**, 707–716 (2019)
8. Dhanala, N.S., Radha, D.: Implementation and testing of a blockchain based recruitment management system. In: Proceeding of the Fifth International Conference on Communication and Electronics Systems (ICCES 2020) IEEE Conference, pp. 583–588. IEEE Xplore (2020). https://doi.org/10.1109/icces48766.2020.9138093
9. Serranito, D., Vasconcelos, A., Guerreiro, S., Correia, M.: Blockchain ecosystem for verifiable qualifications. In: 2020 2nd Conference on Blockchain Research and Applications for Innovative Networks and Services, BRAINS 2020, pp. 192–199 (2020). https://doi.org/10.1109/BRAINS49436.2020.9223305
10. Al Hamrani, N.R., Al Hamrani, A.R.: People of determination (disabilities) recruitment model based on blockchain and smart contract technology. Technol. Invest. **12**, 136–150 (2021)
11. Sarma, J.G., Shrivastava, A.: Enhancing training effectiveness for organizations through blockchain-enabled training effectiveness measurement (BETEM). J. Org. Chang. Manage. **34**, 439–461 (2021)
12. Margherita, E.G., Bua, I.: The role of human resource practices for the development of operator 4.0 in industry 4.0 organisations: a literature review and a research agenda. Businesses **1**, 18–33 (2021)
13. Nasir, S.Z.: Emerging challenges of HRM in 21st century: a theoretical analysis. Int. J. Acad. Res. Bus. Soc. Sci. **7**, 216–223 (2017)
14. Padachi, K., Bhiwajee, S.L.: Barriers to employee training in small and medium sized enterprises: insights and evidences from Mauritius. Eur. J. Train. Dev. **41**, 232–247 (2015)
15. Forbes. Why Business Should Matches Employee Skills with Job Roles. Innovation (2022). https://www.forbes.com/sites/forbestechcouncil/2022/03/18/why-businesses-should-match-employee-skills-with-job-roles/?sh=2d8ec1507784. Accessed 16 Sep 2022
16. B20 Indonesia. 'Link & Match' Education and Industry Solutions for Indonesia's Future of Work (2022). https://b20indonesia2022.org/updates/%27link-&-match%27-education-and-industry-solutions-for-indonesia%27s-future-of-work-48. Accessed 18 Sep 2022
17. Fachrunnisa, O., Hussain, F.K.: Blockchain based human resource management practices for mitigating skills and competencies gap in workforce. Int. J. Eng. Bus. Manage. **12**, 1–11 (2020)
18. Disnakertrans. 6 Situs Loker Terpercaya dan Resmi di Indonesia (2020). https://disnakertrans.ntbprov.go.id/6-situs-lowongan-kerja-terpercaya-dan-resmi-di-indonesia/
19. Dinas Tenaga Kerja dan Transmigrasi Provinsi. E-makaryo. Bursa kerja (2022). https://bursakerja.jatengprov.go.id/

20. Ropper, J.: What do we mean when we talk about talent. HR Mag. (2015)
21. CIPD. Talent Management. Specialist Knowledge (2022). https://peopleprofession.cipd.org/profession-map/specialist-knowledge/talent-management#gref. Accessed 4 Nov 2022
22. Glaister, A.J., Karacay, G., Demirbag, M., Tatoglu, E.: HRM and performance—the role of talent management as a transmission mechanism in an emerging market context. Hum. Resour. Manage. J. **28**, 148–166 (2018)
23. CIPD Asia. Talent management: understand the changing context and business case for talent management, and the key features of a talent management strategy. Knowledge 1–8 (2022). https://www.cipd.co.uk/knowledge/strategy/resourcing/talent-factsheet#gref. Accessed 3 Nov 2022

Thai Word Disambiguation: An Experiment on Thai Language Dataset with Various Deep Learning Models

Nontakan Nuntachit[1,2](✉), Karn Patanukhom[3], and Prompong Sugunnasil[4]

[1] Data Science Consortium, Faculty of Engineering, Chiang Mai University, Chiang Mai, Thailand
nontakan_n@cmu.ac.th
[2] Faculty of Medicine, Chiang Mai University, Chiang Mai, Thailand
[3] Deparment of Computer Engineering, Faculty of Engineering, Chiang Mai University, Chiang Mai, Thailand
karn@eng.cmu.ac.th
[4] College of Arts, Media and Technology, Chiang Mai University, Chiang Mai, Thailand

Abstract. Recent advances in Natural Language Processing (NLP), the transformer and BERT models, have brought us more performance. However, one still challenging task in NLP is the word disambiguation task. Even in English, this task requires complex language model to achieve. Thai is one of the most difficult languages to learn and understand, even for Thai people. To understand the true meaning of an ambiguous word, the context of the sentence is needed. Attention mechanism, which is the core of both transformer and BERT, can be used to understand underlying context of sentences and hence can be used for disambiguation. In this article, we evaluate both unsupervised and supervised BERT based models for disambiguating the Thai word "เขา" (Khao). Zero-shot classification models (unsupervised) can achieve an accuracy of around 60–70%, either Thai language fine-tuning models or other cross languages BERT-based models. On the other hand, fine-tuning supervised BERT-based models can achieve an accuracy of up to 95% from the best model we fine-tuned.

1 Introduction

One of the most important challenges in Natural Language Processing (NLP) is word sense disambiguation (WSD), which resolves natural language ambiguities. It involves choosing the best meaning for an ambiguous term in a given context. For large-scale language understanding applications and related activities like machine translation (MT), information retrieval (IR), and natural language understanding (NLU), word ambiguity resolution is the main bottleneck. These natural language processing applications require word meaning to identify the correct word sense in a situation. One meaning of the Thai term เขา "khao", for instance, can be translated into English as "him", "horn" and another as "mountain."

Thai is one of the most difficult languages to learn and understand even for Thai people. As the language has no space between words, has many vowels, one word can

L. Barolli (Ed.): EIDWT 2023, LNDECT 161, pp. 243–247, 2023.
https://doi.org/10.1007/978-3-031-26281-4_24

be written in many forms, etc. [1, 2]. So, dealing with word disambiguation in Thai is also a challenging task too.

There is recent advance in NLP introduced by Google called "attention" and "transformer" in 2017 [3] and after that they developed deep learning model called "BERT" [4], which utilized the encoder part of the transformer. BERT can be trained with specific task such as classification, question-answering, sentence tagging. After BERT has been released, there are many variants of BERT which tried to improve the performance or do some more specific tasks [5]. In this paper, we do an experiment of BERT-variant models on word sense disambiguation task.

2 Related Works

Since 2001, there have been many approaches that aim for disambiguation of Thai words, including Support Vector Machine (SVM), Sparse Network of Winnow (SNOW), Nave Bayes, and others. The most recent work we found is from 2015 [6].

2.1 TH_WSD [6]

This technique uses cross language knowledge approach by using AsianWordNet (AWN) and PrincetonWordNet (PWN). They compare ambiguous word by translating Thai word to English by AWN and PWN then compare similarity from translated words. This approach uses many resources for comparing and building the database, but the accuracy is around 70%.

2.2 Latent Semantic Indexing [7]

A vector-based semantic model called LSI is based on word cooccurrences [8]. Semantically related words are those that appear together in similar circumstances.

Words that co-occur and have a syntagmatic or thematic relationship are also semantically comparable. The corpus of documents is used to train the LSI algorithm. In this context, a document is any group of words that make up a sentence, a paragraph, an article, etc.

Then, to make the LSI model for the experiments in this paper, a large co-occurrence matrix of documents is made, with words in the vocabulary in the rows and documents in the columns. The weighted frequency of the associated term in the corresponding document is represented by each element in the matrix.

Then the matrix dimension is reduced by using Singular Value Decomposition (SVD). Cosine similarity between each vector corresponding to word context is calculated to determine the correlation between words. This approach requires no model training and depends on the context vector. However, when considering the calculation for matrix reduction and cosine similarity, it requires high computation.

2.3 Decision List Collocation [9]

This is the first work we found to deal with word sense disambiguation. They use the decision list collocation method to create a prototype word sense disambiguation program in Thai. The representative for nouns is 'hua' whereas the representative for verbs is 'kep'. In order to create a manually sense-tagged corpus based on the Thai dictionary of "The Royal Institute" and 1800 samples of 'hua' and 1800 samples of 'kep' obtained from a corpus of Bangkok Business newspaper, we examined the senses of 'hua' and 'kep'.

Nine senses of 'hua' and twenty senses of 'kep' were discovered. To test for the ideal span and the position of sense indicators, we trained the algorithm at twenty spans of collocation and obtained twenty decision lists and tested the algorithm with these choice lists. The sense with the largest collocational weight was the one the algorithm decided to choose. As a result, it appears that the span 2 is adequate for deciphering both words, and the sense indicators for both words are primarily on the right side. The best span for distinguishing between 'hua' and 'kep' is one word to the right and left, with a precision rate of 87%, and two words to the right, with a precision rate of 80.25%. The decision list algorithm utilized in this study may be suitable for the task, based on the high precision rate.

3 Data and Methodology

In this work, we selected Thai word 'khao' which has three meanings.

(Noun) mountain: Class M
(Pronoun) he: Class P
(Noun) horn: Class H

The dataset used in this work was generated by users. Sentences in the dataset have different lengths. But the distribution for each class is balanced. There are 2874 sentences in the training set. For the test set (unseen), the number is 200 sentences.

Deep learning model (BERT-variants) we selected for unsupervised classification are RoBERTa [10, 11], DeBERTa [12] and WangchanBERTa [13]. Criteria for model selection in unsupervised task is the model that contain Thai language (cross language, xlm) and able to do zero-shot classification. For supervised model, we trained RoBERTa-base and Bert-th [14].

4 Results

Table 1 show the result from all models we used and trained in this study. The highest accuracy came from RoBERTa-base (supervised model). Unsupervised model that has highest accuracy is RoBERTa following by DeBERTa and the lowest accuracy is WangchanBERTa.

Table 1. Result from all deep learning model in this study

Model	Accuracy	Precision	Recall	F1-score	Support
Unsupervised Model					
RoBERTa	**0.73**				
Class H		0.72	0.81	0.76	67
Class M		0.84	0.58	0.69	65
Class P		0.68	0.79	0.73	68
DeBERTa	**0.72**				
Class H		0.75	0.63	0.68	67
Class M		0.86	0.57	0.69	65
Class P		0.53	0.79	0.64	68
WangchanBERTa	**0.29**				
Class H		0.34	0.19	0.25	67
Class M		0.23	0.08	0.11	65
Class P		0.29	0.6	0.39	68
Supervised Model					
BERT-th	**0.72**				
Class H		0.73	0.85	0.79	67
Class M		0.7	0.78	0.74	65
Class P		0.73	0.53	0.62	68
RoBERTa-base	**0.94**				
Class H		0.96	0.99	0.97	67
Class M		0.88	0.98	0.93	65
Class P		1.00	0.85	0.92	68

5 Discussion

In this study, we adapted the usability of a deep learning model for a word sense disambiguation task on Thai word. To the best of our understanding, our work is the first to use this approach. The result shows that Supervised-model still outperform Unsupervised-model but considering that Unsupervised-model require no training and can be used right out of the box, this should be an advantage of this type of model. Our study found that RoBERTa is still the best model for classification. The study from Microsoft suggests that the DeBERTa model can overcome RoBERTa in the SuperGLUE benchmark [15], however, in our study, we cannot train DeBERTa to be a supervised model due to different model architecture and software package incompatible. Surprisingly, the WangchanBERTa which is the native Thai transformer model, gives us the lowest accuracy. We tried to investigate this issue and found that this model's vocabulary contained Thai phrases instead of words; this can be explained as to why the model gave such a poor result. The vocabulary in the zero-shot model tends to affect the model's accuracy, as discussed in our previous work [16].

6 Conclusion

This study demonstrates the usability of BERT-variant models on a word sense disambiguation task. The supervised model can achieve an accuracy of 94%. However, if we consider the model that requires no training, the zero-shot model still gives an acceptable result. The vocabulary in the zero-shot model should be considered when we need the model for a classification task.

Acknowledgments. This work was supported by Erawan HPC Project, Information Technology Service Center (ITSC), Chiang Mai University, Chiang Mai, Thailand. The authors would like to thank Faculty of Engineering and ITSC staff for supporting us in this study.

References

1. Is Thai Hard to Learn? Honest Answer with Learning Tips. https://en.amazingtalker.com/blog/en/other/63705/. Accessed 10 Dec 2022
2. (1) Why is the Thai language so difficult? – Quora. https://www.quora.com/Why-is-the-Thai-language-so-difficult. Accessed 10 Dec 2022
3. Vaswani, A., et al.: Attention is all you need. Adv. Neural Inf. Process. Syst. **30**, 5998–6008 (2017)
4. Devlin, J., Chang, M.-W., Lee, K., Toutanova, K.: BERT: pre-training of deep bidirectional transformers for language understanding. In: NAACL HLT 2019 – 2019 Conference of the North American Chapter of the Association for Computational Linguistics: Human Language Technologies – Proceedings of the Conference, vol. 1, pp. 4171–4186 (2018). http://arxiv.org/abs/1810.04805. Accessed 30 Sep 2020
5. BERT Variants and their Differences – 360DigiTMG. https://360digitmg.com/blog/bert-variants-and-their-differences. Accessed 10 Dec 2022
6. Mitrpanont, J.L., Chongcharoen, P.: TH_WSD: Thai word sense disambiguation using cross-language knowledge sources approach. Int. J. Comput. Theory Eng. **7**(6), 428–433 (2015). https://doi.org/10.7763/ijcte.2015.v7.997
7. Pongpinigpinyo, S., Rivepiboon, W.: Distributional Semantics Approach to Thai Word Sense Disambiguation (2008). https://doi.org/10.5281/ZENODO.1079186
8. Deerwester, S.C., Dumais, S.T., Landauer, T.K., Furnas, G.W., Harshman, R.A.: Indexing by latent semantic analysis. J. Am. Soc. Inf. Sci. **41**, 391–407 (1990)
9. Kanokrattananukul, W.: Word sense disambiguation in Thai using decision list collocation (2001)
10. Liu, Y., et al.: "RoBERTa: A Robustly Optimized BERT Pretraining Approach 1 (2019). http://arxiv.org/abs/1907.11692
11. joeddav/xlm-roberta-large-xnli…Hugging Face. https://huggingface.co/joeddav/xlm-roberta-large-xnli. Accessed 26 Mar 2022
12. He, P., Liu, X., Gao, J., Chen, W.: DeBERTa: decoding-enhanced BERT with disentangled attention. In: 2021 International Conference on Learning Representations (2020). http://arxiv.org/abs/2006.03654
13. Lowphansirikul, L., Polpanumas, C., Jantrakulchai, N., Nutanong, S.: WangchanBERTa: Pretraining transformer-based Thai Language Models (2021). http://arxiv.org/abs/2101.09635
14. monsoon-nlp/bert-base-thai…Hugging Face. https://huggingface.co/monsoon-nlp/bert-base-thai. Accessed 10 Dec 2022
15. Microsoft DeBERTa surpasses human performance on the SuperGLUE benchmark – Microsoft Research. https://www.microsoft.com/en-us/research/blog/microsoft-deberta-surpasses-human-performance-on-the-superglue-benchmark/. Accessed 10 Dec 2022
16. Nuntachit, N., Sugunnasil, P.: Do we need a specific corpus and multiple high-performance GPUs for training the BERT model? An experiment on COVID-19 dataset. Mach. Learn. Knowl. Extr. **4**(3), 641–664 (2022). https://doi.org/10.3390/make4030030

Energy-Efficient Locking Protocol in Virtual Machine Environments

Tomoya Enokido[1(✉)], Dilawaer Duolikun[2], and Makoto Takizawa[3]

[1] Faculty of Business Administration, Rissho University, 4-2-16, Osaki, Shinagawa-ku, Tokyo 141-8602, Japan
eno@ris.ac.jp

[2] Department of Advanced Sciences, Faculty of Science and Engineering, Hosei University, 3-7-2, Kajino-cho, Koganei-shi, Tokyo 184-8584, Japan

[3] Research Center for Computing and Multimedia Studies, Hosei University, 3-7-2, Kajino-cho, Koganei-shi, Tokyo 184-8584, Japan
makoto.takizawa@computer.org

Abstract. A huge number of transactions has to be concurrently performed to design and implement current distributed applications. However, each server in a system consumes the large amount of electric energy to perform methods issued by transactions. In this paper, the EE2PL-VM (Energy-Efficient Two-Phase Locking in Virtual Machine environment) protocol is newly proposed to not only serialize conflicting transactions but also reduce the total processing electric energy consumption (PEE) of servers. The evaluation results show the EE2PL-VM protocol can reduce the total PEE of servers compared with the 2PL (Two-Phase Locking) protocol.

Keywords: EE2PL-VM protocol · Energy-efficient systems · Concurrency control · Virtual machine · Object-based systems

1 Introduction

A huge volume and various kinds of data has to be collected by using various types of IoT (Internet of Things) devices [1] and network systems to design and implement current distributed applications. An object [2,3] is an encapsulation of data and methods to manipulate the data like a database system [4,5]. Each object holds a collected data and provides methods to manipulate the data. A transaction [4,5] is defined to be an atomic sequence of methods supported by objects. In object-based systems [2,3], a transaction on a client issues methods to objects in order to use a provided application service. In information systems like a cloud computing system [6–8], a large number of transactions are concurrently performed on each object. Hence, high performance and scalable computing systems are essential to design and implement application services. A server cluster equipped with virtual machines [6–8] is one way to realize a high performance and scalable computing system. Each object is allocated to a virtual machine supported by physical server in a server cluster. On the other

L. Barolli (Ed.): EIDWT 2023, LNDECT 161, pp. 248–255, 2023.
https://doi.org/10.1007/978-3-031-26281-4_25

hand, all the objects have to be kept mutually consistent. Hence, conflicting methods issued by transactions have to be serialized on each object [2,4,5].

The *2PL* (*Two-Phase Locking*) protocol [5,9,10] is so far widely used to serialize conflicting transactions. In the 2PL protocol, the overhead to lock object increases if a large number of transactions is concurrently performed on each object. As a result, the throughput of each application decreases. In addition, a large amount of electric energy is consumed in each server to perform methods issued by transactions on objects stored in virtual machines. In our previous studies, the *EE2PL* (*Energy-Efficient Two-Phase Locking*) protocol [9] is proposed to reduce the total processing electric energy consumption (PEE) of servers. In EE2PL protocol, *meaningless write methods* [9] are not performed on each object. As a result, the total PEE of servers can be reduced. However, the EE2PL protocol assumes each object is directly allocated to a physical server. Hence, the EE2PL protocol cannot be adopted to a server cluster equipped with virtual machines. In this paper, the *EE2PL-VM* (*Energy-Efficient Two-Phase Locking in Virtual Machine environment*) protocol is newly proposed by improving the EE2PL protocol to take into account a server cluster equipped with virtual machines. The EE2PL-VM protocol is compared with the 2PL protocol in a server cluster equipped with virtual machines in the evaluation. In the evaluation, we show the average execution time of each transaction and the total PEE of servers can be reduced in the EE2PL protocol than the 2PL protocol.

The system model, data access model, and power consumption model are presented in Sect. 2. The EE2PL-VM protocol is discussed in Sect. 3. The evaluation of the EE2PL-VM protocol is shown in Sect. 4.

2 System Model

2.1 Servers and Objects

A cluster S is composed of physical servers $s_1, ..., s_n$ $(n \geq 1)$. A term *server* means a physical server in this paper. One multi-core CPU is equipped in each server s_t. A CPU in each server s_t holds the total number nc_t (≥ 1) of cores. Each server s_t holds a set C_t of cores $c_{1t}, ..., c_{nc_t t}$. Each core c_{lt} in a server s_t holds the total number ct_t $(ct_t \geq 1)$ of threads. Hence, a server s_t supports a set TH_t of threads $th_{1t}, ..., th_{nt_t t}$ where $nt_t = nc_t \cdot ct_t$. In this paper, each virtual machine VM_{kt} is performed by occupying one thread th_{kt} in a server s_t. Let NV_t be a set of virtual machines $VM_{1t}, ..., VM_{nt_t t}$ supported by a server s_t. Let O be a set of objects $o_1, ..., o_m$ $(m \geq 1)$ [2,3]. Each object o_h supports *write* and *read* methods. There are *partial* and *full* types of write methods. A partial write method writes a part of data in an object o_h. A full write method writes a whole data in an object o_h. A read method fully reads data in an object o_h. Let $op^i \circ op^j(o_h)$ be a state of an object o_h where a method op^j is performed after a method op^i. A method op^i *conflicts* with another method op^j on an object o_h if and only if (iff) $op^i \circ op^j(o_h) \neq op^j \circ op^i(o_h)$. Otherwise, a pair of methods op^i and op^j are *compatible* on an object o_h. The conflicting relation among methods is symmetric.

2.2 2PL Protocol

Each *transaction* T^i [5] is an atomic sequence of methods supported by objects. A set $\mathbf{T} = \{T^1, ..., T^k\}$ $(k \geq 1)$ of transactions on clients issue methods to manipulate objects. In order to keep all the objects mutually consistent, conflicting transactions have to be *serializable* [4,5]. A notation sc shows a schedule of transactions in a set $\mathbf{T}$. A transaction T^i *precedes* another transaction T^j ($T^i \rightarrow_{sc} T^j$) in a transaction schedule sc iff a method op^i issued by the transaction T^i conflicts with another method op^j issued by the transaction T^j and op^i is performed before op^j. A transaction schedule sc is serializable iff the precedent relation $\rightarrow_{sc}$ is acyclic [5]. In order to make a schedule sc serializable, the $2PL$ (*Two-Phase Locking*) protocol [4,5,10] is proposed. Let $lm(op)$ indicate a *lock mode* of a method op. Each transaction T^i locks an object o_h by a lock mode $lm(op)$ before manipulating the object o_h by a method op in the 2PL protocol. An object o_h is manipulated by a transaction T^i with a method op if the object o_h can be locked in a lock mode $lm(op)$ by the transaction T^i. Locks on each object o_h are released when transaction T^i commits or aborts.

2.3 Data Access

The *DAVM* (*Data Access model for Virtual Machine environments*) model [7] is proposed in our previous studies to perform read and write methods on objects supported by virtual machines in a server s_t. Let $r^i_{kt}(o_h)$ and $w^i_{kt}(o_h)$ be read and write methods, respectively, issued by a transaction T^i to manipulate an object o_h supported by a virtual machine VM_{kt}. Notations $W_t(\tau)$ and $R_t(\tau)$ show sets of write and read methods being performed on every virtual machine in a server s_t at time τ, respectively. Each read method $r^i_{kt}(o_h)$ reads data in an object o_h supported by a virtual machine VM_{kt} at read rate $RR^i_{kt}(\tau)$ [B/sec] at time τ. Each write method $w^i_{kt}(o_h)$ writes data in an object o_h supported by a virtual machine VM_{kt} at write rate $WR^i_{kt}(\tau)$ [B/sec] at time τ. The maximum read rate $maxRR_t$ [B/sec] and the maximum write rate $maxWR_t$ [B/sec] of each server s_t depend on the performance of the server s_t. Here, $RR^i_{kt}(\tau) = fr_t(\tau) \cdot maxRR_t$ $(\leq maxRR_t)$ and $WR^i_{kt}(\tau) = fw_t(\tau) \cdot maxWR_t$ $(\leq maxWR_t)$ where $fr_t(\tau)$ and $fw_t(\tau)$ are degradation ratios. Here, $fr_t(\tau) = 1 / (|R_t(\tau)| + rw_t \cdot |W_t(\tau)|)$ where $0 \leq rw_t \leq 1$. $fw_t(\tau) = 1 / (wr_t \cdot |R_t(\tau)| + |W_t(\tau)|)$ where $0 \leq wr_t \leq 1$.

2.4 Electric Energy Consumption

In our previous studies, we carried out the experiments [7] to measure the electric power of a physical server where read and write methods are concurrently performed on objects supported by virtual machines in the physical server. The *PCDAVM* (*Power Consumption model for Data Access in Virtual Machine environment*) model [7] is proposed based on the experiments. At time τ, the base electric power $BE_t(\tau)$ of a server s_t depends on the number $ac_t(\tau)$ of active cores and the number $at_t(\tau)$ of active threads in the server s_t in the PCDAVM model. The base electric power $BE_t(\tau)$ is given as equation (1):

$$BE_t(\tau) = minE_t + \gamma_t \cdot (minC_t + ac_t(\tau) \cdot cE_t + at_t(\tau) \cdot tE_t) \ [W]. \quad (1)$$

In equation (1), $\gamma_t = 1$ if at least one virtual machine is active in a server s_t. Otherwise, $\gamma_t = 0$ and $BE_t(\tau)$ is the minimum electric power $minE_t$ [W]. The electric power $minC_t$ [W] is consumed by a server s_t if $ac_t(\tau) \geq 1$. Let cE_t and tE_t be the electric power [W] consumed by a server s_t to make one core and thread active, respectively.

At time τ, a server s_t consumes the electric power $E_t(\tau)$ [W] to perform write and read methods on virtual machines. The electric power $E_t(\tau)$ is given as equation (2):

$$E_t(\tau) = \begin{cases} BE_t(\tau) + WE_t & \text{if } |W_t(\tau)| \geq 1 \text{ and } |R_t(\tau)| = 0. \\ BE_t(\tau) + WRE_t(\alpha(\tau)) & \text{if } |W_t(\tau)| \geq 1 \text{ and } |R_t(\tau)| \geq 1. \\ BE_t(\tau) + RE_t & \text{if } |W_t(\tau)| = 0 \text{ and } |R_t(\tau)| \geq 1. \\ BE_t(\tau) & \text{if } |W_t(\tau)| = |R_t(\tau)| = 0. \end{cases} \quad (2)$$

In equation (2), RE_t and WE_t show the electric power [W] consumed by a server s_t to perform only read and write methods on virtual machines in the server s_t. A server s_t consumes the electric power $BE_t(\tau) + RE_t$ [W] if only read methods are performed in the server s_t. A server s_t consumes the electric power $BE_t(\tau)$ + WE_t [W] if only write methods are performed in the server s_t. If read and write methods are concurrently performed in a server s_t, a server s_t consumes the electric power $BE_t(\tau) + WRE_t(\alpha(\tau)) = (1 - \alpha(\tau)) \cdot RE_t + \alpha(\tau) \cdot WE_t$ [W] where $\alpha(\tau) = |W_t(\tau)| \ / \ |(W_t(\tau) + R_t(\tau))|$. Here, $RE_t \leq WRE_t(\alpha(\tau)) \leq WE_t$. Here, the total processing electric energy $TPE_t(\tau_1, \tau_2)$ [J] of a server s_t between time τ_1 and τ_2 is $\Sigma_{\tau=\tau 1}^{\tau_2}(E_t(\tau) - minE_t)$.

3 EE2PL-VM Protocol

A method op^1 *locally precedes* another method op^2 in a local schedule sc_h of an object o_h ($op^1 \rightarrow_{sc_h} op^2$) iff op^1 and op^2 are performed on the object o_h and op^1 $\rightarrow_{sc_h} op^2$. A full write method op^1 absorbs another full or partial write method op^2 in a local schedule sc_h of an object o_h iff (1) $op^2 \rightarrow_{sc_h} op^1$ and there is no read method r such that $op^2 \rightarrow_{sc_h} r \rightarrow_{sc_h} op^1$, or (2) op^1 absorbs op^3 and op^3 absorbs op^2 for some method op^3.

[Definition]. A *meaningless write method op* is a method which is absorbed by another method op' on an object o_h in the local schedule sc_h.

The *EE2PL-VM* (*Energy-Efficient Two-Phase Locking in Virtual Machine environment*) protocol is proposed in this paper to reduce the total PEE of servers. In the EE2PL-VM protocol, meaningless write methods are not performed on each object. Each method op^i issued by a transaction T^i is performed on the object o_h by the **EE2PL-VM** procedure as shown in Algorithm 1:

Algorithm 1. EE2PL-VM procedure

```
procedure EE2PL-VM(op^i)
    if op^i(o_h) = a write method then            ▷ A write method op^i is issued by T^i.
        if o_h.wait = φ then
            o_h.wait = op^i(o_h);
        else
            if op^i(o_h) absorbs o_h.wait then
                perform(op^i(o_h));
            else
                perform(o_h.wait); o_h.wait = op^i(o_h);
            end if
        end if
    else                                           ▷ A read method op^i is issued by T^i.
        if o_h.wait = φ then
            perform(op^i(o_h));
        else
            perform(o_h.wait); o_h.wait = φ; perform(op^i(o_h));
        end if
    end if
end procedure
```

In the Algorithm 1, a variable $o_h.wait$ indicates a write method $w^i_{kt}(o_h)$ issued by a transaction T^i to write data of an object o_h stored in a virtual machine VM_{kt} in a server s_t. A write method shown by a variable $o_h.wait$ is waiting for the next method op' to be performed on the object o_h. Here, if the next method op' absorbs the write method $w^i_{kt}(o_h)$ shown by the variable $o_h.wait$, the method op' is performed on the object o_h without performing the write method $w^i_{kt}(o_h)$ shown by the variable $o_h.wait$ since the write method $w^i_{kt}(o_h)$ is a meaningless write method. By not performing meaningless write methods on each object, the total PEE of servers can be more reduced in the EE2PL-VM protocol than the 2PL protocol.

4 Evaluation

The EE2PL-VM protocol is evaluated in terms of the total PEE of a cluster S of servers and the average execution time of each transaction compared with the 2PL protocol [5,10]. There is a cluster S of five homogeneous servers s_1, ..., s_5. Every server s_t in the cluster S follows the same DAVM model [7]. The maximum read and write rates of each server s_t are 98.5 and 85.3 [MB/sec], respectively. Parameters wr_t and rw_t on degradation ratios of write and read rates are 0.077 and 0.667, respectively. Every server s_t in the cluster S also follows the same PCDAVM model [7]. The minimum electric power $minE_t$ of each server s_t is 17 [W]. $minC_t$ = 1.1 [W], cE_t = 0.6 [W], tE_t = 0.5 [W], WE_t = 4 [W], RE_t = 1 [W], and $maxE_t$ = 24.3 [W]. There are sixty objects o_1, ..., o_{60}. Each object o_h is randomly allocated to a virtual machine VM_{kt} in the cluster S. The size of data

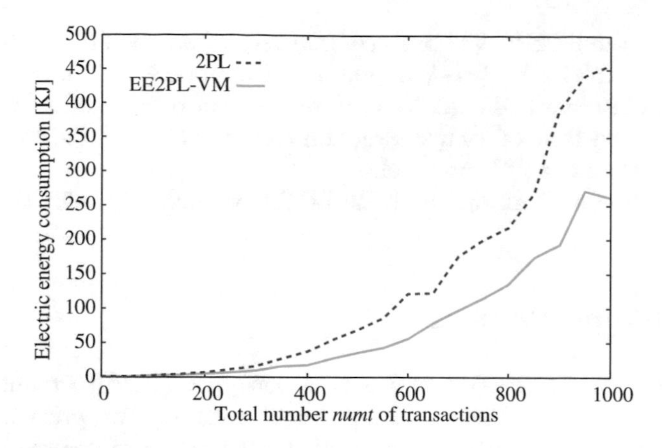

Fig. 1. Total PEE [KJ].

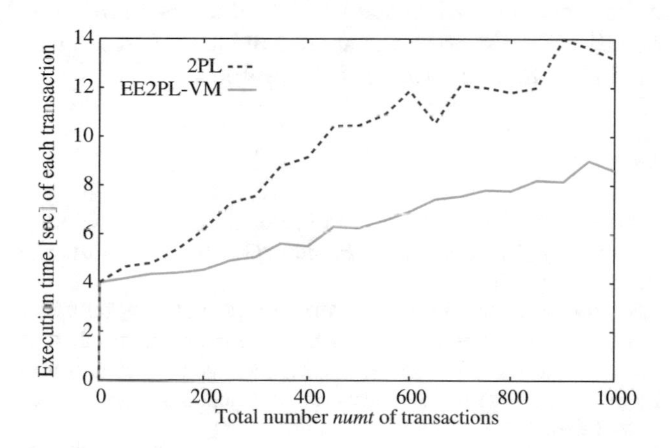

Fig. 2. Execution time [sec] of each transaction.

in each object o_h is randomly selected between 10 and 100 [MByte]. The total number *numt* of transactions are issued to the cluster S ($0 \leq numt \leq 1{,}000$). Each transaction randomly selects five methods on sixty objects and issues the five methods to manipulate objects.

The total PEE [KJ] of the cluster S to perform the number *numt* of transactions in the 2PL and EE2PL-VM protocols is shown in Fig. 1. In the EE2PL-VM protocol, meaningless write methods are not performed on each object. As a result, the total PEE of the cluster S can be more reduced in the EE2PL-VM protocol than the 2PL protocol for $0 \leq numt \leq 1{,}000$.

The average execution time [sec] of the *numt* transactions in the 2PL and EE2PL-VM protocols is shown in Fig. 2. The average execution time of each transaction in the EE2PL-VM protocol is shorter than the 2PL protocol for $0 < numt \leq 1{,}000$. In the EE2PL-VM protocol, computation resources to perform meaningless write methods can be used to perform other methods. Hence, the average execution time of each transaction can be more reduced in the EE2PL-VM protocol than the 2PL protocol.

Following the evaluation, the EE2PL-VM protocol is useful than the 2PL protocol.

5 Concluding Remarks

In this paper, the EE2PL-VM protocol is newly proposed to reduce the total PEE of servers. In the EE2PL-VM protocol, meaningless write methods are not performed on each object. As a result, the EE2PL-VM protocol can reduce not only the total PEE of servers but also the average execution time of each transaction than the 2PL protocol. From the evaluation results, the total PEE of servers can be more reduced in the EE2PL protocol than the 2PL protocol. In addition, the EE2PL-VM protocol can also reduce the average execution time of each transaction compared with the 2PL protocol.

References

1. Nakamura, S., Enokido, T., Takizawa, M.: Information flow control based on capability token validity for secure IoT: implementation and evaluation. Internet of Things **15**, 100423 (2021)
2. Enokido, T., Duolikun, D., Takizawa, M.: Energy consumption laxity-based quorum selection for distributed object-based systems. Evol. Intell. **13**, 71–82 (2020)
3. Enokido, T., Duolikun, D., Takizawa, M.: An energy-efficient quorum-based locking protocol by omitting meaningless methods on object replicas. J. High Speed Netw. **28**(3), 181–203 (2022)
4. Gray, J.N.: Notes on data base operating systems. In: Bayer, R., Graham, R.M., Seegmüller, G. (eds.) Operating Systems. LNCS, vol. 60, pp. 393–481. Springer, Heidelberg (1978). https://doi.org/10.1007/3-540-08755-9_9
5. Bernstein, P.A., Hadzilacos, V., Goodman, N.: Concurrency Control and Recovery in Database Systems. Addison-Wesley, Boston (1987)
6. Enokido, T., Duolikun, D., Takizawa, M.: The improved redundant active time-based (IRATB) algorithm for process replication. In: Proceedings of the 35th IEEE International Conference on Advanced Information Networking and Applications (AINA-2021), pp. 172–180 (2021)
7. Enokido, T. Takizawa, M.: The power consumption model of a server to perform data access application processes in virtual machine environments. In: Proceedings of the 34th International Conference on Advanced Information Networking and Applications (AINA-2020), pp. 184–192 (2020)
8. Duolikun, D., Enokido, T., Takizawa, M.: A negotiation protocol among servers for virtual machines to migrate to reduce the energy consumption. In: Proceedings of the 25th International Conference on Network-Based Information Systems (NBiS-2022), pp. 1–12 (2020)

9. Enokido, T., Duolikun, D., Takizawa, M.: Energy-efficient concurrency control by omitting meaningless write methods in object-based systems. In: Proceedings of the 36th International Conference on Advanced Information Networking and Applications (AINA-2022), pp. 129–139 (2022)
10. Garcia-Molina, H., Barbara, D.: How to assign votes in a distributed system. J. ACM **32**(4), 814–860 (1985)

A Flexible Fog Computing (FTBFC) Model to Reduce Energy Consumption of the IoT

Dilawaer Duolikun[1(✉)], Tomoya Enokido[2], and Makoto Takizawa[1]

[1] RCCMS, Hosei University, Tokyo, Japan
dilewerdolkun@gmail.com, makoto.takizawa@computer.org
[2] Faculty of Business Administration, Rissho University, Tokyo, Japan
eno@ris.ac.jp

Abstract. The FC (Fog Computing) model is proposed to efficiently realize the IoT (Internet of Things), where fog nodes support application processes to handle sensor data in addition to exchanging messages. The IoT consumes a huge amount of energy due to the scalability and we have to reduce the electric energy consumed by nodes in the IoT. In our previous studies, the TBFC (Tree-Based FC) model is proposed to energy-efficiently realize the IoT. Here, fog nodes are hierarchically structured. However, the tree structure is fixed even if some node is overloaded and consumes a larger amount of energy than expected. In this paper, we newly propose a flexible TBFC (FTBFC) model where the tree structure of nodes and processes supported by nodes can be dynamically changed so that the total energy consumption of the nodes can be reduced. In the evaluation, we show the total energy consumption of the FTBFC model is smaller than the TBFC and cloud computing (CC) models.

Keywords: Green computing · FTBFC (Flexible Tree-Based Fog Computing) model · IoT · Fog Computing (FC) model · Energy consumption

1 Introduction

The IoT (Internet of Things) is now one of the most significant infrastructures to realize various applications in our societies. The IoT is so scalable that millions devices like cars and home electric appliances are interconnected in addition to servers and clients and accordingly consumes a huge amount of electric energy [2,3]. In order to decrease carbon footprint on the earth, the total electric energy consumed by the IoT has to be reduced. There are cloud computing (CC) [4] and fog computing (FC) [5,6] models to realize the IoT. In the CC model, devices send sensor data to servers in a cloud through networks and application processes on the servers handle the sensor data to decide on actions to be performed by actuators. Servers and networks are heavily loaded to process and transmit sensor data from a huge number of device nodes.

In the FC model [5,6], application processes are distributed to not only servers in clouds but also fog nodes. Sensor data is processed by fog nodes and

L. Barolli (Ed.): EIDWT 2023, LNDECT 161, pp. 256–267, 2023.
https://doi.org/10.1007/978-3-031-26281-4_26

the processed data is mostly smaller than the sensor data like a maximum value of sensor data from multiple devices. Hence, a smaller amount of data can be processed by servers and transmitted in networks. On the other hand, fog nodes consume energy to process sensor data in addition to servers while servers consume smaller energy than the CC model. Furthermore, it takes time for fog nodes to process and exchange data with other fog nodes. In the TBFC (Tree-Based FC) model [18,19,21–24], fog nodes are structured in a tree whose root node shows a cloud of servers and each leaf node indicates a device node. Since sensor DUs sent by device nodes are in parallel processed by multiple fog nodes, the response time can be reduced as well as the energy consumption. In the TBFC model, the tree structure of fog nodes is fixed even if some node is overloaded and consumes a larger energy to process data. In the FTTBFC (Fault-Tolerant TBFC) model [24–26], if some node$f$ is faulty, the child nodes are reconnected to another node which supports the same processes as the node f. In the DTBFC (Dynamic TBFC) model [19,26], processes of a node f migrate to the parent and child nodes if the node f consumes larger energy to process input data. Processes on fog nodes can migrate to other nodes in a live manner of virtual machines [14–18]. Here, the tree structure of nodes is fixed. A network model of fog nodes is also discussed where each fog node f negotiates with other fog nodes on whether or not the output data of the node f can be processed [28]. A fog node f can dynamically select a fog node which can process the output data of the node f but it takes time to do the negotiation with other fog nodes.

In this paper, we newly propose an FTBFC (Flexible TBFC) model where not only the tree structure of nodes but also processes supported by nodes are changed to reduce the electric energy consumption of nodes and delivery time of sensor data to servers. In the evaluation, we show the total energy consumption and delivery time of the IoT can be reduced in the FTBFC model compared with the CC model in the simulation.

In Sect. 2, we present the FTBFC model. In Sect. 3, we present the power consumption and computation models of fog nodes. In Sect. 4, we discuss operations to realize the flexibility. In Sect. 5, we evaluate the TBFC model.

2 The FTBFC Model

The FC (Fog Computing) model [5,6] of the IoT is composed of fog nodes in addition to clouds of servers and device nodes. Application processes and databases are distributed to not only servers but also fog nodes. Sensor data is processed by fog nodes and the processed sensor data is delivered to servers through fog-to-fog communication. Thus, the traffic of the servers and networks and time to activate actuators can be reduced in the FC model.

In the TBFC model [18–24] proposed to energy-efficiently realize the IoT, nodes are tree-structured where the root node is a cloud of servers, each leaf node shows a collection of devices, and the other non-root and non-leaf nodes are fog nodes. A root node f is interconnected with *child* nodes $f_1, \ldots, f_b$ in networks. Each node f_i is further interconnected with child nodes $f_{i1}, \ldots, f_{i,b_i}$. An index I of a node f_I is a sequence $\langle i_1 i_2 \ldots i_{l-1} \rangle$ $(l \geq 1)$ of numbers, which means a path $\langle f, f_{i_1}, f_{i_1 i_2}, \ldots, f_{i_1 i_2 \ldots i_{l-1}} \rangle$ from the root node f to the node $f_{i_1 i_2 \ldots i_{l-1}}$ $(= f_I)$. Here, $|I|$ shows $l-1$. A node f_I is at level $|I|+1$. For example, a fog node f_{21} is at level 3. Thus, each node f_{Ii} communicates with a parent node f_I and child nodes $f_{Ii1}, \ldots, f_{Ii,b_{Ii}}$ $(b_{Ii} \geq 0)$. An edge node f_I communicates with a leaf node d_I named *device* node. Here, the edge node f_I receives sensor data from the device node d_I and sends actions to d_I. Here, a device node is an abstraction of one or more then one device. In this paper, the total number dv of devices is fixed. Each device node denotes dv/nd devices for number nd of device nodes.

In this paper, an application process P to handle sensor data is assumed to be a sequence $\langle p_1, \ldots, p_m \rangle$ $(m \geq 1)$ of processes. Here, p_1 is the *top* process and p_m is the *tail* process. A data unit (DU) is a unit of data transmission among nodes. The tail process p_m first receives a DU of sensor data from a device node and obtains an output DU dt_m by processing the sensor data. Then, the process p_{m-1} receives dt_m and calculates an output DU dt_{m-1} on the input DU dt_m. Thus, each process p_i receives an input DU dt_{i+1} from the process p_{i+1} and sends an output DU dt_i to the process p_{i-1}. The top process p_1 finally obtains an output DU dt_1 which shows actions to be performed by actuators.

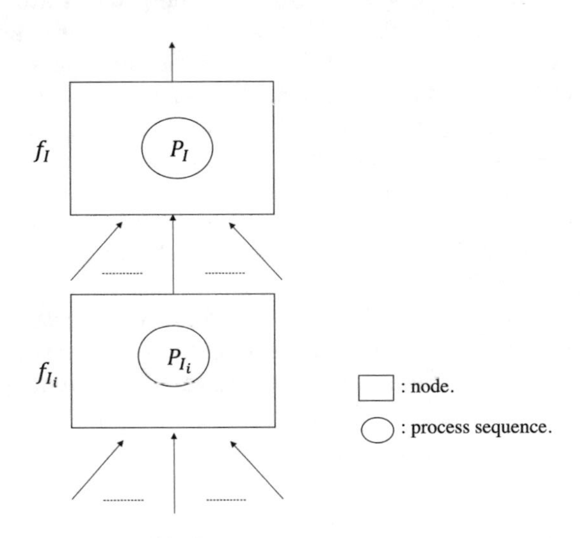

Fig. 1. Node of the TBFC model.

Each node f_I supports a subsequence P_I of P, i.e. $\langle p_{t_I}, p_{t_I+1}, \ldots, p_{l_I} \rangle$ where $1 \leq t_I \leq l_I \leq m$ [Fig. 1]. Here, the process p_{t_I} is the top process and p_{l_I} is the tail process in the node f_I. The node f_I receives an input DU id_{Ii} from each child node f_{Ii} $(i = 1, \ldots, b_I)$. Let ID_I be a set $\{id_{I1}, \ldots, id_{I,b_I}\}$ of the input DUs from the child nodes $f_{I1}, \ldots, f_{I,b_I}$. The tail process f_{l_I} calculates an output DU d_{l_I} on the input DUs ID_I and sends the output DU dt_{l_I} to the process p_{l_I-1}. Then, the process p_{l_I-1} calculates an output DU dt_{l_I-1} and sends dt_{l_I-1} to the process

p_{l_I-2}. Thus, the top process p_{t_I} finally calculates an output DU od_I $(= dt_{t_I})$ on the input data dt_{t_I+1} from the process p_{t_I+1} and sends od_I to the parent node. Each child node f_{Ii} supports a subsequence P_{Ii} $(= \langle p_{t_{Ii}}, \ldots, p_{l_{Ii}} \rangle)$ of processes where $l_I = t_{Ii} - 1.$, i.e. $p_{l_I} = p_{t_{Ii}-1}$. Here, the top process $p_{t_{Ii}}$ of every child node f_{Ii} has to be the same, while the tail process $p_{l_{Ii}}$ may be different from the tail process $p_{l_{Ij}}$ of another child node f_{Ij}.

3 Power Consumption and Computation Models

Each fog node f_I consumes electric energy [J] to communicate with parent and child nodes and to calculate an output DU od_I on input DUs ID_I from the child nodes. In order to discuss the energy consumption of the IoT, it is critical to make a power consumption model to show how much power [W] a node consumes to communicate with child and parent nodes and perform processes. In our previous studies, the SPC (Simple Power Consumption) [8,9] and MLPCM (Multi-Level Power Consumption) [13–15] models are proposed. Here, we consider only the total power [W] consumed by a whole node to perform application processes at macro level and do not discuss how much power each hardware component consumes. In this paper, we consider the SPC model since a small computer like Raspberry Pi3 [29] is used to realize a fog node which follows the SPC model. In addition, more processes than the number of threads are usually performed on each server. Here, the server follows the SPC model. The power consumption NE_I [W] of a fog node f_I is given as follows:

[SPC model] [8,9]

$$NE_I = \begin{cases} maxE_I & \text{if at least one process is active on a node } f_I. \\ minE_I & \text{otherwise.} \end{cases} \quad (1)$$

For example, for a Raspberry Pi3 node f_I, $maxE_I$ and $minE_I$ are 3.7 and 2.1 [W], respectively. On the other hand, $maxE_I$ and $minE_I$ are 301.3 and 126.1 [W] for a server HP DL360 [7] node f_I, respectively.

A node f_I also consumes power RE_I and SE_I [W] to receive and send data, respectively. In this paper, we assume $RE_I = re_I{\cdot}maxE_I$ and $SE_I = se_I{\cdot}maxE_I$ where $re_I(\leq 1)$ and $se_I(\leq 1)$ are constants. For a Raspberry Pi3 node f_I, $se_I =$ 0.68 and $re_I = 0.73$ [23].

Next, we discuss the execution time [sec] of a process p_h $(t_I \leq h \leq l_I)$ on a node f_I. In this paper, we assume the execution time $PT_h(x)$ [sec] of a process p_h to calculate on input data dt_{h+1} of size x is $O(x)$ or $O(x^2)$ as discussed in papers [20–22,24], i.e. $PT_h(x)$ [sec] is $cc_{Ih}{\cdot}x$ or $cc_{Ih}{\cdot}x^2$ where cc_{Ih} [sec/bit] is a constant showing how long it takes for p_h to process one bit of the data on a node f_I. In this paper, we assume $cc_{Ih} = cc_{Ik} = cc_I$ for every pair of processes p_h and p_k on each node f_I. A process p_h is typed $O1$ and $O2$ iff $PT_h(x)$ is $cc_I{\cdot}x$ and $cc_I{\cdot}x^2$, respectively. The *computation ratio* cr_I of a node f_I is the ratio of computation speed of f_I to the root node f. If the same process is performed for the same input data on f and f_I, the computation ratio cr_I is cc/cc_I (≤ 1). The computation ratio cr_I of the Raspberry Pi3 fog node f_I [29] is 0.18 compared with a HP DL360 server node f [7] as discussed [21]. The *computation rate* CR_I [bps] shows how many bits a node f_I processes for one second. CR_I is $CR \cdot cr_I$ (≤ 1). As discussed in papers [12,13,15–17], the computation residue R_h [bit] of a process p_h is x and x^2 if p_h is $O1$ and $O2$ types, respectively, and the execution time T_h [sec] is zero when the process p_h starts processing an input DU of size x. For each second, R_h is decremented by the computation rate CR_I and T_h is incremented by one. If $R_h \leq 0$, the process p_h terminates on the node f_I. Thus, T_h gives the execution time $PT_h(x)$ to calculate on the input DU of size x. The size $|od_h|$ of the output DU od_h of a process p_h is $pr_h \cdot x$ where pr_h is a *reduction ratio*. For example, if a process p_h on a node f_I obtains a maximum value dt_h of a collection of n input DUs, the reduction ratio pr_h is $1/n$ (≤ 1). The total execution time $ET_I(x)$ of the node f_I is $PT_{l_I}(x) + PT_{l_I - 1}(pr_{l_I} \cdot x) + \ldots + PT_{t_I}(pr_{l_t} \cdot \ldots \cdot pr_{t_I + 1} \cdot x)$. Let fr_I be the reduction ratio $pr_{l_t} \cdot \ldots \cdot pr_{t_I}$ of the node f_I. The size $|od_I|$ of the output DU od_I of the node f_I is $fr_I \cdot x$.

A node f_I receives input DUs ID_I from the child nodes and sends an output DU od_I to the parent node. In this paper, it takes time $RT_I(x)$ and $ST_I(x)$ [sec] to receive and send a DU of size x, respectively, i.e. $RT_I(x) = rc_I \cdot x$ and $ST_I(x) = sc_I \cdot x$ where rc_I and sc_I are constants which depend on the transmission rate of the network. For example, $sc_I / rc_I = 0.22$ for a Raspberry Pi3 node f_I with a 100Mbps network.

Thus, it totally takes time $TT_I(x) = RT_I(x) + ET_I(x) + ST_I(fr_I \cdot x)$ [sec] from f_I receives ID_I of size x until f_I sends od_I. The node f_I consumes the total energy $TE_I(x) = RT_I(x) \cdot RE_I + ET_I(x) \cdot NE_I + ST_I(fr_I \cdot x) \cdot SE_I$ [J].

4 Flexible Operations on the FTBFC Model

The more input DUs arrive at a fog node f_I, the longer time it takes for f_I to process input DUs and accordingly the larger amount of energy the fog node f_I consumes. In this paper, we propose an *FTBC* (*flexible TBFC*) model to reduce the total energy consumption of the nodes in the IoT. A fog node f_I supports b_I (≥ 1) child nodes $f_{I1}, \ldots, f_{Ib_I}$ as presented in Sect. 2. Suppose a child node f_{Ii} is overloaded. If a new child node nf is connected to the node f_I and some input DUs to the node f_{Ii} are distributed to the node nf, the node f_{Ii} consumes smaller energy since the node f_{Ii} processes fewer input DUs.

We consider an application process P which is a sequence $\langle p_1, \ldots, p_m \rangle$ of subprocesses. Each node f_{Ii} supports a subsequence $P_{Ii} = \langle p_{t_{Ii}}, p_{t_{Ii}+1}, \ldots, p_{l_{Ii}} \rangle$ of the process sequence P, where $p_{t_{Ii}}$ is a top process and $p_{l_{Ii}}$ is a tail process. The FTBFC model supports a pair of operation types to move processes supported by nodes to other nodes and change the tree structure of nodes. In one type, processes supported by a node are changed by making some of the processes migrate up to the parent node and down to the child nodes. Processes can migrate to nodes in a live manner [1] by taking advantage of virtual machines [14–18]. On receipt of input DUs ID_{Ii} from the child nodes, the tail process

$p_{l_{Ii}}$ is first performed and the output DU $dt_{l_{Ii}}$ is sent to the process $p_{l_{Ii}-1}$. The process $p_{l_{Ii}-1}$ is then performed and sends an output DU $dt_{l_{Ii}-1}$ to the process $p_{l_{Ii}-2}$. Thus, the top process $p_{t_{Ii}}$ finally sends an output DU od_{Ii} $(= dt_{t_{Ii}})$ to the parent node f_I. In order to make processes on nodes migrate to other nodes, the following operations are supported.

[Process migration operations]

1. Migrate_up ($MUP(f_{Ij})$).
2. Migrate_down ($MDW(f_I)$).

In the *migration_up* (MUP) operation $MUP(f_{Ij})$ on a node f_{Ij}, the top process $p_{t_{Ii}}$ of the node f_{Ii} migrates to the parent node f_I and the top process $p_{t_{Ij}}$ of every other child node f_{Ij} is removed if every child node f_{Ii} supports more than one process, i.e. $l_{Ii} - t_{Ii} > 1$. It is noted that the top process $p_{t_{Ii}}$ of every child node f_{Ii} is the same as presented in the FTBFC model. Now, the node f_I and every child node f_{Ii} support subsequences $\langle p_{t_I}, \ldots, p_{l_I}, p_{t_{Ii}} \rangle$ and $\langle p_{t_{Ii}+1}, \ldots, p_{l_{Ii}} \rangle$, respectively, where $t_{Ii} = l_I + 1$.

In the *migration_down* (MDW) operation $MDW(f_I)$ on a node f_I, the tail process p_{l_I} of f_I migrates down to every child node f_{Ii} if the parent node f_I supports more than one process, i.e. $l_I - t_I > 1$. Now, the nodes f_I and f_{Ii} support subsequences $\langle p_{t_I}, \ldots, p_{l_I-1} \rangle$ and $\langle p_{l_I}, p_{t_{Ii}}, \ldots, p_{l_{Ii}} \rangle$, respectively, where $l_I = t_{Ii} - 1$.

In another operation type, child nodes of a node f_I are changed, i.e. the tree structure is changed. In order to change the tree structure of the FTBFC model, we introduce the following operations:

[Structure change operations]

1. $Split(f_{Ii})$.
2. $Merge(f_{Ii}, f_{Ij})$.
3. $Expand(f_I)$.
4. $Shrink(f_{Ii})$.

In the *split* operation $Split(f_{Ii})$ on a child node f_{Ii}, a new node nf is created as a child node of the node f_I. In addition, half of the child nodes of the node f_{Ii} are reconnected to the new node nf if the node f_{Ii} is not an edge node. Hence, the node f_{Ii} is required to have multiple child nodes, i.e. $b_{Ii} > 1$. If the node f_{Ii} is an edge node, one device node is created as a child node of the node nf. Thus, by splitting an edge node, the number *nd* of device nodes increases by one. This means each edge node receives sensor data from a fewer number of devices.

In the *merge* operation $Merge(f_{Ii}, f_{Ij})$ on a pair of child nodes f_{Ii} and f_{Ij}, the node f_{Ij} is merged into the node f_{Ii}. Here, both the nodes f_{Ii} and f_{Ij} have to support the same subsequence of processes, i.e. $P_{Ii} = P_{Ij}$. Every child node f_{Ijk} of the node f_{Ij} is reconnected to the node f_{Ii}. If the nodes f_{Ii} and f_{Ij} are edge nodes, the child device node d_{Ii} of the node f_{Ij} is not connected to the node f_{Ii}. Thus, the number *nd* of device nodes decreases by one. This means each edge node supports more devices.

In the *expand* operation $Expand(f_I)$ on a node f_I, a new node nf is created as a child node of the node f_I and every child node f_{Ii} of the node f_I is reconnected to the new node nf as a child node. Here, the node f_I has only one child node nf

which supports no process. By using the MUP and MDW operations, processes have to migrate from some child node f_{Ii} and the parent node f_I, respectively. The height of the tree increases by one.

In the *shrink* operation $Shrink(f_{Ii})$ on a node f_{Ii}, the node f_{Ii} is merged into the parent node f_I if the node f_I has only one child node f_{Ii}. Here, every process of the node f_{Ii} migrates up to the node f_I. In addition, every child node f_{Iij} is reconnected to the node f_I as a child node. The height of the tree decreases by one.

By using the split and expand operations, the number of nodes can increase. On the other hand, the nodes can be reduced by using the merge and shrink operations.

5 Evaluation

We consider four application processes $P_1, \ldots, P_4$ of five processes $p_1, \ldots, p_5$ as shown in Table 1. Each element ($ptype_i$, pr_i) in Table 1 shows a process type $ptype_i (\in \{O1, O2\})$ and reduction ratio pr_i of each process p_i. In the sequences P_1 and P_2, every process p_i is an $O1$ type. In the sequences P_3 and P_4, every process p_i is an $O2$ type.

Table 1. Parameters of processes

Process	p_1	p_2	p_3	p_4	p_5
P_1	($O1$, 1.0)	($O1$, 0.8)	($O1$, 0.6)	($O1$, 0.4)	($O1$, 0.2)
P_2	($O1$, 0.2)	($O1$, 0.4)	($O1$, 0.6)	($O1$, 0.8)	($O1$, 1.0)
P_3	($O2$, 1.0)	($O2$, 0.8)	($O2$, 0.6)	($O2$, 0.4)	($O2$, 0.2)
P_4	($O2$, 0.2)	($O2$, 0.4)	($O2$, 0.6)	($O2$, 0.8)	($O2$, 1.0)

Every edge node f_I receives an input DU ID_I from the device node d_I at the same time τ every *its* [sec]. Each DU *du* carries time *du.stime* when the DU *du* is issued by a device node. A fog node f_I starts calculating an output DU od_I from the input DUs $ID_I (= \{id_{I1}, \ldots, id_{Ib_I}\})$ only if f_I receives an input DU id_{Ii} from every child node f_{Ii}, where time $id_{Ii}.stime$ is the same $ID_I.stime$. Each output DU od_I which a node f_I creates from the input DUs ID_I carries the ime $ID_I.stime = od_I.stime$. It takes time *tm* [sec] to deliver a DU *du* to a node. The transmission time *tm* is $|du|/tr$ where *tr* is the transmission rate [bps]. In the evaluation, *tr* is 200 [Kbps]. If a child node f_{Ii} sends an output DU od_{Ii} at time τ, the parent node f_I can start calculating an output DU on the input DUs at time $\tau + tm$ where $tm = |od_{Ii}|/tr$. If each node f_I is active, i.e. f_I is calculating preceding input DUs or does not yet receive an input DU from every other child node when an input DU id_{Ii} is delivered to the node f_I, the input DU id_{Ii} is stored in the receipt queue RQ_{Ii}. In the receipt queue RQ_{Ii}, the input DU id_{Ii} is sorted in the time $id_{Ii}.stime$. If the top input DU id_{Ii} in every receipt queue

RQ_{Ii} has the same time, the input DU id_{Ii} is dequeued and then f_I starts to calculate an output DU od_I on the input DUs $ID_I(=\{id_{I1},\ldots,id_{Ib_I}\})$. If a root node f finishes the calculation on an input DU ID at time τ, the delivery time DT is $\tau - ID.stime$.

Table 2 shows the computation ratio cr_I and the maximum power $maxE_I$ and minimum power $minE_I$ of a node f_I. The computation rate CR of the root node f is 800 [Kbps] in the evaluation. The computation rate CR_I of a Raspberry pi3 fog node f_I is 0.185 $\cdot CR$, which is obtained through our experiment [21,23].

Table 2. Parameters of nodes

node f_I	cr_I	$CR_I[Kbps]$	$minE_I$[W]	$maxE_I$[W]
root (server)	1.0	800	126.1	301.3
fog (Raspberry pi)	0.185	148	2.1	3.7

Each device node d_I sends a DU ID_I to the edge node f_I every ist [sec]. Here, the size $|ID_I|$ shows the total amount of sensor data issued by devices denoted by a device node d_I. In the evaluation, the inter-sensing time ist is ten [sec]. Device nodes totally send DUs of size tsd [Kbit] to the edge nodes every ist [sec]. In the evaluation, tsd is 8,000 [Kbit]. If there are nd ($\geq$1) device nodes in the tree T, each edge node f_I receives a sensor DU ID_I of size is ($= tsd/nd$) [Kbit] from a device node d_I. In the FTBFC model, the number nd of device nodes is changed. In the evaluation, each device node sends a DU whose size is is tsd/nd. If the number nd of device nodes increases as presented in Sect. 4, is decreases. Each device node d_I totally sends $xtime/ist$ DUs to the edge node f_I. In the evaluation, $xtime$ = 100 [sec], i.e. each device node sends totally ten DUs to the edge node.

In this paper, the receiving time $RT_I(x)$ and sending time $ST_I(x)$ [sec] to receive and send a DU of size x are assumed to be zero. This means each node is assumed not to consume energy to send and receive DUs.

In the simulation, we evaluate the FTBFC tree in terms of the total energy consumption TE [J] of the nodes and the delivery time DT [sec] of each sensor DU. In the evaluation, nodes and processes of nodes in an FTBFC tree T are manipulated in the following scenario [Fig. 2]:

1. First, there is one root node f_1 and one device node d_2 as shown in Fig. 2 (1). This shows a CC model where the node f_1 is a cloud of servers and the device node d_2 is a collection of all the devices. The root node f_1 supports all the processes $p_1, \ldots, p_5$. The device node d_2 sends a sensor DU of size tsd to the node f_1.
2. By expanding the node f_1, a new node f_3 is created as a child node. Then, a pair of the processes p_5 and p_4 migrate down to the node f_3 [Fig. 2 (2)]. By splitting the node f_3, a new node f_4 is created as a child of f_1 and a device node d_5 is created as a child node of the node f_4 [Fig. 2 (3)]. Here, there are

totally two device nodes d_2 and d_5 where each device node sends a sensor DU of size $tsd/2$.

3. The nodes f_3 and f_4 are expanded where a pair of new nodes f_5 and f_7 are created as child nodes of f_3 and f_4, respectively. The tail process p_5 migrates down to the nodes f_6 and f_7 from the parent nodes f_3 and f_4, respectively. Then, the nodes f_6 and f_8 are split to new nodes f_8 and f_{10} as child nodes of f_3 and f_4, respectively. Here, new device nodes d_9 and d_{11} are created as child nodes of f_8 and f_{10}, respectively. Figure 2 (4) shows the tree. Since there are four device nodes d_2, d_5, d_9, and d_{10} in the tree, each device node sends a sensor DU of size $tsd/4$ to the parent edge node.

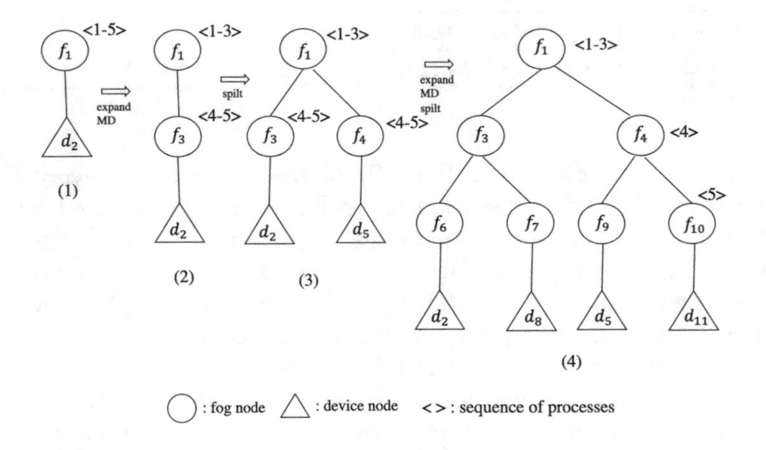

Fig. 2. Flexibility of the TBFC model.

Figure 3 shows the ratio of the total energy consumption TE of a tree obtained at each step to the TE of Fig. 2 (1), i.e. CC model. In the tree shown in Fig. 2 (2), the processes p_5 and p_4 migrate down to the node f_3. The TE is larger than (1). On the other hand, the TE is reduced in the other trees (3), (4), and (5) compared with the tree (1). The more number of nodes processes are performed on, the smaller amount of energy is consumed. Here, sensor data is ditributed to multiple nodes and in parallel processed by multple nodes even if the nodes are less poweful than the root node, i.e. server.

Figure 4 shows the ratio of delivery time DT of each step to the CC model, i.e. Fig. 2 (1). Similarly to Fig. 3, the DT increases just by the migration of processes to child nodes. By increasing child nodes as shown in Fig. 2 (2), (3), and (4), the DT can be reduced.

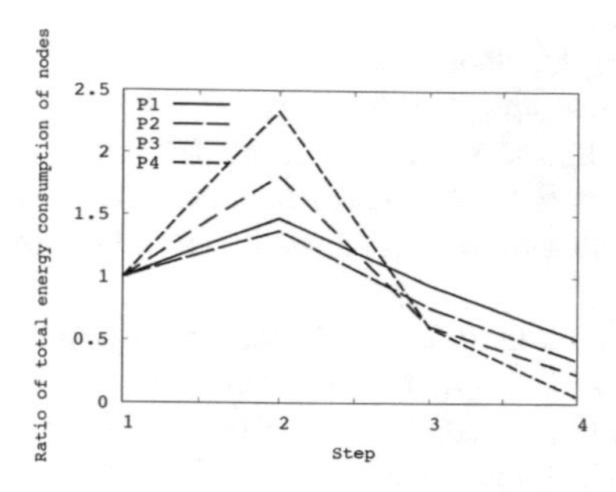

Fig. 3. Ratio of total energy consumption (TE).

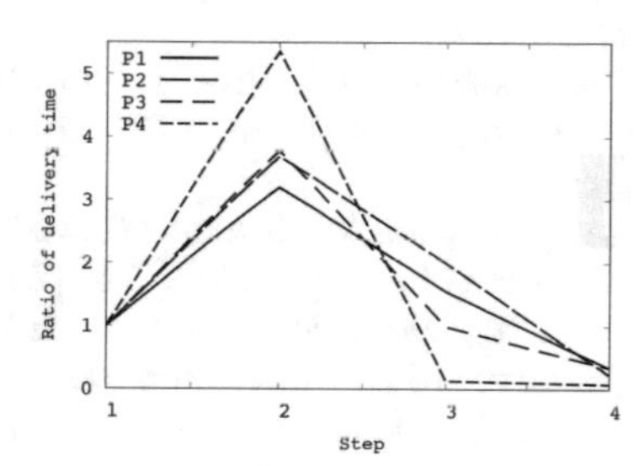

Fig. 4. Ratio of delivery time (DT).

6 Concluding Remarks

In this paper, we newly proposed the FTBFC model to reduce the energy consumption of nodes even if some node is overloaded. Here, processes supported by nodes can migrate up to the parent nodes and down to the child nodes. In addition, a child node can be split to multiple child nodes to increase the number of child nodes and can be expanded to a pair of the parent node and a new child node. In the evaluation, we showed the total energy consumption TE of the nodes and the delivery time DT of each sensor data can be reduced in the FTBFC model than the CC model.

Acknowledgment. This work is supported by Japan Society for the Promotion of Science (JSPS) KAKENHI Grant Number 22K12018.

References

1. KVM: Main Page - KVM (Kernel Based Virtual Machine). http://www.linux-kvm.org/page/Mainx_Page (2015)

2. Dayarathna, M., Wen, Y., Fan, R.: Data center energy consumption modeling: a survey. IEEE Commun. Surv. Tutorials **18**(1), 732–787 (2016)
3. Natural Resources Defense Council: Data center efficiency assessment - scaling up energy efficiency across the data center industry: Evaluating key drivers and barriers. http://www.nrdc.org/energy/files/data-center-efficiency-assessment-IP.pdf (2014)
4. Qian, L., Luo, Z., Du, Y., Guo, L.: Cloud computing: an overview. In: Proceedings of the 1st International Conference on Cloud Computing, pp. 626–631 (2009)
5. Hanes, D., Salgueiro, G., Grossetete, P., Barton, R., Henry, J.: IoT Fundamentals: Networking Technologies, Protocols, and Use Cases for the Internet of Things (First Edition), Cisco Press, Indianapolis, p. 576 (2017)
6. Rahmani, A.M., Liljeberg, P., Preden, J.-S., Jantsch, A. (eds.): Fog Computing in the Internet of Things. Springer, Cham (2018). https://doi.org/10.1007/978-3-319-57639-8
7. HPE: HP server DL360 Gen 9. https://www.techbuyer.com/cto/servers/hpe-proliant-dl360-gen9-server
8. Enokido, T., Aikebaier, A., Takizawa, M.: Process allocation algorithms for saving power consumption in peer-to-peer systems. IEEE Trans. Ind. Electron. **58**(6), 2097–2105 (2011)
9. Enokido, T., Aikebaier, A., Takizawa, M.: A model for reducing power consumption in peer-to-peer systems. IEEE Syst. J. **4**(2), 221–229 (2010)
10. Enokido, T., Aikebaier, A., Takizawa, M.: An extended simple power consumption model for selecting a server to perform computation type processes in digital ecosystems. IEEE Trans. Ind. Inf. **10**(2), 1627–1636 (2014)
11. Enokido, T., Takizawa, M.: Integrated power consumption model for distributed systems. IEEE Trans. Ind. Electron. **60**(2), 824–836 (2013)
12. Kataoka, H., Duolikun, D., Sawada, A., Enokido, T., Takizawa, M.: Energy-aware server selection algorithms in a scalable cluster. In: Proceedings of the 30th International Conference on Advanced Information Networking and Applications, pp. 565–572 (2016)
13. Kataoka, H., Nakamura, S., Duolikun, D., Enokido, T., Takizawa, M.: Multi-level power consumption model and energy-aware server selection algorithm. Int. J. Grid Util. Comput. **8**(3), 201–210 (2017)
14. Duolikun, D., Enokido, T., Takizawa, M.: Energy-efficient dynamic clusters of servers. In: Proceedings of the 8th International Conference on Broadband and Wireless Computing, Communication and Applications, pp. 253–260 (2013)
15. Duolikun, D., Enokido, T., Takizawa, M.: Static and Dynamic Group Migration Algorithms of Virtual Machines to Reduce Energy Consumption of a Server Cluster. Trans. Comput. Collective Intell. **XXXIII**, 144–166 (2019)
16. Duolikun, D., Enokido, T., Takizawa, M.: Simple algorithms for selecting an energy-efficient server in a cluster of servers. Int. J. Commun. Netw. Distrib. Syst. **21**(1), 1–25 (2018). 145–155 (2019)
17. Duolikun, D., Enokido, T., Barolli, L., Takizawa, M.: A monotonically increasing (MI) algorithm to estimate energy consumption and execution time of processes on a server. In: Proceedings of the 24th International Conference on Network-based Information Systems, pp. 1–12 (2021)
18. Duolikun, D., Nakamura, S., Enokido, T., Takizawa, M. : Energy-consumption evaluation of the tree-based fog computing (TBFC) model. In: Proceedings of the 22nd International Conference on Broadband and Wireless Computing, Communication and Applications (2022)

19. Mukae, K., Saito, T., Nakamura, S., Enokido, T., Takizawa, M.: Design and implementing of the dynamic tree-based fog computing (DTBFC) model to realize the energy-efficient IoT. In: Proceedings of the 9th International Conference on Emerging Internet, Data and Web Technologies, pp. 71–81 (2021)
20. Oma, R., Nakamura, S., Duolikun, D., Enokido, T., Takizawa, M.: An energy-efficient model of fog and device nodes in IoT. In: Proceedings of the 32nd International Conference on Advanced Information Networking and Applications Workshops, pp. 301–306 (2018)
21. Oma, R., Nakamura, S., Duolikun, D., Enokido, T., Takizawa, M.: An energy-efficient model for fog computing in the Internet of Things (IoT). Internet of Tings **1–2**, 14–26 (2018)
22. Oma, R., Nakamura, S., Enokido, T., Takizawa, M.: A tree-based model of energy-efficient fog computing systems in IoT. In: Proceedings of the 12th International Conference on Complex, Intelligent, and Software Intensive Systems, pp. 991–1001 (2018)
23. Oma, R., Nakamura, S., Duolikun, D., Enokido, T., Takizawa, M.: Evaluation of an energy-efficient tree-based model of fog computing. In: Proceedings of the 21st International Conference on Network-based Information Systems, pp. 99–109 (2018)
24. Oma, R., Nakamura, S., Duolikun, D., Enokido, T., Takizawa, M.: A fault-tolerant tree-based fog computing model. Int. J. Web Grid Serv. **15**(3), 219–239 (2019)
25. Oma, R., Nakamura, S., Duolikun, D., Enokido, T., Takizawa, M.: Energy-efficient recovery algorithm in the fault-tolerant tree-based fog computing (FTBFC) Model. In: Proceedings of the 33rd International Conference on Advanced Information Networking and Applications (AINA 2019), pp. 132–143 (2019)
26. Oma, R., Nakamura, S., Enokido, T., Takizawa, M.: A Dynamic tree-based fog computing (DTBFC) model for the energy-efficient IoT. In: Proceedings of the 8th International Conference on Emerging Internet, Data and Web Technologies, pp. 24–34 (2020)
27. Chida, R., et al.: Implementation of fog nodes in the tree-based fog computing (TBFC) model of the IoT. In: Proceedings of the 7th International Conference on Emerging Internet, Data and Web Technologies, pp. 92–102 (2019)
28. Guo, Y., Saito, T., Oma, R., Nakamura, S., Enokido, T., Takizawa, M.: Distributed approach to fog computing with auction method. In: Proceedings of the 34th International Conference on Advanced Information Networking and Applications, pp. 268–275 (2020)
29. Raspberry Pi 3 model B. https://www.raspberrypi.org/products/raspberry-pi-3-model-b (2016)

Research on Federated Learning for Tactical Edge Intelligence

Rongrong Zhang(✉), Zhiqiang Gao, and Di Zhou

Information Engineering College, Engineering University of PAP, Xi'an 710086, China
zhangrongrong1109@163.com

Abstract. The gradual maturation of IoT and 5G technologies, as well as edge computing based on cloud-edge-end architecture and the rapid development of artificial intelligence, all provide a very good technical solution to the existing military technology change. At the same time, the information local warfare, equipment modularity, rapid deployment of the development. The network and application layers that connect battlefield information are associated with the efficient transmission of information as well as the effectiveness of information, so transformative improvements must be made to these two major components to suit current wartime and non-wartime tactical command requirements. Therefore, this paper will discuss the possibility of designing a bonding framework that can fully and effectively integrate the network and application layers by using federation learning techniques, and multimodal, multitasking customized edge federation features to expand the network layer and make the application layer intelligent.

1 Introduction

The gradual maturation of IoT and 5G technologies, as well as edge computing [1–3] based on cloud-edge-end architecture and the rapid development of artificial intelligence, all provide a very good technical solution to the existing military technology change. At the same time, the information technology local warfare, equipment is also increasingly to modular, rapid deployment of development. The tactical edge is a very dynamic environment, i.e., the troop location state is constantly changing, the frequency of equipment use changes frequently, network connectivity is intermittent, information flow burst communication is frequent, the schedule mission plan and troop establishment is constantly adjusted, and the tactical edge information shows a multimodal trend. In order to meet the needs of modern warfare information change and tactical edge environment, how to efficiently provide agile, resilient and intelligent information services for tactical forces and how to effectively extract the value of multimodal information are important issues to be solved in the development of tactical edge. Therefore, the integration of artificial intelligence with tactical edge (i.e., tactical edge intelligence) is a key link to win the future distributed joint warfare.

This work was supported by the National Social Science Foundation of China Military Science Youth Program (2022-SKJJ-C-093).

L. Barolli (Ed.): EIDWT 2023, LNDECT 161, pp. 268–274, 2023.
https://doi.org/10.1007/978-3-031-26281-4_27

The network layer and application layer, which are important parts of connecting information on the battlefield, must be transformatively improved to fit current wartime and non-wartime tactical because of the association of how to transmit information efficiently in combat systems and how the effectiveness of information can be truly leveraged on the mission front command requirements. This paper focuses on the design of a cohesive framework that can integrate the network and application layers sufficiently effectively in a tactical edge environment so that the value of information can fully flow through the framework. In this paper, we will use federation learning techniques [4], and multimodal, multi-task [5] customized edge federation features to design the framework with the possibility of expanding the network layer and making the application layer intelligent.

2 Related Work

2.1 Federated Learning

Federated learning [4] is a machine learning setup in which multiple nodes (clients) work together to solve machine learning problems under the unified coordination of an aggregation server or service provider. The raw data of each working node is kept in local storage and is not exchanged or transmitted; this is achieved by using focused updates intended for immediate aggregation to achieve the learning goals. Federated learning consists of two processes, namely the model training process and the model inference process.

As shown in Fig. 1, during the model training process, information related to each party's model can be passed between the parties, while data cannot be passed, and there is no concern about privacy being compromised during the model passing process. The trained model can be placed in any party that joins the federated learning system, or it can be shared among multiple parties. During model inference, the model can be applied to new instances and continue to be trained so that the federation learning process can continue.

Based on the different feature space distribution of each participant and the different data labels, federal learning can be classified into three categories: horizontal, vertical and migration learning. The dataset D_i owned by the participants i is assumed to be represented as a matrix, with the row vector representing each sample and the column vector representing the data features, and the samples involved in the training are denoted as (I, X, Y), where I denotes the identity of the sample, X denotes the feature space of the sample, and Y denotes the label space of the sample.

The existing federation learning generally combines SGD (stochastic gradient descent) to design the algorithm, and the gradient information and parameter information are transmitted between the central server and the sub-node servers. The gradient information refers to the gradient uploaded by the sub-nodes after the initial training to be used as the basis for updating the global model, while the parameter information refers to the gradient used by the central server to update the sub-nodes to learn the training strategy after computing the global model parameters.

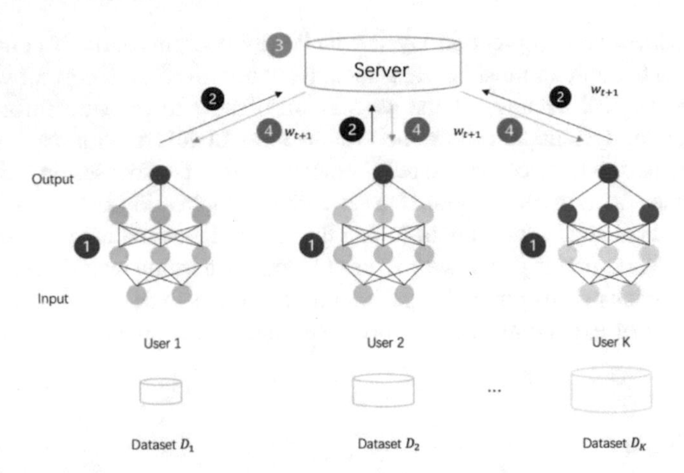

Fig. 1. Federal learning process

Another reason why federated learning outperforms distributed machine learning algorithms is to implement local operations as much as possible. The following mathematical notation is specified for the purpose of explaining this process.

The local update strategy outperforms distributed SGD because it reduces more communication costs, saves sampling and processing time, and the unified planning of tasks is more capable of collaborative resource allocation, allowing relatively critical tasks to be submitted for modification, and reducing the number of communications also reduces the security risk of communications accordingly, thus improving the security strength to a certain extent.

For privacy security protection [4], federated learning is also significantly better than other architectures, specifically in the encryption mechanism based on trust weights. Not only data is kept in local devices and only gradient updates are submitted, which can reduce the risk of data leakage, but also different encryption mechanisms can be used according to different security scenarios.

(1) If the security level of the environment is high at this time, then no encryption at this time.
(2) If there is some security risk in the environment at this time, then encryption is performed at the central server.
(3) If the central server is not trusted at this time, then encryption is performed during the communication process.

2.2 Multidimensional Information Aggregation

1. Multi-task learning.

   Multitask learning [5] refers to learning multiple related tasks simultaneously, allowing these tasks to share knowledge during the learning process, and exploiting the correlation between multiple tasks to improve the performance and generalization ability of the model on each task. Multitask learning can be seen as a kind of inductive transfer learning, i.e., improving generalization ability by using biased information contained in different tasks.

The main issue of multi-task learning is sharing information among tasks. There are four common sharing models: hard sharing model, soft sharing model, hierarchical sharing model, and shared private model.

2. Multimodal Learning

   Multimodal learning [5–8] mainly refers to the ability to process multi-modal data and understand multi-modal information through machine learning methods. Multimodal learning is divided into five main aspects: representation learning, modal transformation, modal alignment, modal fusion, and joint learning.

   Multimodal learning is divided into three categories according to the specific joint approach, (a) mainly joint learning of semantic subspaces of features between different modalities, (b) then mainly feature joint and parallel learning between different modalities under the semantic consistency restriction, and (c) then Encoder-Decoder structure is used to link between different modalities using decoding encoders, thus ensuring semantic consistency under multimodal learning.

3 Designing a Federated Learning Architecture for Tactical Edge Intelligence

The tactical edge is a high-risk information environment located in a poor quality communication environment, frequent disconnections, and a high volume of data with sparse value. The high risk is mainly reflected in the following points; (a) high threat to the information system itself, such as electronic destruction means, the impact of the harsh environment, etc.; (b) high risk of the use of the unit, once the use of the unit has problems, the value of information can not be realized at the end; (c) the system located in the tactical edge must have high availability, high integrity, high transparency.

The design idea of federated learning architecture for tactical edge intelligence is as follows.

1. To solve the problem of poor communication quality and frequent disconnections through asynchronous federation learning techniques [1].
2. To better adapt to the command level information resource aggregation structure, realize hierarchical authority information control and aggregation, as well as security considerations, so the hierarchical federation learning is used to solve the corresponding problems.
3. For better aggregation of information and multi-dimensional access to information value, asynchronous hierarchical federation learning is used as the base architecture, and multimodal and multi-task based joint retrieval is used as the upper layer application structure to design the implementation scheme (Fig. 2).

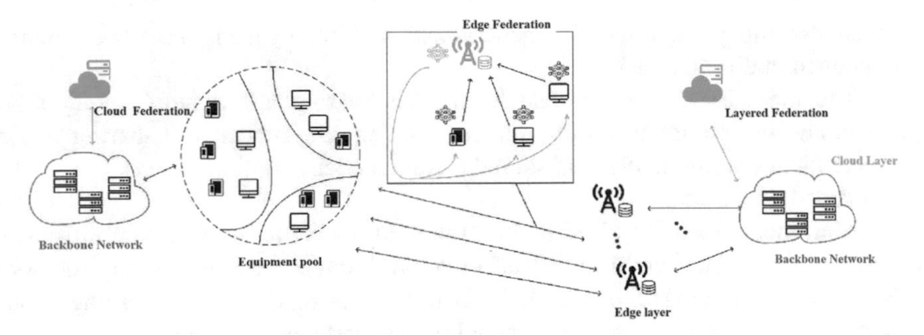

Fig. 2. Tactical edge system architecture design diagram

Asynchronous federation learning is based on a temporally asynchronous federation learning algorithm. In such algorithm design, the simplest and plain temporal decay contribution method is used for parameter weight assignment, i.e., the one with longer update time according to the last model occupies a larger update weight ratio than the one with shorter update time, so as to ensure that a parameter model close to the real data change characteristics can be fitted while taking into account both historical and new information.

Hierarchical federation learning is a federation learning method based on cloud-edge-end architecture. Data from the bottom device does a specified number of rounds of local iterations, after which the initial model parameters are uploaded to the edge for aggregation, and after the edge finishes aggregating the model and performing complex data enhancement training, the model is uploaded to the cloud, which does the final aggregation of data to form a global model, and broadcasts the global model across the network to ensure that every node receives the model parameter push and performs local model update.

The algorithm framework is shown in Fig. 3.

In the tactical edge environment, the information is more dynamic, and the combined complexity of known requirements and unknown situations of a task makes it difficult to fully define the execution details of a task, which leads to an incomplete processing process framework, thus directly leading to a sharp drop in the confidence level of correctness of predicted answers to solve the task, and as a supporting decision. In order to solve the problem, a joint retrieval framework that can solve the fog of war and unclear task requirement orientation is born.

Multimodality mainly focuses on fusion at the feature layer with the neural network layer, and the output layer, waiting for the interrogation request, and then returning the result, while multitasking receives the request at the beginning, performs a relevance prior while processing the requested task, and then synthesizes it, and finally outputs the result directly.

For example, when an enemy unknown threat unit appears, you can quickly estimate the threat threshold of this unit based on the surface characteristics and related behaviors, combined with the previous analysis database, and also do a quick screening of the enemy combat unit type based on the existing unit information, plus some fuzzy features, so as to finally get the result.

Algorithm Asynchronous hierarchical federal learning algorithm

Input: Model $\boldsymbol{w_0}$, number of edge nodes ℓ, total number of worker nodes N, edge cluster worker nodes $i \in \mathcal{C}^\ell$
Output: Model $\boldsymbol{w}$
1: **procedure** ASYNCHRONOUS STRATIFIED FEDERAL AVERAGE
2: Initialize the client parameter $\boldsymbol{w_0}$, while $\alpha_t \leftarrow \alpha, \forall t \in [T]$
3: edge and cloud nodes open Scheduler() threads and Update()
4: **Threads** $Scheduler()$:
5: Start training tasks on a portion of the worker or edge node at regular intervals
6: Pass the timestamp τ down to the global model.
7: **Threads** $Updater()$:
8: **for** *epoch* $t \in [T]$ **do**
9: Receive timestamped data from all worker nodes (w_{new}, τ)
10: Optimization: $\alpha_t \leftarrow \alpha \times s(t - \tau)$, where $s(\cdot)$ is the decay function
11: $w_t \leftarrow (1 - \alpha_t)w_{t-1} + \alpha_t w_{new}$
12: **end for**
13: **Main Process** $Main()$:
14: **for** $k = 1, 2, ..., K$ **do**
15: **for** worker Node $i = 1, 2, ..., N$ **do**
16: **if** Opened by thread $Scheduler()$ **then**
17: $\boldsymbol{w}_i^\ell(k) \leftarrow \boldsymbol{w}_i^\ell(k-1) - \eta \nabla F_i^\ell\left(\boldsymbol{w}_i^\ell(k-1)\right)$
18: $\tau_i \leftarrow t_i$
19: **end if**
20: **end for**
21: **if** $k \mid \kappa_1 = 0$ **then**
22: **for** edge Node $\ell = 1, ..., L$ **do**
23: $\boldsymbol{w}^\ell(k) \leftarrow$ Edge Aggregation $\left(\left\{\boldsymbol{w}_i^\ell(k)\right\}_{i \in \mathcal{C}^{ell}}\right)$
24: **if** $k \mid \kappa_1\kappa_2 \neq 0$ **then**
25: **for** worker Node $i \in \mathcal{C}^\ell$ **do**
26: $\boldsymbol{w}_i^\ell(k) \leftarrow \boldsymbol{w}^\ell(k)$
27: $\tau_\ell \leftarrow t_\ell$
28: **end for**
29: **end if**
30: **end for**
31: **end if**
32: **if** $k \mid \kappa_1\kappa_2 = 0$ **then**
33: $\boldsymbol{w} \leftarrow$ Cloud Aggregation $\left(\left\{\boldsymbol{w}^\ell(k)\right\}_{\ell=1}^L\right)$
34: **for** worker Node $i = 1, 2, ..., N$ **do**
35: $\boldsymbol{w}_i^\ell(k) \leftarrow \boldsymbol{w}^\ell(k)$
36: **end for**
37: **end if**
38: **end for**
39: **end procedure**
40: **function** EDGE AGGREGATION$(\ell, \left\{\boldsymbol{w}_i^\ell(k)\right\}_{i \in \mathcal{C}^{ell}})$
41: $\boldsymbol{w}^\ell(k) \leftarrow \frac{\sum_{i \in \mathcal{C}^\ell} \left|\mathcal{D}_i^\ell\right| \boldsymbol{w}_i^\ell(k)}{|hcal D^\ell|}$
42: **return** $\boldsymbol{w}^\ell(k)$
43: **end function**
44: **function** CLOUD AGGREGATION$(\ell, \left\{\boldsymbol{w}_i^\ell(k)\right\}_{\ell=1}^L)$
45: $\boldsymbol{w}(k) \leftarrow \frac{\sum_{\ell=1}^L \left|\mathcal{D}^\ell\right| \boldsymbol{w}^\ell(k)}{|\mathcal{D}|}$
46: **return** $\boldsymbol{w}(k)$
47: **end function**

Fig. 3. Algorithmic framework

4 Conclusion

Federated Learning greatly reduces the number of communications and improves communication efficiency through the model of local training + upper layer aggregation model + global push update, thus effectively circumventing the problems of poor communication quality and network intermittency. At the same time, by shifting the training task, the number of unnecessary transfers of data in the system architecture is reduced, and the problem of utilizing big data in the tactical edge environment is solved.

In fact, federated learning should be used as a service similar to that as an environment improvement in the tactical edge, as an intermediate layer to optimize the bad environment in the tactical edge into a normal environment, and deliver the confidential tasks from the application layer to the base layer, thus making most of the civilian applications, and various modules can be quickly migrated after a simple functional conversion, which is actually FlaaS here (Federated In fact, this is FlaaS (Federated learning as a service), which encapsulates federated learning as a service black box for third parties to interface with.

References

1. Liang, T., Yong, L., Wei, G.: A hierarchical edge cloud architecture for mobile computing. In: IEEE INFOCOM 2016 – The 35th Annual IEEE International Conference on Computer Communications. IEEE (2016)
2. Shi, W., Zhang, X., Wang, Y., et al.: Edge computing: current status and outlook (2019)
3. Jararweh, Y., Doulat, A., Alqudah, O., et al.: The future of mobile cloud computing: integrating cloudlets and mobile edge computing. In: International Conference on Telecommunications. IEEE (2016)
4. Lu, X., Liao, Y., Lio, P., et al.: Privacy-preserving asynchronous federated learning mechanism for edge network computing. IEEE Access (99), 1–1 (2020)
5. Hu, R., Singh, A.: Transformer is all you need: multimodal multitask learning with a unified transformer (2021)
6. Baltrusaitis, T., Ahuja, C., Morency, L.P.: Multimodal machine learning: a survey and taxonomy. IEEE Trans. Pattern Anal. Mach. Intell. (99), 1–1 (2017)
7. Zhang, C., Yang, Z., He, X., et al.: Multimodal intelligence: representation learning, information fusion, and applications. IEEE J. Sel. Top. Sig. Process. (99), 1–1 (2020)
8. Guo, W., Wang, J., Wanga, S.: Deep multimodal representation learning: a survey. IEEE Access **7**(99), 63373–63394 (2019)

Load Balancing Algorithm for Information Flow Control in Fog Computing Model

Shigenari Nakamura[1](✉), Tomoya Enokido[2], and Makoto Takizawa[3]

[1] Tokyo Metropolitan Industrial Technology Research Institute, Tokyo, Japan
nakamura.shigenari@iri-tokyo.jp
[2] Rissho University, Tokyo, Japan
eno@ris.ac.jp
[3] Hosei University, Tokyo, Japan
makoto.takizawa@computer.org

Abstract. In the IoT (Internet of Things), it is critical to keep devices secure against malicious accesses. In this paper, a device includes data objects where sensor data obtained by sensors of the devices are stored. Subjects like users get sensor data from objects in devices and put the data in objects of other devices. In the IoT, the FC (Fog Computing) model is proposed, which is composed of fog nodes in addition to subjects and devices. Data from objects are processed at fog nodes and processed data are forwarded to subjects. Even if subjects access objects in authorized ways like the CBAC (Capability-Based Access Control) model, the subjects can get data in objects which are not allowed to be gotten. In the FC model, fog nodes execute tasks to process data from objects. In order to avoid concentrating loads at a fog node, an SLB (Source objects-based Load Balancing) algorithm in the CBAC model is newly proposed in this paper. Here, tasks are sent to fog nodes based on not only computation loads but also data locality.

Keywords: Iot (internet of things) · Device security · CBAC (capability-based access control) model · Information flow control · FC (fog computing) model · Load balancing algorithm

1 Introduction

Access control models [3] are widely used to make information systems like database systems secure. The CBAC (Capability-Based Access Control) model [4,6] is considered to be useful to protect devices from malicious accesses in the IoT (Internet of Things) [22]. Here, a capability token is a collection of access rights to manipulate objects in devices. Subjects such as users and applications are issued capability tokens. Only the authorized subjects can manipulate objects in devices only in the authorized operations. Suppose a subject s_{si} gets data from an object o_{oi}^{di} and put the data in another object o_{oj}^{dj}. Here, a subject s_{sj} can get the data via the subject s_{si} even if the subject s_{sj} is granted no access right to get the data, i.e. illegal information flow [10–14]. In addition, a subject s_{sj} might get data which are older than the subject s_{sj} expects to get. In this case, the data come to the subject s_{sj} *late* [15].

L. Barolli (Ed.): EIDWT 2023, LNDECT 161, pp. 275–283, 2023.
https://doi.org/10.1007/978-3-031-26281-4_28

In order to prevent the illegal and late information flows from occurring by interrupting operations implying both information flows in the IoT, the OI (Operation Interruption) [14] and TBOI (Time-Based OI) [15] protocols are proposed and implemented in a Raspberry Pi3 Model B+ [1] with Raspbian [2]. For these protocols, the MRCTSD (Minimum Required Capability Token Selection for Devices) algorithm is also proposed [17]. Here, the request processing time is shorten. In addition, the MRCTSS (MRCTS for Subjects) algorithm is proposed [16]. Here, the communication traffic among subjects and objects is reduced.

In the IoT, it is critical to reduce the electric energy consumed by devices. In order to evaluate the protocols in terms of the electric energy consumption, the power consumption model of a device is proposed [18] by measuring the power consumption of a Raspberry Pi 3 Model B+ equipped with Raspbian. According to the model, electric energy consumption of each protocol is made clear in a simulation evaluation [19]. The request processing time and the electric energy are shown to be reduced in the OI and TBOI protocols with capability token selection algorithms.

In these protocols, the amount of data obtained by entities such as objects and subjects monotonically increases through manipulating objects. If the amount of data kept by entities increases, the numbers of both types of illegal and late information flows also increase. Consequently, the more number of operations are interrupted. Unnecessary data should not be exchanged among entities so that the amount of data obtained by entities does not increase. For this aim, the FC (Fog Computing) model [5] is considered. In the FC model, a fog layer is introduced between devices and subjects. The fog layer is composed of computers named fog nodes. When subjects try to get data from devices, the data from devices arrive at fog nodes before subjects. The data which arrive fog nodes are processed and then the summarized data are sent to subjects. Therefore, subjects can avoid obtaining unnecessary data. In the papers [20,21], the FCOI (FC-based OI) and FCTBOI protocols are proposed so that the number of operations interrupted is reduced by reducing the amount of data kept by entities in the FC model.

In the FC model, fog nodes execute tasks to process data from objects. In order to avoid concentrating loads at a fog node, it is significant to consider how to balance the loads to fog nodes, i.e. load balancing [24]. A lot of load balancing methods have been proposed and discussed so far. Generally, the load of a node is estimated based on the usage of computation resources in the node. For example, an algorithm to select servers to execute computation processes is proposed to reduce the electric energy consumption of a server cluster in our previous studies [8]. On the other hand, the load is estimated based on the features of data sometimes. In a data-aware scheduling [9], while the tasks which do not need large data are sent to other idle nodes, the tasks which need large data are sent to other nodes already having the large portion of the data. In the FC model, since data processing are executed at fog nodes, it is useful to estimate the loads of fog nodes by not only the usage of computation resources of the fog nodes but also the features of data from devices.

In this paper, an SLB (Source objects-based Load Balancing) algorithm to balance the loads to fog nodes is proposed. Here, tasks to process data are sent to fog nodes based on not only computation loads but also data locality. The computation loads are estimated based on the *source* objects. In the FCOI and FCTBOI protocols, every entity holds a set of source objects. A set of source objects indicates which objects the data flow from to its entity.

In Sect. 2, the CBAC model and information flow relations are discussed. In Sect. 3, the SLB algorithm is proposed. In Sect. 4, the FCOI and FCTBOI protocols with the SLB algorithm are discussed to prevent both types of information flows.

2 System Model

In an IoT, subjects $s_1, \ldots, s_{sn}$ ($sn \geq 1$) access to devices $d_1, \ldots, d_{dn}$ ($dn \geq 1$). Each device d_{di} holds the number on^{di} of objects $o_1^{di}, \ldots, o_{on^{di}}^{di}$ ($on^{di} \geq 1$). Here, o_{oi}^{di} indicates an object in the device d_{di}. In this paper, sensors, actuators, and hybrid devices are considered. A sensor d_{di} collects sensor data obtained by sensing events and stores the data in its object o_{oi}^{di}. An actuator d_{di} receives sensor data and stores the data in its object o_{oi}^{di}. The actuator d_{di} performs actions according to the data. A hybrid device d_{di} is equipped with both the sensors and actuators.

Data of objects in devices are manipulated by subjects. In order to make the devices secure, the CBAC model [4,6] is considered. A capability token is a collection of access rights $\langle o, op \rangle$ where o is allowed to be manipulated in op. A subject gets the data from the object and designates actions for actuators by using the data. Here, the data are put to objects of the actuators by the subject. Each subject s_{si} is issued a set CAP^{si} of the number cn^{si} of capability tokens $cap_1^{si}, \ldots, cap_{cn^{si}}^{si}$ ($cn^{si} \geq 1$).

Each capability token cap_{ci}^{si} has validity period. How the subject s_{si} can manipulate which objects is indicated in an access right field of each capability token cap_{ci}^{si}. In addition, a pair of public keys [7] of a subject and an issuer are included. A signature generated with the private key of the issuer is also included.

A subject s_{si} sends an access request with a capability token cap_{ci}^{si}. If the device d_{di} confirms the capability token cap_{ci}^{si}, an operation is performed on an object. Since the device d_{di} just checks the capability token cap_{ci}^{si}, it is easier to adopt the CBAC model to the IoT than the RBAC (Role-Based Access Control) [23] and ABAC (Attribute-Based Access Control) [25] models.

A set of objects whose data a subject s_{si} is allowed to get is $IN(s_{si})$, i.e. $IN(s_{si}) = \{o_{oi}^{di} \mid \langle o_{oi}^{di}, get \rangle \in cap_{ci}^{si} \wedge cap_{ci}^{si} \in CAP^{si}\}$. Through manipulating data of objects in devices, the data are exchanged among subjects and objects. Objects whose data flow into an entity are referred to as source objects for the entity. Let $o_{oi}^{di}.sO$ and $s_{si}.sO$ be sets of source objects of an object o_{oi}^{di} and a subject s_{si}, respectively, which are initially ϕ.

A capability token cap_{ci}^{si} is only valid for some period. Let $gt^{si}.st(o_{oi}^{di})$ and $gt^{si}.et(o_{oi}^{di})$ be the start and end time when a subject s_{si} is allowed to get data from the object o_{oi}^{di}. The generation time is time when data of an object o_{oi}^{di} are generated. Let $minOT_{oi}^{di}(o_{oj}^{dj})$ and $minSBT^{si}(o_{oj}^{dj})$ be the earliest generation times of data of an object o_{oj}^{dj} which flow to an object o_{oi}^{di} and a subject s_{si}, respectively.

In the FC model of the IoT, a fog layer is introduced between devices and subjects as shown in Fig. 1 so that the network traffic is reduced and the time constrains are satisfied. The fog layer F [5] is composed of fog nodes, $f_1, \ldots, f_{fn}$. A fog node f_{fi} does some computation on a collection of input data from sensors. Let $f_{fi}.O$ be a set of objects held by the fog node f_{fi}. Here, if the set $f_{fi}.O$ includes an object o_{oi}^{di}, the fog node f_{fi} holds the object o_{oi}^{di}, i.e. there are the input data from the o_{oi}^{di} in the fog node f_{fi}. The output data are generated by processing the input data.

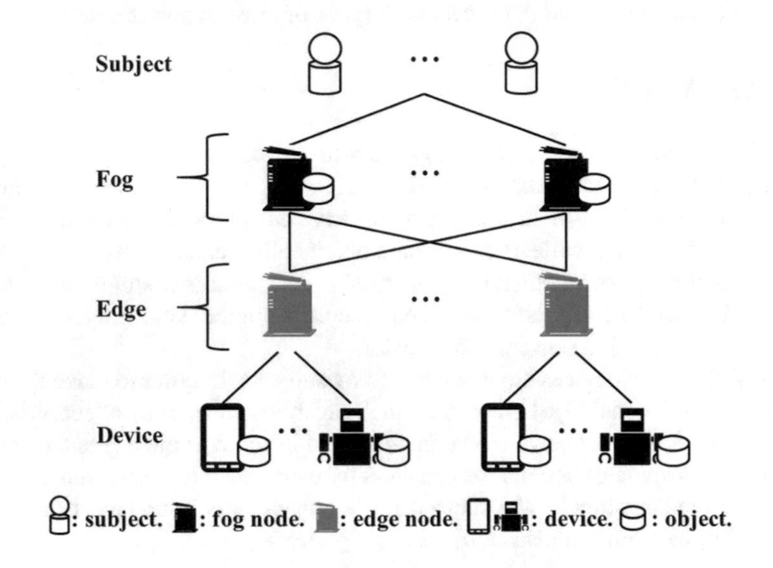

Fig. 1. FC model.

Suppose a subject s_{si} issues a *get* operation on an object o_{oi}^{di} to a device d_{di}. First, the object o_{oi}^{di} arrive at an edge node before being sent to the subject s_{si}. Next, the object o_{oi}^{di} is forwarded to a fog node f_{fi}. Here, the input data held by o_{oi}^{di} are processed and the output data are generated at the fog node f_{fi} for the subject s_{si}. Let $f_{fi}(o_{oi}^{di}.sO)$ be a set of source objects whose data are included in the output data. The set $f_{fi}(o_{oi}^{di}.sO)$ is decided in accordance with the data processing of the fog node f_{fi}. Hence, only the data of objects in $f_{fi}(o_{oi}^{di}.sO)$ flow to the subject s_{si}.

In the CBAC and FC model, following types of information flow relations on objects and subjects are defined:

Definition 1. $o_{oi}^{di} \rightarrow^{f} s_{si}$ *iff* $f_{fi}(o_{oi}^{di}.sO) \neq \phi$ *and* $o_{oi}^{di} \in IN(s_{si})$.

Definition 2. $o_{oi}^{di} \Rightarrow^{f} s_{si}$ *iff* $o_{oi}^{di} \rightarrow^{f} s_{si}$ *and* $f_{fi}(o_{oi}^{di}.sO) \subseteq IN(s_{si})$.

Definition 3. $o_{oi}^{di} \mapsto^{f} s_{si}$ *iff* $o_{oi}^{di} \rightarrow^{f} s_{si}$ *and* $f_{fi}(o_{oi}^{di}.sO) \not\subseteq IN(s_{si})$.

Definition 4. $o_{oi}^{di} \Rightarrow_{t}^{f} s_{si}$ *iff* $o_{oi}^{di} \Rightarrow^{f} s_{si}$ *and* $\forall o_{oj}^{dj} \in f_{fi}(o_{oi}^{di}.sO)$ $(gt^{si}.st(o_{oj}^{dj}) \leq minOT_{oi}^{di}(o_{oj}^{dj}) \leq gt^{si}.et(o_{oj}^{dj}))$.

Definition 5. $o_{oi}^{di} \mapsto_{l}^{f} s_{si}$ *iff* $o_{oi}^{di} \Rightarrow^{f} s_{si}$ *and* $\exists o_{oj}^{dj} \in f_{fi}(o_{oi}^{di}.sO)$ $\neg(gt^{si}.st(o_{oj}^{dj}) \leq minOT_{oi}^{di}(o_{oj}^{dj}) \leq gt^{si}.et(o_{oj}^{dj}))$.

3 Load Balancing Algorithm

In the IoT, subjects issue *get* operation on objects to get data in the objects. In the FC model, the objects in devices are forwarded to fog nodes in the fog layer via edge nodes in the edge layer. Fog nodes access input data of the objects and process the input data in accordance with requirements of the subjects. Output data are generated by fog nodes and the output data are sent to subjects. Here, tasks to process the input data are sent to the fog nodes. In order to avoid concentrating loads at a fog node, it is significant to consider how to balance the loads to fog nodes, i.e. load balancing [24].

A lot of load balancing methods for various types of information systems have been proposed and discussed so far. Generally, the loads of a node are estimated based on the usage of computation resources in the node. For example, an algorithm to select servers to execute computation processes is proposed to reduce the electric energy consumption of a server cluster in our previous studies [8]. On the other hand, the load is estimated based on the features of data sometimes. In the paper [9], a data-aware scheduling is proposed. Here, while the tasks which do not need large data are sent to other idle nodes, the tasks which need large data are sent to other nodes already having the large portion of the data. In the FC model, since data processing are executed at fog nodes, it is useful to estimate the loads of fog nodes by not only the usage of computation resources of the fog nodes but also the features of data from devices.

In this paper, an SLB (Source objects-based Load Balancing) algorithm to balance the loads to fog nodes is proposed. Here, tasks to process data are sent to fog nodes based on not only computation loads but also data locality. Let t_{ti}^{fi} be a task which will be executed at a fog node f_{fi}. Each fog node f_{fi} has a task queue $f_{fi}.Q$ and a task t_{ti}^{fi} is in the queue $f_{fi}.Q$. In the FC model, objects whose data will be processed are forwarded to fog nodes. Hence, each task t_{ti}^{fi} has a set $t_{ti}^{fi}.O$ of objects. In this paper, the meta-data of each object o_{oi}^{di} is considered. Let $o_{oi}^{di}.d$ and $o_{oi}^{di}.o$ indicate the device holding the object o_{oi}^{di} and itself, i.e. $o_{oi}^{di}.d = di$ and $o_{oi}^{di}.o = oi$. $o_{oi}^{di}.ty$ means the type of device d_{di}, i.e. $o_{oi}^{di}.ty = sensor$, *actuator*, or *hybrid device*. $o_{oi}^{di}.sz$ is the size of data the object o_{oi}^{di} holds. $o_{oi}^{di}.sO$ is the set of source objects of the object o_{oi}^{di}. Each task t_{ti}^{fi} has a set $t_{ti}^{fi}.sO$ of source objects which is defined to be $\bigcup_{oi=1}^{|t_{ti}^{fi}.O|} o_{oi}.sO$.

In this paper, it is assumed that a task is generated at an edge node according to an operation issued by a subject s_{si}. After that, the edge node selects a fog node to execute the task. The following SLB algorithm to select a fog node for the task execution is proposed in this paper:

SLB(a task t, a threshold th, meta-data of objects held by fog nodes in F) {
$f = f_1$;
for each object o ($\in t.O$) {
if $o.ty = sensor$ {
$data\_size += o.sz$;
}
}
if $data\_size >= th$ {
for each object o ($\in (t.O \cap f_1.O)$) where $o.ty = sensor$ {
$data\_difference += |o'.sz - o''.sz|$;

/* o' and o'' mean o ($\in t.O$) and o ($\in f_1.O$), respectively. */
}
$min\_data\_difference = data\_difference$;
for each fog node f_{fi} ($\in F$) { /* $fi = 2, \ldots, fn$. */
for each object o ($\in (t.O \cap f_{fi}.O)$) where $o.ty = sensor$ {
$data\_difference += |o'.sz - o''.sz|$;
}
if $min\_data\_difference \geq data\_difference$ {
$min\_data\_difference = data\_difference$;
$f = f_{fi}$;
}
}
return f; /* a fog node f is selected to execute the task t. */
}
else {
for each task t_{ti}^{1} ($\in f_1.Q$) {
$so\_num += |t_{ti}^{1}.sO|$;
}
$min\_so\_num = so\_num$;
for each fog node f_{fi} ($\in F$) { /* $fi = 2, \ldots, fn$. */
for each task t_{ti}^{fi} ($\in f_{fi}.Q$) {
$so\_num += |t_{ti}^{fi}.sO|$;
}
if $min\_so\_num \geq so\_num$ {
$min\_so\_num = so\_num$;
$f = f_{fi}$;
}
}
return f; /* a fog node f is selected to execute the task t. */
}
}

In the SLB algorithm, first, the locality of data needed by the task t is checked. For this aim, the total size of data needed by the task t is calculated. If the total data size is larger than the threshold th, a fog node f_{fi} which already has the large portion of the data is searched. Eventually, such fog node f_{fi} is selected for execution of the task t. Here, the task t is sent to the selected fog node f_{fi} and included in the end of queue $f_{fi}.Q$ of the fog node f_{fi}. At the same time, only the difference between data held by the selected fog node f_{fi} and data held by the task t are sent to the selected fog node f_{fi}. Hence, if the large portion of the data needed by the task t already exist in the selected fog node f_{fi}, the size of data sent with the task t from an edge node becomes smaller. In this procedure, the data locality of objects in only sensor type devices are considered for the simplicity. For hybrid devices, it is difficult to estimate the difference of data size of objects between a fog node f_{fi} and the task t because the data size may be largely changed after the fog node f_{fi} is selected by executing the tasks in the queue $f_{fi}.Q$ of the fog node f_{fi}.

If the total size of data needed by the task t does not exceed the threshold th, the idlest fog node f_{fi} is searched in the SLB algorithm. Here, the loads of each fog node f_{fi} are estimated by the number of source objects of the tasks in its queue $f_{fi}.Q$. The set $t_{ti}^{fi}.sO$ of source objects of a task t_{ti}^{fi} includes objects whose data are obtained by the task t_{ti}^{fi}. Therefore, it is possible to predict the features of data obtained by the task t_{ti}^{fi} by using the set $t_{ti}^{fi}.sO$. For example, the more number of source objects are included in the set $t_{ti}^{fi}.sO$, the more various types of data are obtained by the task t_{ti}^{fi}. Here, the computation complexity of the task t_{ti}^{fi} is generally large. Hence, a fog node f_{fi} whose tasks in its queue $f_{fi}.Q$ hold the smallest number of source objects is regarded as the idlest and selected to execute the new task t.

4 Protocols Based on the FC Model

If subjects try to get data from objects, objects including the data arrive at fog nodes before the data are sent to the subjects. The input data are processed and the output data are generated at fog nodes. Finally, subjects get the output data. As noted above, data are exchanged among subjects and objects in the CBAC model. Hence, subjects may get data which are not allowed to be gotten even if the subjects manipulate objects according to capability tokens issued them. For example, a subject s_{si} can get data of an object o_{oi}^{di} brought to another object o_{oj}^{dj} by accessing the object o_{oj}^{dj} even if the subject s_{si} is not allowed to get the data from the object o_{oi}^{di}, i.e. illegal information flow occurs. In addition, a subject s_{si} can get data which are older than the subject s_{si} expects to get, i.e. information comes to the subject s_{si} *late*. In order to prevent illegal and late types of information flows, the FCOI (FC-based Operation Interruption) and FCTBOI (FC and Time-Based OI) protocols are proposed and evaluated [20,21].

In order to prevent both illegal and late types of information flows, sets of source objects are manipulated in the FCOI and FCTBOI protocols. For example, since data flow from an object o_{oi}^{di} to a subject s_{si} via a fog node f_{fi} in a *get* operation, the set $f_{fi}(o_{oi}^{di}.sO)$ of source objects are added to the set $s_{si}.sO$ of the subject s_{si}. On the other hand, since data flow from a subject s_{si} to an object o_{oi}^{di} in a *put* operation, the set $s_{si}.sO$ are added to the set $o_{oi}^{di}.sO$. In the FCTBOI protocol, the earliest generation time of data of every source object is also updated. The FCOI and FCTBOI protocols perform as follows:

[FCOI protocol] A *get* operation on an object o_{oi}^{di} issued by a subject s_{si} is interrupted if $\exists o_{oj}^{dj} \in t.O \neg(o_{oj}^{dj} \Rightarrow^{f} s_{si})$ holds.

[FCTBOI protocol] A *get* operation on an object o_{oi}^{di} issued by a subject s_{si} is interrupted if $\exists o_{oj}^{dj} \in t.O \neg(o_{oj}^{dj} \Rightarrow_{t}^{f} s_{si})$ holds.

In the FCOI and FCTBOI protocols, tasks to process data from objects are sent to fog nodes in the SLB algorithm. $t.O$ is a set of objects whose data are needed to be processed. In this paper, calculable data are assumed to be exchanged among entities. The set $f_{fi}(o_{oi}^{di}.sO)$ of source objects are decided in accordance with data calculation. For example, if a specific value such as maximum value is required by a subject s_{si}, only the data including the value are extracted from an object o_{oi}^{di} and sent to the subject s_{si}. Here, the set $f_{fi}(o_{oi}^{di}.sO)$ is composed of only one object including the data.

5 Concluding Remarks

The CBAC (Capability-Based Access Control) model is proposed as an access control model of devices in the IoT. The data required by subjects are processed at fog nodes in the FC (Fog Computing) model. Since data are exchanged among entities through manipulating objects, two types of illegal and late information flows occur. In order to prevent both types of illegal and late information flows from occurring, the FCOI (FC-based Operation Interruption) and FCTBOI (FC and Time-Based OI) protocols were proposed. In the FC model, data from devices are processed in a fog layer and the processed data are sent to subjects. Hence, fog nodes execute tasks to process data from objects. In order to avoid concentrating loads at a fog node, it is significant to consider how to balance the loads to fog nodes. In this paper, an SLB (Source objects-based Load Balancing) algorithm to balance the loads to fog nodes is proposed. Here, tasks to process data are sent to fog nodes based on not only computation loads but also data locality. The computation loads are estimated based on the source objects.

Acknowledgments. This work was supported by Japan Society for the Promotion of Science (JSPS) KAKENHI Grant Number JP22K12018.

References

1. Raspberry pi 3 model b+. https://www.raspberrypi.org/products/raspberry-pi-3-model-b-plus/
2. Raspbian, version 10.3, (2020). https://www.raspbian.org/
3. Denning, D.E.R.: Cryptography and Data Security. Addison Wesley, Boston, MA, USA (1982)
4. Gusmeroli, S., Piccione, S., Rotondi, D.: A capability-based security approach to manage access control in the internet of things. Math. Comput. Model. **58**(5–6), 1189–1205 (2013)
5. Hanes, D., Salgueiro, G., Grossetete, P., Barton, R., Henry, J.: IoT Fundamentals: Networking Technologies, Protocols, and Use Cases for the Internet of Things. Cisco Press, Indianapolis, IN, USA (2018)
6. Hernández-Ramos, J.L., Jara, A.J., Marín, L., Skarmeta, A.F.: Distributed capability-based access control for the internet of things. J. Internet Serv. Inf. Secur. **3**(3/4), 1–16 (2013)
7. Johnson, D., Menezes, A., Vanstone, S.: The elliptic curve digital signature algorithm (ECDSA). Int. J. Inf. Secur. **1**(1), 36–63 (2001)
8. Kataoka, H., Nakamura, S., Duolikun, D., Enokido, T., Takizawa, M.: Multi-level power consumption model and energy-aware server selection algorithm. Int. J. Grid Util. Comput. **8**(3), 201–210 (2017)
9. Ke, W., Xraobing, Z., Tonglin, L., Dongfang, Z., Michael, L., Ioan, R.: Optimizing load balancing and data-locality with data-aware scheduling. In: IEEE International Conference on Big Data, pp. 119–128 (2014)
10. Nakamura, S., Duolikun, D., Aikebaier, A., Enokido, T., Takizawa, M.: Read-write abortion (RWA) based synchronization protocols to prevent illegal information flow. In: Proc. of the 17th International Conference on Network-Based Information Systems, pp. 120–127 (2014)
11. Nakamura, S., Duolikun, D., Enokido, T., Takizawa, M.: Influential abortion probability in a flexible read-write abortion protocol. In: Proc. of IEEE the 30th International Conference on Advanced Information Networking and Applications, pp. 1–8 (2016)

12. Nakamura, S., Duolikun, D., Enokido, T., Takizawa, M.: A read-write abortion protocol to prevent illegal information flow in role-based access control systems. Int. J. Space-Based Situated Comput. **6**(1), 43–53 (2016)
13. Nakamura, S., Enokido, T., Takizawa, M.: Information flow control in object-based peer-to-peer publish/subscribe systems. Concurrency and Computation Practice and Experience **32**(8), e5118 (2020)
14. Nakamura, S., Enokido, T., Takizawa, M.: Implementation and evaluation of the information flow control for the internet of things. Concurrency and Computation Practice and Experience **33**(19), e6311 (2021)
15. Nakamura, S., Enokido, T., Takizawa, M.: Information flow control based on capability token validity for secure IoT: Implementation and evaluation. Internet of Things **15**, 100423 (2021)
16. Nakamura, S., Enokido, T., Takizawa, M.: Traffic reduction for information flow control in the IoT. In: Proc. of the 16th International Conference on Broad-Band Wireless Computing, Communication and Applications, pp. 67–77 (2021)
17. Nakamura, S., Enokido, T., Takizawa, M.: Capability token selection algorithms to implement lightweight protocols. Internet of Things **19**, 100542 (2022)
18. Nakamura, S., Enokido, T., Takizawa, M.: Energy consumption model of a device supporting information flow control in the IoT. In: Proc. of the 10th International Conference on Emerging Internet, Data, and Web Technologies, pp. 142–152 (2022)
19. Nakamura, S., Enokido, T., Takizawa, M.: Energy consumption of the information flow control in the IoT: Simulation evaluation. In: Proc. of the 36th International Conference on Advanced Information Networking and Applications, pp. 285–296 (2022)
20. Nakamura, S., Enokido, T., Takizawa, M.: Evaluation of the information flow control in the fog computing model. In: Proc. of the 17th International Conference on Broad-Band Wireless Computing, Communication and Applications, pp. 78–90 (2022)
21. Nakamura, S., Enokido, T., Takizawa, M.: Fog computing model for the information flow control. In: Proc. of the 25th International Conference on Network-Based Information Systems, pp. 25–34 (2022)
22. Oma, R., Nakamura, S., Duolikun, D., Enokido, T., Takizawa, M.: An energy-efficient model for fog computing in the internet of things (IoT). Internet of Things **1–2**, 14–26 (2018)
23. Sandhu, R.S., Coyne, E.J., Feinstein, H.L., Youman, C.E.: Role-based access control models. IEEE. Computer **29**(2), 38–47 (1996)
24. Willebeek-LeMair, M.H., Reeves, A.P.: Strategies for dynamic load balancing on highly parallel computers. IEEE Trans. Parallel Distrib. Syst. **4**(9), 979–993 (1993)
25. Yuan, E., Tong, J.: Attributed based access control (ABAC) for web services. In: Proc. of the IEEE International Conference on Web Services (ICWS'05), p. 569 (2005)

Federated Reinforcement Learning Technology and Application in Edge Intelligence Scene

Xuanzhu Sheng(✉), Zhiqiang Gao, Xiaolong Cui, and Chao Yu

Engineering University of PAP, Xi'an, China
1007939465@qq.com

Abstract. This paper provides a comprehensive review of the emerging technology of federated reinforcement learning in the edge intelligence scenario. Starting from the problems and challenges faced by edge intelligence, we introduce the federated learning framework and use reinforcement learning algorithms to solve the problem of resource scheduling. This paper introduces the generation background, definition and classification of federated reinforcement learning, and focuses on the horizontal and vertical federal reinforcement learning technology and comparison research. Finally, it analyzes and improves the application of federal reinforcement learning in edge intelligence scenarios.

1 Introduction

Intelligent core ability from machine learning, in recent years, reinforcement learning is an important representative of machine learning, however, the promotion of intelligence needs a lot of training data, and data involves user privacy, especially under the edge computing scene, edge computing scene close to the client, for example, mobile phones, video monitoring and other terminal environment. Intelligent edge scenarios, on the other hand, usually involves the remote cloud service nodes, near edge service nodes, intelligent terminal access node distributed components, such as how to cloud-edge-end architecture node intelligent can assign, improve computing, storage, network, intelligent model overall efficiency, and both model training data privacy protection, is the edge of the ground problems to be solved.

Deep reinforcement learning has produced a lot of research results in game scenarios such as Go and game, which provides a reference solution for optimizing the overall performance of multiple resources in edge intelligence.

Integrating the advantages of federated learning and reinforcement learning, forming federated reinforcement learning can solve the problem of optimal resource allocation in both training data privacy and edge intelligence.

2 Edge Intelligence

2.1 The Background and Definition of Edge Intelligence

With the rise of edge computing and the new breakthrough in deep learning algorithms and Moore's Law, artificial intelligence has ushered in a new spring. Edge intelligence is

L. Barolli (Ed.): EIDWT 2023, LNDECT 161, pp. 284–291, 2023.
https://doi.org/10.1007/978-3-031-26281-4_29

combining artificial intelligence with edge computing and deploying [1] in edge devices. Edge intelligence will become the next development stage of edge computing, which can provide advanced service capabilities such as data analysis, scene perception and organizational decision making at the edge [2]. It can also provide offline, privacy-protected smart services. Edge Intelligence is designed to improve data collection, storage, analysis, and processing capabilities. Edge intelligence consists of four parts, namely, edge cache, edge training, edge reasoning, and edge unloading. As an extension of cloud computing, Edge Intelligence pushes cloud services to end-users.

2.2 The Development of Edge Intelligence

In September 2015, ETSI released a Mobile Edge Computing report. Then, at Microsoft Build conference in June 2017, intelligent Cloud and intelligent edge were proposed; in March 2018, Ali Cloud released Link Edge; in August, ICA Alliance released Edge Intelligence White Paper; in November 2018, ECC held Edge Intelligence, Edge Computing Industry Summit; in December, Huawei proposed KubeEdge in Seattle, and released TurboX Edge Platform, an Internet open architecture platform in December 2019.

2.3 Edge Federal Intelligence

In view of the problem of user data security in problem 3, federal learning, as a new type of distributed machine learning, has the role of protecting the privacy and encryption of user data. Federated learning is similar to edge intelligence architecture, which both disperse services and computing resources. Therefore, through the collaboration between terminal devices and edge servers, federated learning and edge intelligence are combined to form edge federal intelligence [3]. Edge federal intelligence can mainly solve the problem between service quality and user data security, and based on the "cloud-edge-end" framework for collaborative computing, in three dimensions, algorithm, task, can

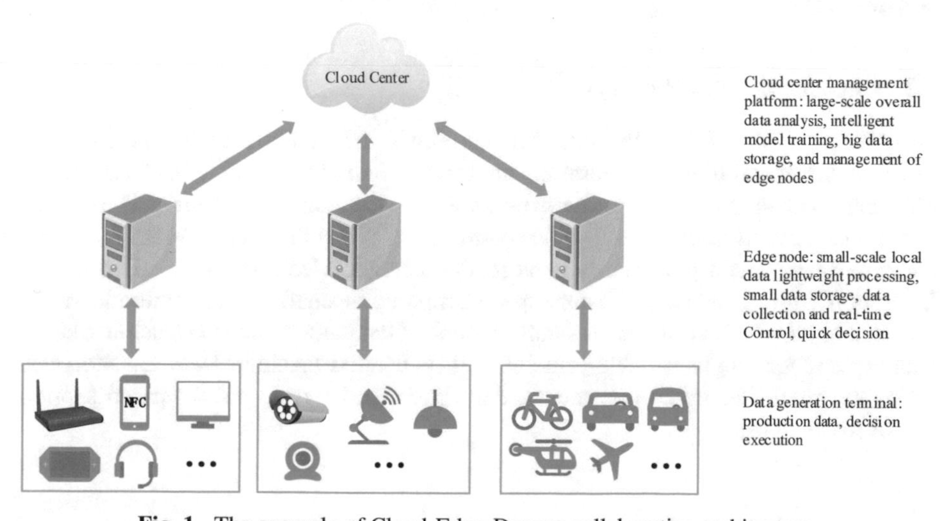

Fig. 1. The example of Cloud-Edge-Devoce collaborative architecture

break the "data island", solve the problem of user data security, collaborative task and provide consistent application scenarios, intelligent, many kinds of services (Fig. 1).

Based on the relationship between the two roles in federated learning and the three levels of edge intelligence, the following three training schemes can be summarized: ①-edge: edge node as server, end as client; ② edge-cloud: cloud as aggregation server, while edge as client participating in federal learning; ③ cloud-edge-end: end and edge as client, and cloud as aggregation server. The third approach is actually a combination of the first two approaches [3].

2.4 Problems Faced with

As can be seen from the development of edge intelligence, edge intelligence is moving from a single agent to the direction of collaborative computing, and the multi-party collaboration mode will inevitably bring new challenges. At present, edge intelligence mainly has four indicators: real-time, precision, energy consumption, and user data security. This can be attributed to the following three main problems: one is the contradiction between accuracy and real-time; the second is the contradiction between accuracy and energy consumption; the third is the contradiction between user data security and algorithm accuracy.

3 Federal Study

3.1 The Definition of Federal Learning

In the data storage and model training stage of machine learning locally, the trained model is passed to the central server and updated together [4]. Federated learning is an algorithmic framework used to build machine learning models, with the following features: multi-party participation, local computing, message encryption, federated performance, and others [5].

3.2 Classification of Federal Learning

From the data point of view, the federated learning is divided into longitudinal federated learning, horizontal federated learning, and transfer federated learning [6] based on the different characteristics of the participants applied to the federated learning. The data set is divided between sample space and feature space. When the feature space similarity and small sample space similarity is large, the federation learning is called horizontal federation learning, when the feature space similarity is small, the federation learning, it can help transfer learning with small amount of data and weak supervision when it cannot meet the single variable principle [7]. Federated machine learning: federated linear algorithm, federated tree model, namely, federal forest, and federated support vector machine [8].

4 Reinforcement Learning

4.1 The Definition of Reinforcement Learning

When supervised and unsupervised learning attempts to enable agents to replicate datasets, i. e., learning from pre-provided samples, reinforcement learning makes the agent gradually enhanced in its interaction with the environment, i. e., generating the samples to learn by themselves [9]. Reinforcement learning (RL) is mainly used to solve the problem of the Markov decision-making process. Reinforcement learning is a hot topic in the field of machine learning. Great progress has been made in many application fields, such as the Internet of Things, autonomous driving and games, etc. [10]. For example, the AlphaGo program developed by DeepMind is a good example [11]. The agent learns from the environment through actual interaction, adjusting its behavior when it receives the results, and ultimately maximizing the reward.

In the game process of interactive learning with the environment in multiple agents, the local performance is still not achieved optimal, but the overall performance is still not maximized [12]. Reinforcement learning is all about solving how to optimize the decision. Resource allocation problems in multi-agents usually contain three basic elements: state space, action space, and reward function.

4.2 The Classification of Reinforcement Learning

Reinforcement learning has the following several algorithms: DQN algorithm, DQN improvement algorithm, policy gradient algorithm, Actor- -Critic algorithm, TRPO algorithm, PPO algorithm, DDPG algorithm, SAC algorithm [13].—— Actor- -Critic algorithm [14] learns both value function and policy function. The Actor- -Critic algorithm itself is a strategy-based function algorithm, whose goal is to optimize a parameter, but it helps the function to learn better by increasing the learning value function [15].

5 Federated Reinforcement Learning

5.1 The Definition of Federted Reinforcement Learning

Federated reinforcement learning consists of four key sub-elements: policy, reward, value function, and environment model. The policy defines the way the agent selects an action in a given state; the reward is the immediate feedback from the environment to the agent; the value function is the long-term reporting reward in the starting state; and the environment model is a virtual model that simulates the environmental action. Four algorithms include federated reinforcement learning: model-based and model-based; value-based and policy-based; Monte Carlo update and time difference update; in policy and off-policy. Among them, value-based and strategy-based algorithms are divided into value-based methods for trying to learn value functions and policy-based methods to search directly from policy parameters. The application of federal reinforcement learning is mainly in two aspects: one is adaptive control and optimal control of —— autonomous driving; the other is discrete and continuous time dynamic feedback control, such as the coal-fired boiler control system. Its important use of reinforcement learning terminal intelligence to obtain observed results, and perform actions, combined with coal-fired boilers to evolve. The difficulties of studying federated reinforcement learning mainly have two aspects: first, the knowledge of the optimal operation of a given state is still limited; second, and the action of the agent will affect the future state of the environment (Fig. 2).

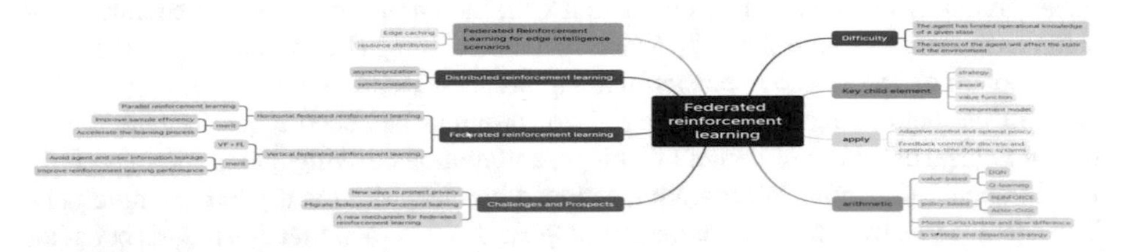

Fig. 2. Federated reinforcement learning framework for edge intelligence scenarios

5.2 Horizontal Federted Reinforcement Learning

Horizontal federated reinforcement learning can be applied to scenarios where agents may be geographically distributed, but they face similar decision-making tasks, with little interactions with each other in the observed environment. Horizontal federated reinforcement learning is the use of reinforcement learning algorithm on the basis of horizontal federated learning. Its classification principle is consistent with the horizontal federated learning, and it can be applied to the same feature space and different sample space, similar to the horizontal division of data in the table. The so-called horizontal, namely "horizontal division", is widely used in the traditional tabular form used to represent data characteristics. Horizontal federated reinforcement learning also has two common system architectures, respectively, called customer-server (clientserver) architecture and peer-to-peer (Peer-to-Peer, P2P) network architecture [17]. The client-server architecture is also known as the primary-slave (master-worker) architecture or the wheel-spoke (hub-and-spoke) architecture [18]. In HFRL problems, the environment, state space, and action space can replace the dataset, feature space, and label space of the basic FL [19].

5.3 Vertical Federted Reinforcement Learning

Vertical federated reinforcement learning is in contrast to horizontal federated reinforcement learning, which is divided by the participants of the same sample space and different feature space, or the federated reinforcement learning by feature [20]. The training process of the VFRL system generally consists of two parts: first, the aligned entities with the same ID but distributed to different participants; then the encrypted (security for user data) model is trained based on these aligned entities [21]. The longitudinal federated reinforcement learning process can be divided into two parts: encrypted sample alignment and encrypted model training [22]. The training process of vertical federated learning generally consists of two parts: first, align the entities with the same ID but that are distributed to different participants, and then perform encrypted or privacy-protected model training based on these aligned entities [23].

6 Federated Reinforcement Learning Applications in Edge Intelligence Scenarios

6.1 Edge Cache

In edge intelligence scenarios, federated reinforcement learning can provide caching services [24]. For edge-initiated Iot of Things, methods of dynamic action caching based on horizontal federated reinforcement learning emerge [16]. Save and efficient cache content by predicting user requests. This horizontal federated reinforcement learning approach defines a header —— for processing with model aggregation in a cache cluster [25]. The method improves the response delay and avoids the drawbacks that may be experienced between small and large clusters [26]. Considering the low latency in the service process and the user's data security problems, the high efficiency and security of the cache method arouse the favor of many researchers [6]. Therefore, the federated reinforcement learning collaborative caching algorithm is used to dynamically determine the content to be replaced or cached, so as to achieve high adaptability and fast convergence in intelligent scenarios [27]. Federal learning can realize the Internet of things application of distributed energy management, the agent through the Actor-Critic algorithm strategy gradient, will interact with the environment Actor strategy network and Critic value function, using the federal gradient descent method, update the deployment model, in convergence speed, equipment energy consumption and the number of agents have obvious advantages in [28].

6.2 Resource Allocation

In order to deal with different types of optimization tasks, resource allocation methods in edge intelligent scenarios have been valued by researchers [29]. To realize the joint collaboration, based on the federated learning framework, the multi-agent participant-criticism algorithm is used, where the edge devices are an interactive learning strategy independently through their own observation, that is, a multi-agent device for modeling the data allocation strategy of the edge devices [30]. In the edge smart scenario, smart homes have undergone a revolution, where smart meters are deployed in a metering infrastructure to analyze and monitor users' energy consumption [30].

7 Summary and Outlook

This paper summarizes the applications and challenges in edge intelligence scenarios, summarizes the definition, connotation, and application of federated learning and reinforcement learning, and focuses on the application of federated reinforcement learning in edge intelligence scenarios. While FRL-based methods have many advantages, some key open issues will need to be considered in future implementations. Therefore, research on addressing these issues provides future directions.

1. Resource consumption of terminal equipment

   Although the existing research work can effectively deal with the resource allocation and optimization problems in edge intelligent scenarios, in practical application, the limited energy reserve of terminal equipment has higher requirements for continuous endurance, thus affecting the allocation effect of resource optimization. At the same time, the energy available for terminal devices to update local models is even more limited. Therefore, how to ensure the resource optimization performance of the edge intelligent system, while guaranteeing the endurance of the terminal equipment, is a future research direction.

2. Data security issues

Although federated learning adopts the "data motionless model movement" method to protect the privacy security of user data, in the decision task of federated reinforcement learning, the training data is generated by the agent interaction with the environment, therefore, may produce data poisoning. Moreover, not only viruses produce internally, but the external environment may also produce various attacks. Therefore, there is a great research space for the threat technology of protecting internal and external attacks and security and privacy.

References

1. Qiao, D., Guo, S., He, J., Zhu, Y.: Edge intelligence: research progress and challenges. Radio Commu. Technol. **48**(01), 34-45 (2022)
2. Chen, M., Zhang, J., Li, T.: Summary of federal learning attack and defense studies. Computer Science **49**(07), 310–323 (2022)
3. Mo, Z., Gao, Z., Miao, D.: Edge intelligence: a new tentacles of artificial intelligence to edge distribution. Data and Computing Development Frontier **2**(04), 16-27 (2020)
4. Zhao, Y., Zhang, H., Wang, S.: The federal learning review. Comp. Program. Tips Maint. (01), 117–119 (2022). https://doi.org/10.16184/j.cnki.comprg.2022.01.063
5. Liang, T., Zeng, B., Chen, G.: Federal learning review: concepts, technologies, applications, and challenges. Computer application 1–13 [2022–10–20]. http://kns.cnki.net/kcms/detail/51.1307.TP.20211231.1727.014.html
6. Zhu, P.: A federated learning model and algorithm based on mobile edge computing. Nanjing University of Posts and Telecommunications (2021). https://doi.org/10.27251/d.cnki.gnjdc.2021.001160
7. Gao, S., Yuan, L., Zhu, J., Ma, X., Zhang, R., Ma, J.: A blockchain-based privacy-preserving asynchronous federated learning. Chinese Sci. Info. Sci. **51**(10), 1755–1774 (2021)
8. Chen, L., Liu, W.: Frontier and application of data security technology from the perspective of compliance. Frontiers of Data and Computing Development **3**(03), 19–31 (2021)
9. Liu, Z.: Channel estimation and resource management research. Beijing Jiaotong University (2021). https://doi.org/10.26944/d.cnki.gbfju.2021.001145
10. Hu, X.: Research and application of target detection algorithm based on border collaboration. Harbin Institute of Technology (2021). https://doi.org/10.27061/d.cnki.ghgdu.2021.003651
11. Anning, Z.Z.: Anti-jamming resource dispatching method for D2D communication network based on deep reinforcement learning. Electric Power Info. Commu. Technol. **20**(09), 108–114 (2022). https://doi.org/10.16543/j.2095-641x.electric.power.ict.2022.09.014
12. Liu, C.: Reinforcement learning algorithm research based on ESN. Hainan University (2021). https://doi.org/10.27073/d.cnki.ghadu.2021.000172
13. Dong, S., Yang, C., Liu, T.: Federated learning-based wireless tasks: does data non-IID necessarily affect performance? Signal Process. **37**(08), 1365–1377 (2021). https://doi.org/10.16798/j.issn.1003-0530.2021.08.003

14. Jiang, Z.: Reinforcement learning-based cloud computing resource scheduling algorithm research. Wuhan University (2021). https://doi.org/10.27379/d.cnki.gwhdu.2021.000128
15. Pu, T., Du, S., Li, Y., Wang, X.: Distributed power supply collaborative optimization strategy based on federated reinforcement learning. Power System Automation: 1–12 [2022–10–31]. http://kns.cnki.net/kcms/detail/32.1180.tp.20221025.1937.013.html
16. Chen, M., Sun, Y., Hu, Y., Liu, C., Pang, P., Xie, Z.: Collaborative training and optimization management method of residential community integrated energy system based on longitudinal federal intensive learning. Chinese J. Elec. Eng. **42**(15), 5535–5550 (2022). https://doi.org/10.13334/j.0258-8013.pcsee.210812
17. Li, J., Shao, J., Lin, J.: Review of the federal reinforcement learning literature. Fintech Era **29**(10), 87–89 (2021)
18. Wang, X., Fan, F., Sun, Y., Wu, Y.: Internet of vehicles resource allocation based on federated deep reinforcement learning. Electr. Measure. Techniq. **44**(10), 114–120 (2021). https://doi.org/10.19651/j.cnki.emt.2106047
19. Zhou, Jr.: Research and application of multi-agent deep reinforcement learning collaborative technology. Univ. Elec. Sci. Technolo. (2021). https://doi.org/10.27005/d.cnki.gdzku.2021.003772
20. Jiang, J.: Research on resource allocation and task scheduling oriented to edge intelligence. Jilin University (2020). https://doi.org/10.27162/d.cnki.gjlin.2020.000122
21. Zhang, X., Liu, Y., Liu, J., Han, Y.: Federated learning review for edge intelligence. Computer Research and Development 1–27 [2022–10–31]. http://kns.cnki.net/kcms/detail/11.1777.TP.20221027.0839.002.html
22. Wanduo, H.M., Xiao, M., Zhang, Y.: Review of heterogeneous parallel computing platforms for edge-oriented intelligent computing. Computer Engineering and Application: 1–12 [2022–10–31]. http://kns.cnki.net/kcms/detail/11.2127.TP.20221006.1805.004.html
23. Guo, Y.: Task unloading and resource allocation based on federated learning. Guangzhou University (2022). https://doi.org/10.27040/d.cnki.ggzdu.2022.001385
24. Cai, X.: Research on resource optimization allocation for distributed edge intelligence. Guangdong University of Technology (2022). https://doi.org/10.27029/d.cnki.ggdgu.2022.000975
25. Wang, R., Qi, J., Chen, L., Yang, L.: Edge collaborative inference review for edge intelligence. Computer Research and Development: 1–17 [2022–10–31]. http://kns.cnki.net/kcms/detail/11.1777.tp.20220426.1612.006.html
26. Yuan, X.: Federated learning node selection mechanism in an edge intelligent network. University of Electronics, Science and Technology (2022). https://doi.org/10.27005/d.cnki.gdzku.2022.000290
27. Shu, Z.: Research on federated learning methods based on edge computation. Nanjing University of Posts and Telecommunications (2021). https://doi.org/10.27251/d.cnki.gnjdc.2021.000897
28. Sun, B., Liu, Y., Wang, T., Peng, S., Wang, G., Jia, W.: Summary of federated learning efficiency optimization in mobile edge networks. Comp. Res. Develop. **59**(07), 1439-1469 (2022)
29. Qi, J., et al.: Federated reinforcement learning: techniques, applications, and open challenges (2021). arXiv preprint arXiv:2108.11887
30. Hu, Y., et al.: Reward shaping based federated reinforcement learning. IEEE Access **9**, 67259–67267 (2021)

Cryptanalysis of a Public Cloud Auditing Scheme

Xu An Wang(✉), Mingyu Zhou, and Wenyong Yuan

Engineering University of People's Armed Police, Xi'an, China
wangxazjd@163.com

Abstract. Cloud data integrity verification is an important means to keep data security. Many integrity verification schemes were proposed for cloud storage. In 2019, Zhang et al. proposed a conditional identity privacy-preserving public auditing (CIPPPA) mechanism. We found that the signature algorithm is insecure because the Cloud Service Provider (CSP) can modify the user's outsourced data, while it can generate a correct proof.

1 Introduction

Today people have to store more and more data. To reduce the burden and cost of physical storage, users prefer to choose cloud storage services, upload their data to cloud service providers (CSP), and hire CSP to store and manage the data. Once the data are uploaded to the CSP, users lose direct control of the data which causes new data security issues. For example, the CSP may change users' data for profit, external attackers may attack the cloud to modify the data, and the integrity of user data is threatened. Therefore maintaining the integrity of outsourced data is an important issue that needs to be solved, and thus the design of schemes that can verify the integrity of cloud data has become an important research topic.

2 Related Work

Since in 2007 Ateniese et al. proposed the provable data possession (PDP) construction [1], many audit schemes were proposed. Nowadays, researchers have proposed cloud audit schemes with more and more interesting properties, and cloud audit schemes are gradually becoming mature. Huang et al. [2] designed a collaborative cloud data verification blockchain framework for cloud data storage. Wang et al. [3] proposed a lightweight certificate-based public/private auditing scheme based on asymmetric bilinear pairing. The solution can change to public and private auditing based on requirements. Rao et al. [4] propose a dynamic outsourced integrity verification scheme. The scheme can protect against any dishonest entity for collusion and support verifiable dynamic updates to outsourced data. They present a new approach to batch-verify multiple leaf nodes and the corresponding indexes based on batch-leaves-authenticated Merkle Hash Tree (MHT). In the process of flight exploration, UAV needs to dynamically update the cloud data frequently when using the cloud storage service, and the privacy of the cloud data must

L. Barolli (Ed.): EIDWT 2023, LNDECT 161, pp. 292–296, 2023.
https://doi.org/10.1007/978-3-031-26281-4_30

be protected. Therefore, Liu et al. [5] proposed a public cloud audit scheme based on distributed string equality verification protocol and Merkle-Hash tree multi-level index structure, which realized efficient dynamic data operation and comprehensive privacy protection. Zhang et al. [6] proposed a new storage audit scheme, which implements efficient user revocation and eliminates the complex certificate management in traditional PKI systems. Yang et al. [7] proposed a stateless cloud audit scheme for privacy protection of non-administrator dynamic group data. The scheme can protect the privacy of user identity and data content, and use Shamir secret sharing technology to divide the re-signature process into several parts, so as to resist the conspiracy attack. Through the binary tree design, group users can track the dynamic changes of data and recover the latest data when the existing data is destroyed. Furthermore, users and TPA do not need to maintain data index information during cloud auditing. Thc solution also implements mutual supervision between users and CSP.

Recently, Zhang et al. proposed a conditional identity privacy-preserving public auditing (CIPPPA) mechanism which could protect users' identity privacy and data privacy [8]. Through analysis, we found that the same random number r was used in the signature process, which made the signature algorithm unsafe. CSP may modify the data, but generate a correct proof to pass the integrity audit.

The rest of the paper is organized as following. In Sect. 3, we review the original protocol of Zhang et al. In Sect. 4, we attack the original protocol to show that it is insecure. Finally, we conclude the paper.

3 Review of Zhang et al.'s Protocol

In the reference [8], the signature algorithm was unsafe, so we mainly reviewed the signature algorithms related with signature. Their scheme is as follows:

Setup: Private key generator (PKG) sets two groups G_1 and G_2, which is an addition cyclic group and a multiplication cyclic group. The order of them is the same large prime order p, where P is a generator of G_1. Then PKG Chooses a random number $s \leftarrow Z_q^*$ as the master secret key, computes $P_{pub} = sP$ as the public key. Finally, it sets five different collision resistant hash functions: $H_1 : G_1 \times \{0, 1\}^* \times G_1 \rightarrow \{0, 1\}^{\rho_1}$, $H_2 : G_1 \times \{0, 1\}^{\rho_1} \rightarrow G_1$, $H_3 : \{0, 1\}^* \rightarrow G_1$, $H_4 : G_1 \rightarrow Z_q$ and $H_5 : \{0, 1\}^* \rightarrow \{0, 1\}^{\rho_2}$.

Keygen: The user with identity $id \in \{0, 1\}^{\rho_1}$, randomly selects $x \leftarrow Z_q^*$ to compute $AID_1 = xP$, and send (id, AID_1) to the PKG. Once receiving (id, AID_1), the PKG computes $AID_2 = id \oplus H_1(P_{pub}||T||sAID_1)$, and outputs the anonymous identity $AID = (AID_1, AID_2)$, Here the valid period of the identity is T. PKG computes corresponding private key as $SK_{AID} = sH_2(AID)$, and sends (AID, T, SK_{AID}) to the user securely.

Siggen: Given a file M, the user splits M into n blocks $M = (m_1, m_2..., m_n) \in Z_q^n$. Then user selects a number $r \leftarrow Z_q^*$, and computes $R = rP$. For m_i, the user generates the signature as following:

$$W_i = m_i SK_{AID} + rH_3(N||i) \tag{1}$$

Denote $\Omega = \{(m_i, W_i)_{1 \leq i \leq n}, N, R\}$.The user uploads $\{\Omega, AID, T\}$ to CSP.

Challenge: Third party auditor (TPA) generates the challenge message after receiving the request for the auditing task. It chooses a subset $I \subseteq \{1, 2, ...n\}$ and computes the challenge subset $Q = (i, v_i)_{i \in I}$. Later it sends the challenge information $Q = (i, v_i)_{i \in I}$ to CSP.

ProofGen: Once receiving challenge from TPA, CSP generates the response auditing proof as following. It chooses a random number $\alpha \leftarrow Z_q^*$, and computes: $V = \alpha H_2(AID)$, $\xi = \alpha^{-1} \sum_{i \in I} v_i m_i + H_4(V)$ and $W = \sum_{i \in I} v_i W_i$.

Finally, CSP returns proof $\{V, \xi, W, N, R\}$ to TPA.

ProofVerify: Upon receiving the auditing proof from CSP, TPA checks Eq. (2) whether or it holds:

$$e(W, P) = e((\xi - H_4(V))V, P_{pub}) \cdot e(\sum_{i \in I} v_i H_3(N||i), R) \tag{2}$$

If the Eq. (2) holds, TPA takes result as accept; otherwise, TPA takes result as reject.

4 Our Attack

In this section, we showed the malicious CSP can modify the data, but can generate a correct response proof to pass the verification Eq. (2). The malicious cloud servers run as following:

1. After receiving $\{\Omega, AID, T\}$, the malicious CSP (adversary) will obtain M, W_i and R, where:$M = (m_1, m_2..., m_n)$,$W_i = m_i SK_{AID} + rH_3(N||i)$ and $R = rP$.
2. The malicious CSP first randomly modifies $M = (m_1, m_2..., m_n) \in Z_q^n$ into $M' = (km_1, km_2..., km_n) = (m_1', m_2'..., m_n')$, $k \in Z_q^*$, $k \in Z_q^*$. Then CSP forges signature by utilizing M', and computes: $W_i' = kW_i = km_i SK_{AID} + krH_3(N||i)$ and $R' = kR = krP$. Denote the new set of signatures by $\Omega' = \{(km_i, W_i')_{1 \leq i \leq n}, N, R'\}$.
3. Once receiving challenge information $Q = (i, v_i)_{i \in I}$ from TPA, CSP forges evidence through the modified signature. CSP chooses a random number $\alpha \leftarrow Z_q^*$, computes:$V = \alpha H_2(AID)$,$\xi' = \alpha^{-1} \sum_{i \in I} v_i k m_i + H_4(V)$ and $W' = \sum_{i \in I} v_i W_i'$. Finally, the auditing proof $\{V, \xi', W', N, R'\}$ is sent to TPA by CSP.
4. Upon receiving the proof from CSP, TPA checks the Eq. (3) whether or it holds:

$$e(W', P) = e((\xi' - H_4(V))V, P_{pub}) \cdot e(\sum_{i \in I} v_i H_3(N||i), R') \tag{3}$$

The forged evidence can pass the integrity verification, which proves as following:

$$
\begin{aligned}
e(W',P) &= e(\sum_{i\in I} v_i W_i', P) \\
&= e(\sum_{i\in I} v_i (km_i SK_{AID} + krH_3(N||i)), P) \\
&= e(\sum_{i\in I} v_i km_i SK_{AID}, P) \cdot e(\sum_{i\in I} krv_i H_3(N||i), P) \\
&= e(\sum_{i\in I} v_i km_i sH_2(AID), P) \cdot e(\sum_{i\in I} krv_i H_3(N||i), P) \\
&= e(\sum_{i\in I} v_i km_i H_2(AID), P_{pub}) \cdot e(\sum_{i\in I} v_i H_3(N||i), R') \\
&= e(\alpha(\xi' - H_4(V))H_2(AID), P_{pub}) \cdot e(\sum_{i\in I} v_i H_3(N||i), R') \\
&= e((\xi' - H_4(V))V, P_{pub}) \cdot e(\sum_{i\in I} v_i H_3(N||i), R')
\end{aligned}
$$

Obviously, CSP can forge signature and pass integrity verification. Therefore, Zhang et al.'s protocol is unsafe. The core reason is that the same random number r was used in the signature process for many data blocks.

5 Conclusion

In this article, we found that Zhang et al.' s protocol was unsafe, and malicious cloud can modify data and forge the signature successfully. We leave how to improve the security of the signature algorithm as an open problem, one way maybe use different random number r for every data block, but this will require R to be different for data blocks, which needs new technique.

Acknowledgement. This work is supported by Engineering University of PAP's Funding for Scientific Research Innovation Team (No.KYTD201805), Engineering University of PAP's Funding for Key Researcher (No. KYGG202011).

References

1. Ateniese, G., et al.: Provable data possession at untrusted stores. In: Proc. Conference on Computer and Communications Security2007, pp. 598–609 (2007)
2. Huang, P., Fan, K., Yang, H., Zhang, K., Li, H., Yang, Y.: A collaborative auditing blockchain for trustworthy data integrity in cloud storage system. IEEE Access **8**, 94780–94794 (2020). https://doi.org/10.1109/ACCESS.2020.2993606
3. Wang, F., Xu, L., Choo, K.-K.R., Zhang, Y., Wang, H., Li, J.: Lightweight certificate-based public/private auditing scheme based on bilinear pairing for cloud storage. IEEE Access **8**, 2258–2271 (2020). https://doi.org/10.1109/ACCESS.2019.2960853
4. Rao, L., Zhang, H., Tu, T.: Dynamic outsourced auditing services for cloud storage based on batch-leaves-authenticated merkle hash tree. In: IEEE Transactions on Services Computing, vol. 13, no. 3, pp. 451–463 (1 May-June 2020). https://doi.org/10.1109/TSC.2017.2708116
5. Liu, J., Wang, X.A., Liu, Z., Wang, H., Yang, X.: Privacy-preserving public cloud audit scheme supporting dynamic data for unmanned aerial vehicles. IEEE Access **8**, 79428–79439 (2020). https://doi.org/10.1109/ACCESS.2020.2991033
6. Zhang, Y., Yu, J., Hao, R., Wang, C., Ren, K.: Enabling efficient user revocation in identity-based cloud storage auditing for shared big data. In: IEEE Transactions on Dependable and Secure Computing, vol. 17, no. 3, pp. 608–619 (1 May-June 2020). https://doi.org/10.1109/TDSC.2018.2829880
7. Yang, X., Wang, M., Wang, X., Chen, G., Wang, C.: Stateless cloud auditing scheme for non-manager dynamic group data with privacy preservation. IEEE Access **8**, 212888–212903 (2020). https://doi.org/10.1109/ACCESS.2020.3039981
8. Zhang, X.J., Zhao, J., Xu, C., Li, H.W., Wang, H.X., Zhang, Y.: CIPPPA: conditional identity privacy-preserving public auditing for cloud-based WBANs against malicious auditors. IEEE Trans. Cloud Comp. **9**(4), 1362–1375 (July 2019)

A Fuzzy-Based Approach for Selection of Radio Access Technologies in 5G Wireless Networks

Phudit Ampririt[1(✉)], Makoto Ikeda[2], Keita Matsuo[2], and Leonard Barolli[2]

[1] Graduate School of Engineering, Fukuoka Institute of Technology, 3-30-1 Wajiro-Higashi, Higashi-Ku, Fukuoka 811-0295, Japan
bd21201@bene.fit.ac.jp

[2] Department of Information and Communication Engineering, Fukuoka Institute of Technology, 3-30-1 Wajiro-Higashi, Higashi-Ku, Fukuoka 811-0295, Japan
makoto.ikd@acm.org, {kt-matsuo,barolli}@fit.ac.jp

Abstract. The 5-th Generation (5G) heterogeneous networks are expected to provide dense network services and a plethora of different networks for fulfilling the user requirements and are supposed to give User Equipment (UE) the ability to connect with the appropriate Radio Access Technology (RAT). However, for the selection of RAT many parameters should be considered, which make the problem HP-hard. For this reason, in this paper, we propose a Fuzzy-based RATs Selection System (FRSS) considering three parameters: Coverage (CV), User Priority (UP) and Spectral Efficiency (SE). From simulation results, we found that when CV, UP and SE are increased, the RAT Decision Value (RDV) parameter value is increased.

1 Introduction

In the 5-th Generation (5G) wireless networks, the unprecedently explosive growth of user devices with erratic traffic patterns will cause a huge volume of data, congesting the Internet and affecting Quality of Service (QoS) [1]. In order to meet the QoS requirements in many application situations, the 5G Wireless Networks will offer enhanced reliability, throughput, latency, and mobility in order to overcome these problems.

For improving the performance of 5G Radio Access Technologies (RATs), multiple base stations (BSs) use heterogeneous RATs (such as GSM, HSPA, LTE, LTE-A, Wi-Fi, and so on) which provide different radio coverages (such as macrocell, microcell, picocell, femtocell, Wi-Fi, etc.) with different transmission power levels in order to provide the mobile users with the best Quality of Experience (QoE), Energy Efficiency (EE), redundancy and reliability [2,3].

In the future, 5G will need to provide a diverse range of service objectives for activities, including lifestyle, working, entertainment and transportation. The 5G will solve the challenges by considering three main different application scenarios: enhanced Mobile Broadband (eMBB), massive Machine Type Communication

L. Barolli (Ed.): EIDWT 2023, LNDECT 161, pp. 297–307, 2023.
https://doi.org/10.1007/978-3-031-26281-4_31

(mMTC) and Ultra-Reliable & Low Latency Communications (URLLC). The eMBB enhances seamless QoE and has good accessibility to services and multimedia information related to the human-essential. The mMTC provides extended battery life despite supporting many connected devices. Finally, by effectively decreasing the latency and enhancing reliability, the URLLC can enable operating in real-time, such as those associated with transport security, remote surgeries and the automation of industrial processes [4–6].

Recently, many research works deal with design of systems appropriate for 5G wireless networks. One example is the utilization of Network Function Virtualization (NFV) and Software-Defined Networking (SDN) for several administrative and technological networks, including massive computing resources [7,8]. Also, the mobile handover approach and SDN are used to minimize processing delays [9–11]. In addition, QoS is improved by implementing Fuzzy Logic (FL) to SDN controllers.

In our previous work [12–16], we presented some Fuzzy-based systems for Call Admission Control (CAC) and Handover in 5G Wireless Networks. In this paper, we propose a Fuzzy-based RATs Selection System (FRSS) considering three parameters: Coverage (CV), User Priority (UP) and Spectral Efficiency (SE).

The rest of the paper is organized as follows. We introduce SDN in Sect. 2. We present 5G Network Slicing in Sect. 3. We describe the proposed Fuzzy-based system and its implementation in Sect. 4. We discuss the simulation results in Sect. 5. Finally, conclusions and future work are presented in Sect. 6.

2 Software-Defined Networking (SDN)

The SDN is one of the most promising methods to make networks programmable and virtualizable by separating the network's data plane from control plane. The SDN structure is shown in Fig. 1.

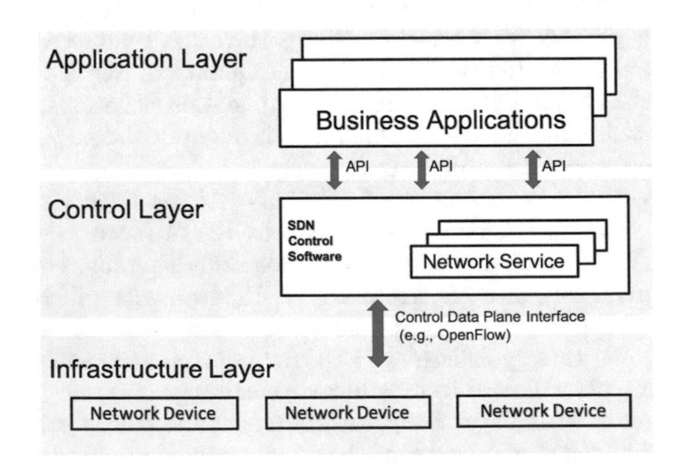

Fig. 1. Structure of SDN.

The Application Layer gathers data from the controller to create an integrated overview of the network for the purpose of making decision. The Northbound Interfaces create a plethora of possibilities for network programming by enabling communication between the Application Layer and the Control Layer. According to the requirements of the application, it will send instructions and data to the control layer, where the controller will establish the best software network feasible with the necessary service quality and security. The Control Layer manages the data plane and transmits various sorts of rules and policies to the Infrastructure Layer via Southbound Interfaces after receiving requests or orders from the Application Layer. The Southbound Interfaces are protocols that allow the controller to set rules for the forwarding plane and offer connectivity and interaction between the Control Plane and the Data Plane. The Infrastructure Layer represents the network's forwarding devices, such as load balancers, switches, and routers, and receives instructions from the SDN controller.

These components of SDN can be efficiently controlled and used by a centralized control plane. The SDN can regulate and modify resources on the control plane effectively in situations of traffic congestion. Forwarding data over many wireless technologies is faster and simpler for mobility management [17,18].

3 5G Network Slicing

The Network Slicing (NS) is a logical network that offers particular network capabilities and characteristics to support users specified purposes. According to network slicing, multiple virtual networks, or "Slices," are constructed on top of a single shared physical infrastructure. Compared with traditional networks, a network with NS can achieve greater performance and be more customizable with individual management and control functions that can be constructed on demand based on different user requirements. Additionally, the network's reliability and security can be enhanced because each slice is logically independent and has no influence on other virtual logical networks [19–22].

The concept for 5G NS was introduced by Next Generation Mobile Networks (NGMN). Three main layers of the network slicing process are the service instance layer, the network slice instance layer, and the resource layer, as illustrated in Fig. 2.

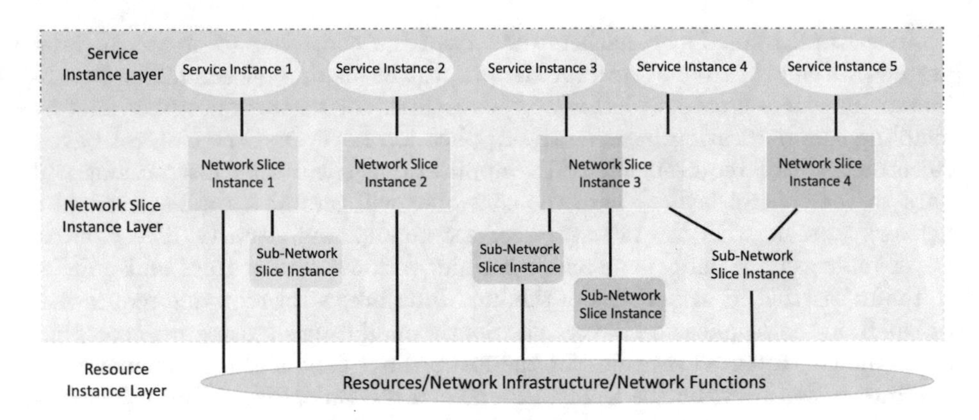

Fig. 2. The concept for 5G NS.

Each service instance in the Service Instance Layer represents a service as an end-user service or a commercial service offered by an aspect of the economy, an application provider, or a mobile network operator. In the Network Slice Instance Layer, the network slice instance is a set of resources designed to meet the specific requirements of a specific service. It includes one or more distinct sub-network instances that are separated from each other or shared between them. For the deployment of slices is used the Resource Layer, which consists of both physical and logical resources [23,24].

4 Proposed Fuzzy-Based System

We use FL to implement the proposed system. In Fig. 3, we present the overview of our proposed system. The SDN controller will provide commands to each evolve Base Station (eBS), allowing them to communicate and transfer data to User Equipment (UE). Additionally, each eBS contains a variety of slices for multiple purposes.

The proposed Fuzzy-based system for selecting a new RAT will be implemented in SDN controller, which will control eBS and other RAT's base stations and collect all the data regarding network traffic situation. The SDN controller will act as a transmission medium between the RAT's base station and the core network. For example, when the UE is connected to Wireless LAN (WLAN) but its QoS is not good, the SDN controller will collect other RAT networks data and decides whether the UE will still be connected with WLAN or connect to other RATs.

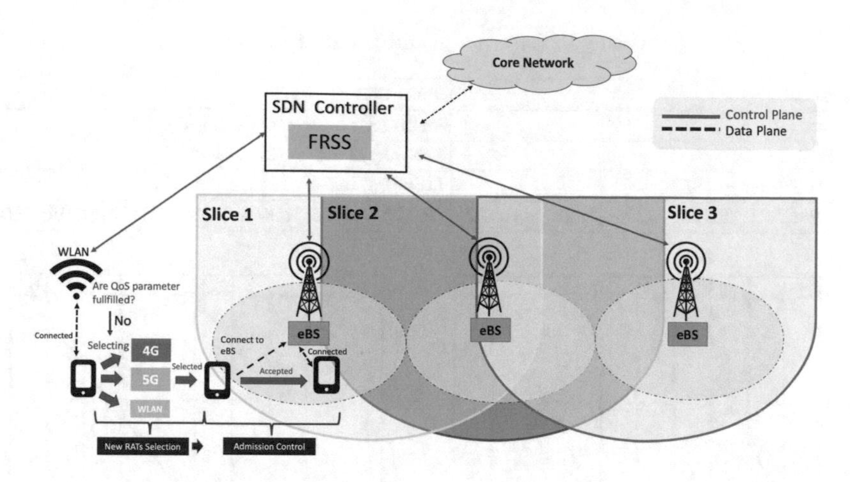

Fig. 3. Proposed system overview.

The proposed system is called Fuzzy-based RATs Selection System (FRSS) in 5G Wireless Networks. The structure of FRSS is shown in Fig. 4. For the implementation of our system, we consider three input parameters: Coverage (CV), User Priority (UP) and Spectral Efficiency (SE) and the output parameter is Radio Access Technology Decision Value (RDV).

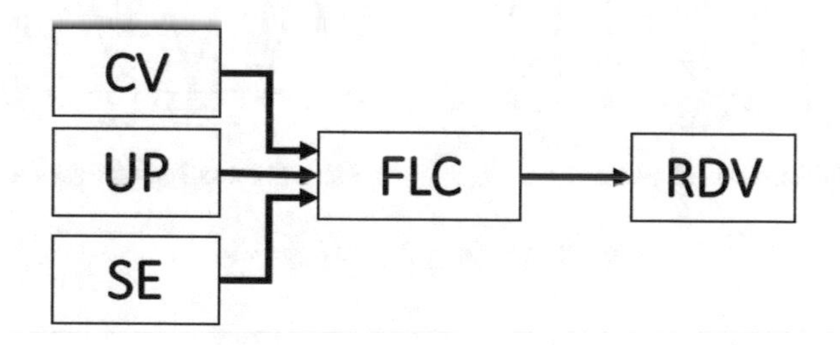

Fig. 4. Proposed system structure.

Coverage (CV): The CV is the cell radius of each RAT. When CV value is high, the RDV is high.

User Priority (UP): The user with high priority have a high possibility of connecting with better RAT.

Spectral Efficiency (SE): The SE shows how many bit rates can be delivered using the current transmission bandwidth (in bps/Hz).

Radio Access Technology Decision Value (RDV): The RDV parameter is the output value decided based on three input parameters.

Table 1. Parameter and their term sets.

Parameters	Term Sets
Coverage (CV)	Small (S), Intermediate (I), Big (B)
User Priority (UP)	Low (L),Medium (M) , High (H)
Spectral Efficiency (SE)	Low (Lo), Medium (Mu), High (Hi)
Radio Access Technology Decision Value (RDV)	RDV1, RDV2, RDV3, RDV4, RDV5, RDV6, RDV7

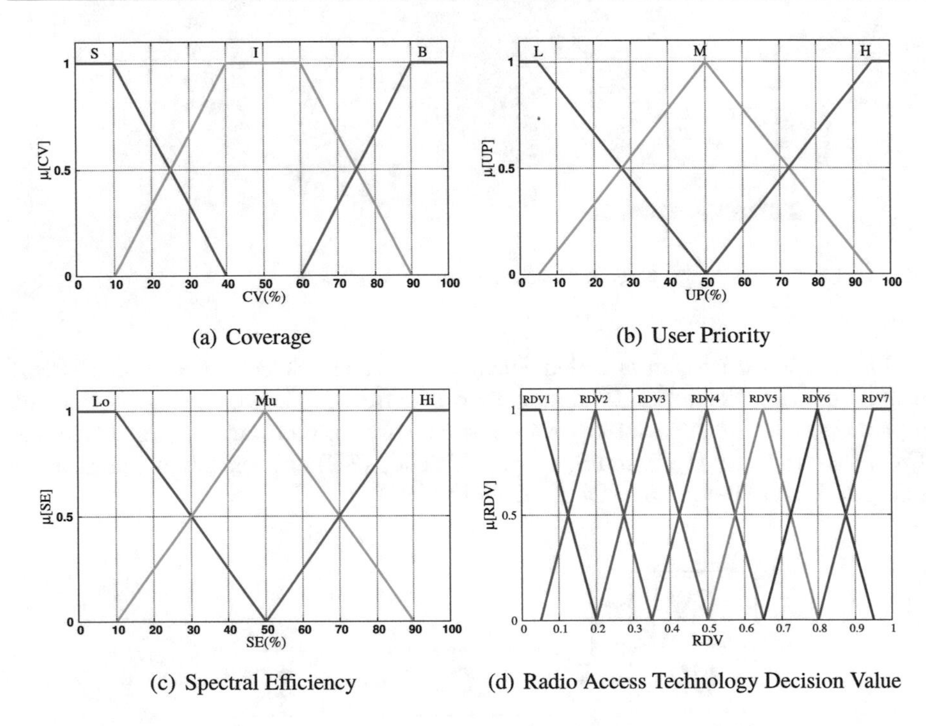

Fig. 5. Membership functions.

Table 1 shows parameters and their term sets. The membership functions are shown in Fig. 5. Due to their better suitability for real-time operations, we employ triangular and trapezoidal membership functions [25–28]. The Fuzzy Rule Base (FRB), which comprises 27 rules, is presented in Table 2. The control rules have the following format: IF "Condition" THEN "Control Action". For instance, Rule 1 can be interpreted as: "IF CV is Sm, UP is Lw and SE is Lo, THEN RDV is RDV1".

Table 2. FRB.

Rule	CV	UP	SE	RDV
1	S	L	Lo	RDV1
2	S	L	Mu	RDV2
3	S	L	Hi	RDV3
4	S	M	Lo	RDV2
5	S	M	Mu	RDV3
6	S	M	Hi	RDV4
7	S	H	Lo	RDV3
8	S	H	Mu	RDV4
9	S	H	Hi	RDV5
10	I	L	Lo	RDV2
11	I	L	Mu	RDV3
12	I	L	Hi	RDV4
13	I	M	Lo	RDV3
14	I	M	Mu	RDV4
15	I	M	Hi	RDV5
16	I	H	Lo	RDV4
17	I	H	Mu	RDV5
18	I	H	Ho	RDV6
19	B	L	Lo	RDV3
20	B	L	Mu	RDV4
21	B	L	Hi	RDV5
22	B	M	Lo	RDV4
23	B	M	Mu	RDV5
24	B	M	Hi	RDV6
25	B	H	Lo	RDV5
26	B	H	Mu	RDV6
27	B	H	Hi	RDV7

5 Simulation Results

In this section, we describe the simulation result of our proposed system. The simulation results are shown in Fig. 6, Fig. 7 and Fig. 8. They present the relation of RDV with SE for various UE values considering CV as a constant parameter.

In Fig. 6, we consider the CV value 10%. When SE is increased from 20% to 40% and 40% to 80% for UP 90%, we see that RDV is increased by 5% and 15%, respectively. When the RAT has a high transmission bandwidth, the RAT decision value will be higher. When we increased the UP value from 10% to 50% and 50% to 90%, both RDV values are increased by 12% when SE is 80%. This indicates that when the user's priority is high, the user has a high possibility of selecting and connecting with a RAT that has better QoS.

We compare Fig. 6 with Fig. 7 to see the effect of CV to RDV. We change the CV value from 10% to 50%. The RDV is increased by 11% when the UP value is 50% and the SE is 50%. When the RAT has better signal coverage, the RAT decision value also is higher.

We increase the value of CV to 90% in Fig. 8. Comparing the results with Fig. 6 and Fig. 7, we can see that the RDV values have been increased significantly. All RDV values for UP 50% and 90% are greater than 0.5. Thus, there is a high probability that the user will select a better RAT with good coverage and bandwidth for satisfying user requirements.

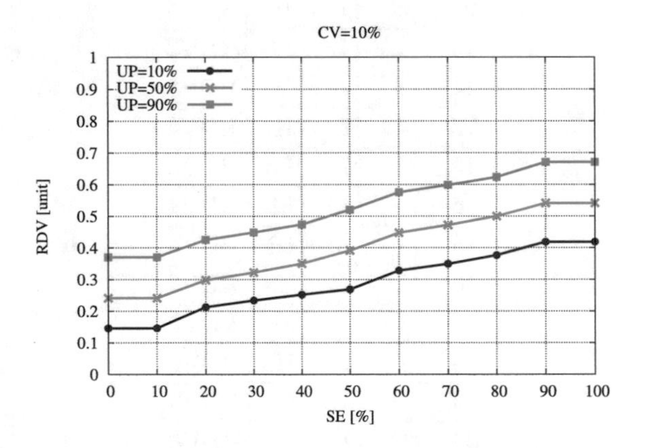

Fig. 6. Simulation results for CV=10%.

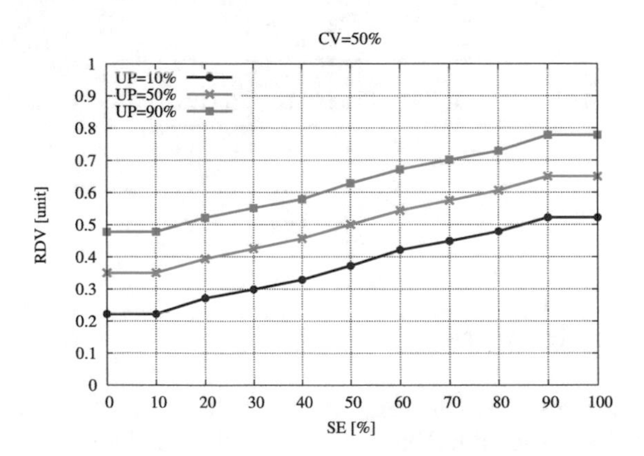

Fig. 7. Simulation results for CV=50%.

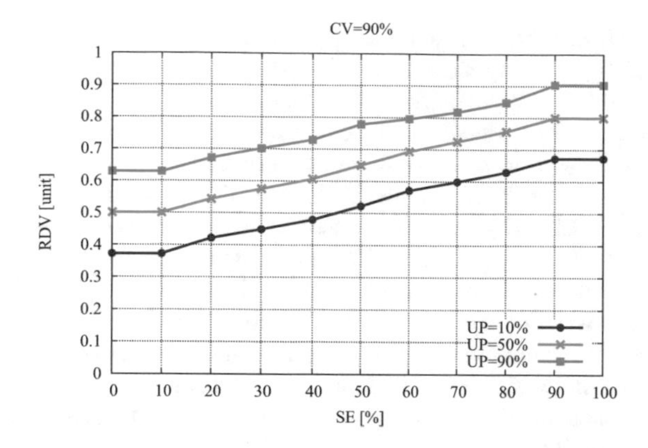

Fig. 8. Simulation results for CV=90%.

6 Conclusions and Future Work

In this paper, we proposed and implemented a Fuzzy-based system for selection of RAT in 5G Wireless Networks. We considered three parameters: CV, UP and SE to decide the RAT decision value. From the simulation results, we found that when CV, UP and SE are increased, the RDV parameter is increased.

In the future work, we will consider different parameters and perform extensive simulations to evaluate the proposed system.

References

1. Navarro-Ortiz, J., Romero-Diaz, P., Sendra, S., Ameigeiras, P., Ramos-Munoz, J.J., Lopez-Soler, J.M.: A survey on 5G usage scenarios and traffic models. IEEE Commun. Surv. Tutorials **22**(2), 905–929 (2020). https://doi.org/10.1109/COMST.2020.2971781
2. Pham, Q.V., et al.: A survey of multi-access edge computing in 5G and beyond: fundamentals, technology integration, and state-of-the-art. IEEE Access 8, 116, 974–117, 017 (2020). https://doi.org/10.1109/ACCESS.2020.3001277
3. Orsino, A., Araniti, G., Molinaro, A., Iera, A.: Effective rat selection approach for 5G dense wireless networks. In: 2015 IEEE 81st Vehicular Technology Conference (VTC Spring), pp. 1–5 (2015). https://doi.org/10.1109/VTCSpring.2015.7145798
4. Akpakwu, G.A., Silva, B.J., Hancke, G.P., Abu-Mahfouz, A.M.: A survey on 5G networks for the internet of things: communication technologies and challenges. IEEE Access **6**, 3619–3647 (2018)
5. Palmieri, F.: A reliability and latency-aware routing framework for 5G transport infrastructures. Computer Networks 179 (9), Article 107365 (2020). https://doi.org/10.1016/j.comnet.2020.107365
6. Kamil, I.A., Ogundoyin, S.O.: Lightweight privacy-preserving power injection and communication over vehicular networks and 5G smart grid slice with provable security. Internet Things **8**(100116), 100–116 (2019). https://doi.org/10.1016/j.iot.2019.100116

7. Hossain, E., Hasan, M.: 5G cellular: key enabling technologies and research challenges. IEEE Instrumentation Measurement Magazine 18, no. 3(3), 11–21 (2015). https://doi.org/10.1109/MIM.2015.7108393
8. Vagionas, C., et al.: End-to-end real-time service provisioning over a SDN-controllable analog mmwave fiber-wireless 5g x-haul network. Journal of Lightwave Technology, pp. 1–10 (2023). https://doi.org/10.1109/JLT.2023.3234365
9. Yao, D., Su, X., Liu, B., Zeng, J.: A mobile handover mechanism based on fuzzy logic and MPTCP protocol under SDN architecture*. In: 18th International Symposium on Communications and Information Technologies (ISCIT-2018), pp. 141–146 (2018). https://doi.org/10.1109/ISCIT.2018.8587956
10. Lee, J., Yoo, Y.: Handover cell selection using user mobility information in a 5G SDN-based network. In: 2017 Ninth International Conference on Ubiquitous and Future Networks (ICUFN-2017), pp. 697–702 (2017). https://doi.org/10.1109/ICUFN.2017.7993880
11. Moravejosharieh, A., Ahmadi, K., Ahmad, S.: A fuzzy logic approach to increase quality of service in software defined networking. In: 2018 International Conference on Advances in Computing, Communication Control and Networking (ICACCCN-2018), pp. 68–73 (2018). https://doi.org/10.1109/ICACCCN.2018.8748678
12. Ampririt, P., Qafzezi, E., Bylykbashi, K., Ikeda, M., Matsuo, K., Barolli, L.: IFACS-Q3S-A new admission control system for 5G wireless networks based on fuzzy logic and its performance evaluation. Int. J. Distrib. Syst. Technol. (IJDST) **13**(1), 1–25 (2022)
13. Ampririt, P., Qafzezi, E., Bylykbashi, K., Ikeda, M., Matsuo, K., Barolli, L.: A fuzzy-based system for handover in 5G wireless networks considering network slicing constraints. In: Computational Intelligence in Security for Information Systems Conference, pp. 180–189. Springer, Cham (2022). https://doi.org/10.1007/978-3-031-08812-4_18
14. Ampririt, P., Qafzezi, E., Bylykbashi, K., Ikeda, M., Matsuo, K., Barolli, L.: A fuzzy-based system for handover in 5G wireless networks considering different network slicing constraints: effects of slice reliability parameter on handover decision. In: International Conference on Broadband and Wireless Computing, Communication and Applications, pp. 27–37. Springer, Cham (2022). https://doi.org/10.1007/978-3-031-20029-8_3
15. Ampririt, P., Ohara, S., Qafzezi, E., Ikeda, M., Matsuo, K., Barolli, L.: An integrated fuzzy-based admission control system (IFACS) for 5G wireless networks: its implementation and performance evaluation. Internet of Things 13, 100, 351 (2021). https://doi.org/10.1016/j.iot.2020.100351
16. Ampririt, P., Qafzezi, E., Bylykbashi, K., Ikeda, M., Matsuo, K., Barolli, L.: Application of fuzzy logic for slice QoS in 5G networks: a comparison study of two fuzzy-based schemes for admission control. Int. J. Mobile Comput. Multimedia Commun. (IJMCMC) **12**(2), 18–35 (2021)
17. Li, L.E., Mao, Z.M., Rexford, J.: Toward software-defined cellular networks. In: 2012 European Workshop on Software Defined Networking, pp. 7–12 (2012). https://doi.org/10.1109/EWSDN.2012.28
18. Mousa, M., Bahaa-Eldin, A.M., Sobh, M.: Software defined networking concepts and challenges. In: 2016 11th International Conference on Computer Engineering & Systems (ICCES-2016), pp. 79–90. IEEE (2016)
19. An, N., Kim, Y., Park, J., Kwon, D.H., Lim, H.: Slice management for quality of service differentiation in wireless network slicing. Sensors **19**, 2745 (2019). https://doi.org/10.3390/s19122745

20. Jiang, M., Condoluci, M., Mahmoodi, T.: Network slicing management & prioritization in 5G mobile systems. In: European Wireless 2016; 22th European Wireless Conference, pp. 1–6. VDE (2016)
21. Chen, J., et al.: Realizing dynamic network slice resource management based on SDN networks. In: 2019 International Conference on Intelligent Computing and its Emerging Applications (ICEA), pp. 120–125 (2019)
22. Li, X., et al.: Network slicing for 5G: challenges and opportunities. IEEE Internet Comput. **21**(5), 20–27 (2017)
23. Afolabi, I., Taleb, T., Samdanis, K., Ksentini, A., Flinck, H.: Network slicing and softwarization: a survey on principles, enabling technologies, and solutions. IEEE Commun. Surv. Tutorials **20**(3), 2429–2453 (2018). https://doi.org/10.1109/COMST.2018.2815638
24. Alliance, N.: Description of network slicing concept. NGMN 5G P 1(1), 7 Pages (2016). https://ngmn.org/wp-content/uploads/160113_NGMN_Network_Slicing_v1_0.pdf
25. Norp, T.: 5G requirements and key performance indicators. J. ICT Stand. **6**(1), 15–30 (2018)
26. Parvez, I., Rahmati, A., Guvenc, I., Sarwat, A.I., Dai, H.: A survey on low latency towards 5G: ran, core network and caching solutions. IEEE Commun. Surv. Tutorials **20**(4), 3098–3130 (2018)
27. Kim, Y., Park, J., Kwon, D., Lim, H.: Buffer management of virtualized network slices for quality-of-service satisfaction. In: 2018 IEEE Conference on Network Function Virtualization and Software Defined Networks (NFV-SDN-2018), pp. 1–4 (2018)
28. Barolli, L., Koyama, A., Yamada, T., Yokoyama, S.: An integrated CAC and routing strategy for high-speed large-scale networks using cooperative agents. IPSJ J. **42**(2), 222–233 (2001)

A Comparison Study of FC-RDVM and LDVM Router Placement Methods for WMNs Considering Uniform Distribution of Mesh Clients and Different Instances

Shinji Sakamoto[1(✉)], Admir Barolli[2], Yi Liu[3], Elis Kulla[4], Leonard Barolli[5], and Makoto Takizawa[6]

[1] Department of Information and Computer Science, Kanazawa Institute of Technology, 7 -1 Ohgigaoka, Nonoichi, Ishikawa 921 -8501, Japan
shinji.sakamoto@ieee.org

[2] Department of Information Technology, Aleksander Moisiu University of Durres, L.1, Rruga e Currilave, Durres, Albania
admirbarolli@uamd.edu.al

[3] Department of Computer Science, National Institute of Technology, Oita College, 1666, Maki, Oita 870-0152, Japan
y-liu@oita-ct.ac.jp

[4] Department of System Management, Fukuoka Institute of Technology, 3-30-1 Wajiro-Higashi, Higashi-Ku, Fukuoka 811-0295, Japan
kulla@fit.ac.jp

[5] Department of Information and Communication Engineering, Fukuoka Institute of Technology, 3-30-1 Wajiro-Higashi, Higashi-Ku, Fukuoka 811-0295, Japan
barolli@fit.ac.jp

[6] Department of Advanced Sciences, Faculty of Science and Engineering, Hosei University, Kajino-Machi, Koganei-Shi, Tokyo 184-8584, Japan
makoto.takizawa@computer.org

Abstract. In this work, we present a hybrid intelligent simulation system based on Particle Swarm Optimization (PSO) and Hill Climbing (HC) called WMN-PSOHC for node placement problem in Wireless Mesh Networks (WMNs). As mesh router replacement methods, we consider Fast Convergence Rational Decrement of Vmax Method (FC-RDVM) and Linearly Decreasing Vmax Method (LDVM). By using WMN-PSOHC system, we carry out a comparison study between FC-RDVM and LDVM considering Uniform distribution of mesh clients and different instances. The simulation results show that FC-RDVM has better behavior compared with LDVM.

1 Introduction

The Wireless Mesh Networks (WMNs) have many advantages such as low-up front cost, easy deployment and high robustness compared with conventional

L. Barolli (Ed.): EIDWT 2023, LNDECT 161, pp. 308–316, 2023.
https://doi.org/10.1007/978-3-031-26281-4_32

Wireless Local Area Networks (WLANs) [1,7]. Also, they can provide better services and can be used in different scenarios and applications [2,6,8]. However, they have many problems and issues such as hidden terminal problem, security, increased workload for each mesh node and lattency issues, which should be considered when designing and engineering WMNs. Therefore, the optimization of mesh routers placement in order to cover many mesh clients is very important problem in WMNs. However, the node placement problem is a NP-hard problem.

In order to deal with these issues, in our previous work, we implemented intelligent systems using single algorithms such as Particle Swarm Optimization (PSO) and Hill Climbing (HC) for solving node placement problem in WMNs [10,11]. We call these intelligent systems as WMN-PSO system and WMN-HC system, respectively. We also implemented a hybrid intelligent simulation system based on PSO and HC [12]. We called this system WMN-PSOHC system. In WMN-PSOHC system, we implemented different mesh router replacement methods [13].

In this paper, we implemented in WMN-PSOHC system the Fast Convergence Rational Decrement of Vmax Method (FC-RDVM) and Linearly Decreasing Vmax Method (LDVM). By using WMN-PSOHC system, we carry out a comparison study between FC-RDVM and LDVM considering Uniform distribution of mesh clients and different instances. The simulation results show that FC-RDVM has better behavior compared with LDVM.

The rest of the paper is organized as follows. In Sect. 2, we present intelligent algorithms. We present the implemented WMN-PSOHC hybrid intelligent simulation system in Sect. 3. The simulation results are shown in Sect. 4. Finally, we give conclusions and future work in Sect. 5.

2 Intelligent Algorithms

2.1 Particle Swarm Optimization

In the Particle Swarm Optimization (PSO) algorithm, a group of simple entities, known as particles, are placed in a search space to evaluate an objective function by using their current location. The objective function is typically minimized, and the search space is explored through the movement of the particles rather than through evolution [9]. Each particle determines its movement through the search space by combining its current and best-fitness locations with those of other particles in the swarm. After all particle have moved, the next iteration begins. Eventually, the movement of the swarm as a whole is likely to converge on an optimum fitness function, much like a flock of birds collectively searching for food.

Each particle in the swarm is composed of three $\mathcal{D}$-dimensional vectors, where $\mathcal{D}$ is the dimensionality of the search space. These vectors are the current position, previous best position, and particle velocity. The swarm is more than just a collection of individual particles, as the progress made in solving a problem emerges from the interactions and behaviors of the particles as a whole. The topology of the population, or how the particles are connected, often resembles a social network, with bidirectional edges connecting pairs of particles such that if particle j is in the particle i's neighborhood, particle i is also in the particle j's neighborhood. Each particle is affected by the best point found by members of its neighborhood, represented by the vector $\boldsymbol{p}_g$. During the PSO process, the velocity of each particle is iteratively adjusted such that the particle oscillates stochastically around the $\boldsymbol{p}_i$ and $\boldsymbol{p}_g$ locations.

2.2 Hill Climbing

Hill climbing (HC) is a heuristic algorithm that relies on a simple concept. The current solution is accepted as the new solution if $\delta \leq 0$, where $\delta = f(s') - f(s)$ and f is the fitness function, which evaluates the current solution s and the next solution s'.

The key factor in HC is the effective definition of a neighbor solution, as this directly impacts the performance of the algorithm. In our WMN-PSOHC system, we use the next step of particle-pattern positions as the neighbor solutions for the HC part.

3 Implemented WMN-PSOHC Hybrid Intelligent Simulation System

The flowchart of WMN-PSOHC system is shown in Fig 1. We combine PSO and HC intelligent algorithms to make a better optimization of mesh routers in order to cover many mesh clients. Thus, the network connectivity and client coverage will be improved.

We present in following some processes, functions and methods used in WMN-PSOHC system such as initialization, particle-pattern, fitness function and router replacement methods.

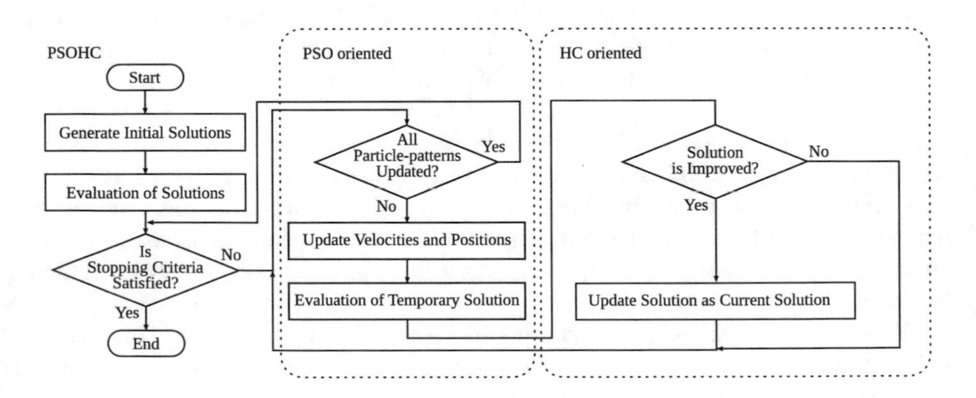

Fig. 1. WMN-PSOHC-flowchart

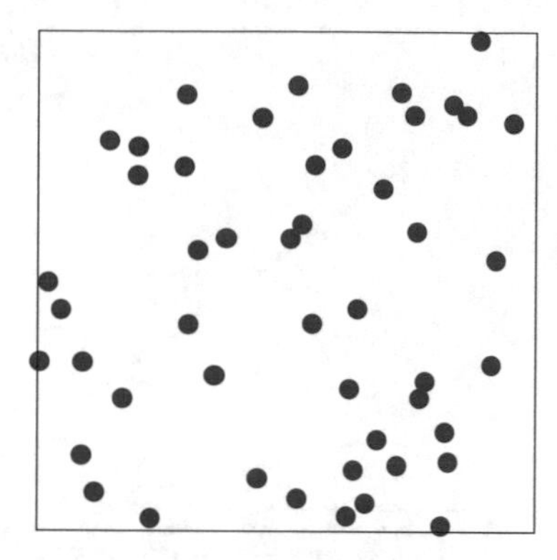

Fig. 2. Uniform distribution of mesh clients

Initialization. The WMN-PSOHC system generates an initial solution randomly by using *ad hoc* methods [17]. The velocity of particles is decided randomly by considering the area size. For example, when the area size is $W \times H$, the velocity is decided randomly from $-\sqrt{W^2 + H^2}$ to $\sqrt{W^2 + H^2}$.

The WMN-PSOHC system can generate many mesh client distributions such as Normal, Uniform, Weibull, Chi-Square, Stadium, Two Islands, Subway and Boulevard distributions. In this paper, as shown in Fig. 2, we consider Uniform distribution of mesh clients.

Particle-Pattern. We consider a particle as a mesh router and the fitness value of a particle-pattern is computed by combination of mesh routers and mesh clients positions. Each particle-pattern is considered as a solution. Therefore, the number of particle-patterns is the number of solutions. The relationship between global solution, particle-patterns and mesh routers is shown in Fig. 3.

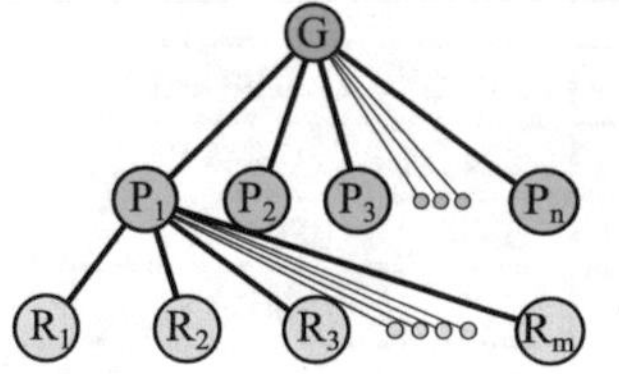

Fig. 3. Relationship between global solution, particle-patterns and mesh routers

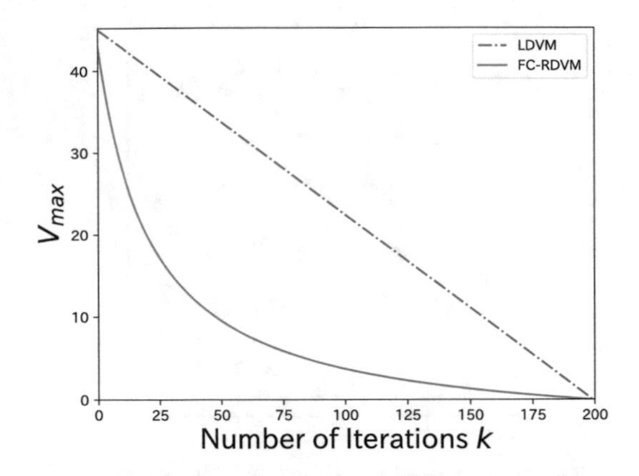

Fig. 4. Relation between V_{max} and nubmer of iterations for LDVM and FC-RDVM.

Fitness Function. In optimization problems the determination of an appropriate objective function and its encoding is a very important issue. In our case, the fitness function follows a hierarchical approach, with the main objective to maximize the SGC and then improving the NCMC parameter. We use weight coefficients α and β in the fitness function, which is defined as follows.

$$\text{Fitness} = \alpha \times \text{SGC}(\boldsymbol{x}_{ij}, \boldsymbol{y}_{ij}) + \beta \times \text{NCMC}(\boldsymbol{x}_{ij}, \boldsymbol{y}_{ij})$$

Router Replacement Methods. The mesh routers movement is based on their velocities. Their placement is done by different router replacement methods [5,14–16]. In this paper, we compare two replacement methods LDVM and FC-RDVM.

Table 1. Parameters for instances.

Parameters	Instance 1	Instance 2
Area size	32×32	64×64
Number of mesh routers	16	32
Number of mesh clients	48	96

Table 2. Parameter settings.

Parameters	Values
Clients distribution	Uniform distribution
Instances	Instance 1, Instance 2
Total iterations	800
Iteration per phase	4
Number of particle-patterns	9
Radius of a mesh router	From 2.0 to 3.0
Fitness function weight-coefficients (α, β)	0.7, 0.3
Curvature parameter (γ)	10.0
Replacement methods	LDVM, FC-RDVM

In LDVM, PSO parameters are set to unstable region ($\omega = 0.9$, $C_1 = C_2 = 2.0$). The V_{max} is the maximum velocity of particles and by increasing of iterations the V_{max} is decreased linearly [3,4,14]. The V_{max} is defined as shown in Eq. (1).

$$V_{max}(k) = \sqrt{W^2 + H^2} \times \frac{T - k}{T} \tag{1}$$

The W and H are the width and height of the considered area, while T and k are the total number of iterations and a current number of iteration, respectively. The k is a variable varying from 1 to T, which is increased by increasing the number of iterations.

In FC-RDVM, the V_{max} decreases with increasing the number of iterations as shown in Eq. (2).

$$V_{max}(k) = \sqrt{W^2 + H^2} \times \frac{T - k}{T + \gamma k} \tag{2}$$

The γ parameter is curvature parameter. When γ is larger, the curvature is larger as shown in Fig. 4. The other parameters are the same with LDVM.

4 Simulation Results

In this section, we show simulation results. For simulations, we consider Uniform distribution of mesh clients and the number of particle-patterns is 9. The total number of iterations is 800 and the iterations per phase is 4.

In order to evaluate the performance of FC-RDVM and LDVM, we consider two instances: Instance 1 and Instance 2 as shown in Table 1. The parameter settings for WMN-PSOHC system are shown in Table 2.

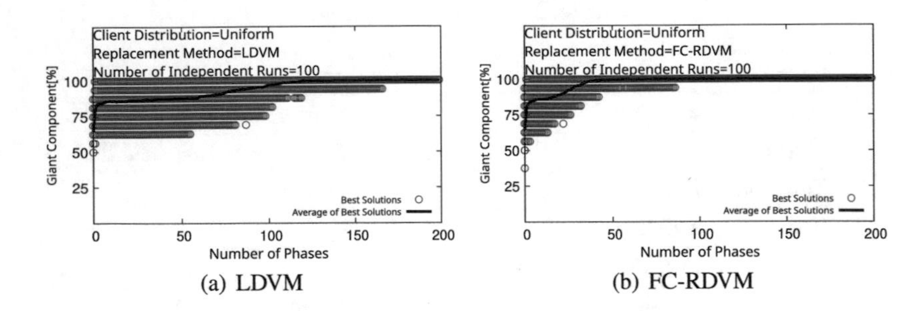

(a) LDVM (b) FC-RDVM

Fig. 5. Simulation results of WMN-PSOHC system for SGC considering Instance 1.

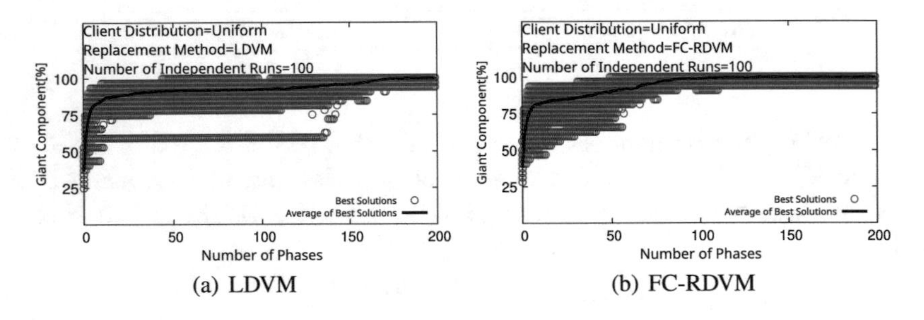

(a) LDVM (b) FC-RDVM

Fig. 6. Simulation results of WMN-PSOHC system for SGC considering Instance 2.

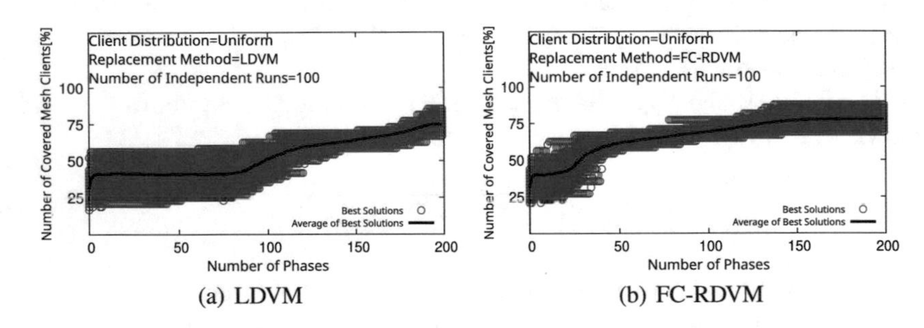

(a) LDVM (b) FC-RDVM

Fig. 7. Simulation results of WMN-PSOHC system for NCMC considering Instance 1.

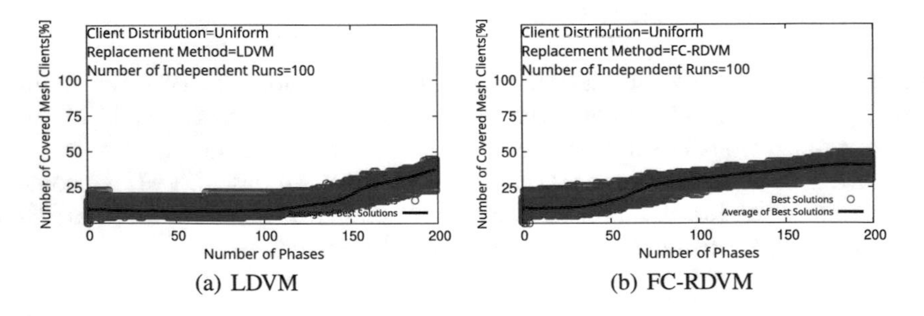

(a) LDVM (b) FC-RDVM

Fig. 8. Simulation results of WMN-PSOHC system for NCMC considering Instance 2.

We show the simulation results from Fig. 5 to Fig. 8. Considering SGC parameter, both replacement methods can reach 100% for both instances as shown in Fig. 5 and Fig. 6. This shows that all mesh routers are connected and the WMN has a good connectivity. However, the LDVM convergence is slower than FC-RDVM and has more oscillations. For NCMC, the FC-RDVM covers more mesh clients than LDVM for both instance as shown in Fig. 7 and Fig. 8. This shows that FC-RDVM has better coverage than LDVM.

5 Conclusions

In this work, we presented WMN-PSOHC hybrid intelligent simulation system. We implemented in WMN-PSOHC system FC-RDVM and LDVM router replacement methods. Then, we compared the performance of FC-RDVM with LDVM considering Uniform distribution of mesh clients and different instances. Simulation results show that FC-RDVM has better behavior compared with LDVM for both instances.

In our future work, we would like to evaluate the performance of WMN-PSOHC system considering different parameters, mesh client distributions, replacement methods and scenarios.

References

1. Akyildiz, I.F., Wang, X., Wang, W.: Wireless mesh networks: a survey. Comput. Netw. **47**(4), 445–487 (2005)
2. Amaldi, E., Capone, A., Cesana, M., Filippini, I., Malucelli, F.: Optimization models and methods for planning wireless mesh networks. Comput. Netw. **52**(11), 2159–2171 (2008)
3. Barolli, A., Sakamoto, S., Ohara, S., Barolli, L., Takizawa, M.: Performance analysis of wmns by wmn-psohc-dga simulation system considering random inertia weight and linearly decreasing vmax router replacement methods. In: Barolli, L., Hussain, F.K., Ikeda, M. (eds.) CISIS 2019. AISC, vol. 993, pp. 13–21. Springer, Cham (2020). https://doi.org/10.1007/978-3-030-22354-0_2
4. Barolli, A., Bylykbashi, K., Qafzezi, E., Sakamoto, S., Barolli, L., Takizawa, M.: A Comparison Study of UNDX and UNDX-m Methods for LDVM and RDVM Router Replacement Methods by WMN-PSODGA Hybrid Intelligent System Considering Stadium Distribution. In: International Conference on Broadband and Wireless Computing, Communication and Applications (BWCCA-2022), Springer, pp 1–14 (2022). https://doi.org/10.1007/978-3-031-20029-8_1
5. Clerc, M., Kennedy, J.: The particle swarm-explosion, stability, and convergence in a multidimensional complex space. IEEE Trans. Evol. Comput. **6**(1), 58–73 (2002)
6. Franklin, A.A., Murthy, C.S.R.: Node Placement Algorithm for Deployment of Two-tier Wireless Mesh Networks. In: Proceedings of the Global Telecommunications Conference, pp. 4823–4827 (2007)
7. Islam, M.M., Funabiki, N., Sudibyo, R.W., Munene, K.I., Kao, W.C.: A dynamic access-point transmission power minimization method using pi feedback control in elastic wlan system for iot applications. Internet Things **8**(100), 089 (2019)

8. Muthaiah, S.N., Rosenberg, C.P.: Single Gateway Placement in Wireless Mesh Networks. In: Proceedings of the 8th International IEEE Symposium on Computer Networks, pp. 4754–4759 (2008)
9. Poli, R., Kennedy, J., Blackwell, T.: Particle Swarm Optimization. Swarm Intell. **1**(1), 33–57 (2007)
10. Sakamoto, S., Lala, A., Oda, T., Kolici, V., Barolli, L., Xhafa, F.: Analysis of WMN-HC Simulation System Data Using Friedman Test. In: The Ninth International Conference on Complex, Intelligent, and Software Intensive Systems (CISIS-2015), IEEE, pp. 254–259 (2015)
11. Sakamoto, S., Oda, T., Ikeda, M., Barolli, L., Xhafa, F.: Implementation and evaluation of a simulation system based on particle swarm optimisation for node placement problem in wireless mesh networks. Int. J. Commun. Netw. Distrib. Syst. **17**(1), 1–13 (2016)
12. Sakamoto, S., Ozera, K., Ikeda, M., Barolli, L.: Implementation of intelligent hybrid systems for node placement problem in wmns considering particle swarm optimization, hill climbing and simulated annealing. Mobile Netw. Appl. **23**(1), 27–33 (2018)
13. Sakamoto, S., Barolli, L., Okamoto, S.: A comparison study of linearly decreasing inertia weight method and rational decrement of vmax method for wmns using wmn-psohc intelligent system considering normal distribution of mesh clients. In: Barolli, L., Natwichai, J., Enokido, T. (eds.) EIDWT 2021. LNDECT, vol. 65, pp. 104–113. Springer, Cham (2021). https://doi.org/10.1007/978-3-030-70639-5_10
14. Schutte, J.F., Groenwold, A.A.: A study of global optimization using particle swarms. J. Global Optim. **31**(1), 93–108 (2005)
15. Shi, Y.: Particle Swarm Optimization. IEEE Connect. **2**(1), 8–13 (2004)
16. Shi, Y., Eberhart, R.C.: Parameter selection in particle swarm optimization. In: Porto, V.W., Saravanan, N., Waagen, D., Eiben, A.E. (eds.) EP 1998. LNCS, vol. 1447, pp. 591–600. Springer, Heidelberg (1998). https://doi.org/10.1007/BFb0040810
17. Xhafa, F., Sanchez, C., Barolli, L.: Ad hoc and Neighborhood Search Methods for Placement of Mesh Routers in Wireless Mesh Networks. In: Proceedings of 29th IEEE International Conference on Distributed Computing Systems Workshops (ICDCS-2009), pp. 400–405 (2009)

Performance Evaluation of FBRD Protocol Considering Transporter Autonomous Underwater Vehicles for Underwater Optical Wireless Communication in Delay Tolerant Networking

Keita Matsuo[1](✉), Elis Kulla[2], and Leonard Barolli[1]

[1] Department of Information and Communication Engineering, Fukuoka Institute of Technology (FIT), 3-30-1 Wajiro-Higashi, Higashi-Ku, 811-0295 Fukuoka, Japan
{kt-matsuo,barolli}@fit.ac.jp

[2] Department of System Management, Fukuoka Institute of Technology (FIT), 3-30-1 Wajiro-Higashi, Higashi-Ku, 811-0295 Fukuoka, Japan
kulla@fit.ac.jp

Abstract. Recently, there are many ways to realize underwater communication. The most important problem is to achieve high speed connection in the underwater environment. For this reason, we focus on Underwater Optical Wireless Communication (UOWC) to solve this issue. However, it is difficult to have high speed data transmission, because of narrow beam of LED optical signals, which may cause intermittent communication. So, we consider to use DTN technologies to make better communication environment. In this paper, we implemented FBRD protocol in The ONE Simulator and evaluate the effects of the transporter AUV using FBRD protocol for UOWC. The performance evaluation show that the VT or HT can increase the delivery probability. In addition, when we use both HT and VT the delivery probability is the highest. But, there is not a big difference with the case of using only VT.

1 Introduction

Recently, there are many technologies to realize communication in underwater environment, such as Wired Communication (WC), Underwater Acoustic Communication (UAC), Underwater Radio-wave Wireless Communication (URWC) and Underwater Optical Wireless Communication (UOWC). Presently, the underwater communication is using communication cables because of the limited development of underwater wireless communications and the high cost of other communication devices [3].

The UAC uses sound signals in water which is able to send the sound signals to longer distances. However, UAC communication suffers from transmission

L. Barolli (Ed.): EIDWT 2023, LNDECT 161, pp. 317–323, 2023.
https://doi.org/10.1007/978-3-031-26281-4_33

losses and time-varying signal distortion due to its dependency on environmental properties [2].

In URWC, radio waves are used for communication. Also, the radio waves for terrestrial wireless communication can be utilized for underwater communication. They can achieve high data rate for short communication range, but suffers from Doppler effect [3]. Furthermore, radio waves propagation through water differs greatly from that through air because of the electrical conductivity and high permittivity of water. In comparison to air, water has a significant plane wave attenuation that rapidly increases with frequency [1].

UOWC technology is gradually maturing, providing a high data rate and transmission bandwidth. The 1.2-m LED system already achieved a data rate of 14.6 Gb/s. However, the UOWC channel has limitations in transmission range and coverage area, which requires the establishment and maintenance of a Line-of-Sight (LOS) link for communication. In the vast ocean environment, it is hard to complete initial location identification for the link establishment [5].

The underwater environment is unstable for data communication. Therefore, we considered to use new technologies such as Delay Tolerant Networking (DTN) for underwater environment.

By combining UOWC and DTN, it can be achieved a high speed communication network and more stable underwater communication. However, in UOWC LED optical signals have narrow beam, which may cause intermittent communication. Also, it is difficult to match the optical axis of the transmitter and receiver. Therefore, we consider to use DTN technology which can provide good underwater environment. But, we have to increase the delivery probability in DTN environment. Therefore, we used Focused Beam Routing (FBR) in DTN. The FBR has shown good results in previous research [4]. In addition, we consider transporter nodes to get high delivery probability of messages.

The transporter nodes move in specified directions such as Horizontal (H) or Vertical (V) axis, and use message forwarding in DTN. We use Focused Beam Routing considering node Direction (FBRD) protocol for UOWC. We evaluate FBRD protocol by simulations. The simulation results show that vertical Transporter (VT) and Horizontal Transporter (HT) can increase the delivery probability. Also, when we use both VT and HT the delivery probability is the highest. But, there is not a big difference with the case of using only VT.

The rest of the paper is structured as follows. In Sect. 2, we explain the proposed approach. In Sect. 3, we show implementation of FBRD in The ONE Simulator. In Sect. 4, we present the simulation results. In Sect. 5 we give conclusions and future work.

2 Proposed Approach

In this section, we explain FBR and FBRD protocols. In Fig. 1 is shown image of UOWC using DTN. Autonomous Underwater Vehicles (AUVs) are able to communicate to each other using DTN. The surface station can receive data from AUVs and sends the data to a surface sink.

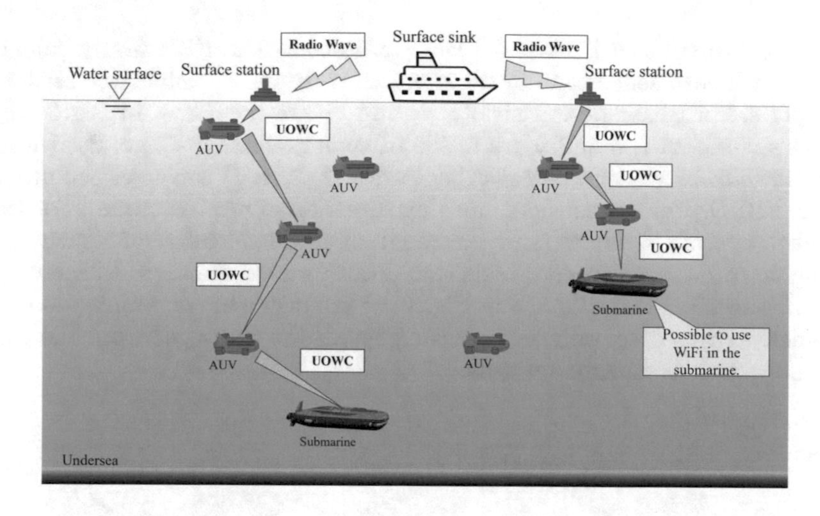

Fig. 1. Image of UOWC of using DTN.

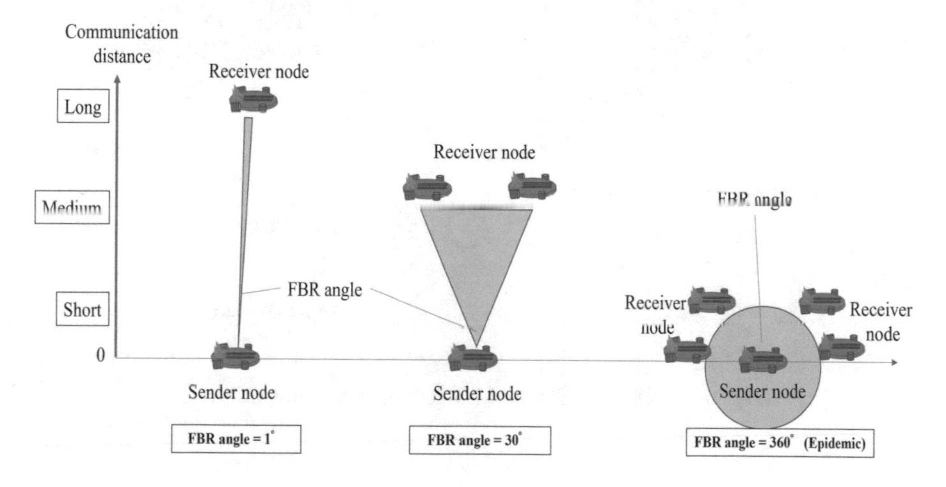

Fig. 2. Image of FBR protocol by using UOWC.

By using DTN can be realized sending of data from bottom of sea to the sea surface. Also, big data size such as videos or high quality photos can be transmitted by high speed connection of UOWC.

In Fig. 2 is shown the image of FBR protocol by using UOWC. If we use narrow FBR angle such as 1°C, the communication distance will be increased, but when the FBR angle increases, the communication distance becomes shorter.

In Fig. 3 are shown FBR and FBRD protocols. The FBR firstly calculates the angle between sender and destination, then θ_1 and θ_2 values are set for left and right side FBR angles. Basically, θ_1 and θ_2 are the same. In Fig. 3 (a) the receiver's numbers 1, 5 and 6 are inside of communication range. So, they will get message from sender whereas other nodes (2, 3, 4, 7) can't get the message.

The FBRD protocol considers moving direction of receiver nodes. For example, when one of the nodes is in the range of communication of sender and is moving downwards, the sender node would not forward messages to the receiver node. The receiver nodes 1, 5, 6 in Fig. 3 (b) are inside of communication range of sender. In this case, only node 5 can receive the messages from the sender node, because it is moving upwards.

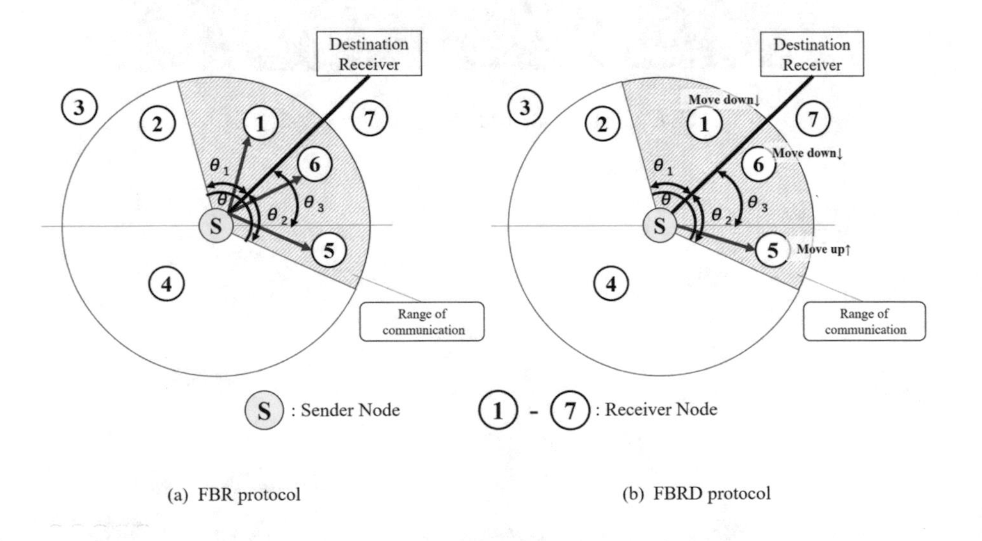

Fig. 3. Image of FBR and FBRD protocol for UOWC.

3 Proposed Transporter AUVs and Simulations Scenario for UOWC Considering FBRD Protocol

The proposed simulation scenario is shown in Fig. 4. We installed three sensors on the undersea ground which make random signals to the surface station. For the simulations, we consider Horizontally Transporter (HT) and Vertically Transporter (VT). Both HT and VT can move linearly along the same course. The aim is to increased node encounters.

For simulations we consider four scenario as following:

- Horizontally Transporter and Vertically Transporter not used (None),
- Only Horizontally Transporter used (HT),

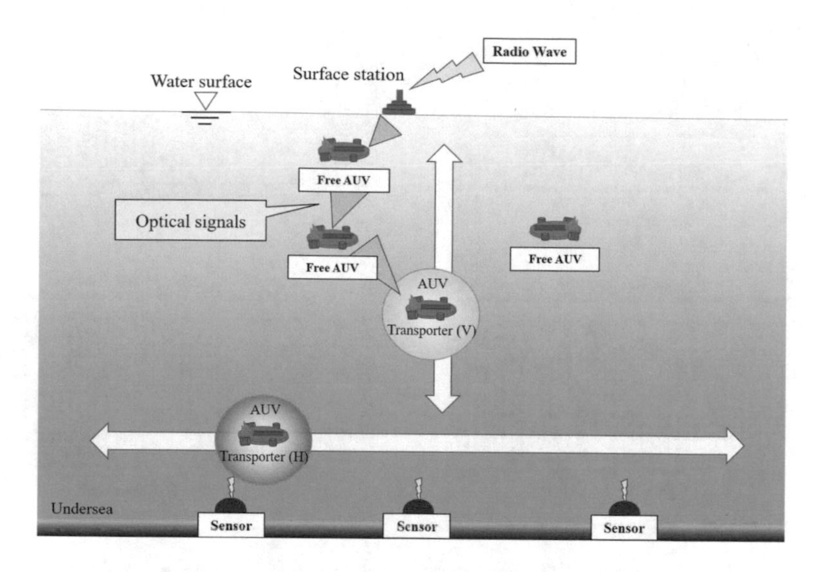

Fig. 4. Proposed scenario for using transporter AUVs.

- Only Vertically Transporter used (VT),
- Horizontally Transporter used and Vertically Transporter used (HT and VT).

The simulation parameters are: Transmission Speed is 2 Mbps, Transmission Range is 50 m, Data Size is between 80 and 120 kB, Event Interval is between 2 and 5 s, Number of Surface Stations is 1 Static (Middle Top), Number of Free Nodes are 20, Nodes Speed is between 0.5 and 2.5 m/s, Movement Model is Random Waypoint, Simulation Time is 1080 s $\times$ 10 Times, Simulation Area is 400 $\times$ 400 m, Buffer Size is 30 MBytes.

4 Simulation Results

We use The ONE simulator to implement FBRD protocol. To simplify the implementation, the following assumptions are made. The simulation environment is considered 2 dimensional. Every node in the simulator knows current position and destination. The nodes in the simulator will move based on Random Waypoint model. Also, all messages are directed towards the surface station. There is only one surface station on the water surface.

First, we evaluated the delivery probability, for each situation (None, HT, VT and using both HT and VT) with different FBRD angles (θ): 10 – 360. The results are shown in Fig. 5. When we did not use HT and VT, the delivered probability is low (see Fig. 5 (a)). However, if we use HT or VT, the delivery probability is increased as show in Fig. 5 (b) and Fig. 5 (c). In addition, when we use both HT and VT the delivery probability is the highest (see Fig. 5 (d)). But, there is not a big difference with the case of using only VT.

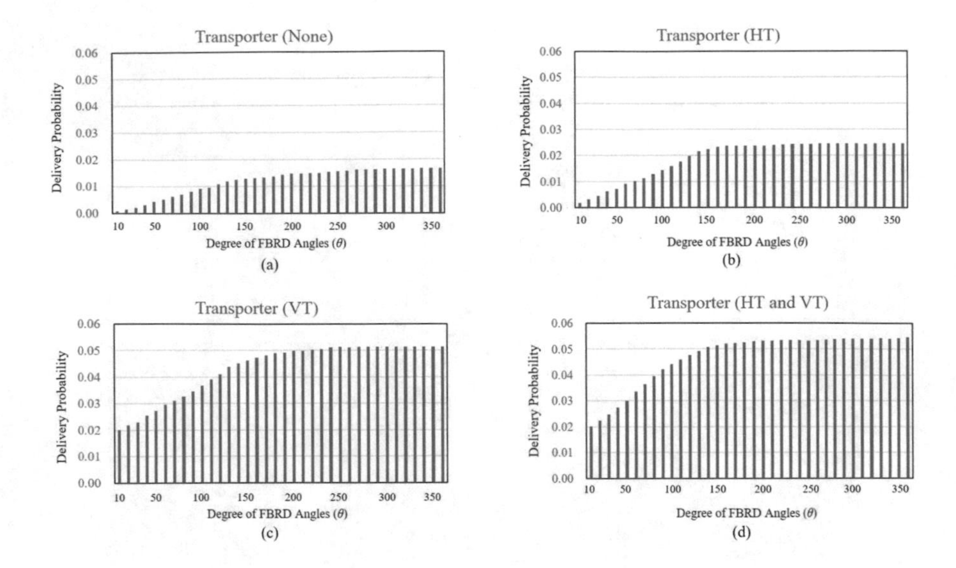

Fig. 5. Relation of between delivery probability and FBRD angles for four situations.

5 Conclusions and Future Work

In this paper, we implemented FBRD protocol in The ONE Simulator and evaluated the effects of the transporter AUV using FBRD protocol for UOWC. The performance evaluation show that the VT or HT can increase the delivery probability. In addition, when we use both HT and VT the delivery probability is the highest. But, there is not a big difference with the case of using only VT.

In the future work, we need to consider various situations in the underwater environment. In addition, we would like to implement other protocols in The ONE simulator and evaluate their performance.

References

1. Ali, M.A., Kaja Mohideen, S., Vedachalam, N.: Current status of underwater wireless communication techniques: A review. In: 2022 Second International Conference on Advances in Electrical, Computing, Communication and Sustainable Technologies (ICAECT), pp. 1–9 (2022)
2. Awan, K.M., Shah, P.A., Iqbal, K., Gillani, S., Ahmad, W., Nam, Y.: Underwater wireless sensor networks: a review of recent issues and challenges. Wireless Commun. Mobile Comput. **2019**(3), 1–20 (2019)
3. Jouhari, M., Ibrahimi, K., Tembine, H., Ben-Othman, J.: Underwater wireless sensor networks: A survey on enabling technologies, localization protocols, and internet of underwater things. IEEE Access **7**, 96879–96899 (2019)

4. Matsuo, K., Kulla, E., Barolli, L.: Evaluation of focused beam routing protocol on delay tolerant network for underwater optical wireless communication. In: International Conference on Emerging Internetworking, Data & Web Technologies, pp. 263–271. Springer (2022). https://doi.org/10.1007/978-3-030-95903-6_28
5. Weng, Y., Pajarinen, J., Akrour, R., Matsuda, T., Peters, J., Maki, T.: Reinforcement learning based underwater wireless optical communication alignment for autonomous underwater vehicles. IEEE J. Oceanic Eng. **47**(4), 1231–1245 (2022)

A Road State Decision Method Based on Roughness by Crowd Sensing Technology

Yoshitaka Shibata[1(✉)] and Yasushi Bansho[2]

[1] Iwate Prefectural University, Sugo 152-89, Takizawa 020-0611, Japan
shibata@iwate-pu.ac.jp
[2] Holonic Systems, Ltd, Kamihirasawa 7-3, Shiwa, Iwate 028-3411, Japan
bansho@holonic-systems.com

Abstract. In this paper, a road state decision method based on roughness by crowd sensing technology is proposed to maintain the safe and reliable driving environment. Various environmental sensers including acceleration, gyro, angular velocity, temperature, humidity, near infrared laser sensors and global positioning system (GPS) are used as crowd sensing. In particular, roughness is precisely studied among road states and evaluated through experimental test on the actual road. Through the field test, the proposed method could provide realtime and effective road state sensing ability.

1 Introduction

In recent, autonomous driving systems have been well developed in industrial countries in the world such as mainly the U.S., Europe, China and Japan. In fact, the practical autonomous cars are commercially available and running on exclusive roads and highway roads at level 3 or 4 in those countries [1]. In order to realize safer autonomous driving, the roads must be on ideal conditions such that road surface is flat, the driving lanes are clear and the same directional driving zones are completely separated from the opposite driving zones. In particular, since road surfaces are occupied with sediment discharge, sand-slide, rock, snow and ice, road conditions must be always monitored and those road states information are always shared. Those problems must be classified into sensing road surface conditions technologies and road states information transmission technologies.

So far, we have developed Vehicle-to-Vehicle (V2V) and Vehicle-to-Roadside (V2R) communications system using N-wave length cognitive wireless network and evaluated those functions and performances [2–6]. Based on the V2V/V2R communication system, we constructed a prototype of road state monitoring system to calculate roughness of running road in realtime by using cost-effective three-axis accelerometer [7] and a GPS sensor embedded in a vehicle. However, it is difficult to understand the vehicle environment and road surface conditions, which vary every moment according to the weather, e.g. dry, wet, snow, freezing, and so on.

In order to resolve those problems, we introduce a new road state decision method using various environmental sensors. As environmental sensors, six-axis sensors including acceleration and gyroscope sensors, near infrared laser sensors and GPS are used

L. Barolli (Ed.): EIDWT 2023, LNDECT 161, pp. 324–330, 2023.
https://doi.org/10.1007/978-3-031-26281-4_34

to determine the correct road state. Among the various road states, we focused on the roughness based on the six-axis sensors and GPS and evaluated its functional and performance. Through the evaluation of the field experiment, the proposed method could provide realtime and effective road state sensing and decision ability.

In the followings, system configuration of our road sensing prototype is explained in Sect. 2. Then, calculation of roughness form 6-axis acceleration sensor and GPS is introduced in Sect. 3. Next, an experimental prototype and performance results are explained in Sect. 4. Finally conclusions and future works are summarized.

2 System Configuration

In our research, we proposed an innovative sensor processing system as shown in Fig. 1 to decide road surface conditions. The sensor processing system is organized by several sensors including a LiDAR, a six-axis accelerometer, a near infrared laser sensor, GPS, data logger and processing server. LiDAR is used to identify the obstacles ahead of vehicle. The 6-axis accelerometer is used to observe the roughness of road surface. The near infrared laser sensor which furthermore includes infrared temperature and humidity sensors is used to decide the road states. GPS is used to precisely to identify the locations and time of the running vehicle. The data logger is used to temporally store the sensor data from the near infrared laser sensor. The processing server is used to process and analyze the sensor data and display the output of analyzed results. In this paper, we focus on the analyzed results of roughness by a 6-axis accelerometer by changing sampling rate ranging from 50 to 2000Hz to know whether the sampling rate gives any influence to accuracy and resolution of roughness.

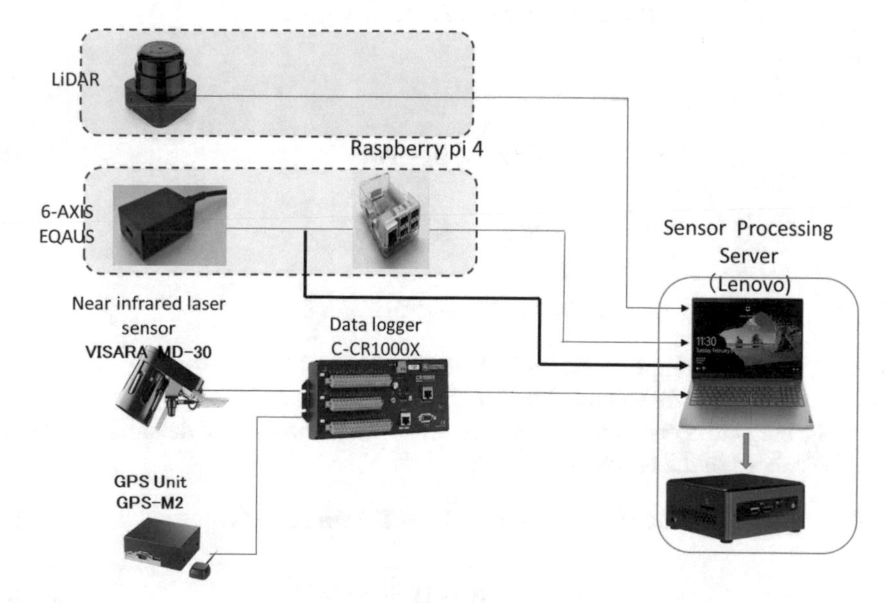

Fig. 1. System configuration of sensor processing system.

3 Observation of Roughness

A set of sensor processing system is installed on the vehicle. The 6-axis acceleration sensor is fixed vertically to the dashboard of the front panel and observes as shown in Fig. 2. Among the accelerations of three coordination (x,y,z), the vertical acceleration, AY is considered as roughness as shown in Fig. 2. However, since the slope angle of the road varies according to the road conditions while running, the coordinates of the sensor is not always identical to the coordinates of the earth, the observed vertical acceleration AY have to be compensated to be identical to the earth coordinates using the following equations,

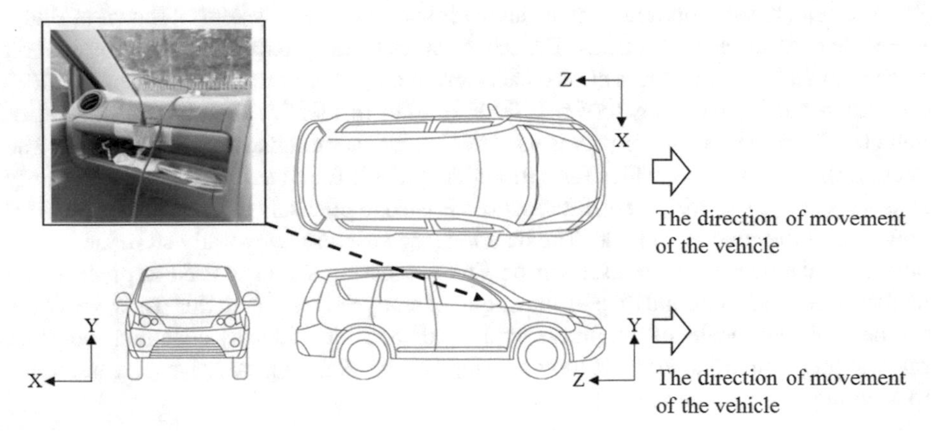

Fig. 2. Installation of 6-Axis acceleration

$$AYz = S_z \cdot AZ \quad (1)$$

$$AYx = S_x \cdot AX \quad (2)$$

$$AYnet = AY - AYz - AYx \quad (3)$$

where AYnet is the compensated vertical acceleration and S_z, S_x are coefficients based on the slop in z and x directions of the road.

Then, using the AYnet, the vertical velocity Vy and the vertical deviation Ly of the can be calculated in the followings,

$$dZ(i) = AYnet(i-1) - \Sigma AYnet\,(j) / N \quad (4)$$

$$Vy(i) = Vy(i-1) + dZ(i) / N \quad (5)$$

$$dVy(i) = Vv(i) - \Sigma Z(j) / N \quad (6)$$

$$Ly(i) = Ly(i-1) + dVy\,(i) / N \tag{7}$$

where N is the sampling frequency. Thus, the deviation of the body of the vehicle is equivalent to the deviation of the road Ly(i), namely roughness of road can be obtained. It is reminded that by changing the sampling frequency, the resolution of the road flatness varies. If frequency is higher, resolution of road sensing become higher, but the computational load become higher. On the contrary, computational load increases, realtime processing and output are influenced (Fig. 3).

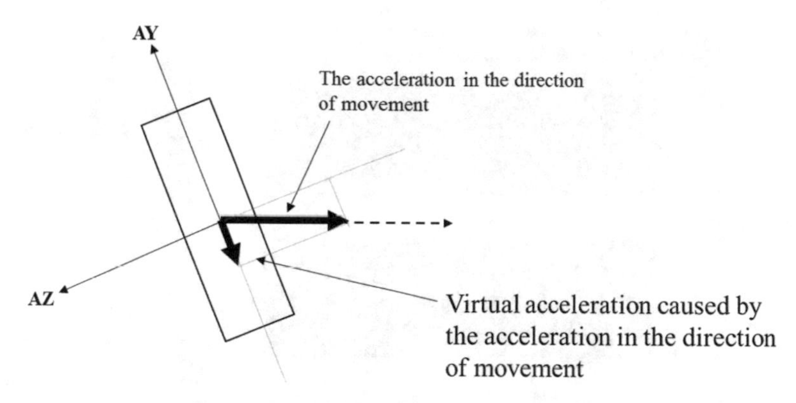

Fig. 3. Compensation of observed vertical acceleration

4 Experimental Result

In order to verify the effects of our proposed sensing system, particularly roughness computing, an experimental prototype is constructed and evaluate its performance as shown in Fig. 4. The six-axis acceleration sensor, EQUAS is fixed on the front dashboard and is connected to the raspberry pi for sampling sensor signals and A/D conversion, and sends to the sensor server PC. In sensor server PC, the sensor data are processed. Figures 5, 6 and 7 show the compensated acceleration (upper parts) and the roughness (lower parts) in (x, y, z) coordination for various sampling frequencies from 50 to 2000 Hz while vehicle's speed is kept at 10 km/h. It is clear that as the sampling frequency is increased, the resolution of both the acceleration and the roughness become higher while the realtime processing and output can be maintained. Thus, our proposed by six-axis acceleration sensor can provide roughness of the road with higher resolution in realtime.

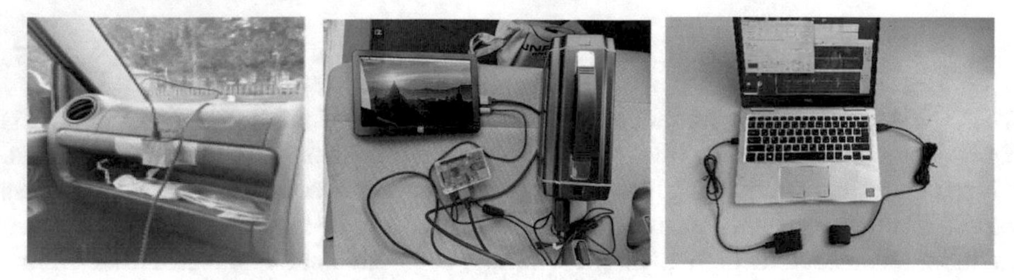

Fig. 4. Experimental prototype system

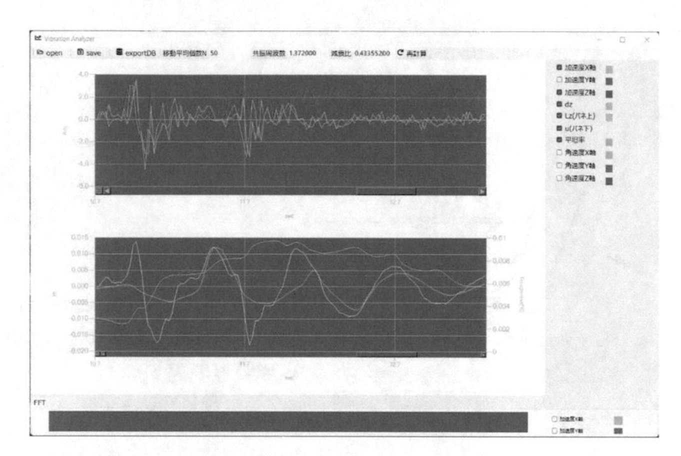

Fig. 5. Acceleration and roughness at sampling rate 50 Hz

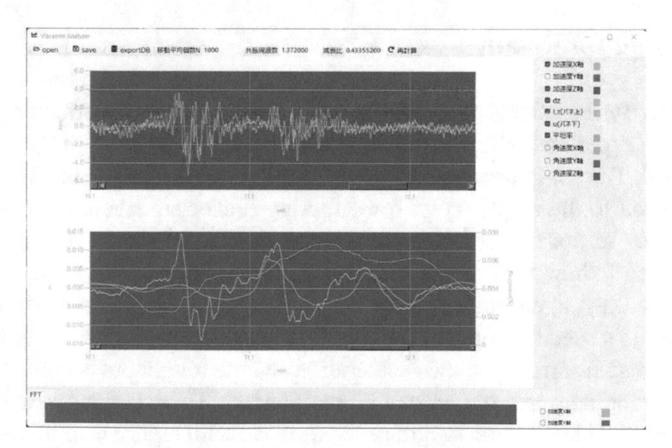

Fig. 6. Acceleration and roughness at sampling rate 1,000 Hz

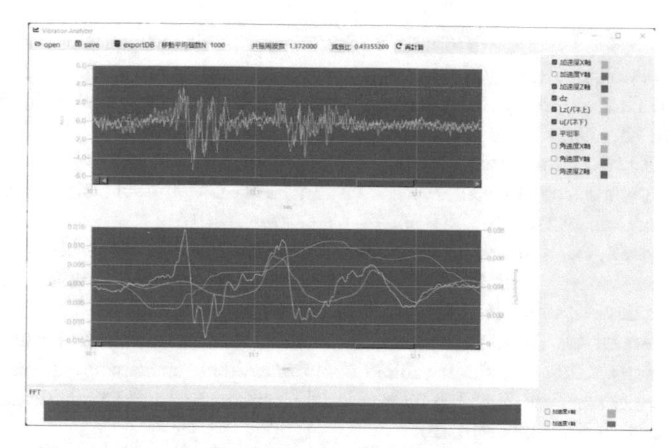

Fig. 7. Acceleration and roughness at sampling rate 2,000 Hz

5 Conclusion and Future Works

In this paper, we proposed a new road state decision system for road surface conditions using various sensor data. Particularly, we introduced system configuration and computation of roughness as index of road surface conditions. We designed and implemented on-board monitoring system using relatively compact and cost-effective sensor devices, in which acceleration and gyroscope sensors are performed to monitor road surface conditions. We actually constructed the experimental prototype to evaluate our system by changing sampling rate. Through our field experiment, it is clear that this six-axis sensor device and road state decision system can provide effective monitoring and realtime decision ability on real running road surface.

Now, we are analyzing the accuracy of the roughness by comparing the results from our decision method and the actual measurement of road surface. As future works, we are going to continue the field experiment by considering surface temperature, humidity and reflection/absorption rates for more general environmental conditions such as dry, rainy, snowy and icy conditions. In addition, we will investigate a study of better road decision algorithm, e.g. the rule-based fuzzy algorithm, nonlinear time series analysis, and deep learning, to analyze and decide road surface conditions more precisely.

Acknowledgments. The research was supported by Japan Keiba Association Grant Numbers 2021M-198, JSPS KAKENHI Grant Numbers JP 20K11773, Strategic Information and Communications R&D Promotion Program Grant Number 181502003 by Ministry of Affairs and Communication.

References

1. SAE International: Taxonomy and Definitions for Terms Related to Driving Automation, Systems for On-Road Motor Vehicles. J3016_201806 (2018)

2. Shibata, Y., Ito, K., Uchida, N.: A new V2X communication system to realize long distance and large data transmission by N-wavelength wireless cognitive network. In: 2018 IEEE 32nd International Conference on Advanced Information Networking and Applications (AINA), pp. 587–592. Krakow, Poland (2018)
3. Uchida, N., Ito, K., Hirakawa, G., Shibata, Y.: Evaluations of wireless V2X communication systems of for winter road surveillance systems. In: 19th International Conference on Network-Based Information Systems (NBiS), pp. 58–63. Ostrava (2016)
4. Ito, K., Hirakawa, G., Shibata, Y.: Estimation of communication range using Wi-Fi for V2X communication environment. In: 10th International Conference on Complex, Intelligent, and Software Intensive Systems (CISIS), pp. 278–283. Fukuoka (2016)
5. Ito, K., Hirakawa, G., Arai, Y., Shibata, Y.: A road alert information sharing system with multiple vehicles using vehicle-to-vehicle communication considering various communication network environment. In: 18th International Conference on Network-Based Information Systems, pp. 365–370. Taipei (2015)
6. Ito, K., Arai, Y., Hirakawa, G., Shibata, Y.: A road condition sharing system using vehicle-to-vehicle communication in various communication environment. In: 9th International Conference on Complex, Intelligent, and Software Intensive Systems, pp. 242–249. Blumenau (2015)
7. Chen, K., Lu, M., Fan, X., Wei, M., Wu, J.: Road condition monitoring using on-board Three-axis Accelerometer and GPS Sensor. In: 6th International ICST Conference on Communications and Networking in China (CHINACOM), pp. 1032–1037. Harbin (2011)

Experimental Results of a Wireless Sensor Network Testbed for Monitoring a Water Reservoir Tank Considering Multi-flows

Yuki Nagai[1], Aoto Hirata[1], Chihiro Yukawa[1], Kyohei Toyoshima[1], Tetsuya Oda[2(✉)], and Leonard Barolli[3]

[1] Graduate School of Engineering, Okayama University of Science (OUS), 1-1 Ridaicho, Kita-ku, 700-0005 Okayama, Japan
{t22jm23rv,t21jm02zr,t22jm19st,t22jm24jd}@ous.jp

[2] Department of Information and Computer Engineering, Okayama University of Science (OUS), 1-1 Ridaicho, Kita-ku, 700-0005 Okayama, Japan
oda@ous.ac.jp

[3] Department of Information and Communication Engineering, Fukuoka Insitute of Technology, 3-30-1 Wajiro-Higashi ku, 811 0295 Fukuoka, Japan
barolli@fit.ac.jp

Abstract. Monitoring, predicting and preventing damages in water reservoir tanks is very important to reduce the damages during bad weather conditions. In our previous work, we developed some sensing devices and proposed a Wireless Sensor Fusion Network (WSFN) to monitor the rivers. In this paper, using the results of our previous work, we implement a Wireless Sensor Network (WSN) testbed and analyze the throughput, delay and jitter of a WSN in an outdoor environment considering two scenarios and multi-flows. The experimental results show that the throughput is almost the same for both scenarios. However, the delay and jitter values are increased with increasing of communication distance. The experimental results show that the proposed system can correctly gather the sensing data and monitor the water reservoir tank in real-time.

1 Introduction

There are different water reservoir tanks such as septic tanks, industrial and agricultural water tanks and fire protection tanks. The are basically used in outdoor environment and the amount of water change depending on weather conditions. During bad weather conditions, there is a risk of flooding because of heavy rains.

It is very important to monitor and predict the water reservoir tank in order to reduce the damages by learning about hazards in their early stage. However, this process should be done in real-time and it is difficult to grasp, monitor and predict all situations.

In our previous work, we proposed a Wireless Sensor Fusion Network (WSFN) [1–7] and implemented some sensing devices in order to monitor rivers [8]. The proposed WSFN can obtain different data and information as real-time streaming and then by analyzing the data was able to make good prediction.

L. Barolli (Ed.): EIDWT 2023, LNDECT 161, pp. 331–340, 2023.
https://doi.org/10.1007/978-3-031-26281-4_35

In this paper, using the results of our previous work, we implement a Wireless Sensor Network (WSN) testbed and analyze the throughput, delay and jitter of a WSN in an outdoor environment considering Line of Sight (LoS) [9–11], two scenarios and multi-flows. The experimental results show that the throughput is almost the same for both scenarios. However, the delay and jitter values are increased with increasing of communication distance. The experimental results show that the proposed system can correctly gather the sensing data and monitor the water reservoir tank in real-time.

The paper structure is as follows. In Sect. 2, we present water reservoir tank monitoring system. In Sect. 3, we describe of the implemented testbed. In Sect. 4, we discuss the experimental results. Finally, in Sect. 5, we conclude the paper.

2 Water Reservoir Tank Monitoring System

The water reservoir tank monitoring system is shown in Fig. 1. This system is called WSFN and is able to gather the sensed data by multiple sensing devices using a WSN. The WSN is based on a Wireless Mesh Network (WMN), where sensor nodes are connected by multiple paths. So, in the case when a sensor node may fail the collection of data will continue.

In the proposed system, we use different sensing devices for measuring the water level, rainfall, soil moisture, temperature, humidity, atmospheric pressure and water temperature. We installed the WSN around the water reservoir tank in order to gather the sensing data and predict water reservoir tank conditions by time-series analysis.

2.1 Sensing Devices for Water Reservoir Tank Monitoring System

In Fig. 2 are shown some sensing devices that are used for different measurements. We used these sensing devices: Water Level Indicator, Indicator of Temperature, Humidity and Atmospheric Pressure, Soil Moisture Indicator and Tipping Bucket Type Rain Gauge. We developed the waterproof mechanism and some parts of the sensing devices with a 3D printer in order to have a low cost [12–15].

In Fig. 2 is showing the water level indicator, which can calculate the water pressure by using the data from water pressure sensor and atmospheric pressure indicator. In Fig. 2 is showing the indicator, which measure temperature, humidity and atmospheric pressure for different ranges. The soil moisture indicator shown in Fig. 2 uses the relative permittivity of water to obtain the soil moisture content. While, the tipping bucket type rain gauge shown in Fig. 2 considers the number of times, the tipping bucket tilts per hour and the measurements per tilt in order to measure the rainfall.

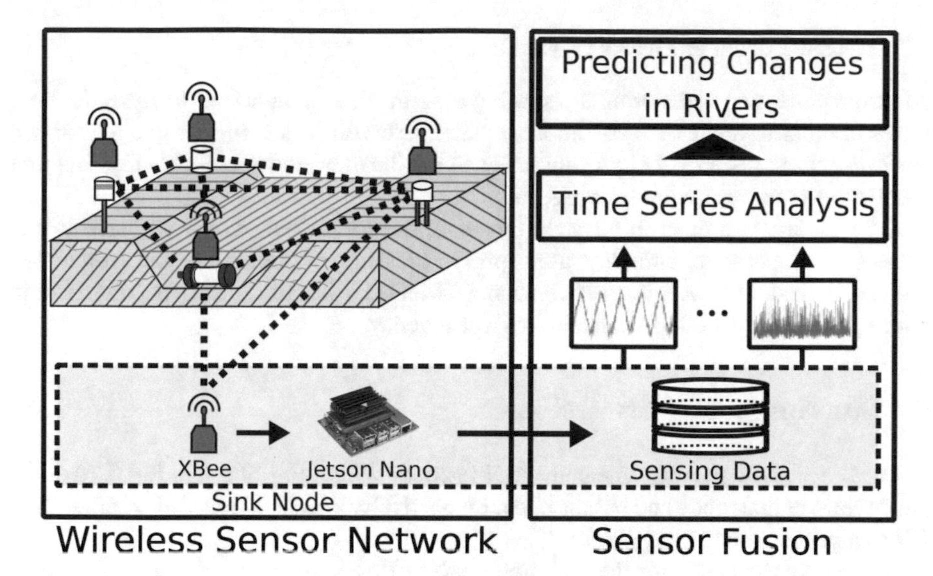

Fig. 1. Proposed system.

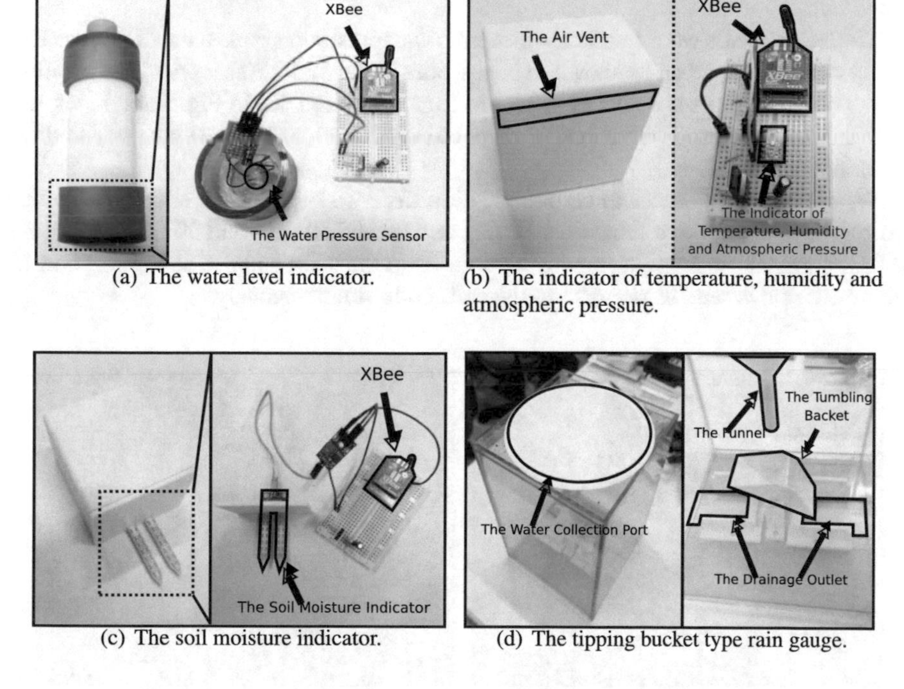

(a) The water level indicator.

(b) The indicator of temperature, humidity and atmospheric pressure.

(c) The soil moisture indicator.

(d) The tipping bucket type rain gauge.

Fig. 2. Sensing devices for water reservoir tank monitoring system.

2.2 Sensor Fusion of Measurement Data

After the collection and storing the sensing data in the sink node, we analyze the time series in order to predict the condition of water reservoir tank [16]. For this reason, we perform sensor fusion [17] by using Long Short-Term Memory (LSTM) [18–21] and predict the water reservoir tank changes.

The LSTM is trained in advance by using observed data, which we have in the dataset of temperature, humidity, atmospheric pressure and rainfall of the Automated Meteorological Data Acquisition System (AMeDAS) installed near the observation points provided by the Japan Meteorological Agency.

3 Description of WSN Testbed

In Fig. 3, we show the implemented WSN testbed. While, Fig. 3(a) and Fig. 3(b) show the devices of sink node and sensor node, respectively. We use XBee with ZigBee [22–27] as a device to build the WSN.

Jetson Nano is used for the sink node, while Raspberry Pi for the sensor node to control XBee. The Raspberry Pi is equipped with indicators of temperature, humidity and atmospheric pressure. The testbed equipment is stored in a waterproof case for protection.

The experiments were conducted around a water reservoir tank at the Okayama University of Science, Japan as shown in Fig. 4 considering a LoS scenario. The locations of sink node and sensor nodes as shown in Fig. 5. The red dot in Fig. 5 and black dots show the sink node and sensor nodes, respectively. We conisder one sink node and two sensor nodes.

We evaluate three cases of communication distance between the sink node and the two sensor nodes for two scenarios: 10 [m] and 20 [m], 10 [m] and 30 [m], 20 [m] and 30 [m]. The sink node requests sensing data from the sensor nodes at 1 [$sec.$] and the two sensor nodes send sensor data to the sink node simultaneously.

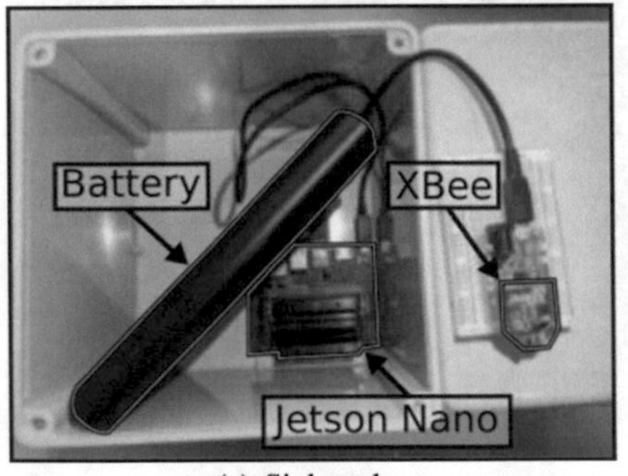

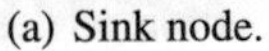
(a) Sink node.

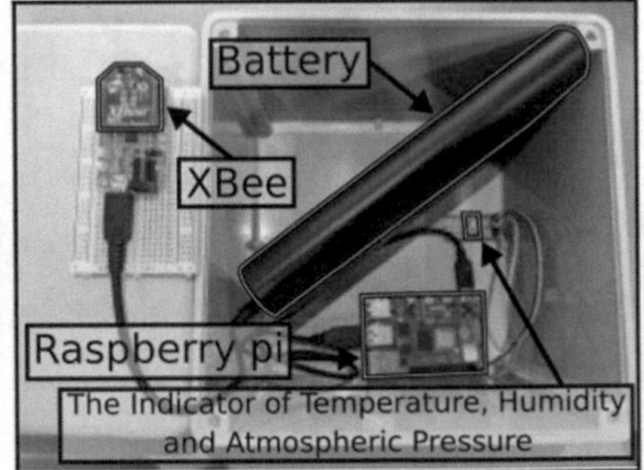

(b) Sensor node.

Fig. 3. Devices of WSN testbed.

Fig. 4. Water reservoir tank at Okayama University of Science used as experimental environment.

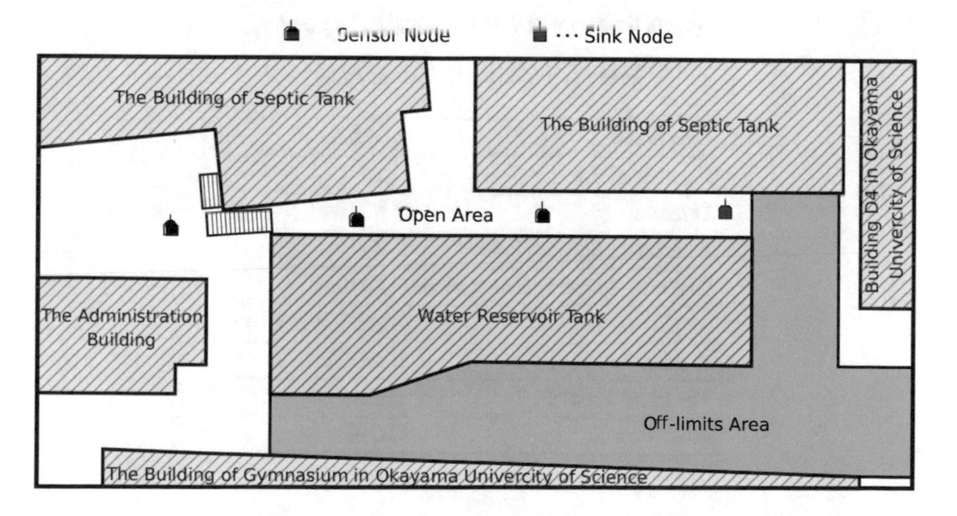

Fig. 5. Location of sink node and sensor nodes.

Figure 6 shows the Snapshot of each node. The experimental parameters are shown in Table 1. The sensing data transmitted by the sensor node is the value measured by the indicator of temperature, humidity and atmospheric pressure.

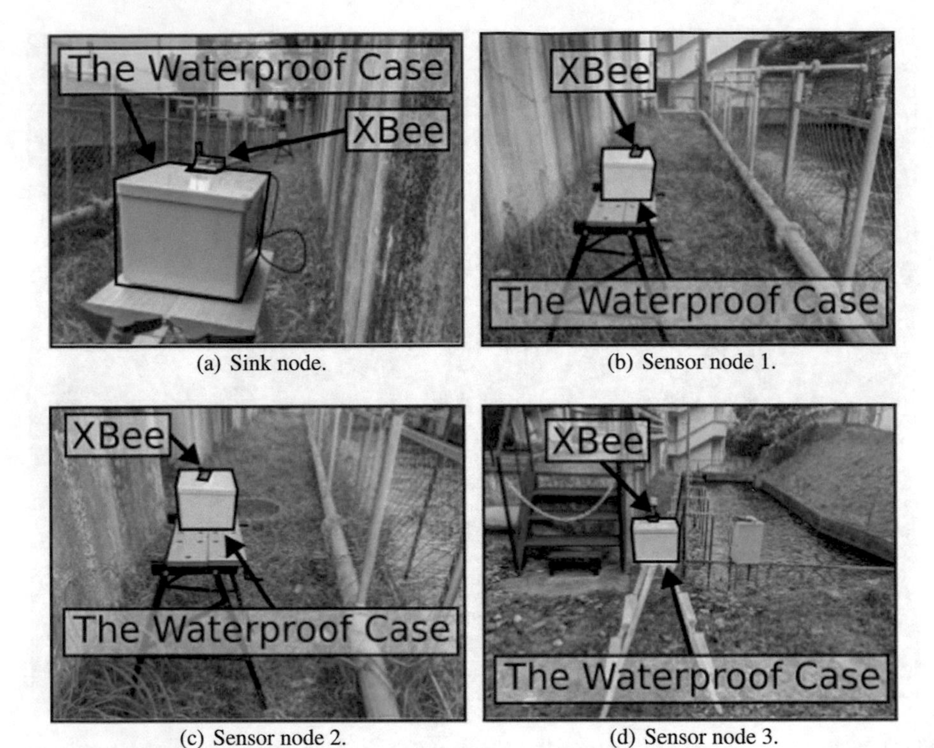

(a) Sink node. (b) Sensor node 1.

(c) Sensor node 2. (d) Sensor node 3.

Fig. 6. Snapshot of sink node and sensor nodes.

Table 1. Experimental parameters.

Functions	Values
Number of Trials	10
Duration	60 [*sec*]
Number of Sensor Nodes	2
Distance Between Sensor Nodes	10 [*m*]
MAC	IEEE 802.15.4
Routing Protocol	AODV
Transport Protocol	UDP
Flow Type	CBR
Sensor Data	52 [*bps*]

4 Experimental Results

For evaluation, we used multi-flows from the sensor node to the sink node. Figures 7, 8 and 9 show the experimental results of average throughput, average delay and average jitter, respectively. The throughput, the delay and the jitter are the average value of 10

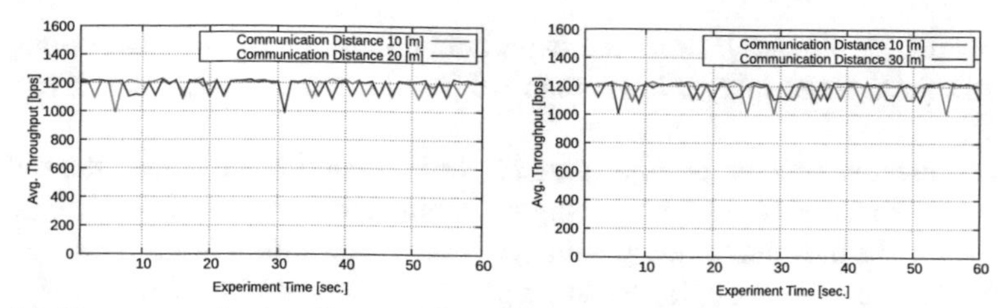

(a) The communication distance 10 and 20 [*m*]. (b) The communication distance 10 and 30 [*m*].

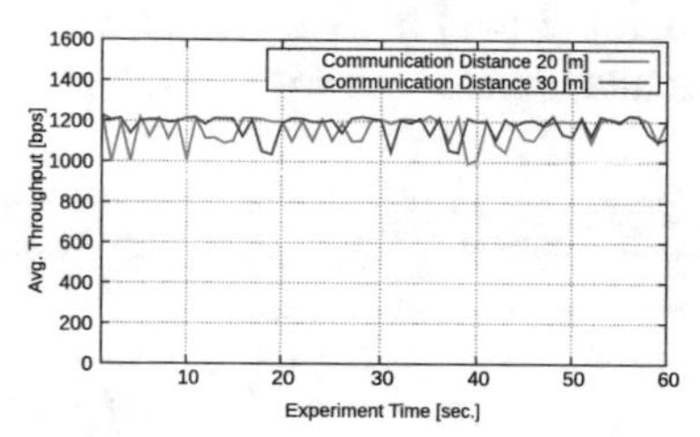

(c) The communication distance 20 and 30 [*m*].

Fig. 7. Experimental results of throughput.

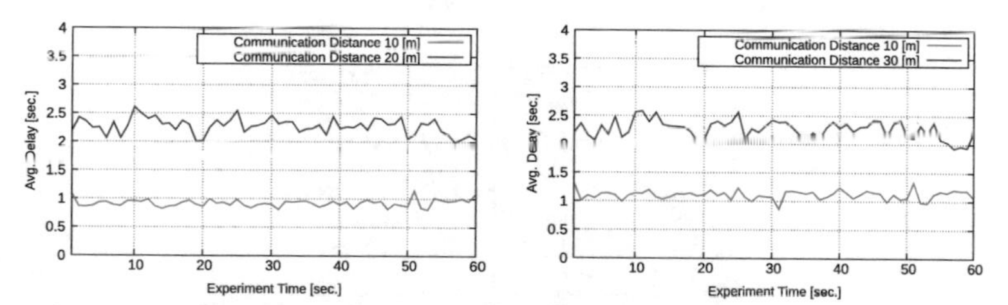

(a) The communication distance 10 and 20 [*m*]. (b) The communication distance 10 and 30 [*m*].

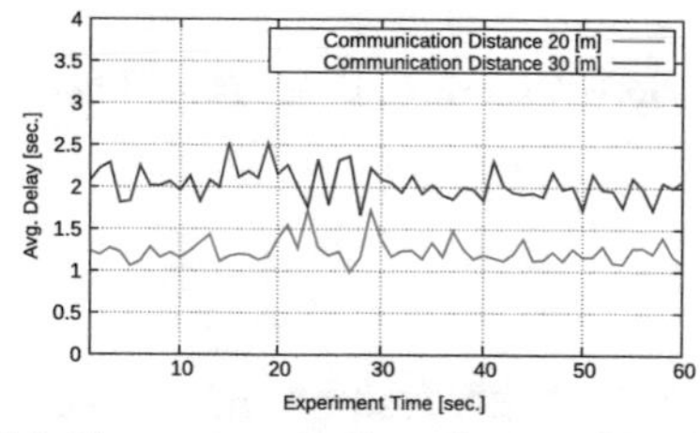

(c) The communication distance 20 and 30 [*m*].

Fig. 8. Experimental results of dclay.

measurements for 60 [*sec.*]. From the experimental results, the average throughput is 1178 [*bps*], the average delay is lower than 3.0 [*sec.*] and the average jitter is lower than 1.0 [*sec.*].

The throughput was almost the same for both scenarios. However, the delay and jitter values are increased with increasing of communication distance. Also, the water level indicator, the indicator of temperature, humidity and atmospheric pressure, the soil moisture indicator and the tipping bucket type rain gauge send 24 [*bit*], 52 [*bit*], 12 [*bit*] and 1 [*bit*] sensing data to the sink node via WSN. The delay time of 52 [*bit*] for the indicator of temperature, humidity and atmospheric pressure with the largest sensing data size is less than 3.0 [*sec.*]. Therefore, we conclude tha that the proposed system can collect the sensing data and monitor the water reservoir tank in real-time.

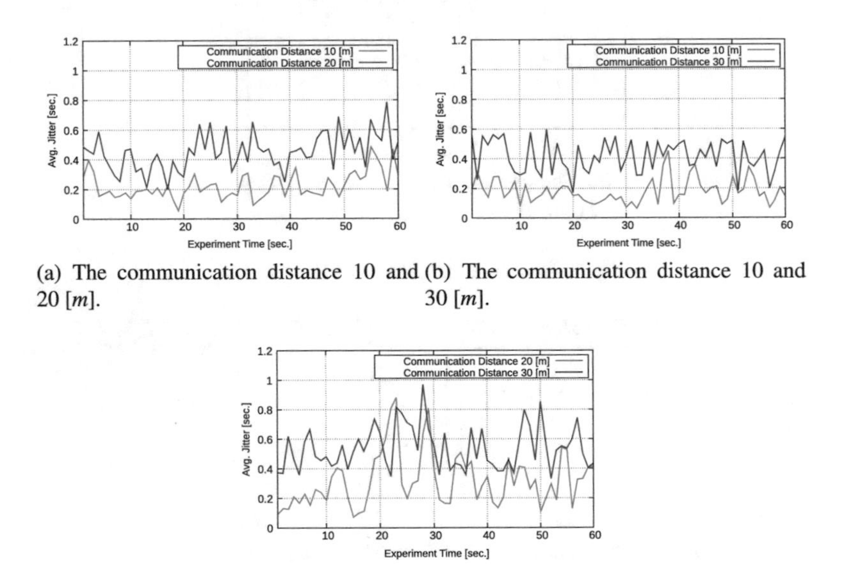

(a) The communication distance 10 and 20 [*m*].

(b) The communication distance 10 and 30 [*m*].

(c) The communication distance 20 and 30 [*m*].

Fig. 9. Experimental results of jitter.

5 Conclusions

In this paper, we implemented a WSN testbed and carried out some experiments. We analyzed the throughput, delay and jitter. From the experimental results, we found that the average throughput is 1178 [*bps*], the average delay is lower than 3.0 [*sec.*] and the average jitter is lower than 1.0 [*sec.*]. These results show that the proposed system can collect the sensing data and monitor the water reservoir tank in real-time.

In the future, we would like to conduct experiments with various node configurations and different scenarios.

Acknowledgement. This work was supported by JSPS KAKENHI Grant Number JP20K19793.

References

1. Lewi, T., et al.: Aerial sensing system for wildfire detection. In: Proceeding of The 18-th ACM Conference on Embedded Networked Sensor Systems, pp. 595-596 (2020)
2. Mulukutla, G., et al.: Deployment of a large-scale soil monitoring geosensor network. SIGSPATIAL Special **7**(2), 3–13 (2015)
3. Gayathri, M., et al.: A low cost wireless sensor network for water quality monitoring in natural water bodies. In: The IEEE Global Humanitarian Technology Conference, pp. 1-8 (2017)
4. Gellhaar, M., et al.: Design and evaluation of underground wireless sensor networks for reforestation monitoring. In: Proceeding of the 41-st International Conference on Embedded Wireless Systems and Networks, pp. 229-230 (2016)
5. Yu, A., et al.: Research of the factory sewage wireless monitoring system based on data fusion. In: Proceeding of The 3rd International Conference on Computer Science and Application Engineering, No. 65, pp. 1-6 (2019)
6. Suzuki, M., et al.: A high-density earthquake monitoring system using wireless sensor networks. In: Proceeding of The 5-th International Conference on Embedded Networked Sensor Systems, pp. 373-374 (2007)
7. Oda, T., et al.: Design and implementation of a simulation system based on deep Q-network for mobile actor node control in wireless sensor and actor networks. In: Proceeding of the IEEE 31st International Conference on Advanced Information Networking and Applications Workshops, pp. 195-200 (2017)
8. Nagai, Y., Oda, T.: A river monitoring and predicting system considering a wireless sensor fusion network and LSTM. In: Proceeding of the 10-th International Conference on Emerging Internet, Data and Web Technologies, Okayama, Japan, pp. 283-290, March (2022)
9. Oda, T., et al.: Implementation and experimental results of a WMN testbed in indoor environment considering LOS scenario. In: Proceeding of the IEEE 29-th International Conference on Advanced Information Networking and Applications, pp. 37-42 (2015)
10. Nagai, Y., et al.: A wireless sensor network testbed for monitoring a water reservoir tank: experimental results of delay. In: Proceeding of the 16-th International Conference on Complex, Intelligent, and Software Intensive Systems, pp. 49-58 (2022)
11. Nagai, Y., et al.: A wireless sensor network testbed for monitoring a water reservoir tank: experimental results of delay and temperature prediction by LSTM. In: Proceeding of the 25-th International Conference on Network-Based Information Systems, pp. 392-401 (2022)
12. Saito, N., et al.: Design and implementation of a DQN Based AAV. In: Proceeding of the 15-th International Conference on Broad-Band and Wireless Computing, Communication and Applications, pp. 321-329 (2020)
13. Saito, N., et al.: Simulation results of a DQN Based AAV testbed in corner environment: a comparison study for normal DQN and TLS-DQN. In: Proceeding of the 15-th International Conference on Innovative Mobile and Internet Services in Ubiquitous Computing, pp. 156-167 (2021)
14. Saito, N., et al.: A tabu list strategy based DQN for AAV mobility in indoor single-path environment: implementation and performance evaluation. Internet Things **14**, 100394 (2021)
15. Saito, N., et al.: A LiDAR based mobile area decision method for TLS-DQN: improving control for AAV mobility. In: Proceeding of the 16-th International Conference on P2P, Parallel, Grid, Cloud and Internet Computing, pp. 30-42 (2021)
16. Sharma, P., et al.: A machine learning approach to flood severity classification and alerting. In: Proceeding of the 4-th ACM SIGSPATIAL International Workshop on Advances in Resilient and Intelligent, pp. 42-47 (2021)

17. Hang, C., et al.: Recursive truth estimation of time-varying sensing data from online open sources. In: The 14-th International Conference on Distributed Computing in Sensor Systems, pp. 25-34 (2018)
18. Hochreiter, S., et al.: Long short-term memory. Neural Comput. **9**, 1735–1780 (1997)
19. Karevan, Z., et al.: Transductive LSTM for time-series prediction: an application to weather forecasting. Neural Netw. **125**, 1–9 (2019)
20. li, Y., et al.: Hydrological time series prediction model based on attention-LSTM neural network. In: Proceeding of the 2nd International Conference on Machine Learning and Machine, pp. 21-25 (2019)
21. Toyoshima, K., et al.: Proposal of a haptics and LSTM based soldering motion analysis system. In: Proceeding of the IEEE 10-th Global Conference on Consumer Electronics, pp. 1-2 (2021)
22. Oda, T., et al.: A genetic algorithm-based system for wireless mesh networks: analysis of system data considering different routing protocols and architectures. Soft. Comput. **20**(7), 2627–2640 (2016)
23. Oda, T., et al.: Evaluation of WMN-GA for different mutation operators. Int. J. Space-Based Situated Comput. **2**(3), 149–157 (2012)
24. Oda, T., et al.: WMN-GA: a simulation system for WMNs and its evaluation considering selection operators. J. Ambient Intell. Human. Comput. **4**(3), 323–330 (2013)
25. Nishikawa, Y., et al.: Design of stable wireless sensor network for slope monitoring. In: The IEEE Topical Conference on Wireless Sensors and Sensor Networks, pp. 8-11 (2018)
26. Hirata, A., et al.: A coverage construction method based hill climbing approach for mesh router placement optimization. In: Proceeding of the 15-th International Conference on Broad-Band and Wireless Computing, Communication and Applications, pp. 355-364 (2020)
27. Hirata, A., et. al.: A coverage construction and hill climbing approach for mesh router placement optimization: simulation results for different number of mesh routers and instances considering normal distribution of mesh clients. In: Proceeding of the 15-th International Conference on Complex, Intelligent and Software Intensive Systems, pp. 161-171 (2021)

A Depth Camera Based Soldering Motion Analysis System for Attention Posture Detection Considering Body Orientation

Kyohei Toyoshima[1], Chihiro Yukawa[1], Yuki Nagai[1], Nobuki Saito[1], Tetsuya Oda[2(✉)], and Leonard Barolli[3]

[1] Graduate School of Engineering, Okayama University of Science (OUS), 1-1 Ridaicho, Kita-ku, 700–0005 Okayama, Japan
{t22jm24jd,t22jm19st,t22jm23rv,t21jm01md}@ous.jp

[2] Department of Information and Computer Engineering, Okayama University of Science (OUS), 1-1 Ridaicho, Kita-ku, 700–0005 Okayama, Japan
oda@ous.ac.jp

[3] Department of Informention and Communication Engineering, Fukuoka Insitute of Technology, 3-30-1 Wajiro-Higashi-ku, 811-0295 Fukuoka, Japan
barolli@fit.ac.jp

Abstract. Recently, governments are promoting employment of persons with physical, intellectual, mental or other disabilities, but the employers should ensure safety in the workplace. However, these persons the technical training requires explanation of detailed work procedures and the process monitoring to prevent accidents. In this paper, in order to solve these problems, we present a depth camera based soldering motion analysis system for attention posture detection. We carried out many experiments using the implemented system. We present and discuss experimental results of the posture detection according to the orientation of the body during soldering. From experimental results, we found that the proposed system can detect the posture from the three-dimensional coordinates of the body and the accuracy for three patterns of body orientation is about 96 [%].

1 Introduction

There are many accidents during the industrial production process and the cost resulting from safety-related injuries is very high. Also, many governments are promoting now employment of persons with physical, intellectual, mental or other disabilities, but the employers should ensure safety in the workplace.

On the factory floor, the soldering is considered relatively easy to learn and in Japan the soldering training is provided in junior high school, high school and university classes. But, this operation is prone to human error. For instance, the beginner soldering students are inexperienced, which may lead to different accidents.

It is very important to reduce the accidents by giving instruction during dangerous actions or inappropriate postures during soldering iron work [1–6]. Also, the beginner

L. Barolli (Ed.): EIDWT 2023, LNDECT 161, pp. 341–349, 2023.
https://doi.org/10.1007/978-3-031-26281-4_36

soldering persons should take attention of their posture (attention posture) during soldering process. Thus, the detection of body orientation is expected to reduce mistakes and accidents.

In this paper, we present a soldering motion analysis system based on a depth camera for attention posture detection considering body orientation. Our proposed and implemented system considers three-dimensional coordinates of the left and right shoulders for attention posture detection. Using the implemented system, we carry out the analysis of soldering operation based on object detection [7–12], posture estimation [13–17] and smart speaker. The experimental results show that the proposed system can detect the posture from the three-dimensional coordinates of the body and the accuracy for three patterns of body orientation is about 96 [%].

The paper is organized as shown in following sections. We describe the proposed and implemented system in Sect. 2. Then, in Sect. 3, we present and discuss the experimental results. Finally, conclusions and future work are given in Sect. 4.

2 Proposed and Implemented System

In this section, we present the proposed and implemented system. We show in Fig. 1 and Fig. 2 the proposed system structure and and system environment, respectively. The proposed system can recognize in real time Soldering Iron, Soldering Iron Tip, Hand, Upper Body, Eyes and Legs By using these data, the system can determine dangerous operations during soldering iron process and the smart speaker notifies the soldering iron user of the danger.

In the proposed system, we use a stereo camera, which can calculates the distance between camera and object by triangulation from the parallax of images captured by two cameras. The soldering iron tip and hand tracking indicate the distance from camera to finger tip obtained by triangulation. If the finger is close to the BoundingBox of dangerous area, it is considered as a dangerous situation. We use `YOLOv5` [18–20] and `MediaPipe` [25] for detection of soldering iron and soldering iron tip, and soldering iron users hand by hand tracking [21–24], respectively.

The pose of the entire body is estimated by using UVC camera [26]. We use three JetsonNanos and three UVC cameras for camera diversification in order to avoid blind spots. The data received from each camera is aggregated on the Jetson Nano server and is used to determine the danger detection.

In order to detect whether or not the upper body posture during soldering is dangerous, the height of each camera is kept constant and cameras are positioned at an angle of ±45 [*deg.*] with respect to the camera positioned in front. In this way, the system improves the recognition accuracy of the entire body.

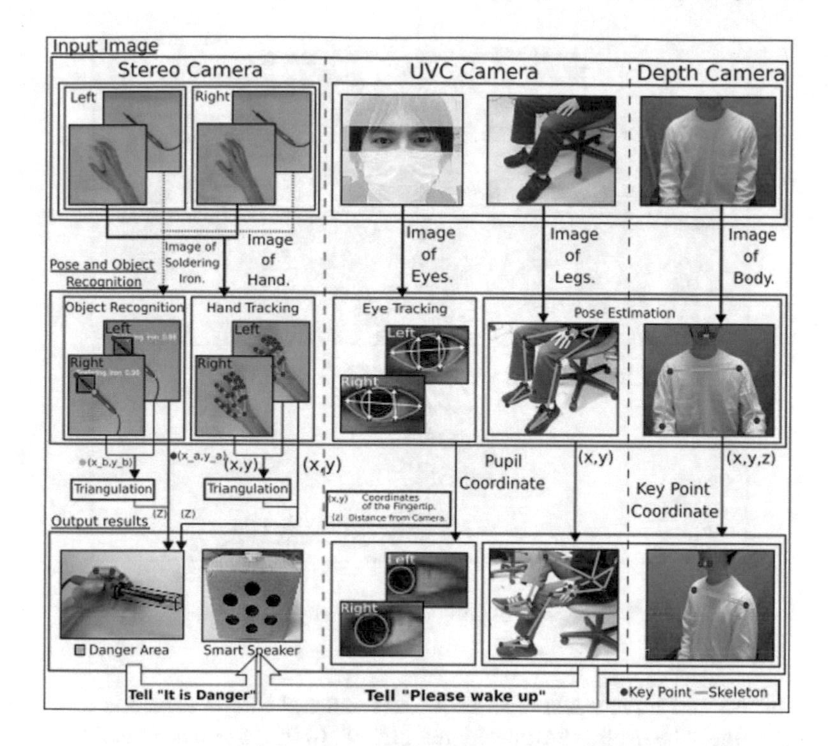

Fig. 1. Proposed system structure.

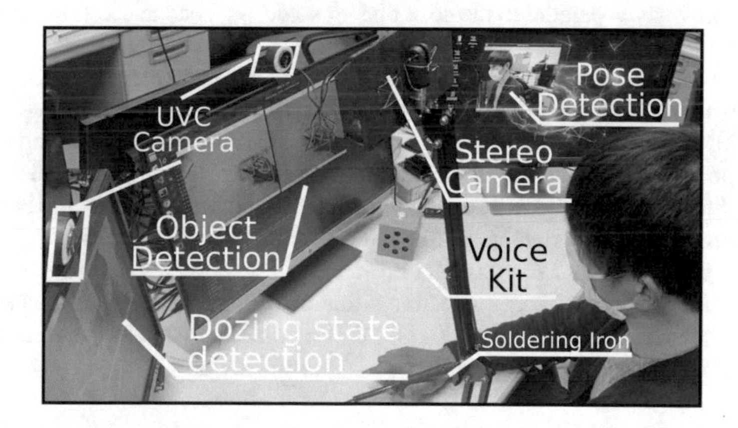

Fig. 2. System environment.

We use a front UVC camera and consider the 2 dimensional landmark coordinates of the left and right eyes for detecting dozing conditions. The left and right Eye Aspect Ratio (EAR) [27] can be decided by using vertical and horizontal Euclidean distances of 2 dimensional landmark coordinates. The eyelids are considered to be closed when

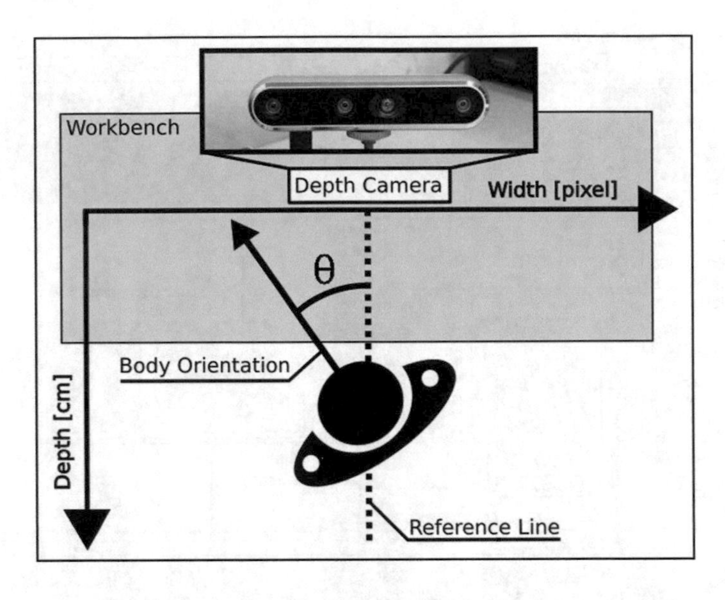

Fig. 3. Measuring of body orientation by depth camera.

the sum of left and right When the sum of left and right EAR valuse is less than 0.4 (threshold value), the eyelids are considered closed. In the case when eyelids are closed for less than 5 [*sec.*], they are considered awake. Finally, the dozing state is when the eyelids are closed for more than 5 [*sec.*].

When the body is parallel to the desk during soldering is considered a correct pose. But, when the body is tilted forward or tilted more than a certain angle to the left or right, it is considered a dangerous pose. The poses are estimated based on the key points (x, y) obtained from skeletal estimation using `MediaPipe` and the distance z from the depth camera [28] to each key point. The body orientation is estimated by deciding the angle from the perpendicular line to the line connecting the left and right shoulder, and the reference line shown in Fig. 3. The attention body orientation is decided based on the body orientation with a certain angle to the depth camera.

By using `Google AIY VoiceKit V2`, it is generated a voice output as shown in Table 1 in the case of a danger detection situation. For instance, when the user finger tip is close to dangerous area, `Google AIY VoiceKit V2` tells the soldering iron user "Please turn the body to the front".

3 Experimental Results

This section describes the experimental results. Figure 4, Fig. 5, Fig. 6 and Fig. 7 show the recognition patterns during soldering. Figure 4a, Fig. 5a, Fig. 6a and Fig. 7a show full color images. While Fig. 4b, Fig. 5b, Fig. 6b and Fig. 7b show depth images.

Table 1. Examples of voice output for danger detection.

Conditions	Content of Output
Angle between the reference line and the body is more than 30 [deg.]	"Please turn the body to the front."
Chin is lower than the shoulders	"Please correct your posture."
Eyelids closed for more than 5 [sec.]	"Please wake up."
Finger tip close to soldering iron dangerous area	"It's a dangerous way to hold it."
Soldering iron is away from hand	"Please hold the soldering iron."
Upper body facing sideways	"It's a dangerous posture."
Crossed legs	"Please put your legs down."
Eyes looking sideways	"Don't look away."

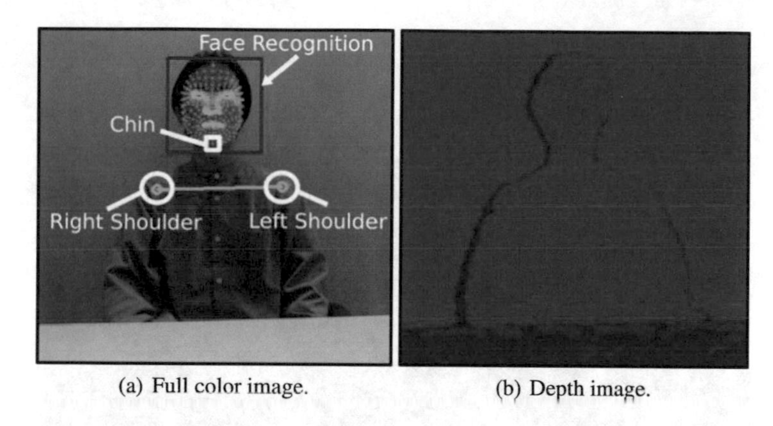

(a) Full color image. (b) Depth image.

Fig. 4. Correct body orientation and posture during soldering.

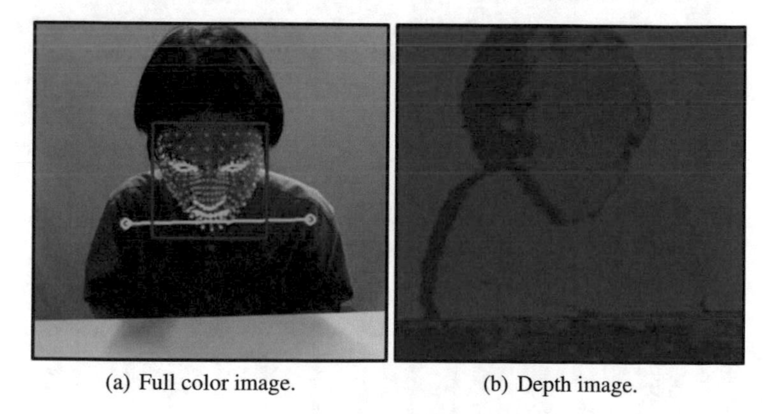

(a) Full color image. (b) Depth image.

Fig. 5. Incorrect posture during soldering.

In the Depth camera-based body direction estimation, the angle between the reference line and the body orientation is decided by the workbench as the front. The attention body orientation is defined as an angle between the reference line and the

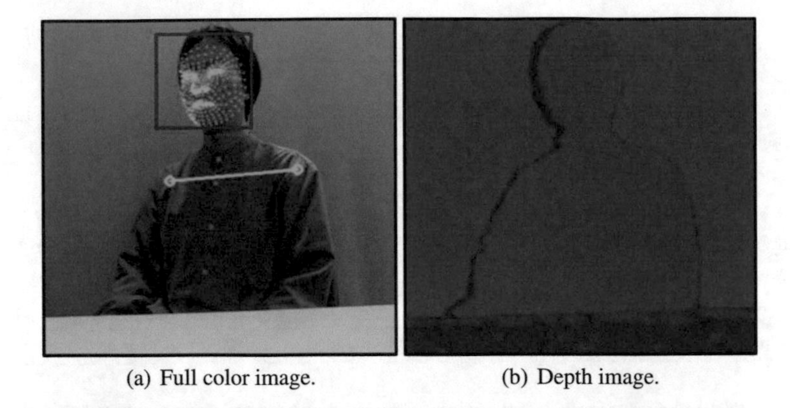

(a) Full color image. (b) Depth image.

Fig. 6. Case of body orientation 30 degrees to the right of the depth camera.

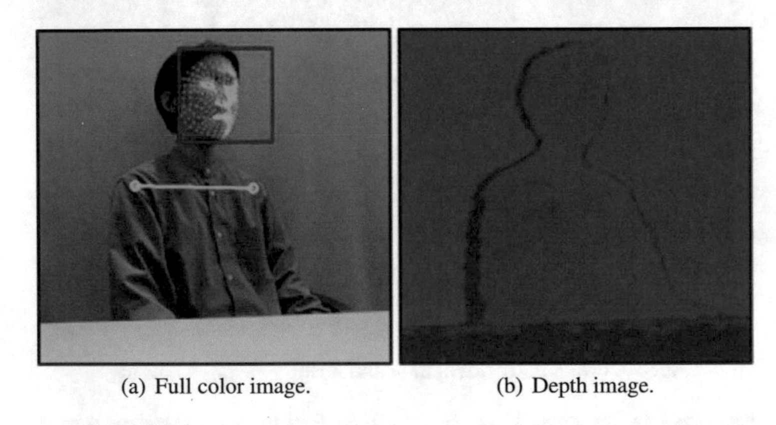

(a) Full color image. (b) Depth image.

Fig. 7. Case of body orientation 30 degrees to the left of the depth camera.

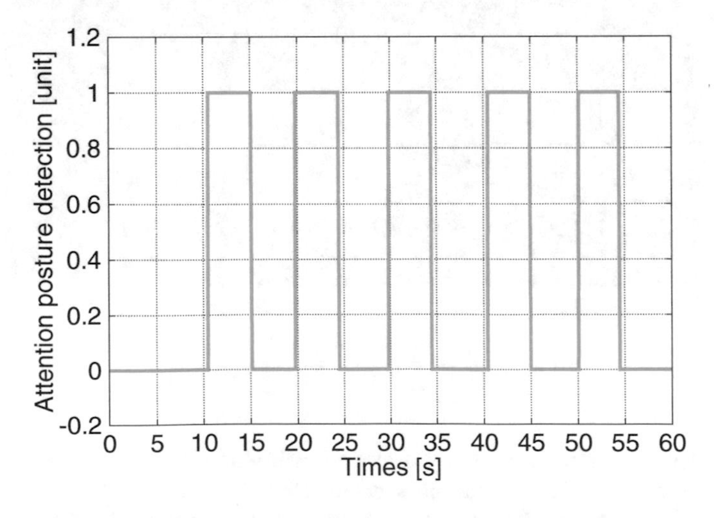

Fig. 8. Visualization results of detecting attention body orientation.

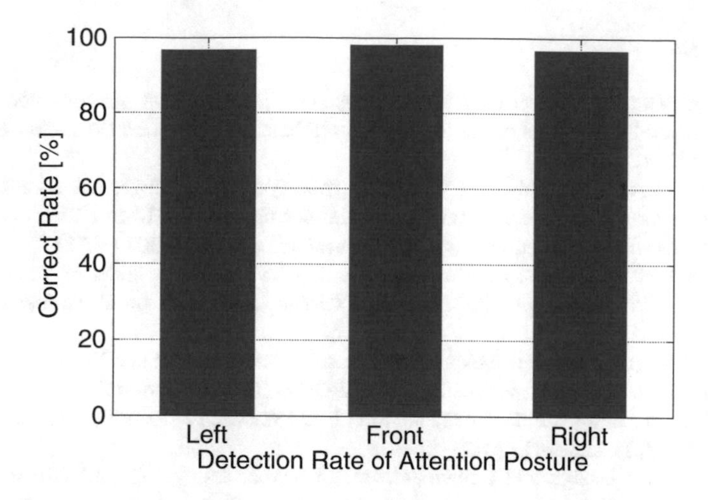

Fig. 9. Experimental results of upper body pose.

body orientation of 30 [*deg.*] or more. In the experiment, the body is turned to the front for 5 [*sec.*] and then turned to the left or right for 5 [*sec.*] more than 30 [*deg.*]. This process is repeated five times. The experimental time is 60 [*sec.*].

Figure 8 shows the visualization results during detection of the attention body orientation. It can be seen that the attention body orientation is detected at about 10 [*sec.*], 25 [*sec.*], 40 [*sec.*] and 55 [*sec.*]. This shows that an accurate attention body orientation is detected considering body orientation. Figure 9 shows the experimental results for each body orientation by depth camera. Also, for all three patterns of body orientation the accuracy is about 96 [%].

4 Conclusions

In this paper, we presented a depth camera based soldering motion analysis system for attention posture detection. We carried out many experiments using the implemented system. We presented and discussed experimental results of the posture detection according to the orientation of the body during soldering. The experimental results have shown that the proposed system can detect the posture from three-dimensional coordinates of the body and the accuracy for three patterns of body orientation is about 96 [%].

In the future, we would like to carry out extensive experiments by using the proposed system in different scenarios.

Acknowledgement. This work was supported by JSPS KAKENHI Grant Number JP20K19793.

References

1. Yasunaga, T., et al.: Object detection and pose estimation approaches for soldering danger detection. In: Proceedings of the IEEE 10-th Global Conference on Consumer Electronics, pp. 776–777 (2021)
2. Yasunaga, T., et al.: A soldering motion analysis system for danger detection considering object detection and attitude estimation. In: Proceedings of the 10-th International Conference on Emerging Internet, Data & Web Technologies, pp. 301–307 (2022)
3. Toyoshima, K., et al.: Proposal of a haptics and LSTM based soldering motion analysis system. In: Proceedings of The IEEE 10-th Global Conference on Consumer Electronics, pp. 1–2 (2021)
4. Hirota, Y., et al.: Proposal and experimental results of an ambient intelligence for training on soldering iron holding. In: Proceedings of BWCCA-2020, pp. 444–453 (2020)
5. Oda, T., et al.: Design and implementation of an IoT-based E-learning testbed. Int. J. Web Grid Serv. **13**(2), 228–241 (2017)
6. Liu, Y., et al.: Design and implementation of testbed using IoT and P2P technologies: improving reliability by a fuzzy-based approach. Int. J. Commun. Netw. Distrib. Syst. **19**(3), 312–337 (2017)
7. Papageorgiou, C., et al.: A general framework for object detection. In: The IEEE 6th International Conference on Computer Vision, pp. 555–562 (1998)
8. Felzenszwalb, P., et al.: Object detection with discriminatively trained part-based models. The IEEE Trans. Pattern Anal. Mach. Intell. **32**(9), 1627–1645 (2009)
9. Obukata, R., et al.: Design and evaluation of an ambient intelligence testbed for improving quality of life. Int. J. Space-Based Situated Comput. **7**(1), 8–15 (2017)
10. Oda, T., et al.: Design of a deep Q-Network based simulation system for actuation decision in ambient intelligence. In: Proceedings of AINA-2019, pp. 362–370 (2019)
11. Obukata, R. et al.: Performance evaluation of an am i testbed for improving QoL: evaluation using clustering approach considering distributed concurrent processing. In: Proceedings of IEEE AINA-2017, pp. 271–275 (2017)
12. Yamada, M., et al.: Evaluation of an IoT-based e-learning testbed: Performance of OLSR protocol in a NLoS environment and mean-shift clustering approach considering electroencephalogram data. Int. J. Web Inf. Syst. **13**(1), 2–13 (2017)
13. Toshev, A., Szegedy, C.: DeepPose: human pose estimation via deep neural networks. In: Proceedings of the 27-th IEEE/CVF Conference on Computer Vision and Pattern Recognition (IEEE/CVF CVPR-2014), pp. 1653–1660 (2014)
14. Haralick, R., et al.: Pose estimation from corresponding point data. The IEEE Trans. Syst. **19**(6), 1426–1446 (1989)
15. Fang, H., et al.: Rmpe: regional multi-person pose estimation. In: Proceedings of the IEEE International Conference on Computer Vision, pp. 2334–2343 (2017)
16. Xiao, B., et al.: Simple baselines for human pose estimation and tracking. In: Proceedings of the European Conference on Computer Vision (ECCV), pp. 466–481 (2018)
17. Martinez, J., et al.: A simple yet effective baseline for 3D human pose estimation. In: Proceedings of the IEEE International Conference on Computer Vision, pp. 2640–2649 (2017)
18. Redmon, J., et al.: You Only Look Once: unified, real-time object detection. In: Proceedings of The 29-th IEEE/CVF Conference on Computer Vision and Pattern Recognition (IEEE/CVF CVPR-2016), pp. 779–788 (2016)
19. Zhou, F., et al.: Safety helmet detection based on YOLOv5. In: The IEEE International Conference on Power Electronics, Computer Applications (ICPECA), pp. 6–11 (2021)
20. Yu-Chuan, B., et al.: Using improved YOLOv5s for defect detection of thermistor wire solder joints based on infrared thermography. In: The 5th International Conference on Automation, Control and Robots (ICACR), pp. 29–32 (2021)

21. Zhang, F., et al.: MediaPipe Hands: on-device real-time hand tracking. arXiv preprint arXiv:2006.10214 (2020)
22. Shin, J., et al.: American sign language alphabet recognition by extracting feature from hand pose estimation. Sensors **21**(17), 5856 (2021)
23. Hirota, Y., et al.: Proposal and experimental results of a DNN based real-time recognition method for ohsone style fingerspelling in static characters environment. In: Proceedings of the IEEE 9-th Global Conference on Consumer Electronics, pp. 476–477 (2020)
24. Erol, A., et al.: Vision-based hand pose estimation: a review. Computer Vision and Image Understanding, pp. 52–73 (2007)
25. Lugaresi, C., et al.: MediaPipe: a framework for building perception pipelines. arXiv preprint arXiv:1906.08172 (2019)
26. Antonio, M., et al.: Real-time upper body detection and 3D pose estimation in monoscopic images. In: European Conference on Computer Vision, pp. 139–150 (2006)
27. Soukupova, T., et al.: Real-Time eye blink detection using facial landmarks. Proceedings of the 21st computer vision winter workshop, Rimske Toplice, Slovenia (2016)
28. Andriyanov, N., et al.: Intelligent system for estimation of the spatial position of apples based on YOLOv3 and real sense depth camera D415. Symmetry 14(1) (2022)

Effect of Lighting of Metal Surface by Different Colors for an Intelligent Robotic Vision System

Chihiro Yukawa[1], Nobuki Saito[1], Aoto Hirata[1], Kyohei Toyoshima[1], Yuki Nagai[1], Tetsuya Oda[2(✉)], and Leonard Barolli[3]

[1] Graduate School of Engineering, Okayama University of Science (OUS), 1-1 Ridaicho, Kita-ku, Okayama 700-0005, Japan
{t22jm19st,t21jm01md,t21jm02zr,t22jm24jd,t22jm23rv}@ous.jp

[2] Department of Information and Computer Engineering, Okayama University of Science (OUS), 1-1 Ridaicho, Kita-ku, Okayama-shi 700-0005, Japan
{oda,ueda}@ice.ous.ac.jp

[3] Department of Informention and Communication Engineering, Fukuoka Insitute of Technology, 3-30-1 Wajiro-Higashi-ku, Fukuoka 811-0295, Japan
barolli@fit.ac.jp

Abstract. By recent advances in manufacturing industry will be achieved a higher level of operational productivity and automatization, which will improve the efficiency of production processes. Also, the measurement at nano-level on the target object surface by a machine or a robot has been considered for automation, but has a high cost and need a long time for measurement. In this paper, we present an intelligent robot vision system for recognizing micro-roughness of arbitrary surfaces. In order to increase the recognition rate of the proposed system, we implement a ring lighting to illuminate the arbitrary target with appropriate light. We consider different colors (red, green, blue and white) and create the dataset for lighting design of metal surface. We carried out experiments with different colors. The experimental results show that the lighting of metal surface by blue color has the highest recognition rate.

1 Introduction

The automation is a very important in manufacturing industry and the measurement of micro-roughness on the target surface by machines or robots is being considered for to improve the efficiency of production processes [1–8]. But the measurement has a high cost and it requires a long time. Also, most of the processes are done by craftsmen who have good skills and experience.

It is important to learn and transmit the skills of craftsmen and artisans to young people, but it takes a lot of time and experience and also young people are not interested to become artisans. Thus, it is difficult to hire young people, resulting in a chronic labor shortage. On the other hand, the skills of artisans can be technically reproduced by applying current machine learning and control technologies. Also, the robotization will become possible by implementing intelligent algorithms that can imitate the skills of artisans [9–12].

L. Barolli (Ed.): EIDWT 2023, LNDECT 161, pp. 350–356, 2023.
https://doi.org/10.1007/978-3-031-26281-4_37

In this paper, we present a robot vision system based on an intelligent algorithm for recognizing micro-roughness of arbitrary surfaces. In order to increase the recognition rate of the proposed system, we implement a ring lighting to illuminate the arbitrary target with appropriate light. We consider different colors (red, green, blue and white) and create the dataset for lighting design of metal surface. We carried out experiments with different colors. The experimental results show that the lighting of metal surface by blue color has the highest recognition rate.

The structure of the paper is as follows. In Sect. 2, we present the proposed system. In Sect. 3, we introduce the lighting design of the metal surface by different colors. In Sect. 4, we discuss the experimental results. Finally, conclusions and future work are given in Sect. 5.

2 Proposed System

In this section, we present the implemented robotic vision system for recognizing micro-roughness on the arbitrary target surface as shown in Fig 1. In the robot arm, in order to improve the recognition rate of micro-roughness, we propose a vibration suppression method based on fuzzy inference [13–17]. Also, we use Hill Climbing (HC) for the reduction of movement vibration of the robot arm.

The proposed system can process images from different angles with 4 degrees of freedom. Also, an electron microscope is mounted to the robot arm edge and the images received by the electron microscope are reconstructed by stitching. Then, the stitched images are sent to a Deep Learning Network (Conventional Neural Network: CNN) for object recognition.

2.1 Servo Motors Vibration Reduction Method

There are some problems with vibrations and low accuracy of movement of the robot arm. In a previous work, the authors use a training dataset for image recognition in order to improve the recognition rate by suppressing the movement vibration of the robot arm [18, 19].

In this research work, we propose a new method for reduction of vibration of the robot arm considering three modules: sensing module, fuzzy inference, and servomotor control to improve the recognition rate of micro-roughness.

The sensing module sends to the `Jetson` acceleration values received from the accelerometer `GY-521` which is mounted to the robot arm edge. Then the sensed data are sent to the controller via serial transmission.

The fuzzy inference considers two input parameters: error of acceleration and angle robot arm. We use Interval Type-2 Fuzzy Sets (IT2FS) [20, 21] and the output is decided by the Enhanced Iterative Algorithm with Stop Condition (EIASC) [22, 23]. The IT2FS output is the gain of servo motor.

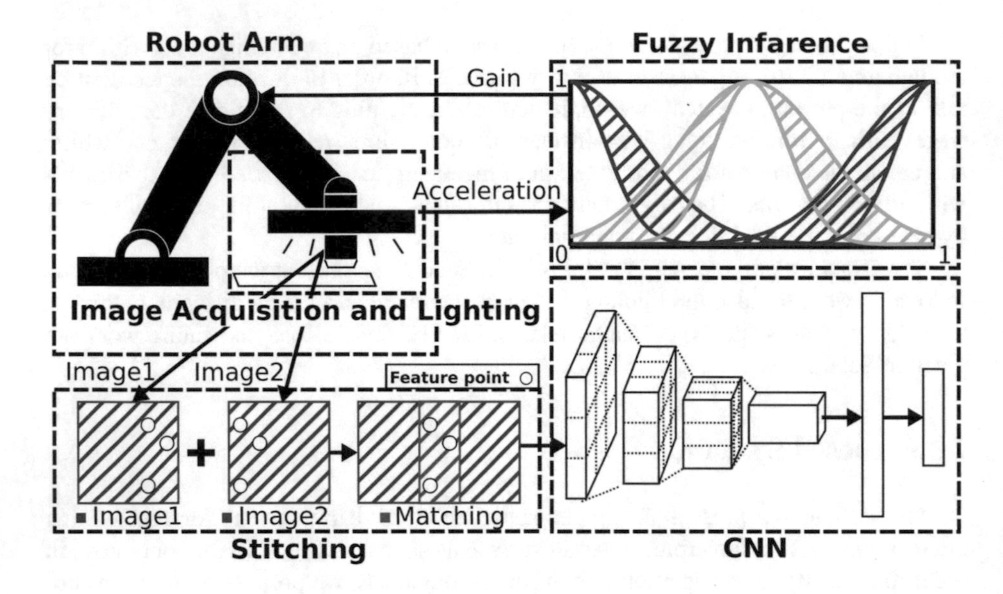

Fig. 1. Proposed system

The servo motor controls the angle based on the gain of fuzzy inference output and the gain is gradually decreased from the starting position to the specified angle to suppress the vibration.

2.2 Robot Arm Movement, Image Stitching and Object Detection

In order to optimize the robot arm movement, we implement a new method that determines the robot arm movement by using image acquisition points. We implement a HC-based method to reduce the movement vibration of robot arm.

In image stitching, by using an electron microscope the feature points are detected in multiple images and a single high-resolution image of the entire acquired object is obtained. Thus, the prediction of object detection time can be reduced.

In the proposed system, the micro-roughness on surface is recognize by `YOLOv5` and a deep learning model is used to perform object detection and class classification. In order to train the model are needed a large number of images containing micro-roughness, but unfortunately there are only few factories or production sites that deal with micro-roughness. For this reason, we created our dataset of images containing micro-roughness for training `YOLOv5`. We consider transfer learning approach [24,25], which can transfer a model already been trained for other problems. Thus, improving the recognition rate of micro-roughness and reduce the training time.

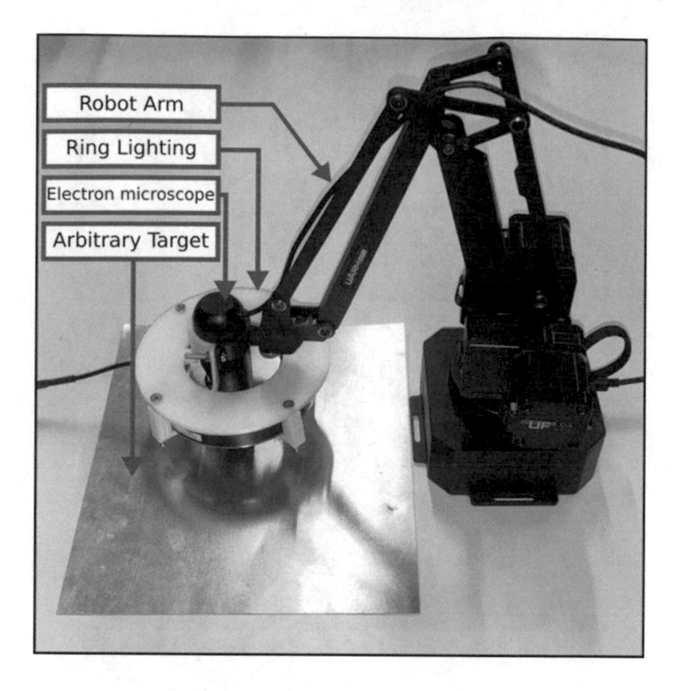

Fig. 2. Experimental environment.

3 Lighting Design

The implemented robot arm, ring lighting, electron microscope and arbitrary target are shown in Fig. 2. It important to design a suitable lighting of surfaces or objects in order to increase the recognition rate. For this reason, we use ring lighting to illuminate the arbitrary target with appropriate light. We consider different colors (red, green, blue and white) as shown in Fig. 3 and we created the dataset for lighting design of metal surface.

4 Experimental Results

We carried out experiments by using the experimental environment shown in Fig. 2. The experiments are performed on the metal surface using the dataset of four lighting colors shown in Fig. 3.

We used uArm Swift Pro Standard with 4 Degrees of Freedom as the robot arm, 5 MP Digital Microscope USB 2.0 as microscope and HPR2 as the ring lighting. The experimental results are shown in Fig. 4. From the experimental results, we found that recognition rate is different for different lighting colors and the lighting of metal surface by blue color has the highest recognition rate.

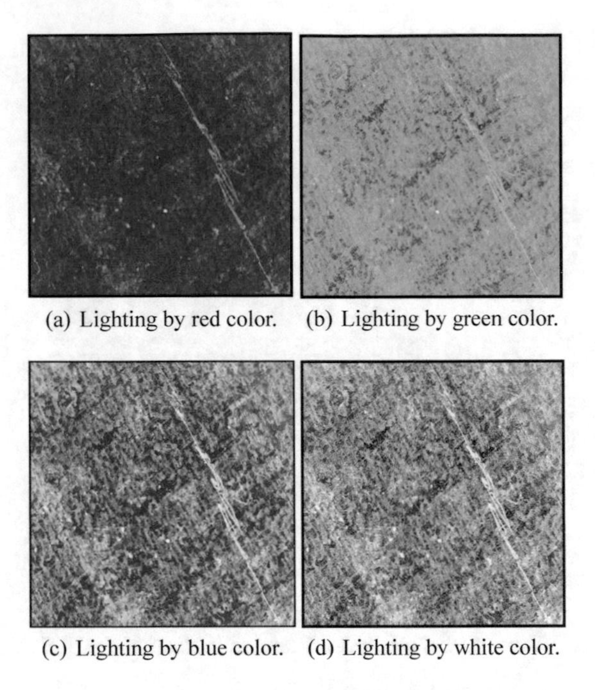

(a) Lighting by red color. (b) Lighting by green color.

(c) Lighting by blue color. (d) Lighting by white color.

Fig. 3. Lighting of metal surfaces by different colors.

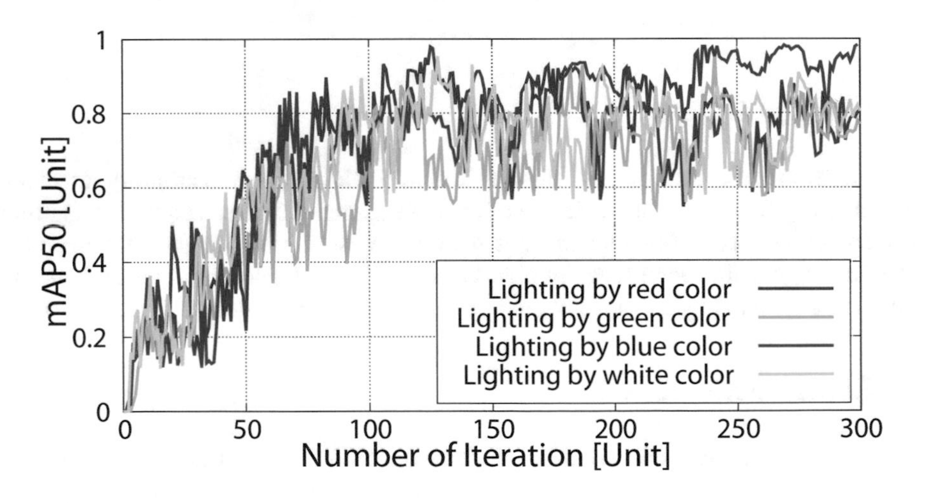

Fig. 4. Experimental result.

5 Conclusions

In this paper, we presented our implemented robot vision system. In order to increase the recognition rate of the proposed system, we implemented a ring lighting to illuminate the arbitrary target with appropriate light. We considered different colors (red, green, blue and white) and created the dataset for lighting design of metal surface. We carried out experiments with different colors. The experimental results show that the lighting of metal surface by blue color has the highest recognition rate.

In the future, we would like to increase the number of image acquisition points and carry out extensive experiments to evaluate the proposed system.

Acknowledgement. This work was supported by JSPS KAKENHI Grant Number 20K19793.

References

1. Dalenogare, L., et al.: The expected contribution of industry 4.0 technologies for industrial performance. Int. J. Prod. Econ. (IJPE-2018), **204**, 383–394 (2018)
2. Shang, L., et al.: Detection of rail surface defects based on CNN image recognition and classification. In: The IEEE 20th International Conference on Advanced Communication Technology (ICACT), pp. 45–51 (2018)
3. Li, J., et al.: Real-time detection of steel strip surface defects based on improved yolo detection network. IFAC-PapersOnLine **51**(21), 76–81 (2018)
4. Oda, T., et al.: Design and implementation of a simulation system based on deep Q-network for mobile actor node control in wireless sensor and actor networks. In: Proceedings of The IEEE 31st International Conference on Advanced Information Networking and Applications Workshops, pp. 195–200 (2017)
5. Saito, N., Oda, T., Hirata, A., Hirota, Y., Hirota, M., Katayama, K.: Design and implementation of a DQN based AAV. In: Barolli, L., Takizawa, M., Enokido, T., Chen, H.-C., Matsuo, K. (eds.) BWCCA 2020. LNNS, vol. 159, pp. 321–329. Springer, Cham (2021). https://doi.org/10.1007/978-3-030-61108-8_32
6. Saito, N., Oda, T., Hirata, A., Toyoshima, K., Hirota, M., Barolli, L.: Simulation results of a DQN based AAV testbed in corner environment: a comparison study for normal DQN and TLS-DQN. In: Barolli, L., Yim, K., Chen, H.-C. (eds.) IMIS 2021. LNNS, vol. 279, pp. 156–167. Springer, Cham (2022). https://doi.org/10.1007/978-3-030-79728-7_16
7. Saito, N., et al.: A Tabu list strategy based DQN for AAV mobility in indoor single-path environment: implementation and performance evaluation. Internet Things **14**, 100394 (2021)
8. Saito, N., Oda, T., Hirata, A., Yukawa, C., Kulla, E., Barolli, L.: A LiDAR based mobile area decision method for TLS-DQN: improving control for AAV mobility. In: Barolli, L. (ed.) 3PGCIC 2021. LNNS, vol. 343, pp. 30–42. Springer, Cham (2022). https://doi.org/10.1007/978-3-030-89899-1_4
9. Wang, H., et al.: Automatic illumination planning for robot vision inspection system. Neurocomputing **275**, 19–28 (2018)
10. Zuxiang, W., et al.: Design of safety capacitors quality inspection robot based on machine vision. In: 2017 First International Conference on Electronics Instrumentation & Information Systems (EIIS), pp. 1–4 (2017)
11. Li, J., et al.: Cognitive visual anomaly detection with constrained latent representations for industrial inspection robot. Appl. Soft Comput. **95**, 106539 (2020)
12. Ruiz-del-Solar, J., et al.: A survey on deep learning methods for robot vision. arXiv preprint arXiv:1803.10862 (2018)
13. Matsui, T., et al.: FPGA implementation of a fuzzy inference based quadrotor attitude control system. In: Proceedings of IEEE GCCE-2021, pp. 691–692 (2021)

14. Saito, N., et al.: Approach of fuzzy theory and hill climbing based recommender for schedule of life. In: Proceedings of LifeTech-2020, pp. 368–369 (2020)
15. Ozera, K., et al.: A fuzzy approach for secure clustering in MANETs: effects of distance parameter on system performance. In: Proceedings of IEEE WAINA-2017, pp. 251–258 (2017)
16. Elmazi, D., et al.: Selection of secure actors in wireless sensor and actor networks using fuzzy logic. In: Proceedings of BWCCA-2015, pp. 125–131 (2015)
17. Elmazi, D., et al.: Selection of rendezvous point in content centric networks using fuzzy logic. In: Proceedings of NBiS-2015, pp. 345–350 (2015)
18. Zaeh, M.F., et al.: Improvement of the machining accuracy of milling robots. Prod. Eng. Res. Devel. **8**(6), 737–744 (2014)
19. Yukawa, C., et al.: Design of a fuzzy inference based robot vision for CNN training image acquisition. In: Proceedings of IEEE GCCE-2020, pp. 871–872 (2021)
20. Liang, Q., et al.: Interval type-2 fuzzy logic systems: theory and design. IEEE Trans. Fuzzy Syst. **8**(5), 535–550 (2000)
21. Mendel, J.M.: Interval type-2 fuzzy logic systems made simple. IEEE Trans. Fuzzy Syst. **14**(6), 808–821 (2006)
22. Dongrui, W., et al.: Comparison and practical implementation of type-reduction algorithms for type-2 fuzzy sets and systems. In: 2011 IEEE International Conference on Fuzzy Systems (FUZZ-IEEE 2011), pp. 2131–2138 (2011)
23. Mendel, J.M.: On KM algorithms for solving type-2 fuzzy set problems. IEEE Trans. Fuzzy Syst. **21**(3), 426–446 (2012)
24. Yosinski, J., et al.: How transferable are features in deep neural networks?. arXiv preprint arXiv:1411.1792 (2014)
25. Zhuang, F., et al.: A comprehensive survey on transfer learning. Proc. IEEE **109**(1), 43–76 (2020)

A Design and Implementation of Dynamic Display Boards in a Virtual Pavilion Based on Unity3D

Zimin Li(✉) and Feng Pan

Engineering University of PAP, Xi'an, Shanxi, China
870221920@qq.com

Abstract. Based on the Unity3D development platform, this research designs and implements a dynamic display board in a virtual exhibition hall, which can be used in various virtual exhibition halls to display information in different formats. The use of this dynamic exhibition board can reduce the complex operation of some display function requirements, and can be better integrated into the virtual exhibition hall and become a part of it. Finally, the design content is run and tested. The results show that this research can help developers to simplify part of the development content, has good operability and repeatability, achieves the expected design goals, and can meet the design requirements.

1 Introduction

With the continuous development of computer technology, the virtual simulation design technology has also made great progress today, especially in the virtual exhibition hall, the virtual simulation design technology has been fully used. In order to realize the dynamic and updateable display of the content to be presented in the virtual pavilion, this research utilizes the integrated PhysX physics engine and the central processing unit-graphics processing unit (Central Processing Unit-Graphics Processing Unit, CPU-GPU) parallel computing capability The Unity3D platform, designed a virtual dynamic update display board based on Unity3D [1].

1.1 Reasons for Designing Dynamic Display Boards

Virtual pavilions have been developed by leaps and bounds in various fields. Compared with traditional physical exhibition halls, it has the characteristics of low development cost, less time and space constraints, and easy operation and use. Virtual exhibition hall technology is widely used by people in various fields. A traditional display board refers to a plate-like medium used for publishing and displaying information. Within the limited screen, the basic elements that make up the display board are as follows: text, lines, pictures and colors. Arrange and combine the content that needs to be highlighted according to the theme, and use the design principles of modeling elements to express the conceived design scheme in an intuitive form. The size of the picture that constitutes one of the basic elements of the display board has a great influence on the overall performance

L. Barolli (Ed.): EIDWT 2023, LNDECT 161, pp. 357–364, 2023.
https://doi.org/10.1007/978-3-031-26281-4_38

of the display board design, and will directly affect the communication of the display board design concept and the performance of the theme [2].

In the space of the virtual exhibition hall, the current mainstream display board method is still to design the traditional display board. Although the information to be conveyed can be expressed, the biggest problem is the irreplaceability of the content and the complicated and difficult operation. This research is to optimize the main problem above. The designed dynamic display board has achieved the expected function through testing [3, 4].

1.2 Our Contributions

The main contributions of the dynamic display board we designed can be summarized as follows:

Firstly, we designed a dynamic display board model that is easy to operate and implement, which simplifies the design process of a major function in the virtual pavilion.

Secondly, this dynamic display board design realizes the display of various types of information, making it more flexible and practical than traditional display boards.

Finally, we have successfully implemented the designed model in the Unity3D development platform, which can be used for various types of virtual exhibition halls in the Unity3D development platform. Although there are still imperfections, this design still proves to be feasible.

2 Features of Dynamic Display Boards

The dynamic exhibition board can be compatible to display various types of information, has a convenient and operable information replacement method, can adjust the size of the exhibition board, and can also change the rotation frequency of the displayed information. Copying some components and arranging display panels in any location and direction in the exhibition hall has strong operability and flexibility [5, 6] (Fig. 1).

2.1 Picture Display Function Introduction

On the 'Game'interface, users can display images in JPEG format by clicking the 'Picture Mode' button next to the dynamic display boards. After the image information is imported into the 'Assets'folder, you can arbitrarily select the image to be displayed in Unity3D, and adapt the image size to the size of the display board. Users can change the speed of the image carousel in the Visual Studio software interface of the MainScene script. Each display board can display any number of pictures in rotation according to demand.

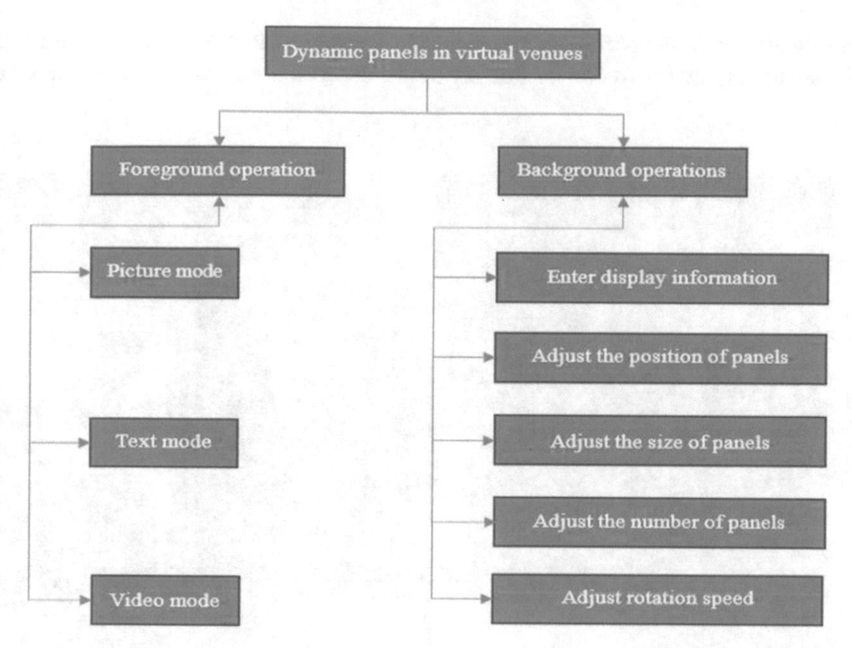

Fig. 1. This picture shows the functional structure design ideas of the dynamic display boards.

2.2 Text Display Function Introduction

On the 'Game'interface, users can display text in TXT format by clicking the 'Text Mode' button next to the dynamic display board. In the text interface, users can adjust the content, font, font, font size, color, alignment, line spacing, etc. of the text to be displayed. In the Visual Studio software interface of the MainText script, users can increase or decrease the number of text paragraphs to be displayed in rotation and their rotation speed. Similarly, the number of text paragraphs that need to be displayed in rotation can also be set to multiple.

2.3 Video Display Function Introduction

On the 'Game'interface, users can display videos in MP4 format by clicking the 'Video Mode' button next to the dynamic display board. After the video information is imported into the 'Assets'folder, you can choose the video you want to display in Unity3D. In the 'Video Player' interface, users can arbitrarily adjust the video playback speed, and can also perform some common operations such as setting the video to loop playback.

2.4 Introduction to the Multi-point Layout of Dynamic Display Boards

When there are multiple display panels that need to be arranged in the venue, you can quickly meet the requirements by duplicating the components that have been produced. As shown in the Fig. 2, on the left side of the large display board, by duplicating the display board and adjusting the size of the display board, two small display boards can be placed to complement each other to highlight the content to be displayed. The content of the small display board can be flexibly set in the format of pictures, texts or videos [7–9].

Fig. 2. This picture shows the effect of arranging dynamic exhibition boards of different sizes at multiple points.

3 Dynamic Display Board Structure

3.1 MainScene Class

The picture display implementation class is mainly responsible for displaying all the picture information to be displayed on the exhibition board in a certain frequency. You can change the rotation frequency by adjusting the parameters in 'WaitForSeconds()', and use the characteristics of the array to expand the number of displayed pictures [10–12].

3.2 MainText Class

The text display implementation class is mainly responsible for displaying the text information to be displayed on the display board at a certain frequency. Different from the picture display, adding the rotating display content requires adding an array in the code, and the last accessed array should point to the first accessed array, otherwise the text display content will not be displayed in a cycle.

3.3 MainVideo Class

The video display implementation class is mainly responsible for displaying the video information to be displayed on the dynamic exhibition board. The adjustment of the video content needs to be done before importing into the background. The dynamic display board is only responsible for displaying the content here, and does not have other playback options.

3.4 UIControl Class

User graphical interface control class. Various components for creating user interface control objects, such as buttons, static text boxes, pop-up menus, etc., and specifying callback functions for these components.

4 Dynamic Display Boards Operation Status Display

4.1 Operational Display in Picture Mode

The picture display interface is to realize the function of displaying a group of pictures at a certain rotation speed on the dynamic exhibition board. In the Picture Mode, the number of pictures that the user can import is almost unlimited, and the user can independently adjust the rotation speed of the picture display. As long as the picture resource type is adjusted to JPEG format, it can be displayed in the dynamic exhibition board. The picture display interface is shown in the figure (Fig. 3).

Fig. 3. This picture shows the display effect of the dynamic exhibition board in Picture Mode.

4.2 Operational Display in Text Mode

The text display interface is to realize the function of displaying multiple sets of text in different forms on the dynamic display board in rotation. In the text mode, the user can adjust the content of the text, the size of the text, the font of the text, the spacing of the text and the color of the text, etc., and the change of text information is relatively easy to achieve, only the number of text groups needs to be increased or decreased in the Visual Studio software. Interface operation, other changes have relatively strong operability. The text display interface is shown in the figure (Fig. 4).

Fig. 4. This picture shows the display effect of the dynamic exhibition board in Text Mode.

4.3 Operational Display in Video Mode

The video display interface is to realize the function of playing imported video information on the dynamic display board. In the video mode, the dynamic display board is more like a player with simple functions, which can only play the pre-made video information, and does not retain other complex playback functions. The adjustment and activation of the video information needs to be completed by the user before importing the video information into the dynamic display board. The video display interface is shown in the figure (Fig. 5).

Fig. 5. This picture shows the display effect of the dynamic exhibition board in Video Mode.

5 Conclusion

This research greatly improves the efficiency of information display and interaction, and users can display information in multiple formats on the same virtual display board. Information exhibition virtual venues can use this research to save development costs, save development resources, and display as much information as possible in a relatively small space designed, and the information replacement operation of this virtual exhibition board is simple and easy to understand, and does not require users to master Too complex knowledge of computer programming and operation.

Acknowledgments. The author would like to thank Dr. Fuqiang Di for his pertinent suggestions and careful guidance on structural framework design and requirements analysis implementation.

References

1. Chen, X., Lu, W., Li, M.: Talking about the modeling technology in Web3D. Computer Engineering Applied Technology, 5082–5083 (2012)
2. Ni, L., Qi, P., Yu, L., Wang, J.: Research and application of Unity3D product virtual display technology. Digital technologies and applications, 54–55 (2010)
3. Zhu, Z.: Design and Application of Virtual Experiment System Based on Unity3D, 32–54 (2012)
4. Xiong, Y.: Research on 3D film and television special effects development based on Unity3D particle system. Software Guide, 134–136 (2011)
5. Jiang, W.: User Interface Design Research, 24–35 (2016)
6. Liu, Z., Jiang, Y., Ding, H.: Virtual Reality Production and Development, 42–56 (2012)
7. Johnson, J.: Cognition and Design - Understanding UI Design Guidelines, 33–37 (2011)

8. Xu, Y., Yang, J., Zhang, A.: Virtools Virtual Interactive Design Example Analysis, 23–32 (2012)
9. Hu, X., Yu, T.: Distributed Virtual Reality Technology, 52–62 (2012)
10. Ou, Y., Li, Q., Lu, X.: Research and Implementation of Virtual Campus Development Based on Unity3D. Modern Electronic Technology, 19–22 (2013)
11. Zhu, H.: A Virtual Tour System Based on Unity3D. Computer System Applications, 36–39 (2013)
12. Cai, Y.: Research and Implementation of Virtual Tourism System of JinSiXia Geopark Based on Unity3D, 33–39 (2012)

A Comparative Study of Several Spatial Domain Image Denoising Algorithm

Rui Deng(✉), Yanli Fu, and Shuyao Li

Engineering University of PAP, Shaanxi, China
dengrui_0120@163.com

Abstract. With the development of science and technology, images have become an essential source for people to obtain external information, but they are easily disturbed by noise in the process of image transmission, which degrades image quality. In this paper, the advantages and disadvantages of the mean filter, median filter, Gaussian filter, and bilateral filter in removing pepper noise and Gaussian noise are investigated through qualitative and quantitative analyses using MATLAB software. The experimental results show that for images containing pepper noise, median filtering has the best denoising effect while retaining the original image information to the greatest extent; for images containing Gaussian noise, mean filtering and Gaussian filtering have better denoising effects; bilateral filtering, although not the best denoising effect, has better retention of edge information.

1 Introduction

Images have a unique insight into the information they convey and are used everywhere, such as photographs, maps, faxes, and satellite cloud maps [1]. However, during transmission and compression, images can be affected by the external environment, such as lighting and motion, which can deteriorate the quality of the image and make it difficult to distinguish, making it impossible to analyze and process the image subsequently. Image denoising aims to remove the effects of this noise and improve the quality of the image so that it can be better processed for the next step of image processing. Therefore, in-depth research into image-denoising techniques to reduce the impact of image noise on image quality so that it can both reduce noise and retain more useful detail information so that the transmitted image is close to the original image is crucial in the image processing process. At present, there has been a lot of research into image-denoising methods. The author selects several classical spatial domain image-denoising methods and compares and analyses their advantages and disadvantages to provide a reference for future image-denoising research [2].

2 Image Quality Evaluation

When transmitting image information, the image goes through a series of processes such as denoising, compression, recording, transmission, and enhancement. The effectiveness of this process depends on the final evaluation of the image quality by the user. Therefore, evaluating the image quality is critical for us to judge the denoising effect of the denoising algorithm [3].

L. Barolli (Ed.): EIDWT 2023, LNDECT 161, pp. 365–374, 2023.
https://doi.org/10.1007/978-3-031-26281-4_39

2.1 Subjective Evaluation of Image Quality

Observers' subjective assessment of the denoised images is based on the quality of the denoised images as observed by the naked eye and judged by different people and then based on each person's feedback. Although subjective evaluation is not reliable due to the various subjective judgments of people, it is the most common and reliable evaluation to improve image quality.

2.2 Objective Evaluation of Image Quality

There are various methods for objective evaluation of image quality. Still, this paper adopts the most authoritative and widely used criteria in objective evaluation, namely: peak signal-to-noise ratio (*PSNR*) and mean square error (*MSE*), which are calculated as follows: assume that $f(i,j)$ is the standard image of $M \times N$, $g(i,j)$ is the processed image, and MSE is the mean squared error. The $f_{\max}$ and $f_{\min}$ in Eq. (1) represent the image's maximum and minimum grey values, which are usually taken as 255 and 0 in commonly used 8bit greyscale images. The formula for the peak denoising ratio is Eq. (1), and the formula for the mean squared error is Eq. (2).

$$PSNR = 10log_{10}\frac{(f_{max} - f_{min})^2}{MSE} \tag{1}$$

$$MSE = \frac{1}{M \times N}\sum_{x=0}^{M-1}\sum_{y=0}^{N-1}\left[g(i,j) - f(i,j)\right]^2 \tag{2}$$

3 Spatial Domain Filtering Image Denoising Algorithm

3.1 Mean Filtering

It is a low-pass filter that removes high-frequency signals from an image. The basic principle of mean filtering is as follows: Frist, select a mask matrix for the image's pixel points $f(x,y)$, which combines with several adjacent elements according to the demand. Then, find the average of all values in the mask matrix, which is the value of each pixel point $f(x,y)$, so the grey scale value $g(x,y)$ of the image after the mean filtering process is the value of the newly generated pixel point.

As shown in Eq. (3), M is the total number of all pixel points in the mask matrix.

$$g(x, y) = \frac{1}{M}\sum f(x, y) \tag{3}$$

3.2 Median Filtering

Median filtering protects edge detail and removes noise when denoising an image [4]. It has the advantage that the algorithm is straightforward, easy to implement, does not require the statistical properties of the image, and is convenient and fast. The basic principle is that first, select an image window, at the same time, arrange the corresponding pixel values in the image window in order of size, and finally choose the median of the pixel points as the new pixel value $y(i,j)$ [5]. This aim is to keep the pixel values close to the neighboring central points close to the pixel values of the real points, thus removing more isolated noise points in the image. The formula for median filtering is Eq. (4), and $f(i,j)$ represents a two-dimensional data sequence of image pixel points [6].

$$y(i,j) = Med\{f(i,j)\} \tag{4}$$

3.3 Gaussian Filtering

Gaussian filtering is a linear smoothing filter, similar to mean filtering, and Gaussian filtering is a process of weighted averaging of the whole image, where each pixel value is obtained by a weighted average of its value and the values of other pixels in its neighborhood. The difference is that mean filtering gives equal weight to each pixel in its neighborhood [7]; on the other hand, Gaussian filtering increases the weight of the central point and decreases the weight of the points away from the central point, on the basis of which calculating the sum of the different weights of each pixel value in the neighborhood [8].

3.4 Bilateral Filtering

Compared with Gaussian filtering, bilateral filtering has two mask matrices, the spatial domain matrix and the value domain matrix [9]. The spatial domain matrix is similar to Gaussian filtering and can blur and denoise the image; the value domain matrix can protect the image edges according to the pixel brightness value of each point [10].

The spatial domain matrice is Eq. (5), and the value domain matrice is Eq. (6). The bilateral filter weight matrix is equal to the product of the spatial domain matrix and the value domain matrix, as Eq. (7). Replace the coordinate value of the center point with the weighted average calculated after denoising, as Eq. (8).$g(i,j)$ denotes the output point, s is the $(2N+1)(2N+1)$ centered on (i,j)size range, $f(k,l)$ refers to the image containing multiple output points, and $w(i,j,k,l)$ represents the final bilateral filtering weight matrix.

$$w_s(i,j,k,l) = e^{-\frac{(i-k)^2+(j-l)^2}{2\sigma_s^2}} \tag{5}$$

$$w_r(i,j,k,l) = e^{\frac{-f(k,l)-f(i,j)^2}{2\sigma_r^2}} \tag{6}$$

$$w_r(i,j,k,l) = e^{-\frac{(i-k)^2+(j-l)^2}{2\sigma_s^2}+\frac{-f(k,l)-f(i,j)^2}{2\sigma_r^2}} \tag{7}$$

$$g(i,j) = \frac{\sum_{(k,l)\in s} f(k,l)w(i,j,k,l)}{\sum_{(k,l)\in s} w(i,j,k,l)} \tag{8}$$

4 Simulation Experiments and Analysis

This chapter will use MATLAB software to simulate the denoising process of two images containing Pepper or Gaussian noise using these spatial domain filtering algorithms. At the same time, make a comprehensive evaluation of the processing results and comprehensively analyze the advantages and disadvantages of these spatial domain filtering algorithms [11].

4.1 Simulation Experiments and Analysis of Spatial Domain Denoising Algorithms Containing Pepper Noise

Qualitative Analysis. In order to investigate the effect of spatial domain denoising algorithms on the denoising of pepper noise, the experiments were conducted by first adding a density of 0.05 pepper noise to the original image, then applying four spatial domain denoising algorithms, namely mean filter, median filter, Gaussian filter, and bilateral filter, to denoise the images containing pepper noise, and finally conducting qualitative and quantitative analyses on the processed images.

In order to avoid the singularity of the judgment results and increase the contrast of the experiment, choose Beauty, Hill, and Cat as the original images for the experiment. Beauty image simulation results, as shown in Fig. 1; Hill image simulation results, as shown in Fig. 2; Cat image simulation results, as shown in Fig. 3.

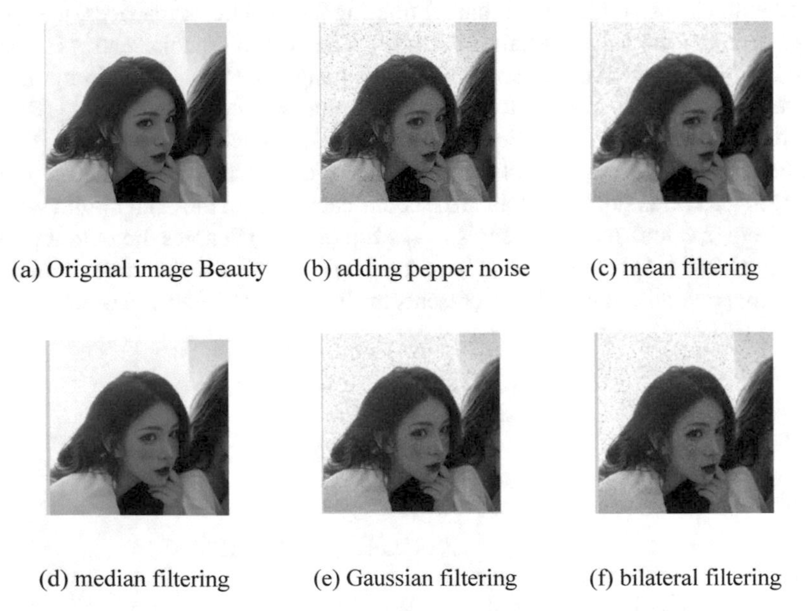

(a) Original image Beauty (b) adding pepper noise (c) mean filtering

(d) median filtering (e) Gaussian filtering (f) bilateral filtering

Fig. 1. Simulation results of the spatial domain denoising algorithm with pepper noise (Beauty image)

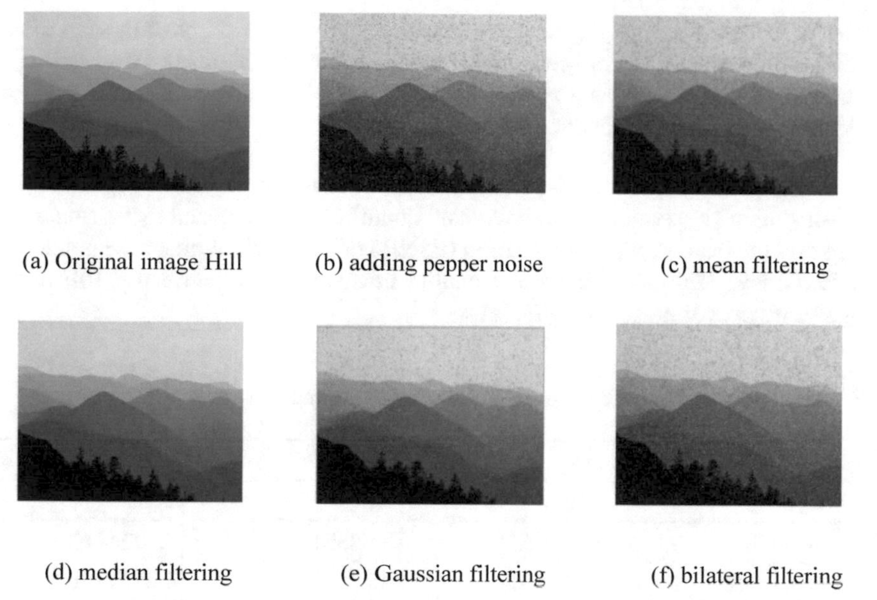

Fig. 2. Simulation results of the spatial domain denoising algorithm with pepper noise (Hill image)

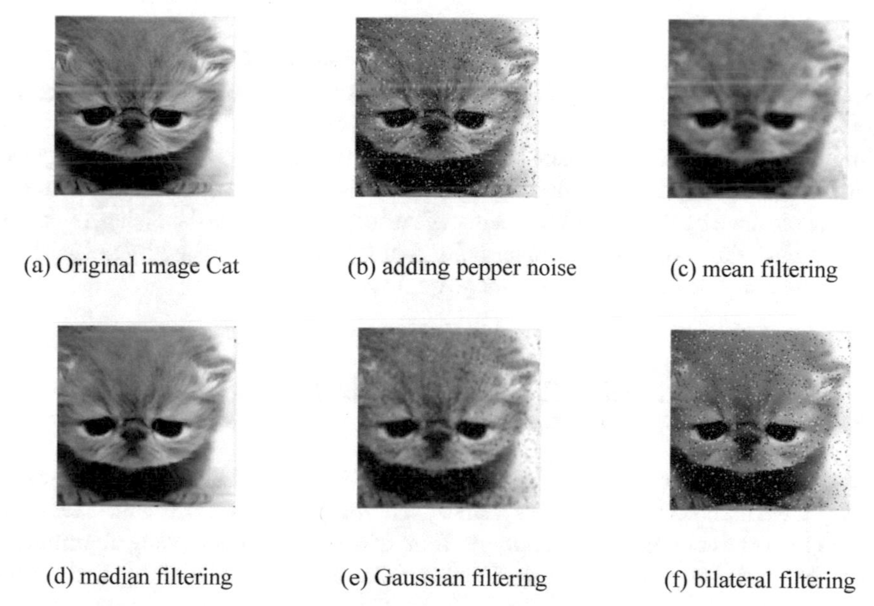

Fig. 3. Simulation results of the spatial domain denoising algorithm with pepper noise (Cat image)

Figure 1, 2 and 3 show that after adding pepper noise, the original image contains many black-and-white noise points. After the denoising process of the four spatial domain filtering algorithms: mean filtering, median filtering, Gaussian filtering, and bilateral filtering, the image quality has been improved. The effect of median filtering is more obvious, and the image is clearly visible after denoising; mean filtering and Gaussian filtering are the next most effective; bilateral filtering is less effective, but it protects edge information better.

Quantitative analysis. The advantages and disadvantages of various spatial-domain denoising algorithms can be objectively and quantitatively analyzed by calculating and comparing the peak signal-to-noise ratio (PSNR) of the original image with that of the denoised image. The calculated peak signal-to-noise ratios for the Beauty, Hill, and Cat images with pepper noise are as Table1.

Table 1. Output peak signal-to-noise ratio for images with pepper noise

	Image		
Filter type	Beauty	Hill	Cat
Mean filtering	32.6230 db	34.8654 db	32.2244 db
Median filtering	40.2587 db	45.5008 db	38.9549 db
Gaussian filtering	31.5518 db	32.7322 db	31.7274 db
Bilateral filtering	22.2264 db	22.7791 db	22.1717 db

A comparison of the data in Table1 shows that the signal-to-noise ratios of the four spatial domain filtering algorithms are consistent with the observed experimental phenomena when adding pepper noise to the different pictures.

Comprehensive qualitative analysis and quantitative analysis, when adding pepper noise into the image, the median filtering should be selected in the denoising algorithm, which is determined by the principle of median filtering, which, when denoising, replaces the pixel point with a surrounding pixel point that is not contaminated by noise, thus removing the noise and improving the image quality.

4.2 Simulation Experiments and Analysis of Spatial Domain Denoising Algorithms with Gaussian Noise

Qualitative analysis. In order to investigate the denoising effect of the spatial domain denoising algorithm on Gaussian noise, firstly, adding Gaussian noise with a density of 0.01 to the original image, then applying four spatial domain denoising algorithms, namely mean filtering, median filtering, Gaussian filtering, bilateral filtering, to denoise the image.

In order to make the experimental results more comparable and reduce the chance of the experiment, choose Beauty, Hill, and Cat as the original images for the experiment. Beauty image simulation results, as shown in Fig. 4; Hill image simulation results, as shown in Fig. 5; and Cat image simulation results, as shown in Fig. 6.

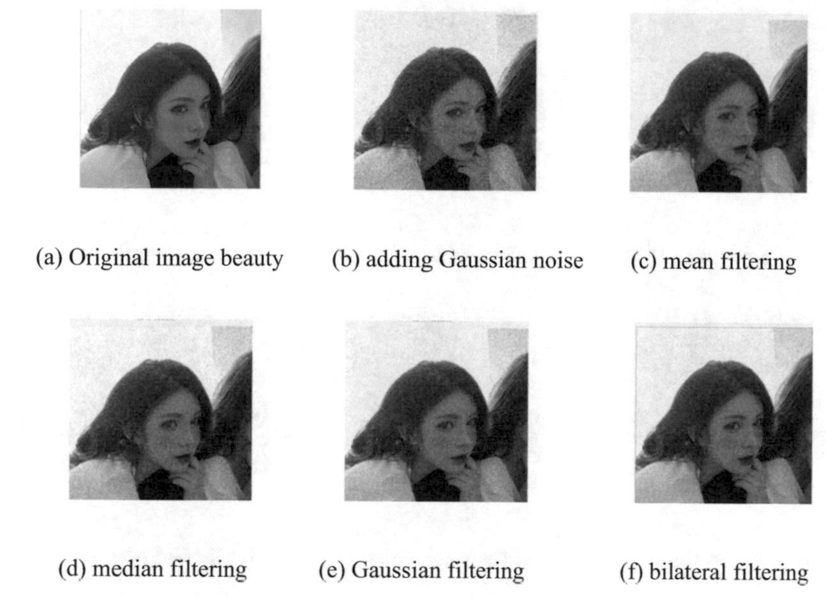

(a) Original image beauty (b) adding Gaussian noise (c) mean filtering

(d) median filtering (e) Gaussian filtering (f) bilateral filtering

Fig. 4. Simulation results of the spatial domain denoising algorithm with Gaussian noise (Beauty image)

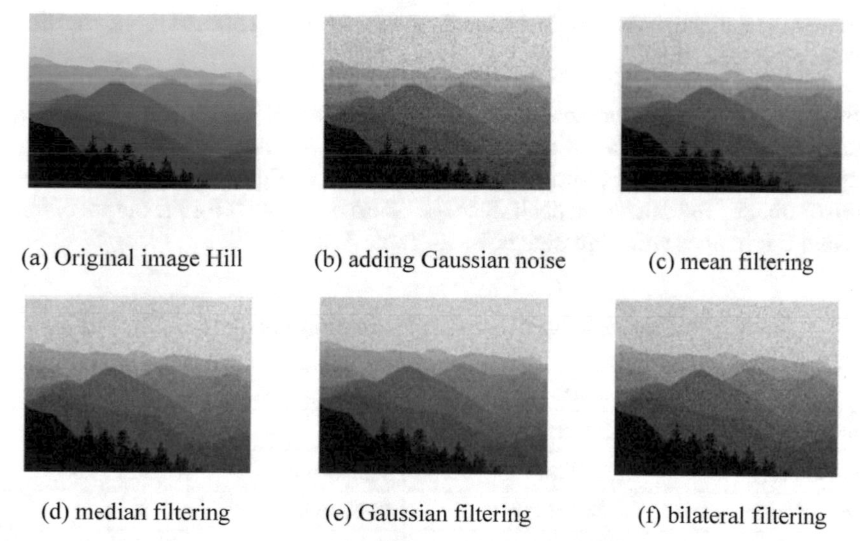

(a) Original image Hill (b) adding Gaussian noise (c) mean filtering

(d) median filtering (e) Gaussian filtering (f) bilateral filtering

Fig. 5. Simulation results of the spatial domain denoising algorithm with Gaussian noise (Hill image)

Figure 4, 5, and 6 show that after adding Gaussian noise, the original image contains many blurred points. After the denoising process of the four spatial domain filtering algorithms: mean filtering, median filtering, Gaussian filtering, and bilateral filtering, the image quality has been improved. The mean and Gaussian filters are more effective

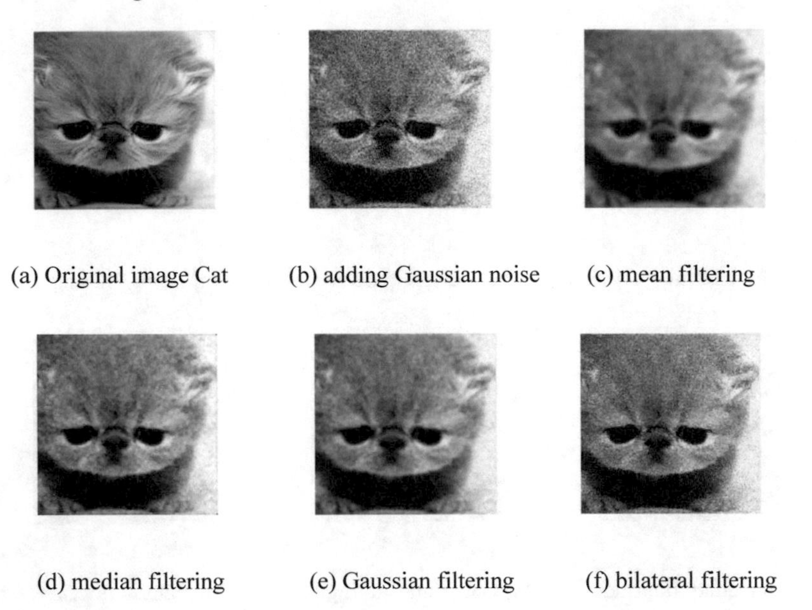

(a) Original image Cat (b) adding Gaussian noise (c) mean filtering

(d) median filtering (e) Gaussian filtering (f) bilateral filtering

Fig. 6. Simulation results of the spatial domain denoising algorithm with Gaussian noise (Cat image)

in removing Gaussian noise, while the median is the next most effective, and the bilateral filters are less effective.

Quantitative analysis. The advantages and disadvantages of various spatial-domain denoising algorithms can be objectively and quantitatively analyzed by calculating and comparing the peak signal-to-noise ratio (PSNR) of the original image with that of the denoised image. The calculated peak signal-to-noise ratio values for the output of beauty, Hill, and Cat images with Gaussian noise as Table 2.

Table 2. Output peak signal-to-noise ratio for images with Gaussian noise

	Image		
Filter type	Beauty	Hill	Cat
Mean filtering	33.8303 db	37.2104 db	31.9121 db
Median filtering	32.1911 db	32.3746 db	30.8354 db
Gaussian filtering	33.7055 db	35.3057 db	32.5278 db
Bilateral filtering	28.6054 db	28.3378 db	27.7535 db

A comparison of the data in Table2 shows that when adding Gaussian noise to different images, mean filtering and Gaussian filtering are more effective, with median filtering coming second and bilateral filtering removing noise insignificantly. Thus, combining the qualitative and quantitative analyses, the peak signal-to-noise ratios of the four spatial domain filtering are consistent with the observed experimental phenomena when adding Gaussian noise to the images.

By comparing the results of the above two sets of simulation experiments, qualitative analysis, and quantitative analysis of the images, it shows that, for the pepper noise, it can be visually seen that the median filtered denoised image is nearly indistinguishable from the original image and has the best denoising effect, and the output peak signal-to-noise ratio of the median filter is the highest and its stability is the strongest, which can better adapt to different noises; for Gaussian noise, the mean filter, and Gaussian filter have better denoising effect. Bilateral filtering is better for edge protection of the image, but bilateral filtering removes the valuable high-frequency detail part in removing noise, which will make the image poorly defined.

5 Conclusion

This paper focuses on images containing pepper noise and Gaussian noise and uses four spatial domain denoising algorithms, namely mean filtering, median filtering, Gaussian filtering, and bilateral filtering, to denoise the images by MATLAB software. A subjective qualitative analysis of the denoising results and a quantitative analysis of the signal-to-noise ratio lead to the following conclusions: for images containing pepper noise, median filtering has the best denoising effect while retaining the original image information to the greatest extent; for images containing Gaussian noise, mean filtering and Gaussian filtering have better denoising effects; bilateral filtering, although not the best denoising effect, has better retention of edge information. Through the comparative study of several spatial domain image-denoising algorithms, their advantages and disadvantages are analyzed, providing data and a theoretical basis for future research on image-denoising methods.

References

1. Makoto, S., et al.: Improved iterative reconstruction method for compton imaging using median filter. PloS ONE **15**(3), 22–36 (2020)
2. Rajeshkannan, R., Punith, N.S., Ashraf, A.K., Gautham, S.: Image denoising using a combination of spatial domain filters and convolutional neural networks. In. J. Recent Technol. Eng. (IJRTE) **7**(6), 1836–1841 (2019)
3. Rajabi, M., Hasanzadeh, R.P.R.: A modified adaptive hysteresis smoothing approach for image denoising based on spatial domain redundancy. Sens. Imaging **22**(1), 42 (2021)
4. Wang, Y.B., Huang, H.L.: Image denoising based on adaptive sector rotation median filter. J. Phys. Conf. Ser. **1769**(1), 012056 (2021)
5. Li, N., Liu, T., Li, H.: An improved adaptive median filtering algorithm for radar image co-channel interference suppression. Sensors **22**(19), 7573 (2022)
6. Yuan, S., Qing, X., Hang, B., Qu, H.: Quantum color image median filtering in the spatial domain: theory and experiment. Quantum Inf. Process. **21**(9), 321 (2022). https://doi.org/10.1007/s11128-022-03660-0
7. Tang, K., Zhou, X.: Evolution algorithm of parametric active contour model based on Gaussian smoothing filter. Mach. Vis. Appl. **33**(6), 83 (2022)

8. Piao, W., Yuan, Y., Lin, H.: A digital image denoising algorithm based on gaussian filtering and bilateral filtering. ITM Web Conf. **17**, 01006 (2018)
9. Jin, L., Xiong, C., Liu, H.: Improved bilateral filter for suppressing mixed noise in color images. Dig. Signal Process. **22**(6), 903–912 (2012). https://doi.org/10.1016/j.dsp.2012.06.012
10. Babu, Md.S.I., Ping, P.: Bilateral filter based image denoising. Int. J. Comput. Commun. **11**, 34–37 (2017)
11. Huang, Z.-K., Li, Z.-H., Huang, H., Li, Z.-B., Hou, L.-Y.: Comparison of different image denoising algorithms for Chinese calligraphy images. Neurocomput **188**, 102–112 (2016). https://doi.org/10.1016/j.neucom.2014.11.106

A Pedestrian Avoidance System for Visual Impaired People Based on Object Tracking Algorithm

Rui Shan[1](✉), Wei Shi[2], Zhu Teng[3], and Yoshihiro Okada[4]

[1] Graduate School and Faculty of Information Science and Electrical Engineering, Kyushu University, Fukuoka, Japan
shan.rui.292@s.kyushu-u.ac.jp
[2] Research Institute for Information Technology, Kyushu University, Fukuoka, Japan
shi.wei.243@m.kyushu-u.ac.jp
[3] School of Computer and Information Technology, Beijing Jiaotong University, Beijing, China
zteng@bjtu.edu.cn
[4] Innovation Center for Educational Resources, Kyushu University, Fukuoka, Japan
okada@inf.kyushu-u.ac.jp

Abstract. Avoiding obstacles and pedestrians during the movement is a challenge for the visual impaired people. With the advancement of computer vision, some new technologies can be applied to improve this problem. In this paper, we proposed a new framework to realize such function based on the object tracking algorithm, FairMOT. Our framework detects the pedestrians around users and alarm users to avoid the collision between them. In our system, we use the 360 degree camera as the system input to record the users' surrounding situations, and use the VR headset as the output device with a feedback application, that is realized using WebVR technologies. In this paper, we also discuss some experiments of the current system and some further improvement methods according to the results of experiments.

1 Introduction

Nowadays, many people still suffer from visual impairment. According to statistics from the World Health Organization (WHO), there are 188.5 million mild visual impaired people, 217 million moderate or severe visual impaired people and 36 million blind people in the world [1]. Such people are surviving great inconvenience in their daily life. Especially, when they want to go outside, there are many dangers around them. Obstacle avoidance in daily movement is a difficult activity since they are hard to perceive the complex environment around them by vision. There have been some solutions proposed to deal with such problem. White canes and guide dogs are two well-known methods. The white cane is low cost and easy to use, yet it can only detect the stationary and fixed obstacles. The guide dogs can detect moving obstacles and interact with user by sound. However, the long training periods cause the guide dogs to be very expensive, and the amount of guide dogs can not satisfy the needs at all.

L. Barolli (Ed.): EIDWT 2023, LNDECT 161, pp. 375–385, 2023.
https://doi.org/10.1007/978-3-031-26281-4_40

During recent decades, computer vision has great advancement and many new technologies have been proposed with help of machine learning and deep learning. Some of these technologies have been applied in real scenarios such as avoiding obstacles in automated driving by using the object organizing and tracking algorithms, and such kind of technologies can also be used to improve the daily travels of the visual impaired people. To realize this target, our team have made some previous works [2] which makes use of a famous object detection technology called YOLO [3] and 360-degree camera to prevent the blind from colliding with obstacles by audio alert, yet the proposed system has some drawbacks need to be improved, as the object detection method cannot re-identify the object and is not efficient to track the objects in real-time live streaming. Moreover, a lot of people are partly visual impaired and many of them can perceive the light and color, hence it is more comfortable for them to receive alerts through various kinds of perceptions.

In this paper, continuing to our previous works, we propose a system which integrates the object tracking technology and the WebVR technology [4] to help visual impaired people avoid colliding with pedestrians during their outdoor movements. The system utilizes the RICOH R Development Kit 360-degree camera to acquire real time video streaming as the system input. The input video streaming will be processed using FairMOT [5] which is an efficient object tracking model to conduct pedestrian tracking. Then the system will determine whether to alert user according to the tracking results. Finally, the system will alert the user when collision is highly possible to occur with Meta Quest 2 VR device via a WebVR application. The backend of system is constructed by python while the front end is constructed by JS + HTML, which means the system can be accessed by multi-platforms (VR/AR devices and smart phones which have the web browser).

In the Sect. 2, we will first discuss some related works of this paper. We will introduce some related research, WebVR technology, and the object tracking model, FairMOT. In the Sect. 3, we will elaborate the design of our system in details and some experiments' results of the methods utilized in our system. In the Sect. 4 we will discuss some drawbacks of the current system and improvement solutions. Finally, we will conclude our work in the Sect. 5.

2 Related Work

In this section, we will first introduce the FairMOT which is an object tracking algorithm. It is used to detect and tracking pedestrians around the users. We also introduce several research related to this paper.

2.1 FairMOT

FairMOT is an approach proposed by Yifu Zhang in 2021, which is on top of CenterNet [6]. It can realize the multiply object tracking (MOT). Comparing to other MOT algorithms, this approach treats detection and re-ID tasks equally. In FairMOT, these two tasks are realized in one single network to improve the data processing speed. In our framework, we choose this method to process the video data recorded by a 360-degree camera.

We choose to implement the FairMOT in our system, because this approach has a good performance on the MOT challenge (https://motchallenge.net/). According to its developers, they "rank first among all the trackers on 2DMOT15, MOT16, MOT17 and MOT20. The tracking speed of the entire system can reach up to 30 FPS." We suppose such a result can satisfy our system's requests of object tracking.

2.2 WebVR

WebVR is a technology which allows users to experience virtual reality (VR) on the web browser. Currently, Most VR devices including smartphone has supported this technology, which means it is easier for people to get into the VR world with help of WebVR no matter what devices they use.

There have been some WebVR APIs that help developers construct WebVR application. We choose to use Three.js to develop our application due to its simplicity and integrity in the development.

2.3 Related Research

We found some researchers already noticed this movement trouble of the visual impaired people. They also proposed several effective systems. Tapu [7] introduced their navigating application. In their paper, they focus on their detecting and tracking algorithm, occlusion detecting and handling methods, and develop a system to prove their theory. In this paper, authors gave the solutions of most possible problems when creating such kind of applications. Comparing to this paper, we focus on how to use 360-degree cameras to detect the dangers from all directions of the users. And our framework will use be designed in a generic method to support different object detecting and tracking algorithms.

Joshi [8] proposed an artificial intelligence-based fully automatic assistive technology to help navigate the visual impaired people. Their system focuses on helping the visual impaired people understand their surroundings by improving the object detection and recognition algorithm. Although their system gains high accuracy and speed in recognition, the whole system is customized with many sensors. Comparing to this system, our framework uses less customized sensors which is designed for high replaceable and scalable for further development.

3 Pedestrian Avoidance System

In this section, we will introduce of our proposed framework and the implemented system.

3.1 Overview of Our Navigation Framework for Visual Impaired People

In previous research, we also already proposed a navigation framework [2]. The overview of this framework is shown in Fig. 1. The framework can be divided into 3 main parts, the input part, the data processing part, and the output part.

Our framework aims to realize to navigate visual impaired people and to assist them to avoid obstacles and pedestrians during their movements. Our framework uses 360-degree video as the input to record users' surroundings. Comparing to other systems, which only record the front of users, our system can more effectively keep users' safety. The recorded video will be sent to the processing part in real time through a desktop or a smart phone. In the processing part, we implement the machine learning algorithm to realize the object recognition and object tracking. Based on the analyzing result, our framework will use our danger judging model to find out all objects which may collide with users. Next, our system will generate feedbacks according to the judging result and send these feedbacks to the output part. In the output part, we use audio, lights, and vibration to alert users. In this step, the route information will also be integrated and guide users to their destinations.

To realize the above function, our framework is designed to be composed by different components. The communication among these components is realized by exchanging of JSON formatted messages in a pre-defined message standard. Such kind of design can ensure the generic of our framework. We can easily update our framework by replacing necessary components, and our framework can also use different input and output devices by developing their adapter components. In the output part, our application is developed based on HTML5. According to users' situation, we design different feedbacks.

Currently, our framework has following components:

(1) Input component: Converting the recorded video into the format our system can process.
(2) Object tracking component: recognizing obstacles and pedestrians.
(3) Danger judgement component: finding out the potential danger to the user.

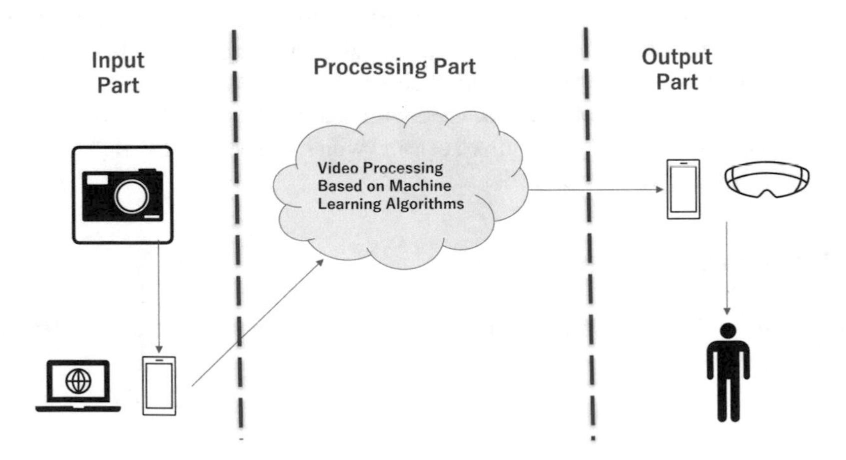

Fig. 1. Overview of our navigation framework

(4) Map component: determining the route according to the user's destination and provide audio route guidance.
(5) Alert generation component: generating alerts and sending back to users.
(6) Feedback component: using sound, lights, vibration and other methods to feedback the user.

3.2 A Pedestrian Avoidance System for Visual Impaired People

For demonstrating our framework design, we develop a pedestrian avoidance system by implementing our proposed framework. Currently, as shown in Fig. 2, our system is composed by following devices: (1) we use the RICOH R Development Kit 360-degree camera as the input device. This device can record the user's surrounding in 360 degrees simultaneously. Such kind of method can detect the potential dangers from the back and front of users. The camera will be set to a helmet to ensure its stability. (2) A PC is used as the processing part. And users are requested to carry the PC. This PC has two functions. One is to provide electric power to the camera. Because the RICOH R Development Kit only be used as a USB camera for continuously record videos. Second, the the video will be processed. Our system will use FairMot to find out all the pedestrians around the users. Using our danger judgement method, our system will filter out the pedestrians who are impossible to collide with the users. Based on the analyzing result, our system will generate alert message and send it to the output Part. (3) A Meta Quest 2 will be used as the output device. Comparing to other systems that only use sound as the feedbacks, our system will use both sound and lights as the feedbacks. Because a great part of the visual impaired people can perceive change of lights, and they depend on their auditory sense to obtain the feedback of the real world, so we use Meta Quest 2, which can provide sound, light, and vibration feedbacks, as the output.

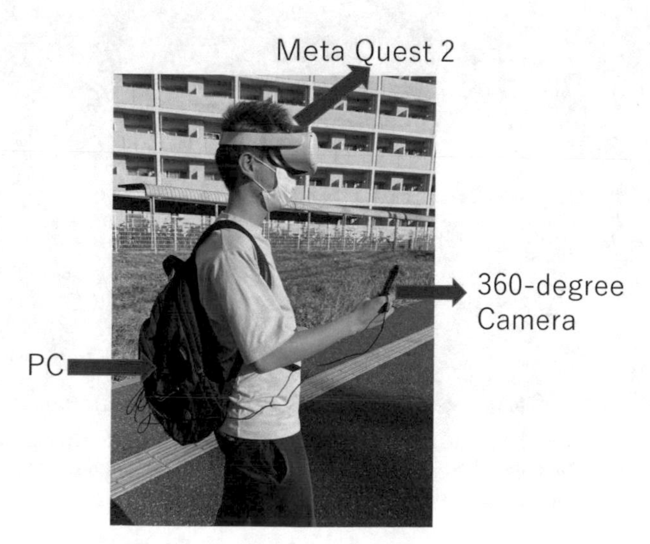

Fig. 2. Devices that are used to compose our system

3.3 Details of Processing the Video Streaming

In current system, we directly use the PC carried by users to process data. There are three main tasks of processing videos.

First, we use the object tracking algorithm FairMOT, which has already been trained, to tracking pedestrians. Figure 3 shows an example of the recognizing and tracking result. The input is a video, which transformed from two videos taken by two fish-eye cameras. The two sub figures in the Fig. 3 are separately taken by the 360-degree camera and converted into a transformed rectangular picture to show all information. The above picture is taken earlier than the below one. Two people are tracked. The changes of their positions are shown by the arrows. Although the shapes of the objects in each picture is transformed, FairMOT still can work.

Fig. 3. Organizing and tracking example by FairMOT (Two people are moving from the positions in the above figure to the below figure. They are separately marked by colored boxes. The boxes with the same color in the two figures indicate the same person.)

Second, we need to judge if the pedestrian may collide with the users or not. After we find out all pedestrian people around the users, we need to calculate their walking directions and speeds. As shown in Fig. 4, we define a danger determining rule in our danger judgement component. We suppose a polar coordinates system whose origin is the user. Other tracked pedestrians' moving direction and speed will be calculated according to their distances in the continuous frames. If the distance of some pedestrian and the user is less than 1 m, and in the following 5 frames, the pedestrian is still moving to the user, our system will generate an alert message. The parameter (alert distance, etc.) of the danger judgement component can modified according to the real situation. The distance between the pedestrians and the users is estimated based on the similar triangle rule. We suppose an adult person's width is 0.39 m. If we get a box width of the person in the picture, we can easily get the distance D by the Eq. (1). In the equation, w is the width of a person, f is the focus of the camera, and p is the pixel width of the person's image in the picture which is the width of the box on the person. We also pre-calculate the focus of each used camera using the method introduced in these two sites [9, 10]. The distance estimating result is shown in Table 1. The estimating distance is shorter than the real distance for ensuring the safety of the users.

$$D = (w \times f)/p \tag{1}$$

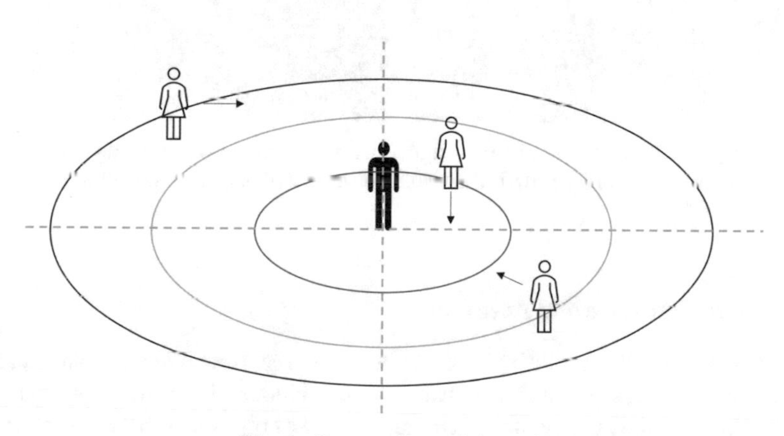

Fig. 4. Danger determining rule

Table 1. The result of the distance measurement method

Target position	Real distance (meters)	Estimating distance (meters)	Error
Front	1.5	1.39	−0.073
Front	3	2.24	−0.253
Left	1.5	1.44	−0.040
Left	3	2.25	−0.250

3.4 Feedback Application

In our system, we develop a VR application as the feedback application. In current step, we use the Meta Quest 2 because it can combine the sound, lights, vibration, and other methods as the feedbacks to users. Our system provides the sound feedback to the people who totally lose their sights, and lights feedbacks to the people who can feel the changes of the light and color. Our feedback application is developed by HTML5, ThreeJS and WebVR technologies. Besides the Meta Quest 2, other VR glasses and smartphone also can be used as the output devices. In our system, the pedestrians around the users are divided into four directions, front left, front right, back left and back right. For the people who can feel the color, we will use the red lights to alert the user that a person appearing at the front of the user. Figure 5 is an example of our application to alert the user.

Fig. 5. Example of alarm in VR space (the figure in left is a person who moving closely from the left of camera, the figure in the right is the visual alarm in the WebVR application.)

3.5 Computation Time and Accuracy

In order to evaluate whether the system can be run in real time, we make an experiment to record the time cost of back-end part which includes object tracking and danger determination, and the latency of the data transmission between WebVR application and back end. The result is shown in Table 2, which shows the system can be run in real time since the average latency of the whole system is about 650.1ms. We also make some experiments to evaluate whether the system can send alerts accurately. We test our system in 8 different kinds of scenarios: the first and the second scene are the user standing in place while the pedestrian moving closely to the user from the right front and left front respectively; the third and the forth scene are the pedestrian standing in place while the user moving closely to the pedestrian from right front and left front respectively; the fifth and sixth scene are the user and the pedestrian approaching each other while the pedestrian is moving from right and left; the seventh and eighth scene are pedestrian standing in place while the pedestrian moving closely from the right back and left back. The result is shown in the Table 3. In Scence 6 and 7, the system cannot give any alerts. The problem only occurred when the user and pedestrian were approaching each other, which might be caused by the speed of the object in the camera was too fast to compare the changing of the distance.

Table 2. The result of the computation time

Processing part	Average time (ms)
Back end	180.5
Data transmission	469.6
Total	650.1

Table 3. The result of the system accuracy

Scene	Accuracy
1	Correct
2	Correct
3	Correct
4	Correct
5	No alert
6	No alert
7	Correct
8	Correct

4 Discussion

Currently, our framework still can not to ensure the safety of the user, because our system only can detect and track pedestrians and generate alarms based on the analyzing result. First, we plan to retrain the FairMOT model to make the system can recognize and track multi-class objects. Besides, to ensure that users walk on the sidewalk, we will use some road segmentation methods to recognize the sidewalk on the video, and alert users when they have high possibility to walk out of sidewalk, currently we are considering using the pretrained segmentation model of OpenVINO [11] which is called road-segmentation-adas-001 [12]. Then, to realize the function of route guidance, we need to complete our map component. This component can generate the routes according to the user's destination. According to the position of the user, this component should provide the audio guidance to the user. Besides, we will use a remote server as our processing part. The recorded video will be sent to the server for future processing in the RTMP. In this method, the best merit is that the weight of the devices carried by the user. Users only need to carry a camera and a smartphone. However, for improving our system, the following problems should be solved. First, we should reduce the processing time. Conducting object tracking on the video streaming from RTMP message has long latency. Besides, the alert distance of the current system is 1 m, we will modify the alert distance of the system according to the result of accuracy experiment, we plan to make some experiments to compare the accuracy of the alert in different alert distances and allow the system send the alert without comparing the distance's changing when the object is very close to the user to improve the performance of danger sending.

5 Conclusion and Future Work

In this paper, we introduce a framework which can realize the navigation for visual impaired people. This framework has input part, processing part and output part. It is composed by components to realize different functions. This framework is defined based on object tracking algorithm to detect the potential dangers around the user and send various types of alerts as the feedbacks to users. After integrating the route information from the map application, our framework can guide the user and ensure the users' safety.

We developed a pedestrian avoiding system by implementing our framework. This system can realize 3 main tasks: (1) detecting pedestrians around the user; (2) detecting the roadside to keep the user in safe areas; (3) using sound and lights as the feedbacks to alert the user when some pedestrians may collide with the user. Our system uses RICOH R Development Kit as the input device, FairMOT to tracking pedestrians, and Meta Quest2 as the output device. The application which provides alters to users is developed using HTML5 and WebVR technologies, which can easily be used in other VR devices or smartphones. Next, we use experiments to show the performance of our system.

In future, we will consider improving our framework in following aspects. First, we will retrain the FairMOT model for recognizing people and other objects on the roads and implement road segmentation method to recognize the sidewalk. Then, our system can keep the users away from the people and objects, such as cars, which may collide with them while keep them walking on the sidewalk. Next, we will set a remote server to processing the videos. Comparing to current method, it allows users not to carry a PC with him. It can greatly reduce the total weight of our system. Third, we will complete the map component to realize route generation and audio guidance.

References

1. Bai, J., Liu, Z., Lin, Y., Li, Y., Lian, S., Liu, D.: Wearable travel aid for environment perception and navigation of visually impaired people. Electronics **8**(6), 697 (2019)
2. Shi, W., Shan, R., Okada, Y.:A navigation system for visual impaired people based on object detection. In: 2022 12th International Congress on Advanced Applied Informatics (IIAI-AAI), pp. 354–358. Kanazawa, Japan (2022)
3. Nepal, U., Eslamiat, H.: Comparing YOLOv3, YOLOv4 and YOLOv5 for autonomous landing spot detection in faulty UAVs. Sensors **22**(2), 464 (2022)
4. Vukicevic, V., Jones, B., Gilbert, K., Wiemeersch, C.V.: "WebVR", Editor's Draft (2017). https://immersive-web.github.io/webvr/spec/1.1/
5. Zhang, Y., Wang, C., Wang, X., Zeng, W., Liu, W.: FairMOT: on the fairness of detection and re-identification in multiple object tracking. Int. J. Comput. Vision **129**(11), 3069–3087 (2021). https://doi.org/10.1007/s11263-021-01513-4
6. Zhou, X., Wang, D., Krähenbühl, P.: Objects as Points. ArXiv, abs/1904.07850 (2019)
7. Tapu, R., Mocanu, B., Zaharia, T.: DEEP-SEE: joint object detection, tracking and recognition with application to visually impaired navigational assistance. Sensors **17**, 2473 (2017). https://doi.org/10.3390/s17112473

8. Joshi, R.C., Yadav, Y.S., Dutta, M.K., Travieso-Gonzalez, C.M.: Efficient multi-object detection and smart navigation using artificial intelligence for visually impaired people. Entropy **22**(9), 941 (2022)
9. Wu, J.: 单目测距算法 (2020). https://www.cnblogs.com/wujianming-110117/p/12901246.htm
10. Lee, S., Hayes, M.H., Paik, J.: Distance estimation using a single computational camera with dual off-axis color filtered apertures. Opt. Express **21**, 23116–23129 (2013)
11. https://www.intel.com/content/www/us/en/developer/tools/openvino-toolkit/overview.html
12. https://docs.openvino.ai/latest/omz_models_model_road_segmentation_adas_0001.html

Web-Based Collaborative VR System Supporting VR Goggles for Radiation Therapy Setup Training

Yuta Miyahara[1], Kosuke Kaneko[2], Toshioh Fujibuchi[3], and Yoshihiro Okada[1,4](✉)

[1] Graduate School of ISEE, Kyushu University, Fukuoka, Japan
okada@inf.kyushu-u.ac.jp
[2] AiRIMaQ, Kyushu University, Fukuoka, Japan
[3] Faculty of Medical Sciences, Kyushu University, Fukuoka, Japan
[4] ICER (Innovation Center for Educational Resources), Kyushu University Library, Fukuoka, Japan

Abstract. This paper proposes a new web-based collaborative VR system for the radiation therapy setup training that solved the following problems: in the authors' previous system, the laser representation was not the same as the real one and the replay function was not enough. In addition, the system configuration was complicated because in the previous system, two sets of a goggle with a smartphone were used as output devices, and a game pad and a hand tracking device were used as input devices. Contrarily, the system configuration of the new version proposed in this paper has become simpler by only supporting VR goggles such as Meta Quest 2 with their hand tracking functionality. Medical students need to take the radiation therapy setup training before practicing such therapies. Because the radiation therapy devices are very expensive and dangerous, we need a training system without using the real devices. Evaluation by medical students proved that the new system proposed in this paper is useful for such the training.

1 Introduction

Currently, three-dimensional computer graphics (3D-CG) technology is widely used in smartphones' applications besides personal computers' desktop applications. 3D-CG technology is also popular in a wide range of application fields, from the entertainment to the practical such as the medical applications. Virtual Reality (VR) and Augmented Reality (AR) are examples of the technologies based on 3D-CG technology. VR is often used in the entertainment field [1, 2], and there are also researches on various fields like education/training [3–7] besides physics, medicine, and architecture. In the research of this paper, we propose a web-based collaborative VR system for the radiation therapy setup training. As shown in Fig. 1, several medical procedures including the setup operation of a radiation therapy device such as TrueBeam [8] sometimes need to be carried out by multiple persons. Because the radiation therapy devices are very expensive and dangerous, we need a training system without using the real devices.

L. Barolli (Ed.): EIDWT 2023, LNDECT 161, pp. 386–400, 2023.
https://doi.org/10.1007/978-3-031-26281-4_41

Therefore, in the researches of our past papers [9–11, 13], we proposed a web-based collaborative VR training system. However, there have remained several functionalities to be improved as follows: the laser representation was not the same as real one and the replay function was not enough. In addition, the system configuration was complicated because two sets of a goggle with a smartphone were used as output devices, and a game pad and a hand tracking device were used as input devices in the system. This time, we developed a new version to improve these problems. The new version, the latest system fully supports VR goggles such as Meta Quest 2 and their hand tracking functionality. As a result, its system configuration has become simpler than the previous one. Also, we evaluated the latest system to justify its usefulness for the radiation therapy setup training and obtained good results.

The remainder of this paper is organized as follows: The next Sect. 2 introduces our first system and the other related work. We describe our previous system and its details in Sect. 3. After that, we describe our latest system and its improved points in Sect. 4. The evaluation results by several medical students are also shown in Sect. 5. Finally, we conclude the paper in Sect. 6.

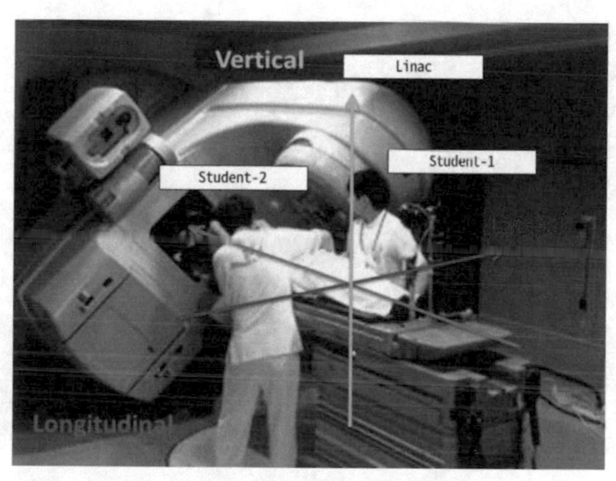

Fig. 1. Image of TrueBeam, here student-1 moves the bed and rotates the linac of TrueBeam via its controller and student-2 adjusts the orientation of a patient.

2 Related Work

As related work, there is a prototype system of a linear accelerator simulator using VR for oncological radiotherapy training [14]. The paper [14] discusses the main challenges and features of the VR prototype, including the system design and implementation. A key factor for trainees' access and usability is the user interface, particularly tailored in the prototype to provide a powerful and versatile friendly user interaction. The purpose of this system is for the training of the operations of radiotherapy devices, but not for the patient setup training. As another related work, there is LINACVR, a collaborative VR radiation therapy simulation prototype that provides an immersive training solution including the patient setup training [15]. The authors of the paper [15] evaluated LINACVR with 15 radiation therapy students and educators. The results indicated that LINACVR would be an effective and cost-effective alternative solution for radiation therapy compared to state-of-the-art simulators. Our latest system proposed in this paper is very similar to LINACVR. The difference between them is their system configurations. LINACVAR is provided as a dedicated software developed using Unity 3D, but our system is provided as a web-application, easier to start the training.

In the following, as related work, we describe our first system proposed by Imura, et al. [12] for the radiation therapy setup training targeting a radiation therapy device called TrueBeam. In our training system, two students collaboratively move a bed and a patient to master the skill that adjusts the position and orientation of the patient. One student operates a smartphone device in which a certain application is installed for moving a bed. The other student operates a real patient model, an artificial chest model on whose surface a gyroscope sensor device is put. The operation data of the two students are sent from each device to a Unity application running on a server for checking setup operation results. The Unity application displays a 3D virtual world reflected to the operation results, and the students can check it on a monitor screen as shown in Fig. 2.

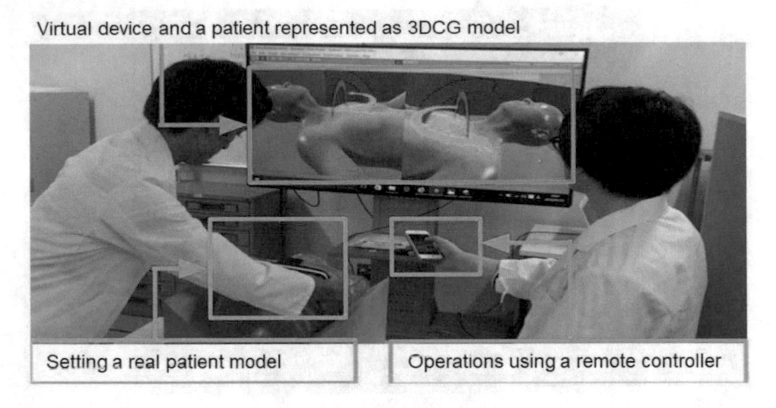

Fig. 2. Image of operation training of TrueBeam on the system of Imura, et al.

Figure 3 shows the components of Imura's training system. Imura's system had some problems as follows: before using the system, users needed to install the applications on each device, and the developers of the system needed to upload the applications on the application stores.

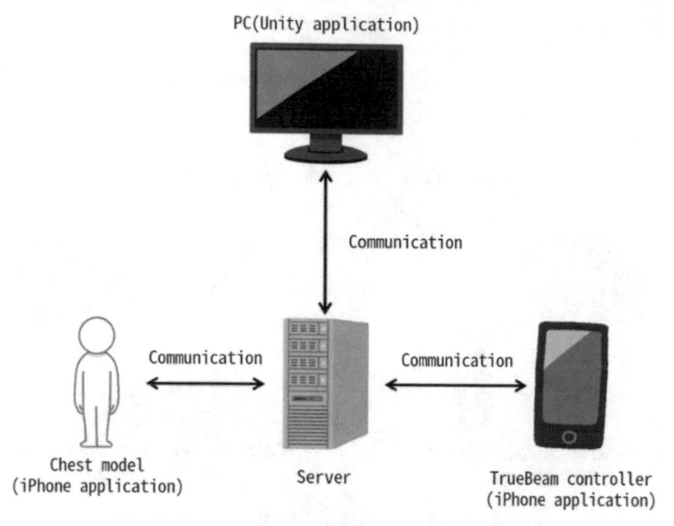

Fig. 3. Components of Imura's system. The server synchronizes the status of the other three components using real-time bi-directional communication.

3 Previous System and Improvements

For solving the above problems of Imura's system, we developed a web-based collaborative VR system [13] for the setup training of TrueBeam. In this section, we describe this previous system and its details.

3.1 System Architecture

The system is divided into the following three components as shown in Fig. 4.

- Head Mounted Display (HMD) and a game pad as a remote controller for the bed movement of TrueBeam.
- HMD and a hand tracking device for the patient setup.
- A server that synchronizes two virtual worlds displayed on two HMDs by web socket communications.

Two trainees can see the 3D virtual world representing a treatment room through a smartphone within an HMD as shown in Fig. 4. However, whenever the trainees wear the HMD, they cannot see any physical objects in the real world because the 3D virtual world spreads before their eyes. For compensating this defeat, we need intuitive input devices easy to operate. As a result, we employed a game pad as the remote controller for the bed movement of TrueBeam, and Leap Motion Controller as the hand tracking device for the patient setup. We used Leap.js to manage hand tracking data sent from Leap Motion Controller on the web application.

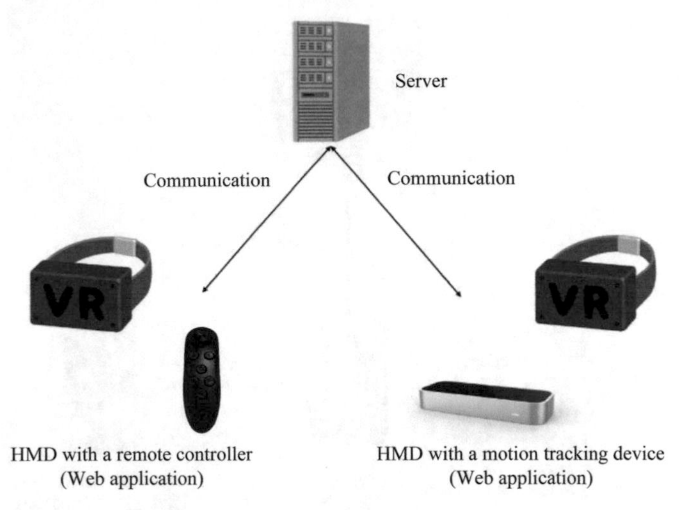

Fig. 4. Components of the previous system.

3.2 Bed Operation

The degrees of freedom (DoF) of the bed and the Linac of TrueBeam are shown in the right part of Fig. 5, those are as follows.

- LAT: The bed moves in the z-direction.
- LNG: The bed moves in the x-direction.
- VRT: The bed moves in the y-direction.

The one of the two trainees can operate the bed with these degrees of freedom by pressing the corresponding buttons on the remote controller as shown in the left part of Fig. 5. The remote controller is connected to the smartphone via Bluetooth.

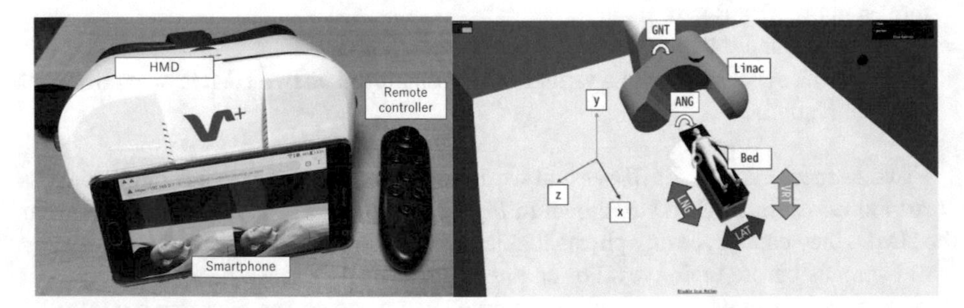

Fig. 5. Degrees of freedom of the bed and the Linac of TrueBeam.

3.3 Patient Setup

Interactions between the other trainee and the 3D patient object are necessary for the training of the patient setup. For the patient setup in the real world, radiation therapists touch the patient body directly by their two hands. In the training system, the hand tracking device attached to the HMD senses the position of the palm and direction of the fingers for each of the trainee's two hands as shown in the left part of Fig. 6, and by the sense data, two 3D hand objects are drawn as shown in the right part of Fig. 6. The rotation angle of the 3D patient object obeys the direction vector between the trainee's both two hands while the both hands are grabbled.

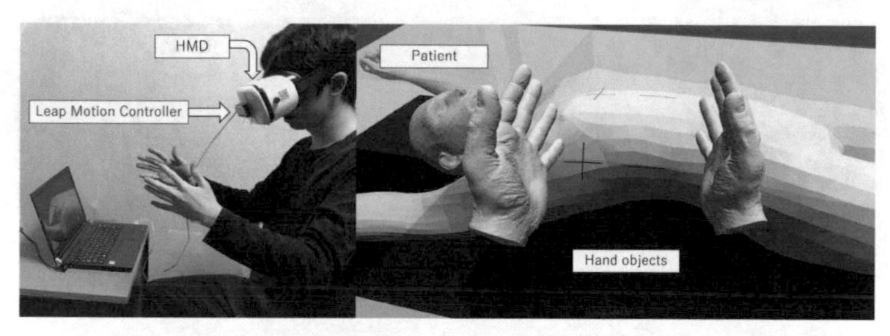

Fig. 6. Training of the patient setup in the system (left), 3D hand objects in the virtual world (right).

3.4 Server Process

For the two users, i.e., two trainees, to share the same 3D virtual world on their HMDs, the following JavaScript code runs on the server.

```
operationNameSpace.on('connect', (socket) => {
    socket.on('operation', (object) => {
        logArray.push({event: 'operation', object: object});
        socket.broadcast.emit('operation', object);
    });
});
```

The "connect" event can establish a connection between one of the two users and the server. After the connection, a certain callback function already registered will become invoked whenever the "operation" event will be fired for the bed setup. As shown in Fig. 7, the "operation" event has the object work as an argument of the callback function. The object includes a name attribute for specifying the 3D object to be operated and a value attribute required for the operation (e.g., a sign corresponding to degrees of freedom of TrueBeam). In each callback function, the object sent from the user is transferred to the other user through the server, and the callback function corresponding to the event will be executed so that the users' HMDs can share the same 3D virtual world. In the same way, another callback function already registered will become invoked whenever the "orientation" event will be fired for the patient setup, so that the users' HMDs can share the same 3D virtual world as well.

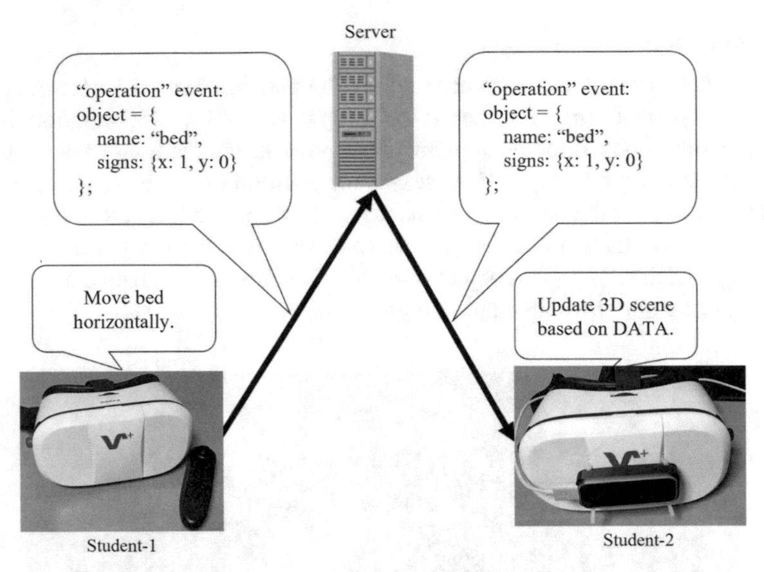

Fig. 7. Example communication for sharing the 3D virtual world.

3.5 Training for Radiation Therapy Setup

Three markers of the cross are drawn on the patient's body, and half-transparent plane objects are placed around the patient object, as shown in Fig. 8. One of the three markers is on the chest, and the others are on the two body sides. The plane objects are substituted as lasers used for positioning in the actual treatment. The users need to match all the markers with the plane objects by moving the bed and the patient in the training.

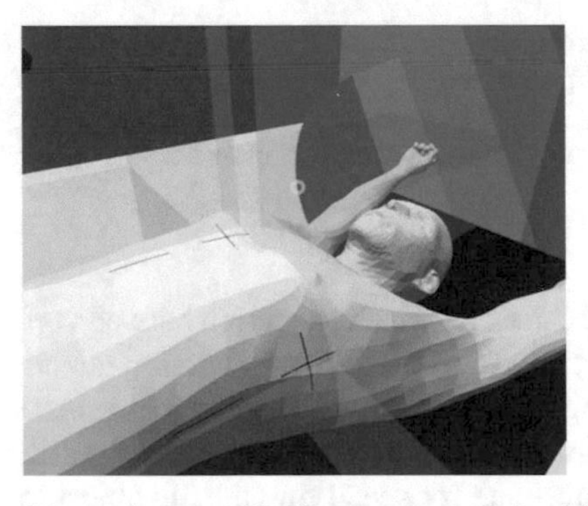

Fig. 8. Three skin markers, and half-transparent plain objects as lasers.

Our system checks for every frame whether all the markers are within error ranges of the lasers or not. When all the markers match the lasers, the patient object will change its color, and the duration time of the training and the number of times the patient object was touched are displayed in the 3D virtual world as feedback to the traineers.

3.6 Log Recording and Replay Functions

Our previous system has another function that records a log of the operations of trainees and runs a replay so that the trainees can confirm their operations after the training. The server stores the operation event data received from the users in an array as shown in Sect. 3.4. When the training is over, the server writes the operation log to a JSON file. When the replay is requested by the user, the server reads the JSON file and broadcasts the operation log to each client, i.e., each user. After that, clients initialize their own 3D virtual world and execute the functions corresponding to each event name included in the received operation log. The recorded operation events do not have any time information when the operation was executed. Therefore, each operation function is executed at 9-ms intervals to prevent the replay from ending in a too short time.

4 Latest System Fully Supporting VR Goggles

The users can operate the bed and patient objects in our training system. However, there were problems with the system, those were the laser representation different from one in the real world and the lack of a full replay function. Therefore, we developed a new version that solved these problems. The new version, the latest system fully supports VR goggles such as Meta Quest 2 and their hand tracking functionality.

4.1 Supporting VR Goggles with Hand Tracking

We decided to discontinue the development of the smartphone edition and proceed with the development of the Meta Quest 2 edition only. Quest is a 6DoF stand-alone HMD marketed by Meta. The smartphone edition required an additional tracking device for hand tracking. On the other hand, Quest has a hand tracking function and does not require an additional tracking device, making the system simpler than ever. Moreover, the users can walk around in the virtual world because Quest has 6DoF. Since our system is a web application and Quest also has a web browser, we were able to reuse most of the previous system's programs for developing the latest system.

In order to employ Quest's hand tracking function to operate the system, it was necessary to support the recognition of a grasping gesture. Each joint data of fingers sent from the hand tracking, as shown in Fig. 9, is used to decide whether the hand is grasping or not. Specifically, the decision is made by calculating whether the distance between the palm and fingertip (e.g., #1 circular joint and #1 square joint) is less than a threshold value. For the index, middle, ring, and little fingers, the distance between pairs of joints is calculated, and if any one pair is less than the threshold, it is considered to be grasping.

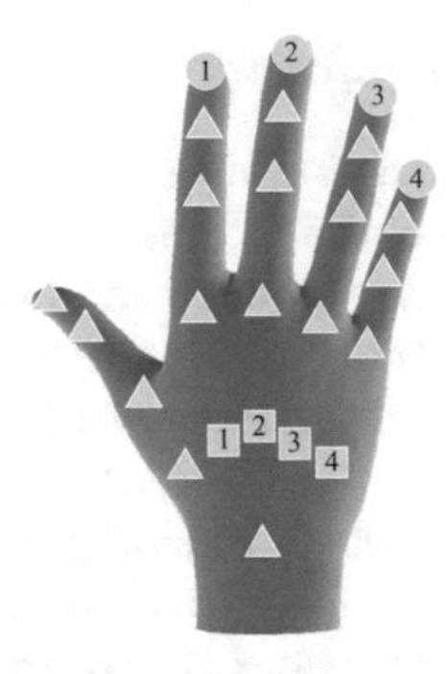

Fig. 9. Joints of the 3D hand model.

4.2 Laser Representation

We already represented the lasers with the half-transparent plane objects in the previous system. However, in a real treatment environment, the lasers can only be seen on the patient's body surface, so representations using such plane objects are far from reality and can detract from the sense of immersion. Therefore, we changed the laser representation to reproduce a more realistic training environment as shown in Fig. 10. We duplicate the patient object and use one for laser drawing. We controlled the appearance of the object using a fragment shader. The fragment shader is a computer program that processes each pixel of a device's screen for rendering 3D-CG. Specifically, it determines what color each pixel is drawn. In the latest system, the processing is described in a shading language called GLSL. As shown in the middle of the figure, only the laser intersection trajectory is colored, and the other parts are completely transparent. By superimposing the original patient object and the laser-drawn object, a more realistic laser expression can be achieved.

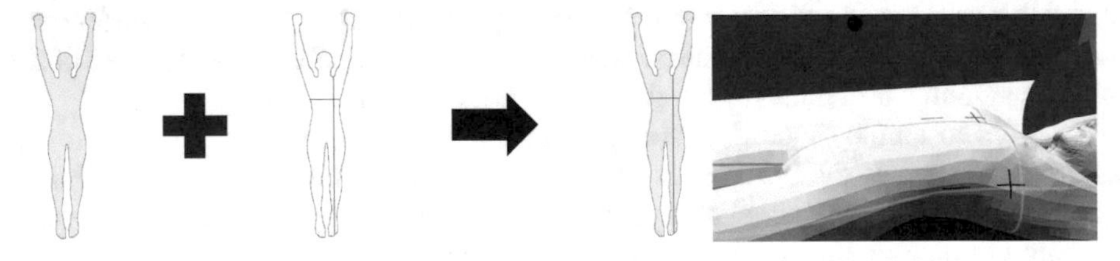

Fig. 10. Method of new laser representation (left) and result image of new laser representation (right).

4.3 Modified Bed Operation Method Supporting Movement Sound

In the previous system, a Bluetooth remote controller was used to operate the bed, however, due to the small number of buttons, there was a problem that the operation method was difficult. The user needs to put the bed or the patient object appropriately in the center of the screen by moving the bed horizontally or vertically. For example, if the user wants to move the bed horizontally, the patient object should be in the center of the screen. Contrary, if the user wants to move the bed vertically, the bed object should be in the center of the screen. This operation procedure is not intuitive and may confuse the users, so we changed the operation method to solve this problem. Figure 11 shows the 3D model used in the new operation method. This model imitates the remote controller used in actual radiation therapy and was created using Blender, a 3D-CG software. Because the program detects collisions between a tiny sphere set at the tip of the index finger and the button, the user can operate the bed by pressing the button on the model.

Moving the bed of TrueBeam makes a very loud noise in the real world. Such a sound is considered to be one of the important factors to enhance the reality of a radiation therapy device in the virtual world. Therefore, we have changed the system to play a monotone sound in conjunction with the operation of the bed.

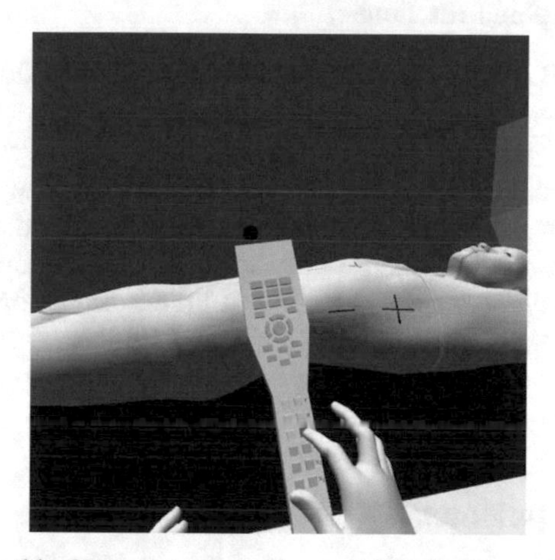

Fig. 11. 3D remote controller model for bed operation.

4.4 Patient Setup with Collision Detection

The user can rotate the patient object by moving both grasped hands while aligning it to the center of the screen in the previous system. However, in actual treatment, the user directly touches the patient's body to move it, so the operation method was uncomfortable. Therefore, we have changed the program so that it uses collision detection, and the patient object can be rotated only when the user is regarded as touching the patient object by collision detection as shown in the right part of Fig. 12.

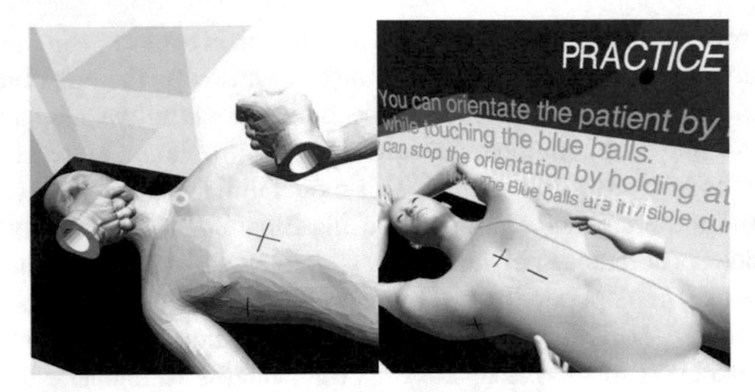

Fig. 12. Patient setup in the previous system (left) and patient setup in the latest system (right).

4.5 Extension of Replay Functions

As explained in Sect. 3.6, our previous system has a replay function that allows the user to look back on the operations after the training. However, the recorded operation events do not have any time information when the operation was executed. Therefore, each operation function is executed at 9-ms intervals to prevent the replay from ending in a too short time. In the latest system, this problem was solved by that the timing when an operation occurred is saved in a log file, and the operation function is executed based on that timing in the replay. This makes the replay of the executed operations appear to be the same as the actual operation executed. In addition, it supports pause, playback at 1.5x and 2x speed, and 10-s rewind and fast-forward, so that the user can more conveniently look back on the operations.

5 Evaluation Experiment

We conducted an experiment to evaluate the latest system. Six medical students of the laboratory of one of the authors participated in this experiment and answered a questionnaire consisting of open-ended questions and five-point scale questions. After we explained the basic usage of the system to the participants beforehand, they actually used the system in pairs. The participants were free to practice bed operation and patient setup before being trained in a sequence of radiation therapy procedures and alternated between bed operator and patient setup roles for the training. After the training, we discussed our system with the participants.

5.1 Questionnaire and Results

Q1, Q2, Q16, and Q17 are open-ended questions as follows and the others are five-point scale questions. The five-point scale questions and their answer results are shown in Table 1.

- Q1: What experience do you have with VR (HTC Vive, Meta Quest, etc.)?
- Q2: What experience (including hands-on training) do you have with radiation therapy?

- Q16: What are the differences from the real world treatment environment?
- Q17: Please let us know what you think of the system.

Five participants had no experience with VR and only one had experience with it in his/her research. Five participants had experience in radiation therapy practice, and only one had only observed clinical practice.

We rated the scores from 1 to 5 on a scale from strongly disagree to strongly agree and calculated the means. Q3–Q5 about the representation of VR space received positive answers. Especially, the result of the question about the laser representation is good and this means that the implementation in the latest system worked well.

Table 1. Results of the five-point scale questions

Question	Number of Respondents					Mean
	Strongly disagree.	Disagree.	Neither.	Agree.	Strongly agree.	
Q3. The treatment environment is well replicated.	0	0	1	3	2	4.2
Q4. The laser representation for positioning is effective.	0	0	0	2	4	4.7
Q5. The representation of bed operation sound is effective.	0	0	1	4	1	4.0
Q6. The bed operation is comfortable.	0	1	1	3	1	3.7
Q7. The bed operation is close to reality.	0	0	0	6	0	4.0
Q8. The patient setup is comfortable.	0	1	3	2	0	3.2
Q9. The patient setup is close to reality.	0	0	1	4	1	4.0
Q10. This system is effective for learning the technique of bed operation.	0	0	0	5	1	4.2
Q11. This system is effective for learning the technique of patient setup.	0	0	0	5	1	4.2
Q12. You felt a synchronization gap.	1	4	0	1	0	2.2
Q13. Collaboration is smooth.	0	0	2	3	1	3.8
Q14. You felt sickness.	3	1	1	1	0	2.0
Q.15 You would like to use this system in the future.	0	0	1	5	0	3.8

It can be said that both bed operation and patient setup have higher reality from the answers to Q7 and Q9. However, while there are positive answers to the comfort level question (Q6 and Q8), some participants did not feel comfortable. The buttons on the 3D remote controller model for the bed operation were sometimes pressed incorrectly due to the small size of the buttons and the narrow spacing between each button. Regarding patient setup, it is thought that it was difficult for participants to operate using the system's unique collision detection, and they had to become accustomed to this.

The questions regarding the effectiveness of the program in acquiring radiotherapy skills (Q10 and Q11) were rated highly. As mentioned above, bed operation and patient setup have the higher reality, and the participants seemed to be able to satisfactorily perform the task of matching all the lasers with the skin markers.

Most participants did not feel a synchronization gap, but only one participant felt it. Some participants indicated that they were able to work together smoothly, but two participants felt that they could say neither. Most participants did not feel sickness, but only one participant answered as feeling so. In order to prevent VR sickness, it is important to take breaks as needed and avoid long training sessions. We received positive feedback that the participants would like to continue to use our training system in the future.

For Q16, some answers point out that the patient object has no weight. In a real-world patient setup, it is not easy to move the patient's body because the human body is heavy. This is the obvious difference from the real-world treatment environment. However, it is very difficult to present weights to the users in VR to solve this problem, and further research is needed. We received a wide variety of comments from Q17. In this paper, we describe some of them. Some participants commented that it would be good to make the 3D remote controller model more realistic, but it would be better to focus on operability rather than realism. As mentioned above, there is room for improving the 3D remote controller. In addition, there are comments that the patient setup seems confusing for novice trainees so any help system will be useful for them. Some participants could feel improvement about the operation times between the first and second sessions because the system could inform the elapsed times taken for the training operations.

5.2 Discussion

We could configure the system more simply by using Meta Quest 2 instead of a smartphone and a hand tracking device, which allows the users to walk around in the virtual world. However, VR-specific operations are not always easy for the users, and the system needs to provide more comfortable operability. It is also necessary to provide information to assist users when they have questions while working in the virtual world, and one such method is any help system. The results of the evaluation experiment suggest that the setup training environment was mostly well reproduced, including improvements in the laser representation. Although we did not evaluate the replay function in this research, it would be a useful educational tool if the replay function could be further expanded for allowing teachers of radiation therapy to always check students' performance in their training sessions later.

6 Conclusions

In this paper, we introduced the latest version of our web-based collaborative VR system for the radiation therapy setup training. This latest system supports fully VR goggles like Meta Quest 2 with their hand tracking function to provide a more useful and realistic training environment. We also conducted experiments to evaluate the latest system and found out what is good about the system and what needs to be improved. Modification based on the evaluation results obtained in this study will make the system more practical for many trainees performing radiation therapy. We have been modifying the system and will introduce it in the near future.

Acknowledgements. This research was partially supported by JSPS KAKENHI Grant Number JP22H03705.

References

1. Okada, Y.: 3D visual component based-approach for immersive collaborative virtual environments. In: Proceedings of the 2003 ACM SIGMM Workshop on Experiential Telepresence, ETP'03, pp. 84–90 (2003)
2. Okada, Y.: IntelligentBox as component based development system for body action 3D games. In: ACM SIGCHI International Conference on Advances in Computer Entertainment Technology (ACE 2005), pp. 454–457 (2005)
3. Okada, Y., Ogata, T., Matsuguma, H.: Component-based approach for prototyping of Tai Chi-based physical therapy game and its performance evaluations. ACM Comput. Entertainment **14**(1), 1–20 (2016)
4. Akase, R., Okada, Y.: Automatic 3D furniture layout based on interactive evolutionary computation, VENOA 2013. In: Proceedings of the 7th International Conference on Complex, Intelligent, and Software Intensive Systems (CISIS-2013), IEEE CS Press, pp. 726–731 (2013)
5. Yuuta, K., Yoshihiro, O.: 3D visual component based development system for medical training systems supporting haptic devices and their collaborative environments, VENOA 2012. In: Proceedings of the 6th International Conference on Complex, Intelligent, and Software Intensive Systems (CISIS-2012), IEEE CS Press, pp. 687–692 (2012)
6. Miyahara, K., Okada, Y.: Collada-based file format for various attributes of realistic objects in networked VR applications supporting various peripherals. J. Mobile Multimedia **6**(2), 128–144 (2010)
7. Miyahara, K., Okada, Y., Niijima, K.: A cloth design system using haptic device and its collaborative environment. In: Proceedings of the 2006 IEEE International Workshop on Haptic Audio Visual Environments and Their Applications, HAVE 2006, pp. 48–53 (2006)
8. TrueBeam: https://www.varian.com/ja/oncology/products/treatment-delivery/truebeam-radiotherapy-system. 8 Apr 2018
9. Kuroda, K., Kaneko, K., Fujibuchi, T., Okada, Y.: Web-based collaborative vr training system for operation of radiation therapy devices. In: Barolli, L., Hussain, F.K., Ikeda, M. (eds.) CISIS 2019. AISC, vol. 993, pp. 768–778. Springer, Cham (2020). https://doi.org/10.1007/978-3-030-22354-0_70
10. Kuroda, K., Kaneko, K., Fujibuchi, T., Okada, Y.: Web-based operation training system of medical therapy devices using VR/AR devices. In: Proceedings of the 13-th International Conference on Broadband and Wireless Computing, Communication and Applications (BWCCA-2018), pp. 250–259 (2018)
11. Kuroda, K., Kaneko, K., Fujibuchi, T., Okada, Y.: Web-based VR system for operation training of medical therapy devices. In: Barolli, L., Javaid, N., Ikeda, M., Takizawa, M. (eds.) CISIS 2018. AISC, vol. 772, pp. 948–957. Springer, Cham (2019). https://doi.org/10.1007/978-3-319-93659-8_88

12. Imura, K., Fujibuchi, T., Kaneko, K., Hamada, E., Hirata, H.: Evaluation of the normal tissues dose and exposure efficiency in lung-stereotactic body radiation therapy. In: the 44th Autumn Scientific Congress, Japanese Society of Radiological Technology, 13–15 Oct 2016. (in Japanese)
13. Miyahara, Y., Kaneko, K., Fujibuchi, T., Okada, Y.: Web-based collaborative VR training system and its log functionality for radiation therapy device operations. In: Barolli, L., Yim, K., Enokido, T. (eds.) CISIS 2021. LNNS, vol. 278, pp. 734–746. Springer, Cham (2021). https://doi.org/10.1007/978-3-030-79725-6_74
14. Chan, V.S., et al.: Virtual reality prototype of a linear accelerator simulator for oncological radiotherapy training. In: Groen, D., de Mulatier, C., Paszynski, M., Krzhizhanovskaya, V.V., Dongarra, J.J., Sloot, P.M.A. (eds.) Computational Science, ICCS 2022. ICCS 2022. Lecture Notes in Computer Science, vol. 13352. Springer, Cham (2022). https://doi.org/10.1007/978-3-031-08757-8_56
15. Bannister, H., Selwyn-Smith, B., Anslow, C., Robinson, B., Kane, P., Leong, A.: Collaborative VR simulation for radiation therapy education. In: Uhl, J.-F., Jorge, J., Lopes, D.S., Campos, P.F. (eds.) Digital Anatomy. HIS, pp. 199–221. Springer, Cham (2021). https://doi.org/10.1007/978-3-030-61905-3_11

Development Framework Using 360VR Cameras and Lidar Scanners for Web-Based XR Educational Materials Supporting VR Goggles

Yoshihiro Okada[1(✉)], Kosuke Kaneko[2], and Wei Shi[3]

[1] ICER (Innovation Center for Educational Resources), Kyushu University Library, Fukuoka, Japan
okada@inf.kyushu-u.ac.jp
[2] AiRIMaQ, Kyushu University, Fukuoka, Japan
[3] Research Institute for Information Technology, Kyushu University, Fukuoka, Japan

Abstract. This paper proposes a development framework using 360VR cameras and lidar scanners for web-based XR educational materials that supports VR goggles. Recently, the demand of XR (VR/AR/MR) applications has become increased due to the advancement of XR technologies, e.g., high specification VR goggles, 360VR cameras and lidar scanners have become available at a cheaper price rather than ever. Although more than ten years ago, many entertainment XR applications have been produced, XR applications in other fields have not been so. Besides the entertainment field, educational materials using XR have become in great demand. Especially, XR seems useful for educational materials of experiment and exercise subjects. However, the development of such educational materials is not easy and needs much time. Therefore, in this paper, the authors propose the development framework for them. In this paper, the authors also show several material examples to justify the usefulness of the proposed framework.

Keywords: e-Learning · 3D Graphics · Educational materials · 360VR

1 Introduction

In this paper, we propose a development framework using 360VR cameras and lidar scanners for web-based XR educational materials that supports VR goggles such as Meta Quest 2, and show several educational materials developed using the framework to justify its usefulness. This research is one of the activities of our center called ICER (Innovation Center for Educational Resources) [1] belonging to Kyushu University Library of Kyushu University, Japan because the mission of ICER is to provide students and teachers in the university with educational materials using ICT like 3D-CG.

Effective education needs to catch students' interest to their learning subjects by providing attractive educational materials. Such materials would be realized by the most recent ICT like 3D-CG because current students sometimes called video game generation or Z generation are used to operate such 3D applications [2, 3]. However, it is difficult

L. Barolli (Ed.): EIDWT 2023, LNDECT 161, pp. 401–412, 2023.
https://doi.org/10.1007/978-3-031-26281-4_42

for teachers to develop such 3D educational materials because it requires much technological knowledge and programming skills that common teachers do not have. Teachers need any tools that make it easier to develop such attractive 3D educational materials. Then, we have already proposed several frameworks so far that make the development easier [4–7]. Furthermore, for developing educational materials that have more educational effectiveness using Virtual Reality (VR) and Augmented Reality (AR), we also added functionalities to the already proposed frameworks for supporting VR goggle with a smartphone [8, 9]. Recently, many types of 360VR cameras have been released from many companies and 360VR images/videos have become popular. The use of 360VR images/videos makes it easier to develop immersive environments that have higher educational efficiency than the use of 3D-CG. Therefore, we introduced new functionalities that support 360VR images/videos into the proposed frameworks [10, 11] and developed educational materials for IoT security education using the framework [12, 13].

At present, the demand of XR applications has become increased more and more because the advancement of XR technologies, e.g., high specification VR goggles, 360VR cameras and lidar scanners have become available at a cheaper price rather than ever. Therefore, as new activity of ICER, we made another development framework for developing web-based XR educational materials that supports VR goggles such as Meta Quest 2, 360VR camera images/videos and Point Cloud DAT(PCD) by extending the already proposed frameworks [10, 11]. We describe details of this new development framework, and show a couple of example materials developed using the framework for justifying its usefulness.

The remainder of this paper is organized as follows: next Sec. 2 describes related work. We briefly explain the previously proposed framework [10, 11], its functional components in Sec. 3. In Sec. 4, we introduce the framework proposed in this paper and Sec. 5 shows several educational material examples. Finally, we conclude the paper and discuss about future work in Sec. 6.

2 Related Work

Usually, we have to use any toolkit systems when creating web-based interactive 3D educational materials. There is *IntelligentBox* [14] that is a development system for interactive 3D graphics desktop applications. There have been many applications actually developed using *IntelligentBox* so far. Although there are web-based 3D graphics applications developed using the web-version of *IntelligentBox* [15], those cannot support VR goggles such as Meta Quest 2. Unity 3D, a game engine is one of the most popular toolkit systems in the world [16], that enables creating web contents those support VR goggles such as Meta Quest 2. The use of Unity 3D requires programming knowledge and skills of the operations for it. Therefore, it is not easy for standard end-users like teachers to use Unity 3D. There are many commercial services for creating interactive web contents using 360VR camera images [17, 18]. The service of RICHO [17] does support 360VR images but not 360VR videos nor PCD. The service of Matterport [18] does support 360VR images and PCD but not 360VR videos. Contrarily, our framework consists of three main systems those are for walkthrough contents supporting 360VR images, for navigation contents supporting 360VR videos and for walkthrough contents supporting PCD.

3 Web Based Interactive 3D Educational Material Development Framework Supporting VR/AR and 360VR Images/Videos

First of all, we introduce our previous development frameworks in this section. First one is a web based interactive 3D educational material development framework supporting VR/AR and the other one is a web based interactive 3D educational material development framework supporting 360VR images/videos.

3.1 Web Based Interactive 3D Educational Material Development Framework Supporting VR/AR

In this framework, there are following functionalities to support VR/AR those are a stereo view, touch interfaces, device orientation/motion interfaces, geolocation interface and a camera interface for tablet devices and smart phones as shown in Fig. 1. See the paper [10] for more details.

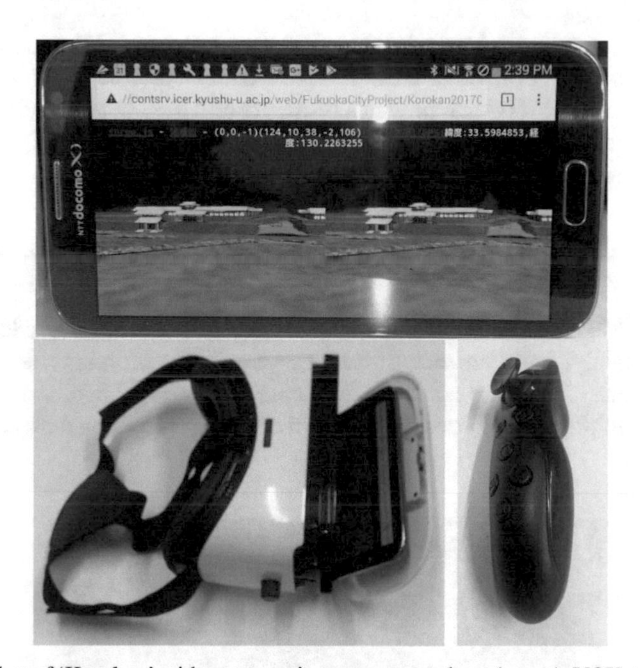

Fig. 1. Snap shot of 'Korokan' with a stereo view on a smart phone(upper), VOX + 3DVR goggle (lower left) and Poskey blue-tooth gamepad (lower right).

1) Stereo view support
   StereoEffect.js is a dedicated JavaScript program for the stereo view of Three.js. This can be read in a HTML file and can call effect.render(scene, camera) for the stereo view instead of calling renderer.render(scene, camera) for the standard view.

2) Touch interfaces
   The implementation for the touch interfaces is also simple because HTML5 supports 'touchstart', 'touchend', 'touchmove' events. The user can access x and y positions of his/her touch fingers by event.touches[*].pageX and event.touches[*].pageY, here, * means the indices of his/her fingers.
3) Device orientation/motion interfaces
   The implementation for the device orientation/motion interfaces is simple because HTML5 supports 'deviceorientation' and 'devicemotion' events. The user can access the device orientation/motion of his/her smart phone by event.alpha, event.beta, event.gamma, event.acceleration.x, event.acceleration.y and event.acceleration.z, respectively.
4) Geolocation interface
   The implementation for geolocation interface is also simple because HTML5 supports navigator.geolocation variable and navigator.geolocation.getCurrentPosition( ) function. The user can access the device geolocation of his/her smart phone by position.coords.latitude and position.coords.longitude.
5) Camera interface
   For creating web-based AR applications, we have to manage video camera images in real-time on a web browser. It is possible to manage video camera images on a web browser using WebRTC that provides browsers and mobile applications with Real-Time Communications (RTC) capabilities including multimedia communications via simple APIs.

3.2 Web Based Interactive 3D Educational Material Development Framework Supporting 360VR Images/videos

The followings are functionalities for supporting 360VR images/videos based on WebGL. Originally, there are many example JavaScript programs developed using Three.js and some of them are examples for displaying 360VR images/videos so that we can extend those programs to make our own web 3D applications that support 360VR images/videos.

Fig. 2. Equirectangular projection image/video frame.

For displaying 360VR images/videos, we have to prepare equirectangular projection images/videos like those shown in Fig. 2. Using a special camera, e.g., RICHO Theta, it is possible to take images/videos shown in Fig. 3 and by converting them, it is possible to generate equirectangular projection images/videos using the dedicated software provided by RICHO. Then, by the texture mapping of such projection image/video frame onto the surface inside of a sphere in a virtual 3D space of WebGL. Figure 4 shows one frame of 360VR video.

In 360VR video of Fig. 4, one of the authors walked from one location to another location with bringing a RICHO Theta. In any video frame, it is possible to see 360-degree image. Therefore, if it is possible to adequately display any video frame anytime, we can walk through freely between the position of a first video frame and the position of a last video frame as shown in Fig. 5.

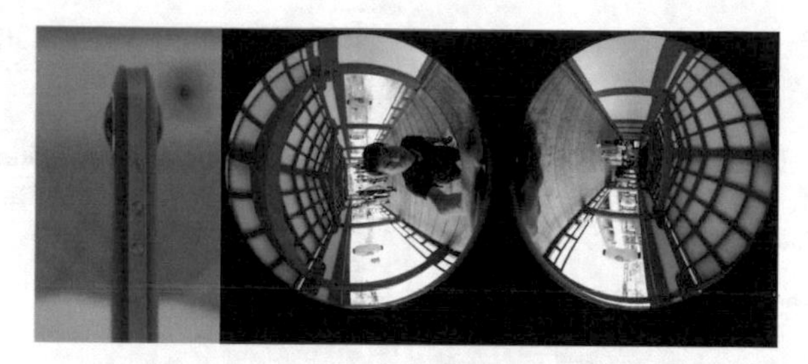

Fig. 3. 360VR camera, e.g., RICHO Theta and its captured image.

Fig. 4. One frame of a certain 360VR video on WebGL based viewer.

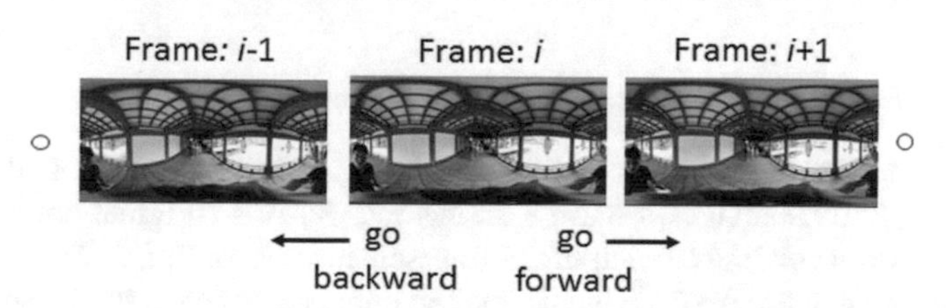

Fig. 5. Several frames in backward (left) and forward (right) direction of 360VR video.

4 Proposed Development Framework

Our framework proposed in this paper consists of three individual systems those are for walkthrough contents supporting 360VR images, for navigation contents supporting 360VR videos and for walkthrough contents supporting Point Cloud Data. The all JavaScript and HTML files of the three systems are shown in Fig. 6.

walk_map.html	JSM {
video_map.html	Several JS files from Three.js library
pcd_map.html	}
JS {	Assets
walk_map.js,	Images { 360VR images, optional images }
walk_map_controls.js,	Videos { 360VR videos }
video_map.js,	Models { PCD files }
video_map_controls.js,	Movies { optional movie files }
pcd_map.js,	Sounds { optional sound files }
pcd_map_controls.js	
}	

Fig. 6. Files for the three systems

Each of the three html files and each pair of the two JavaScript files, i.e., {walk_map.html, walk_map.js and walk_map_controls.js}, {video_map.html, video_map.js and video_map_controls.js} and {pcd_map.html, pcd_map.js and pcd_map_controls.js} are each system for walkthrough content of 360VR images, for navigation content of 360VR videos and for navigation content of PCD, respectively. Walk_map.js, video_map.js and pcd_map.js are main JavaScript programs of the three systems that work with several functions derived from Three.js library [19]. When creating new contents, you do not need to change these programs. Only you need to prepare required media files, i.e., 360VR images, 360VR videos and PCD files included in Assets directory, and just change walk_map_controls.js, video_map_controls.js and pcd_map_controls.js those are subsidiary JavaScript programs to appropriately read required media files into each of the three main JavaScript programs.

Actually, we made VR tour contents of the university library building for our open campus event to appeal that the building is one of the biggest university libraries in Japan. In the followings, we explain each system individually using these VR tour contents.

4.1 Walkthrough Contents Supporting 360VR Images

Figure 7 shows four screen shots of the walkthrough content, 3rd floor of the library building using 360VR images. Figure 8 shows four 360VR (original equirectangular) images, each corresponding to each of the four screen shots of Fig. 7. The center image of Fig. 8 is the map of 3rd floor appears in the left upper part of each screen shot of Fig. 7. On the map, there are many orange dots totally over 100 each indicates the location at where each 360VR image was taken by 360VR camera, Insta 360 Pro.

Fig. 7. Four screen shots of the walkthrough content, 3rd floor of the library building.

By clicking on the orange dot, its corresponding 360VR image will be loaded to display it as the next 360VR scene. Similarly, by clicking on one of the thin grey cylinders in a 360VR scene, the corresponding 360VR image taken at the same location will be loaded to display it as the next 360VR scene.

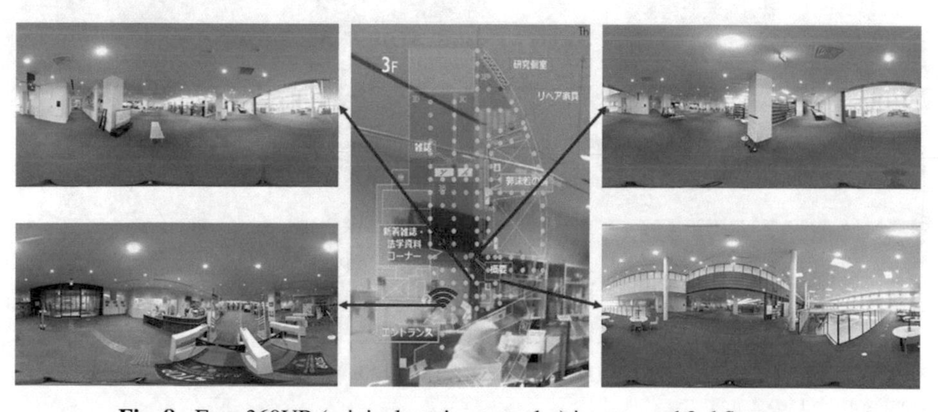

Fig. 8. Four 360VR (original equirectangular) images and 3rd floor map.

The framework supports multimedia like standard image files and movie files. For example, one of the standard movies appears in the screen shot at the upper left part of Fig. 7 and one of the standard images, an explanation panel appears in the screen shot at the upper right part of the figure by clicking on the corresponding red sphere. Also, as shown in the right lower part, there is a yellow sphere close to the stairs and by clicking on it, you can go to downstairs or upstairs.

4.2 Navigation Contents Supporting 360VR Videos

Figure 9 shows three screen shots of the navigation content, 3rd floor of the library building using 360VR video. The lower middle part of each of the three screen shots is a control panel for the 360VR video, i.e., play, pause, backward, forward, etc. The upper left part of the figure is the map of 3rd floor displayed on the left upper part of each of the three screen shots. There is a cyan color closed polyline that means the moving path actually the person wearing a 360VR camera on the top of his/her head moved through when taking this 360VR video. By clicking on the polyline, the 360VR video moves to the corresponding 360VR scene and it will appear. Similar to the system for walkthrough contents, this system also supports multimedia like standard image files. For example, in each of the two screen shots of the right part of Fig. 9, there is an image file displayed on its center. The 360VR scene automatically changes according to the current playing point of the 360VR video. Therefore, we call this type of contents are navigation ones.

Fig. 9. Three screen shots of the navigation content, 3rd floor of the library building using 360VR video and 3rd floor map.

4.3 Walkthrough Contents Supporting PCD

Figure 10 shows four screen shots of the navigation content, 3rd floor of the library building using PCD taken by a lidar scanner, Leica BLK 360. Each of these four screen shots corresponds to each of the four screen shots of Fig. 7. The operations for this content of Fig. 10 are almost the same as those of the content of Fig. 7. The difference between them is supporting data, i.e., 360VR images of Fig. 7 and PCD of Fig. 10. This system also supports multimedia shown in the upper part screen shots, standard movies and images.

Fig. 10. Four screen shots of the navigation content, 3rd floor of the library building using PCD.

5 Educational Material Examples

Figure 11 shows two screen shots, one is the walkthrough content of 360VR images (left) and the other is the walkthrough content of PCD (right) of the wind tunnel experiment room. The upper part of Fig. 12 shows two screen shots, one is the walkthrough content of 360VR images (left) whose WebXR mode shown in the lower part of the figure that supports VR goggles like Meta Quest 2, and the other one is the navigation content of 360VR videos of Ito isotope experiment center.

Besides the above contents introduced in this paper, we created walkthrough contents of 360VR images and PCD for six other center, laboratory, workshop rooms. Although we needed a couple of hours for taking 360VR images, capturing PCD and making a floor map image of one room or one floor, just after that, we could create these contents quickly in less than one hour by using our proposed framework. Therefore, it can be said that the proposed framework is useful for developing these kinds of XR contents.

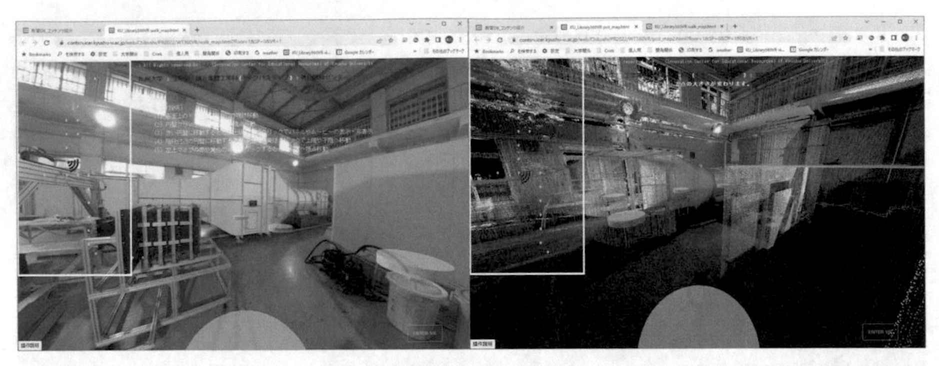

Fig. 11. Two screen shots, one is the walkthrough content of 360VR images (left) and the other is the walkthrough content of PCD (right) of a wind tunnel experiment room.

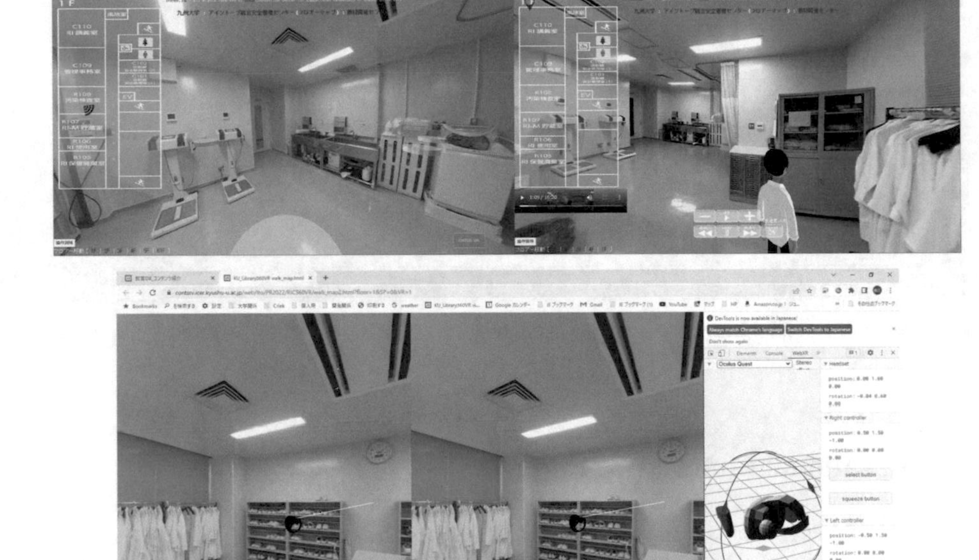

Fig. 12. Two screen shots, one (left) is the walkthrough content of 360VR images (left) whose WebXR mode shown in the lower part, and the other one (right) is the navigation content of 360VR videos of Ito isotope experiment center.

6 Conclusions

In this paper, we proposed the development framework using 360VR cameras and lidar scanners for web-based interactive XR educational materials. Recently, the demand of XR(VR/AR/MR) applications has become increased due to the advancement of XR technologies. Besides entertainment fields, educational materials using XR are also demanded, especially, for experiment and exercise subjects. However, the development of such educational materials is not easy and needs much time. Therefore, we made the development framework that supports 360VR camera images/videos and Point Cloud Data(PCD) of lidar scanners. We showed several material examples to justify the usefulness of the proposed framework.

As future work, we will try to ask teachers to use the proposed framework for actually creating more web-based interactive XR educational materials to clarify the usefulness of the framework. Furthermore, we will create several educational materials with the framework and ask students to learn using the materials. After that, we will consult the students for evaluating educational efficiency of the materials.

Acknowledgements. This research was partially supported by JSPS KAKENHI Grant Number JP22H03705 and VISION EXPO Project of Kyushu University.

References

1. ICER: http://www.icer.kyushu-u.ac.jp/en. 1 Feb 2023
2. Sugimura, R., et al.: Mobile game for learning bacteriology. In: Proceedings of IADIS 10th International Conference on Mobile Learning 2014, pp.285–289 (2014)
3. Sugimura, R., et al.: Serious games for education and their effectiveness for higher education medical students and for junior high school students. In: Proceedings of 4th International Conference on Advanced in Information System, E-Education and Development(ICAISEED 2015), pp. 36–45 (2015)
4. Okada, Y., Nakazono, S., Kaneko, K.: Framework for development of web-based interactive 3D educational contents. In: 10th International Technology, Education and Development Conference, pp. 2656–2663 (2016)
5. Okada, Y., Kaneko, K., Tanizawa, A.: Interactive educational contents development framework based on linked open data technology. In: 9th annual International Conference of Education, Research and Innovation, pp. 5066–5075 (2016)
6. Hirayama, D., Shi, W., Okada, Y.: Web-based interactive 3D educational material development framework and its authoring functionalities. In: Barolli, L., Nishino, H., Enokido, T., Takizawa, M. (eds.) NBiS - 2019 2019. AISC, vol. 1036, pp. 258–269. Springer, Cham (2020). https://doi.org/10.1007/978-3-030-29029-0_24
7. Yamamura, H., et al.: A development framework for RP-type serious games in a 3D virtual environment. In: Barolli, L., Poniszewska-Maranda, A., Enokido, T. (eds.) CISIS 2020. AISC, vol. 1194, pp. 166–176. Springer, Cham (2021). https://doi.org/10.1007/978-3-030-50454-0_16
8. Okada, Y., Kaneko, K. ,Tanizawa, A.: Interactive educational contents development framework and its extension for web-based VR/AR applications. In: Proceedings of the GameOn 2017, Eurosis, pp. 75–79 (2017). ISBN: 978-90-77381-99-1

9. Chenguang, M., Kulshrestha, S., Wei, S., Yoshihiro, O., Ranjan, B.: E-learning material development framework supporting VR/AR based on linked data for IoT security education. In: Proceedings of 6th International Conference on Emerging Internet, Data & Web Technologies (EIDWT 2018), pp. 479–491. Tirana/Albania (2018). ISBN: 978-3-319-75928-9
10. Yoshihiro, O., Akira, H., Wei, S.: Web based interactive 3D educational material development framework supporting 360VR images/videos and its examples. In: Proceedings of the 21st International Conference on Network-Based Information Systems (NBiS-2018), pp. 395–406 (2018). ISBN 978-3-319-98529-9
11. Akira, H., Wei, S., Ginpei, H., Yoshihiro, O.: Web based interactive viewer for 360VR images/videos supporting VR/AR devices with human face detection. In: Proceedings of the 13th International Conference on Broadband and Wireless Computing, Communication and Applications (BWCCA-2018), pp. 260–270 (2018). https://doi.org/10.1007/978-3-030-02613-4_23
12. Okada, Y., Haga, A., Wei, S., Ma, C., Kulshrestha, S., Bose, R.: E-Learning Material Development Framework Supporting 360VR Images/Videos Based on Linked Data for IoT Security Education. In: Barolli, L., Xhafa, F., Khan, Z.A., Odhabi, H. (eds.) EIDWT 2019. LNDECT, vol. 29, pp. 148–160. Springer, Cham (2019). https://doi.org/10.1007/978-3-030-12839-5_14
13. Shi, W., Gao, T., Kulshrestha, S., Bose, R., Haga, A., Okada, Y.: A Framework for Automatically Generating IoT Security Quizzes in 360VR Images/Videos Based on Linked Data. In: Barolli, Leonard, Okada, Yoshihiro, Amato, Flora (eds.) EIDWT 2020. LNDECT, vol. 47, pp. 259–267. Springer, Cham (2020). https://doi.org/10.1007/978-3-030-39746-3_28
14. Okada, Y., Tanaka, Y.: IntelligentBox: a constructive visual software development system for interactive 3D graphic applications. In: Proceedings of Computer Animation'95, pp. 114–125. IEEE CS Press (1995)
15. Okada, Y.: Web version of IntelligentBox (WebIB) and its integration with Webble world. In: Arnold, O., Spickermann, W., Spyratos, N., Tanaka, Y. (eds.) WWS 2013. CCIS, vol. 372, pp. 11–20. Springer, Heidelberg (2013). https://doi.org/10.1007/978-3-642-38836-1_2
16. Unity 3D: https://unity3d.com/jp. 1 Feb 2023
17. Richo Theta360.zi: https://www.theta360.biz/. 1 Feb 2023
18. Matterport: https://matterport.com/ja. 1 Feb 2023
19. Three.js: https://threejs.org/. 1 Feb 2023

A Comparison Study of LDVM and RDVM Router Replacement Methods by WMN-PSODGA Hybrid Simulation System Considering Two Islands Distribution of Mesh Clients

Admir Barolli[1], Kevin Bylykbashi[2], Leonard Barolli[2], Ermioni Qafzezi[3](✉), Shinji Sakamoto[4], and Makoto Takizawa[5]

[1] Department of Information Technology, Aleksander Moisiu University of Durres, L.1, Rruga e Currilave, Durres, Albania
admirbarolli@uamd.edu.al

[2] Department of Information and Communication Engineering, Fukuoka Institute of Technology, 3-30-1 Wajiro-Higashi, Higashi-Ku, Fukuoka 811-0295, Japan
kevin@bene.fit.ac.jp, barolli@fit.ac.jp

[3] Graduate School of Engineering,Fukuoka Institute of Technology, 3-30-1 Wajiro-Higashi, Higashi-Ku, Fukuoka 811-0295, Japan
bd20101@bene.fit.ac.jp

[4] Department of Information and Computer Science, Kanazawa Institute of Technology, 7-1 Ohgigaoka Nonoichi, Ishikawa 921-8501, Japan
shinji.sakamoto@ieee.org

[5] Department of Advanced Sciences, Faculty of Science and Engineering, Hosei University, 3-7-2, Kajino-machi, Koganei-shi, Tokyo 184-8584, Japan
makoto.takizawa@computer.org

Abstract. In this paper, different from our previous work, we consider Two Islands distribution of mesh clients and optimize the number of mesh routers in Wireless Mesh Networks (WMNs) using WMN-PSODGA hybrid intelligent simulation system. For the evaluation of the implemented system, we consider three parameters: network connectivity, mesh client coverage and load balancing. We carry out a comparison study between Linearly Decreasing Vmax Method (LDVM) and Rational Decrement of Vmax Method (RDVM). The simulation results show that RDVM has better load balancing than LDVM.

1 Introduction

The Wireless Mesh Networks (WMNs) can be used as last mile networks. They are the most suitable networks to link the devices at the edge of the networking where the data is generated, due to the cost-effective scalability, fault tolerance and load distribution. Edge devices cooperate with each other for sharing data and computation which results to higher analytics capability, better response time, better performance and higher security and privacy. Scalability is one of the most important requirements in Internet of Things (IoT). A centralized network deals with limited bandwidth and therefore can

L. Barolli (Ed.): EIDWT 2023, LNDECT 161, pp. 413–422, 2023.
https://doi.org/10.1007/978-3-031-26281-4_43

not support the increasing number of devices. On the other hand, edge mesh does not rely in one single device but is based in distributed intelligence. In this way edge mesh resolve bottleneck issue and support large scale networks.

The WMNs have many advantages such as low cost, easy scalability and they are resistant to node failure problems. The mesh nodes are self organizing and requires minimal manual settings for operation. Also, WMNs offer good reliability and freedom for mobile users of the system with minimum configurations. However, they have different problems such as hidden terminal problem, security, increased workload for each mesh node and lattency issues. Therefore, the optimization of WMNs is a very important issue.

For designing and engineering WMNs are needed many parameters which may be uncorrelated with each other. Therefore a trade-off on the solutions is required in the optimization process. Mesh node placement in WMNs is an NP-Hard problem [2,6,13]. For the optimization process, we consider a grid area and deploy a number of mesh router nodes and mesh client nodes of fixed positions. The problem is where to deploy mesh routers in the grid area in order to achieve the maximum network connectivity and client coverage while balancing the load among mesh routers.

We consider three parameters: Size of Giant Component (SGC), Number of Covered Mesh Clients (NCMC) and Number of Covered Mesh Clients per Router (NCMCpR). The network connectivity is measured by SGC, while the client coverage is the number of mesh client nodes that fall within the radio coverage of at least one mesh router node and is measured by the NCMC. For load balancing, we use NCMCpR parameter.

Node placement problems are known to be computationally hard to solve [4,5,14]. In previous works, some intelligent algorithms have been recently investigated for the node placement problem [1,3,7,8].

In [10], we implemented a Particle Swarm Optimization (PSO) based simulation system, called WMN-PSO. Also, we implemented another simulation system based on Genetic Algorithms (GA), called WMN-GA [9]. Then, we designed and implemented a hybrid simulation system based on PSO and Distributed GA (DGA). We call this system WMN-PSODGA.

In this paper, we compare the simulation results for Linearly Decreasing Vmax Method (LDVM) and Rational Decrement of Vmax Method (RDVM) for the Two Islands distribution of mesh clients. The simulation results show that RDVM has better load balancing than LDVM.

The rest of the paper is organized as follows. In Sect. 2, we introduce intelligent algorithms. In Sect. 3 is presented the implemented hybrid simulation system. The simulation results are given in Sect. 4. Finally, we give conclusions and future work in Sect. 5.

2 Intelligent Algorithms for Proposed Hybrid Simulation System

2.1 PSO

The PSO is a population based stochastic optimization algorithm inspired by social behavior of bird flocking or fish schooling. The PSO is initialized with a group of random particles (solutions) and then searches for optimal solution by updating generations. In every iteration, each particle is updated by following two "best" values. The first one is the best solution (fitness) that has achieved so far. This value is called pbest. Another "best" value that is tracked by the particle swarm optimizer is the best value, obtained by any particle in the population. This best value is a global best and called gbest. When a particle takes part of the population as its topological neighbors, the best value is a local best and is called lbest.

The velocities of particles on each dimension are clamped to a maximum velocity Vmax. If the sum of accelerations would cause the velocity on that dimension to exceed Vmax, then the velocity on that dimension is limited to Vmax.

2.2 GA and DGA

The GA is robust search algorithm that searches a population of individuals and can operate on various representations. However, the design of objective function and the representation of operators are difficult. The GAs have been used successfully in a variety of optimization problems in various domains. However, the drawback of GA is that it is computationally expensive and time consuming. The GA has Selection, Crossover and Mutation operators. In this paper, we use the random selection method, Unimodal Normal Distribution Crossover (UNDX) method and the uniformly random mutation method.

GA has the ability to avoid falling to local optima and can escape from them during the search process. DGA has one more mechanism to escape from local optima by considering some islands. Each island computes GA for optimizing and they migrate their genes to provide the ability of avoiding local optima.

2.3 Comparison Between GA and PSO

Both algorithms start with a group of a randomly generated population and they have fitness values to evaluate the population. The algorithms update the population and search for the optimium with random techniques. However, PSO does not have genetic operators like crossover and mutation. Particles update themselves with the internal velocity. They also have memory, which is important to the algorithm.

Compared with GA, the information sharing mechanism in PSO is significantly different. In GAs, chromosomes share information with each other. So the whole population moves like a group towards an optimal area. In PSO, only gbest (or lbest) gives out the information to others. It is a one way information sharing mechanism. The evolution only looks for the best solution. Compared with GA, all particles tend to converge to the best solution quickly in most cases.

3 Proposed and Implemented WMN-PSODGA Hybrid Intelligent Simulation System

In this section, we present the proposed WMN-PSODGA hybrid intelligent simulation system. The structure of WMN-PSODGA is shown in Fig. 1. The implemented system uses Migration function to swaps solutions among lands included in PSO part. The flowchart of WMN-PSODGA is shown in Fig. 2. In following, we describe the initialization, particle-pattern, gene coding, fitness function, mesh clients distribution and replacement methods.

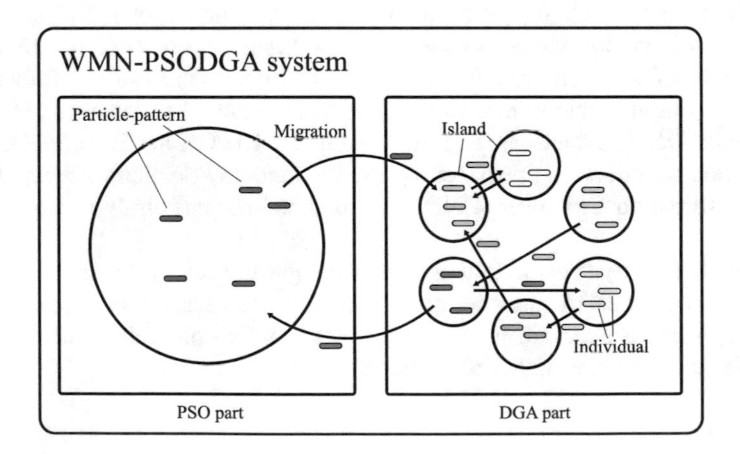

Fig. 1. Structure of WMN-PSODGA system.

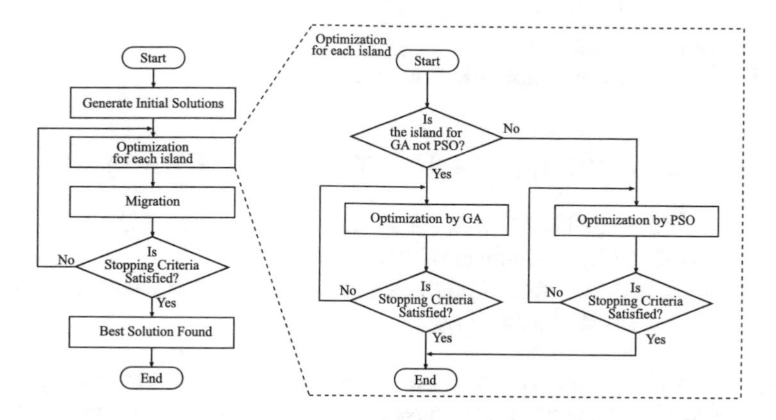

Fig. 2. Flowchart of WMN-PSODGA system.

Initialization

We decide the velocity of particles by a random process considering the area size. For instance, when the area size is $W \times H$, the velocity is decided randomly from $-\sqrt{W^2+H^2}$ to $\sqrt{W^2+H^2}$.

Particle-pattern

A particle is a mesh router. A fitness value of a particle-pattern is computed by combination of mesh routers and mesh clients positions. In other words, each particle-pattern is a solution as shown is Fig. 3.

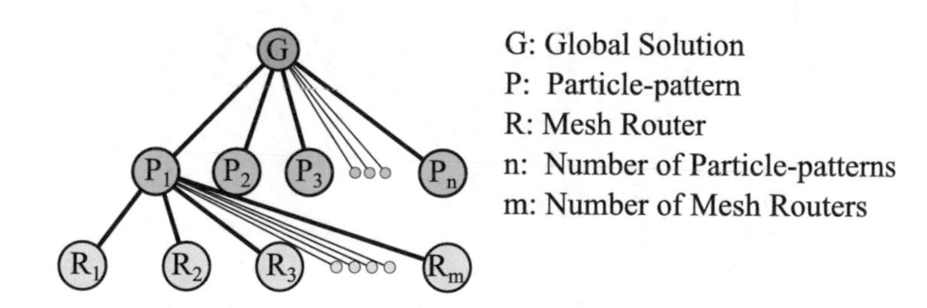

Fig. 3. Relationship between global solution, particle-patterns and mesh routers.

Gene Coding

A gene describes a WMN. Each individual has its own combination of mesh nodes. In other words, each individual has a fitness value. Therefore, the combination of mesh nodes is a solution.

Fitness Function
WMN-PSODGA fitness function is defined as:

$$Fitness = \alpha \times NCMC(\mathbf{x}_{ij}, \mathbf{y}_{ij}) + \beta \times SGC(\mathbf{x}_{ij}, \mathbf{y}_{ij}) + \gamma \times NCMCpR(\mathbf{x}_{ij}, \mathbf{y}_{ij}).$$

- The NCMC is the number of mesh clients covered by routers in SGC.
- The SGC is the maximum number of connected routers.
- The NCMCpR is the number of mesh clients covered by each router. The NCMCpR indicator is used for load balancing.

WMN-PSODGA maximizes the value of fitness function in order to optimize the placement of mesh routers using the above three indicators. Weight-coefficients of the fitness function are α, β, and γ for NCMC, SGC, and NCMCpR, respectively. Moreover, the weight-coefficients are implemented as $\alpha + \beta + \gamma = 1$.

Mesh Clients Distribution
In this work, we implement Two Islands distribution of mesh clients, which can generates locations of mesh clients scattered over two islands. The Two Islands distribution is shown in Fig. 4.

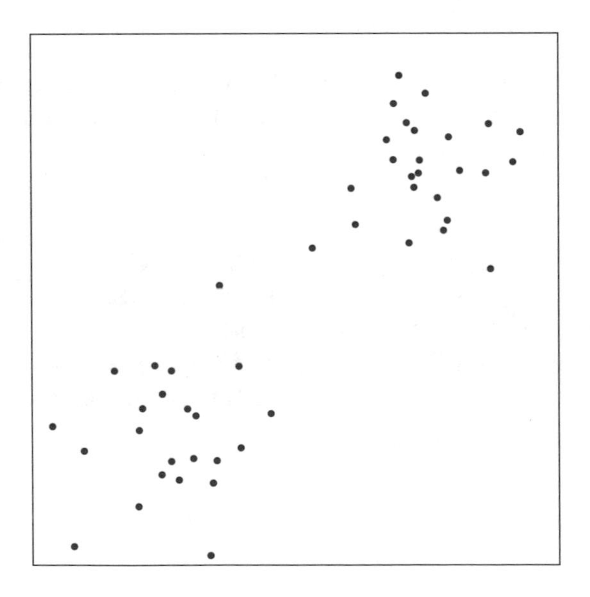

Fig. 4. Two Islands distribution of mesh clients.

Router Replacement Methods
There are many router replacement methods. In this paper, we consider the Linearly Decreasing Vmax Method (LDVM) and Rational Decrement of Vmax Method (RDVM).

Table 1. Simulation parameters.

Parameters	Values
Distribution of mesh clients	Two islands
Number of mesh clients	48
Number of mesh routers	17
Radius of a mesh router	2.0–3.5
Number of GA islands	16
Number of migrations	200
Evolution steps	9
Selection method	Random
Crossover method	UNDX
Mutation method	Uniform
Crossover rate	0.8
Mutation rate	0.2
Replacement method	LDVM, RDVM
Area size	32.0×32.0

Linearly Decreasing Vmax Method (LDVM)
In LDVM, PSO parameters are set to an unstable region ($\omega = 0.9$, $C_1 = C_2 = 2.0$). The V_{max} is the maximum velocity of particles, which decreases linearly after each iteration [12].

Rational Decrement of Vmax Method (RDVM)
In RDVM, PSO parameters are set to tan unstable region ($\omega = 0.9$, $C_1 = C_2 = 2.0$). The V_{max} decreases after each iteration as shown in the following equation.

$$V_{max}(x) = \sqrt{W^2 + H^2} \times \frac{T - x}{x}.$$

Where W and H are the width and height of the considered area, respectively. While, T and x are the total number of iterations and the current number of iterations, respectively [11].

4 Simulation Results

In this section, we present and compare the simulation results of LDVM and RDVM router replacement methods for the Two Islands distribution of mesh clients. The weight-coefficients are $\alpha = 0.6$, $\beta = 0.3$, $\gamma = 0.1$. The number of mesh routers and mesh clients are 16 and 48, whereas the selection, crossover and mutation methods are Random, UNDX and Uniform, respectively. Table 1 presents the common parameters used for the simulations.

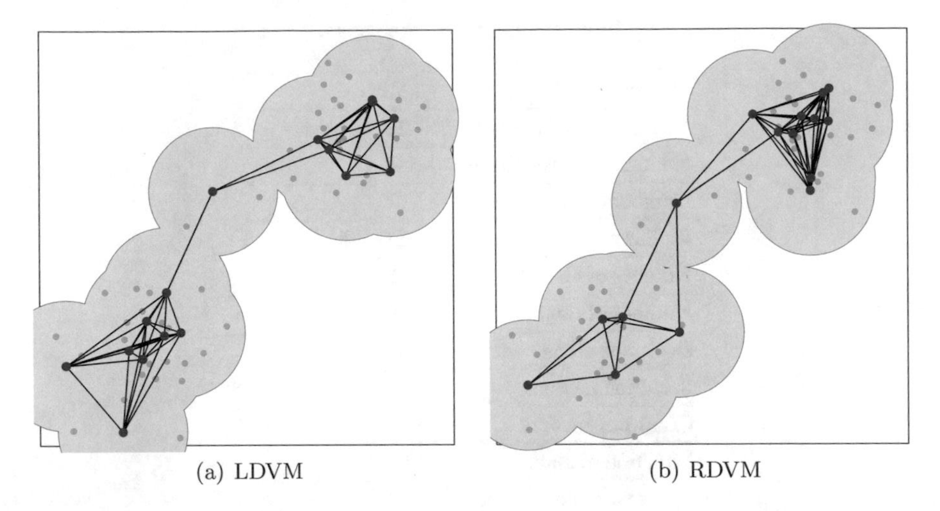

Fig. 5. Visualization results after optimization.

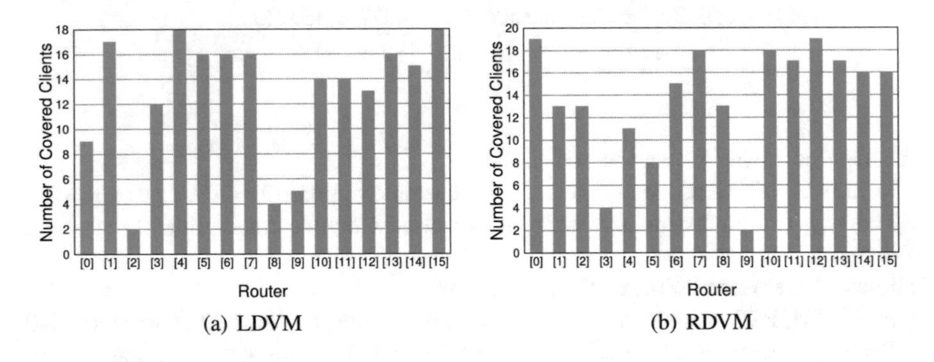

Fig. 6. Number of covered clients by each router after optimization.

In Fig. 5 are shown the visualization results after the optimization, with Fig. 5(a) and Fig. 5(b) showing the results for LDVM and RDVM, respectively. Figure 6 shows the number of covered mesh clients by each router for each replacement method, whereas Fig. 7 the standard deviation where r is the correlation coefficient. Both router replacement methods achieve full coverage and high connectivity. But, they have different performance on load balancing.

In Fig. 6(a) and Fig. 6(b), we see that in each simulation scenario each mesh router covers at least two mesh clients and the majority of mesh routers cover more than ten mesh clients. However, the clear indicator of load balancing is shown in Fig. 7(a) and Fig. 7(b). We compare the standard deviations and their correlation coefficients. When standard deviation is a decreasing line, the number of mesh clients for each router tends to be close to each other. The standard deviation is an increasing line for LDVM and a decreasing line for RDVM. Therefore, a better load balancing is achieved by RDVM.

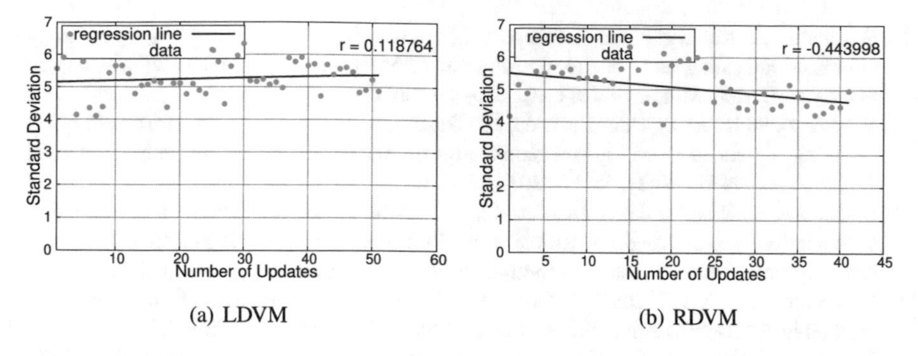

(a) LDVM

(b) RDVM

Fig. 7. Transition of standard deviations.

5 Conclusions

In this paper, we considered Two Islands distribution of mesh clients and optimized the number of mesh routers in WMNs using WMN-PSODGA hybrid intelligent simulation system. For the evaluation we considered three parameters: SGC, NCMC and NCMCpR. We carried out a comparison study between LDVM and RDVM methods. The simulation results show that RDVM had better load balancing than LDVM.

In future work, we will consider different crossover methods, mutation methods and other router replacement methods. Also, we will consider other mesh clients distributions.

References

1. Barolli, A., Sakamoto, S., Ozera, K., Barolli, L., Kulla, E., Takizawa, M.: Design and Implementation of a Hybrid Intelligent System Based on Particle Swarm Optimization and Distributed Genetic Algorithm. In: Barolli, L., Xhafa, F., Javaid, N., Spaho, E., Kolici, V. (eds.) EIDWT 2018. LNDECT, vol. 17, pp. 79–93. Springer, Cham (2018). https://doi.org/10.1007/978-3-319-75928-9_7
2. Franklin, A.A., Murthy, C.S.R.: Node placement algorithm for deployment of two-tier wireless mesh networks. In: Proceedings of Global Telecommunications Conference, pp 4823–4827 (2007)
3. Girgis, M.R., Mahmoud, T.M., Abdullatif, B.A., Rabie, A.M.: Solving the wireless mesh network design problem using genetic algorithm and simulated annealing optimization methods. Int. J. Comput. Appl. **96**(11), 1–10 (2014)
4. Lim, A., Rodrigues, B., Wang, F., Xu, Z.: k-Center problems with minimum coverage. Theor. Comput. Sci. **332**(1–3), 1–17 (2005)
5. Maolin, T., et al.: Gateways placement in backbone wireless mesh networks. Int. J. Commun. Netw. Syst. Sci. **2**(1), 44–50 (2009)
6. Muthaiah, S.N, Rosenberg, C.P.: Single gateway placement in wireless mesh networks. In: Proceedings of 8th International IEEE Symposium on Computer Networks, pp 4754–4759 (2008)
7. Naka, S., Genji, T., Yura, T., Fukuyama, Y.: A hybrid particle swarm optimization for distribution state estimation. IEEE Trans. Power Syst. **18**(1), 60–68 (2003)

8. Sakamoto, S., Kulla, E., Oda, T., Ikeda, M., Barolli, L., Xhafa, F.: A comparison study of simulated annealing and genetic algorithm for node placement problem in wireless mesh networks. J. Mob. Multimedia **9**(1–2), 101–110 (2013)
9. Sakamoto, S., Kulla, E., Oda, T., Ikeda, M., Barolli, L., Xhafa, F.: A comparison study of hill climbing, simulated annealing and genetic algorithm for node placement problem in WMNs. J. High Speed Netw. **20**(1), 55–66 (2014)
10. Sakamoto, S., Oda, T., Ikeda, M., Barolli, L., Xhafa, F.: Implementation and evaluation of a simulation system based on particle swarm optimisation for node placement problem in wireless mesh networks. Int. J. Commun. Netw. Distrib. Syst. **17**(1), 1–13 (2016)
11. Sakamoto, S., Oda, T., Ikeda, M., Barolli, L., Xhafa, F.: Implementation of a new replacement method in WMN-PSO simulation system and its performance evaluation. In: The 30th IEEE International Conference on Advanced Information Networking and Applications (AINA-2016), pp 206–211 (2016)
12. Schutte, J.F., Groenwold, A.A.: A study of global optimization using particle swarms. J. Global Optim. **31**(1), 93–108 (2005)
13. Vanhatupa, T., Hannikainen, M., Hamalainen, T.: Genetic algorithm to optimize node placement and configuration for WLAN planning. In: Proceedings of The 4th IEEE International Symposium on Wireless Communication Systems, pp 612–616 (2007)
14. Wang, J., Xie, B., Cai, K., Agrawal, D.P.: Efficient mesh router placement in wireless mesh networks. In: Proceedings of IEEE Internatonal Conference on Mobile Adhoc and Sensor Systems (MASS-2007), pp 1–9 (2007)

FBCF: A Fuzzy-Based Brake-Assisting Control Function for Rail Vehicles Using Type-1 and Type-2 Fuzzy Inference Models

Mitsuki Tsuneyoshi[1], Makoto Ikeda[2](✉), and Leonard Barolli[2]

[1] Graduate School of Engineering, Fukuoka Institute of Technology,3-30-1 Wajiro-higashi, Higashi-ku, Fukuoka 811-0295, Japan
mgm21106@bene.fit.ac.jp

[2] Department of Information and Communication Engineering, Fukuoka Institute of Technology, 3-30-1 Wajiro-higashi, Higashi-ku, Fukuoka 811-0295, Japan
makoto.ikd@acm.org, barolli@fit.ac.jp

Abstract. In railway networks, the driver expertise such as accuracy, physical and experience condition has a major impact on the quality of comfort for rail car passengers. In our previous work, we have proposed a Fuzzy-based brake-assisting function considering velocity, brake level and environment to assist the rail drivers. In this paper, we consider the gravitational acceleration level and slope angle as additional parameters to control the brake-assisting function. We present a comparison study for Type-2 and Type-1 inference models. The proposed system, called Fuzzy-based Brake-assisting Control Function (FBCF), provides intelligent braking for passengers and train drivers by considering five input parameters. The simulation results show that the proposed FBCF provides soft brake assistance considering various situations created by the train and rail environment in order that the passengers have a comfortable feeling.

Keywords: Brake-assisting function · Fuzzy logic · Rail vehicle

1 Introduction

To realize Sustainable Development Goals (SDGs) and solve environmental challenges, many intelligent vehicles are being designed [1]. The new products different from old models should have electric power and automatic emergency brake [23]. In the field of railroads, Automatic Train Stop (ATS), Automatic Train Control (ATC) and Automatic Train Operation (ATO) have been developed, and the application of Advanced Driver Assistance System (ADAS) in rail vehicles is also attracting attention. In ATS, a bell in the driver cab gives an alert when a train approaches a stop sign within a predetermined distance and the brakes are automatically applied to stop the train if the driver does not execute the required inspections. The ATC is a control function that ensures the safe operation of trains by comparing the current train velocity to the maximum permissible velocity and automatically reducing the train's velocity to the limit if the maximum velocity is exceeded. During ATO, the driver just presses a button to start the train and

L. Barolli (Ed.): EIDWT 2023, LNDECT 161, pp. 423–431, 2023.
https://doi.org/10.1007/978-3-031-26281-4_44

the system thereafter handles acceleration, cruising, braking operations and fixed-point stops automatically.

Unlike automobiles, railroad vehicles did not require advanced control. But now, when the train is operated automatically in a hazardous region, it is crucial to find the nearby obstructions rapidly [3,6,8,10,26]. Even if the brakes are in good condition, the train may still collide with obstructions if cameras mounted in train do not detect them. As a hazard region can be considered pedestrians, animals and automobiles that can enter the railway tracks. Consequently, some transportable devices or roadside units such as relay equipments used to monitor the area surrounding the tracks and transmit emergency information to trains would be necessary.

In several parts of the world, deceleration and acceleration operations on trains are performed primarily by the driver. During strong g-force deceleration, it is hard for old aged or sick passengers to keep their balance in the train car. This paper focuses on deceleration support for railway vehicles using Fuzzy Logic (FL). In a variety of fields, intelligent approaches such as various learning techniques and FL enable humans to make good decisions [4,5,9,12,14–18,20,24].

In our previous work [22], we proposed a FL-based brake-assisting control function. The proposed method considered velocity, brake and environment status. In this paper, we improve the previous function and propose a new Fuzzy-based Brake-assisting Control Function (FBCF) for railway vehicles. The FBCF provides an intelligent braking assist considering velocity, brake level, gravitational acceleration level, slope angle and environment status. We evaluate two Fuzzy inference types. The evaluation results show that Type-2 FLC (T2FLC) can support a better brake-assisting function than Type-1 FLC (T1FLC).

The structure of the paper is as follows. In Sect. 2, we explain the FBCF for railway vehicles. In Sect. 3, we describe the Input and Output (IO) parameters for FL. In Sect. 4, we explain the simulation results. Finally, conclusions and future work are given in Sect. 5.

2 Proposed FBCF for Railway Vehicles

Uncertainty and fuzziness in the inference process of an intelligent system such a logical control system or a knowledge-based system inspired the development of FL [25] and Fuzzy sets [2,7,11,13,19].

Both Mamdani Fuzzy (MF) inference model [13] and the Sugeno Fuzzy (SF) inference model [21] are classified as Type-1 inference model. In order to implement the SF inference model, we consider a linear function. In this technique, we are able to decrease the complexity of the computations compared to the min-max gravity model.

We developed an integrated environment for Type-1 SF inference system and an interval Type-2 SF inference system in MATLAB. The proposed brake assist control function consists of two Fuzzy Logic Controllers (FLCs: T1FLC and T2FLC). For T1FLC, we use trapezoidal and triangular MFs. While for T2FLC, we use sigmoidal and gaussian MFs. For output values of both T1FLC and T2FLC, we use a linear function, because it is good for real time applications.

We use a Karnik-Mendel (KM) algorithm as type-reduction method for T2FLC. The input MFs of T1FLC and T2FLC are shown in Fig. 1. In Fig. 1 at right side, we show Lower MF (blue line), Upper MF (red line) and Footprint of Uncertainty (FOU) for the Type-2 MFs.

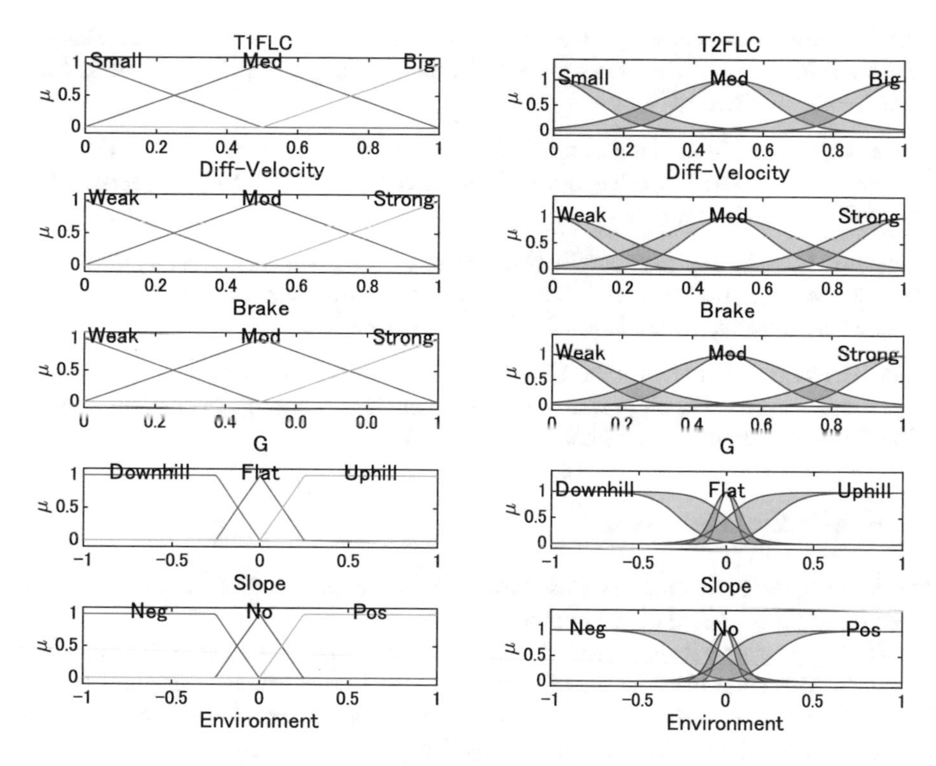

Fig. 1. Input MFs for T1FLC and T2FLC. (Color figure online)

3 IO Parameters

We present in following Diff-Velocity, current brake level, gravitational acceleration, slope angle and environment input parameters and U output parameter used for our T1FLC and T2 FLC.

Diff-Velocity: We consider three Diff-Velocity conditions. If the allowable speed parameter is available at the train cab, camera-equipped speed-reading subsystem should be implemented.

Current Brake Level: We consider the current train brake pressure. There are three conditions of the Brake (Weak, Moderate (Mod) and Strong).

Gravitational acceleration: We consider the current Gravitational (G) acceleration conditions in the train. The G has three different acceleration conditions (Weak, Moderate (Mod) and Strong).

Slope Angle: The Slope indicates the slope angle of the train on the rail. The value is from -1 to 1. The Slope has three different conditions that can be interpreted as: Downhill, Flat and Uphill.

Environment: The Environment (Env) indicates the environment status on the rail vehicles. The value is from -1 to 1. The Env has three different conditions that can be assigned as: Negative (Neg), Normal (No) and Positive (Pos).

Unit: The system can control the Unit (U) value in order to improve the comfortable level for passengers. The output has eleven different levels. For this evaluation, we setup 243 rules in Fuzzy rule-base by combining the IO parameters.

4 Evaluation Results

In this paper, wc evaluate the proposed FBCF, which consists of T1FLC and T2FLC to assist the braking control of rail vehicles.

The relationship between selected input parameters and the output results are shown in Fig. 2. The color-bar indicates the level of the output value U. Figure 2(a) and Fig. 2(b) show the relationship between U output parameter with Diff-velocity and Brake for T1FLC and T2FLC, respectively. For T1FLC, we observe that the output U value is increased linearly with the increase of Diff-Velocity and Brake values. While in T2FLC, the control is improved. Figure 2(c) and Fig. 2(d) show the relationship between U output parameter with gravitational acceleration and slope angle for T1FLC and T2FLC, respectively. For T1FLC, we observe that the output U value is increased linearly with the increase of G values. Also, it has a stepwise braking control with the increase of Slope values. While in T2FLC, we noticed that there are increasing numbers of stepwise braking control. Figure 2(e) and Fig. 2(f) show the relationship between U output parameter with slope angle and Environment for T1FLC and T2FLC, respectively. We see that the results are close, but in T2FLC the stepwise control is enhanced.

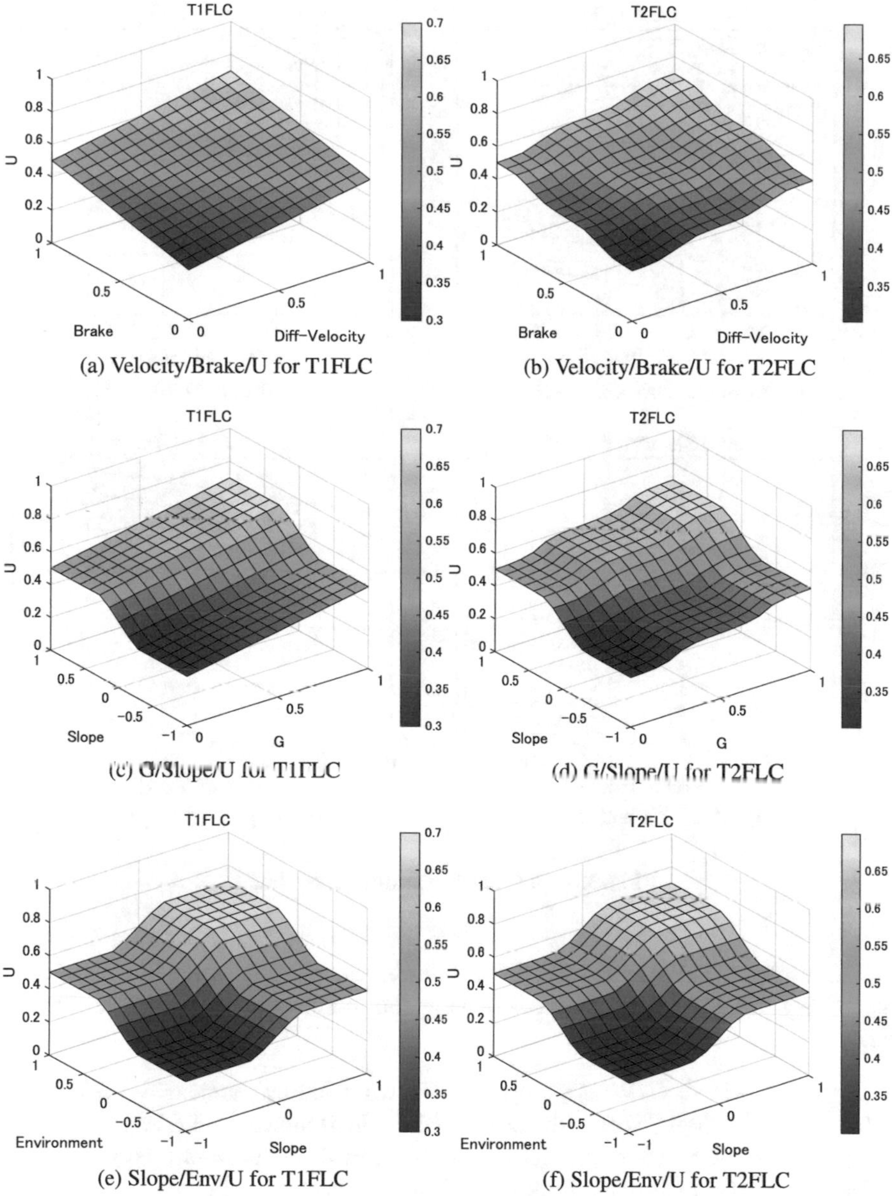

T1FLC
T2FLC
U
Brake
Diff-Velocity
Slope
G
Environment
(a) Velocity/Brake/U for T1FLC
(b) Velocity/Brake/U for T2FLC
(c) G/Slope/U for T1FLC
(d) G/Slope/U for T2FLC
(e) Slope/Env/U for T1FLC
(f) Slope/Env/U for T2FLC

Fig. 2. Output results for different selected parameters.

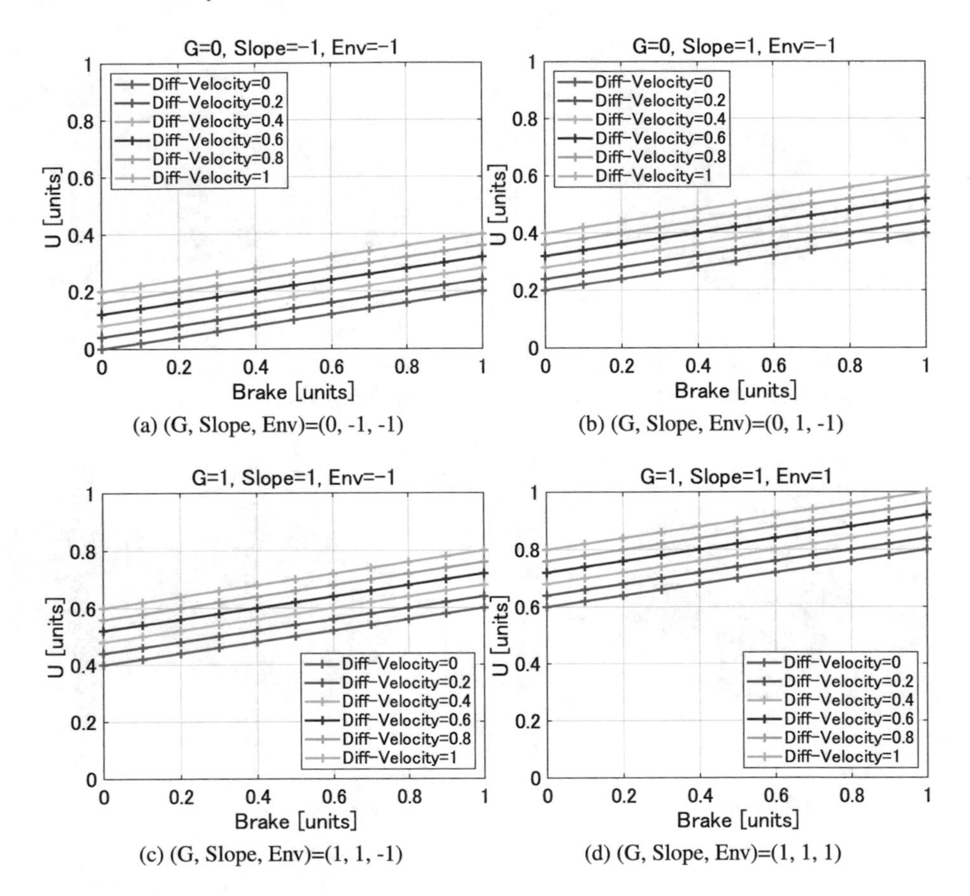

(a) (G, Slope, Env)=(0, -1, -1)

(b) (G, Slope, Env)=(0, 1, -1)

(c) (G, Slope, Env)=(1, 1, -1)

(d) (G, Slope, Env)=(1, 1, 1)

Fig. 3. T1FLC results for different cases.

The results of T1FLC and T2FLC for different cases are shown in Fig. 3 and Fig. 4, respectively. Here, we consider the following four combinations: (G, Slope, Environment):

1. (a) $(0, -1, -1)$: G=0 (Weak), Slope=-1 (Downhill), Environment=-1 (Neg)
2. (b) $(0, 1, -1)$: G=0 (Week), Slope=1 (Uphill), Environment=-1 (Neg)
3. (c) $(1, 1, -1)$: G=1 (Strong), Slope=1 (Uphill), Environment=-1 (Neg)
4. (d) $(1, 1, 1)$: G=1 (Strong), Slope=1 (Uphill), Environment=1 (Pos)

From Fig. 3(a) to Fig. 3(d), we observe that the output U value is increased linearly with the increase of Brake values. In addition, the output values present a constant increase for G from Weak to Strong, Slope from Downhill to Uphill and Environment from Neg to Pos. On the other hand, the T2FLC shows a wave-like increase. We conclude that the proposed FBCF can provide a soft brake assistance considering various situations created by the train and rail environment in order that the passengers have a comfortable feeling.

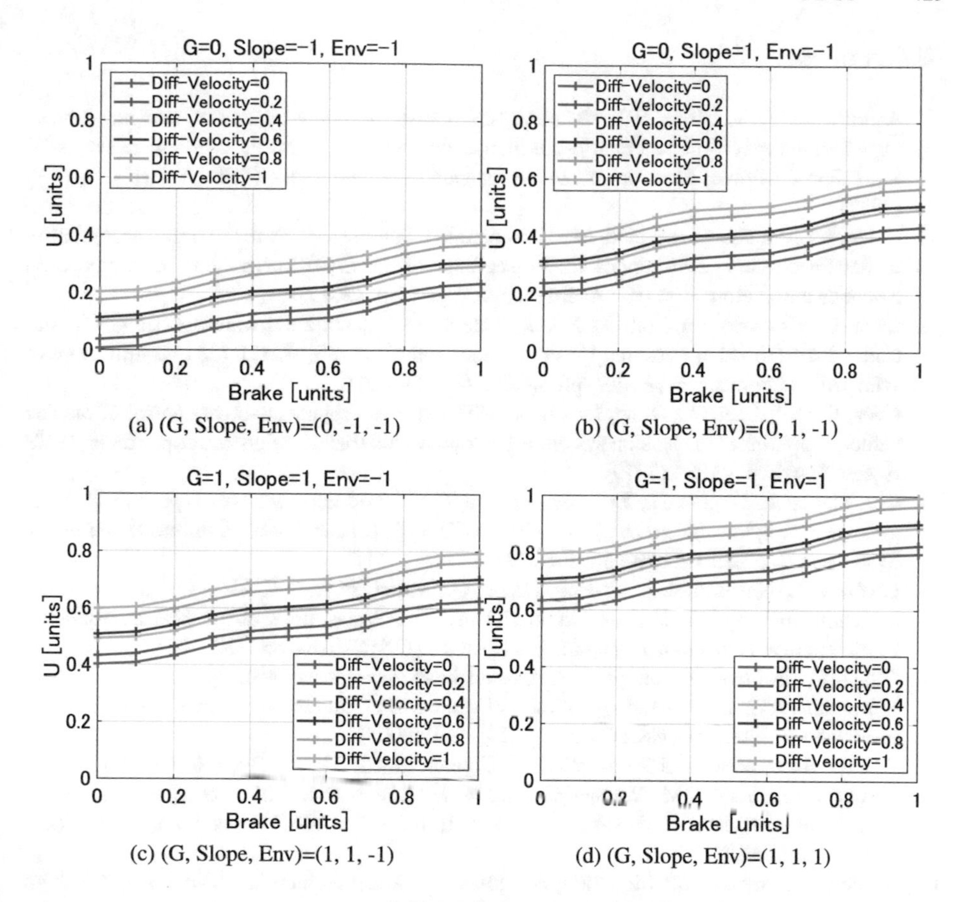

(a) (G, Slope, Env)=(0, -1, -1)

(b) (G, Slope, Env)=(0, 1, -1)

(c) (G, Slope, Env)=(1, 1, -1)

(d) (G, Slope, Env)=(1, 1, 1)

Fig. 4. T2FLC results for different cases.

5 Conclusions

In this paper, we presented a new FBCF for railway vehicles. The proposed FBCF can support an intelligent braking assist considering velocity, brake level, gravitational acceleration level, slope angle and environment on the railway. We used two types of FL inference models. The simulation results have shown that T2FLC can support enhanced brake assisting function than T1FLC. We conclude that the proposed FBCF can provide a soft brake assistance considering various situations created by the train and rail environment in order that the passengers have a comfortable feeling.

In the future work, we will extend our FL-based system considering other functions and other parameters. We plan to develop the testbed system to evaluate the accuracy of our simulation system.

References

1. Asprilla, A.M., Martinez, W.H., Munoz, L.E., Cortes, C.A.: Design of an embedded hardware for motor control of a high performance electric vehicle. In: Proceedings of the IEEE Workshop on Power Electronics and Power Quality Applications (PEPQA-2017), pp. 1–5 (2017)
2. Balan, K., Manuel, M.P., Faied, M., Krishnan, M., Santora, M.: A fuzzy based accessibility model for disaster environment. In: Proceedings of the IEEE International Conference on Robotics and Automation (ICRA-2019), pp. 2304–2310 (2019)
3. Cena, G., Cibrario Bertolotti, I., Hu, T., Valenzano, A.: On a software-defined CAN controller for embedded systems. Comput. Stan. Interfaces **63**, 43–51 (2019). https://www.sciencedirect.com/science/article/pii/S0920548918302101
4. Chen, C., Xu, J., Ji, W., Rong, L., Lin, G.: Sliding mode robust adaptive control of maglev vehicle's nonlinear suspension system based on flexible track: design and experiment. IEEE Access **7**, 41874–41884 (2019)
5. Chimatapu, R., Hagras, H., Kern, M., Owusu, G.: Hybrid deep learning type-2 fuzzy logic systems for explainable AI. In: Proceedings of the IEEE International Conference on Fuzzy Systems (FUZZ-IEEE-2020), pp. 1–6 (2020)
6. Ghallabi, F., Nashashibi, F., El-Haj-Shhade, G., Mittet, M.A.: LIDAR-based lane marking detection for vehicle positioning in an HD map. In: Proceedings of the 21st International Conference on Intelligent Transportation Systems (ITSC-2018), pp. 2209–2214 (2018)
7. Gupta, I., Riordan, D., Sampalli, S.: Cluster-head election using fuzzy logic for wireless sensor networks. In: Proceedings of the 3rd Annual Communication Networks and Services Research Conference (CNSR-2005), pp. 255–260 (2005)
8. Iwnicki, S.: Handbook of Railway Vehicle Dynamics. CRC Press, Boca Raton (2006)
9. Jammeh, E.A., Fleury, M., Wagner, C., Hagras, H., Ghanbari, M.: Interval type-2 fuzzy logic congestion control for video streaming across IP networks. IEEE Trans. Fuzzy Syst. **17**(5), 1123–1142 (2009)
10. Knothe, K., Grassie, S.: Modelling of railway track and vehicle/track interaction at high frequencies. Veh. Syst. Dyn. **22**(3–4), 209–262 (1993)
11. Li, T.S., Chang, S.J., Tong, W.: Fuzzy target tracking control of autonomous mobile robots by using infrared sensors. IEEE Trans. Fuzzy Syst. **12**(4), 491–501 (2004)
12. Liu, Z., Wang, Y., Liu, S., Li, Z., Zhang, H., Zhang, Z.: An approach to suppress low-frequency oscillation by combining extended state observer with model predictive control of emus rectifier. IEEE Trans. Power Electron. **34**(10), 10282–10297 (2019)
13. Mamdani, E.H.: Application of fuzzy algorithms for control of simple dynamic plant. In: Proceedings of the Institution of Electrical Engineers. 12, vol. 121, pp. 1585–1588 (Dec 1974)
14. Noori, K., Jenab, K.: Fuzzy reliability-based traction control model for intelligent transportation systems. IEEE Trans. Syst. Man Cybern.: Syst. **43**(1), 229–234 (2013)
15. Petrakis, E.G.M., Sotiriadis, S., Soultanopoulos, T., Renta, P.T., Buyya, R., Bessis, N.: Internet of things as a service (iTaaS): challenges and solutions for management of sensor data on the cloud and the fog. Internet Things **3–4**, 156–174 (2018)
16. Ruan, J., Jiang, H., Li, X., Shi, Y., Chan, F.T.S., Rao, W.: A granular GA-SVM predictor for big data in agricultural cyber-physical systems. IEEE Trans. Ind. Inf. **15**(12), 6510–6521 (2019)
17. Sekine, S., Nishimura, M.: Application of fuzzy neural network control to automatic train operation. In: Proceedings of 1995 IEEE International Conference on Fuzzy Systems. vol. 5, pp. 39–40 (1995)

18. Silver, D., et al.: Mastering the game of Go without human knowledge. Nature **550**, 354–359 (2017)
19. Su, X., Wu, L., Shi, P.: Sensor networks with random link failures: distributed filtering for T-S fuzzy systems. IEEE Trans. Ind. Inf. **9**(3), 1739–1750 (2013)
20. Sun, Y., Xu, J., Qiang, H., Lin, G.: Adaptive neural-fuzzy robust position control scheme for maglev train systems with experimental verification. IEEE Trans. Ind. Electron. **66**(11), 8589–8599 (2019)
21. Takagi, T., Sugeno, M.: Fuzzy identification of systems and its applications to modeling and control. IEEE Trans. Syst. Man Cybern. SMC **15**(1), 116–132 (1985)
22. Tsuneyoshi, M., Ikeda, M., Barolli, L.: A brake assisting function for railway vehicles using fuzzy logic: A comparison study for different fuzzy inference types. In: Proceedings of the 17th International Conference on Broadband and Wireless Computing, Communication and Applications (BWCCA-2022). pp. 301–311 (Oct 2022)
23. Wehner, P., Schwiegelshohn, F., Gohringer, D., Hubner, M.: Development of driver assistance systems using virtual hardware-in-the-loop. In: Proceedings of the International Symposium on Integrated Circuits (ISIC-2014). pp. 380–383 (Dec 2014)
24. Yazdani, S., Montazeri-Gh, M.: A novel gas turbine fault detection and identification strategy based on hybrid dimensionality reduction and uncertain rule-based fuzzy logic. Computers in Industry 115 (2020), https://www.sciencedirect.com/science/article/pii/S0166361519303811
25. Zadeh, L.: Fuzzy logic, neural networks, and soft computing. ACM Communications pp. 77–84 (1994)
26. Zhai, W., Wang, K., Cai, C.: Fundamentals of vehicle-track coupled dynamics. Veh. Syst. Dyn. **47**(11), 1349–1376 (2009)

A Memetic Approach for Classic Minimum Dominating Set Problem

Peng Rui, Wu Xinyun(✉), and Xiong Caiquan

School of Computer Science, Hubei University of Technology, Wuhan 430068, China
xinyun@hbut.edu.cn

Abstract. Solving the minimum dominating set problem is an NP-hard problem, which cannot be solved in a polynomial time. In this paper, a memetic algorithm is proposed to solve the minimum dominating set problem. We transform the original problem into a sequence of decision problems, i.e., the k-dominating set problem. The algorithm tackling the k-dominating set problem consists of a local search process embedded in a genetic optimization framework. As an optimization algorithm with a parallel structure, the memetic algorithm speeds up the convergence speed and improves the solution efficiency. The results from the memetic algorithm are compared with those of other algorithms, showing that the memetic algorithm is better.

1 Introduction

Given an undirected graph G = (V, E), if there is a vertex set S ⊆ V and the vertex of its complement is adjacent to at least one vertex of S, the set S is called a dominating set (DS) of the graph G. The vertices in the dominating set are dominant points, while the vertices adjacent to them in the complement set are dominated ones. The domination set with the smallest number of all domination sets is called the minimum dominating set (MDS). The minimum dominating set problem mainly applies to AD-hoc networks [1] and broadcast routing [2, 3].

Since the minimum dominating set problem is np-hard [4], no algorithm can guarantee the dominating set to be minimum with a large-scale graph instance. To make it as small as possible, most of the research conducts using heuristic algorithms [5], such as particle swarm optimization algorithm [6], tabu search algorithm [7], ant colony algorithm [8], genetic algorithm [9], and so on. Comparing with traditional algorithms such as the greedy algorithm [10] to solve the minimum dominated set problem, the advantage of the heuristic algorithm is that the solving time is more efficient, and the solution result is closer to the optimal. For example, the particle swarm optimization algorithm, its main characteristics are easy to implement, fast convergence speed, etc. The Tabu search algorithm prevents the searching process from falling into the local optimum prematurely. Ant colony algorithm has the characteristics of robustness. The genetic algorithm does not depend on the initial solution and enhances the diversification of the algorithm.

As a population optimization algorithm, genetic algorithm can iteratively evolve and mutate individuals, and adopt the survival of the fittest to ensure the excellence of the

L. Barolli (Ed.): EIDWT 2023, LNDECT 161, pp. 432–440, 2023.
https://doi.org/10.1007/978-3-031-26281-4_45

population, which greatly solves the problem of initial independence and ensures the diversity of the population. Nowadays, genetic algorithm provides many good ideas for solving a series of dominating set problems. For example, two population optimization algorithms for minimum weight connected dominating set problem are proposed in reference [11]. The first algorithm combines a greedy heuristic algorithm and a genetic algorithm to solve minimum weight connected dominating set problem. The second algorithm optimizes the population by destroying and greedily reconstructing individuals. These two algorithms have achieved good results in solving minimum weight connected dominating set problem. A hybrid algorithm based on a biased random key genetic algorithm and an exact solver is proposed in reference [12]. It solves the problem of minimum capacitated dominating set problem. It is a variant of MDS problem, which is different from MDS problem in that it limits the number of vertices that can be dominated by each vertex in the dominating set. The core idea of the algorithm is that in the iterative process, firstly, the existing individuals in the population are used to generate a sub-instance to be solved, then the exact solver is used to solve this sub-instance, and finally the result is transformed into individuals and then incorporated into the population. In reference [13], a hybrid genetic algorithm is proposed, which sets 0 and 1 attributes of individuals, and these individuals form a population. The adaptability of the individual is judged by setting an efficient fitness function, and MDS is searched by retaining excellent individuals.

Compared with the above algorithm, this paper proposes a memetic approach (MA), which divides the individuals in the population into sets, which can be more efficient, quickly judge whether the individual is a legal dominating set and reduce the search time. In addition, this paper uses a step-by-step method to solve MDS problem, which divides the population into a sub-problem, each of which searches for legal individuals with k vertices of the dominant set in the current population. After a sub-problem is solved, a new population will be regenerated. The new sub-problem is to search for legitimate individuals in the population whose dominant set vertex is k-1. Unlike the traditional genetic algorithm, MA does not search MDS in a completely random way, but directionally searches MDS, which greatly improves the search efficiency. Finally, by comparing with other algorithms, the results verify the superiority of MA.

2 Main Algorithm Framework

As a parallel structure algorithm, MA inherits in the way of survival of the fittest to get the expected result. In this paper, based on MA, we constantly adjust the fitness of the dominant set population to obtain a feasible dominating set. The core idea of the algorithm is to solve a k-DS problem continuously and reduce the dominating set size k when a feasible dominating set if found. The algorithm terminates when no feasible dominating set can be found for the current k. The final feasible dominating set is the minimum dominating set desired. For all the dominating sets of individuals with size k in one population, if one feasible individual appears, the process ends with this feasible dominating set of size k. The algorithm then creates a new population with all individuals with a candidate dominating set with k-1 and repeats the optimization process. The algorithm terminates if it is impossible to find a feasible dominating set individual with size k in this population, and the last feasible solution recorded is considered the best result. As shown in algorithm 1, given G = (V, E), the entire undirected graph is the initial solution that all vertices of the current graph G are dominant points. By iterating to the termination condition and ending the program, the final minimum dominant set is produced.

Algorithm 1 Algorithm framework

```
Input:G=(V,E)
Output:MDS
begin
k ← |V|
iter ← 0
while iter<Max do
Try to solve k-DS problem using MA
if find DS in the population that satisfies the conditions then
k ← k-1
end if
iter ← iter+Δiter
end while
return MDS
end
```

3 Solving a k-Dominating Set Problem

3.1 Idea of the Algorithm

MA is an optimization algorithm derived from the way of biological evolution. It hybridizes individuals among populations to make them evolve and design variation criteria to make individuals mutate positively. The idea of MA designed in this paper is to find a legitimate dominant set individual in a population consisting of dominant set individuals with a dominating number of k and to make the population evolve continuously through hybridization, mutation, and other operations, so as to retain high-quality individuals, eliminate inferior individuals and maintain a constant number of individuals in the population. As shown in Fig. 1, it represents a population evolution process in Algorithm 1. The specific meaning of this process will be introduced in detail in the following section.

3.2 Genetic Division of the Dominant Set Individuals

To find feasible domination sets, the algorithm divides the whole vertices set into three different subsets. Set S is the control set, where vertices are control vertices, trying to ensure all vertices in its complementary are adjacent to one of its elements. The next is the vertex set S+ which contains the vertices dominated by S, each vertex of S+ is strictly adjacent to at least one vertex of the set S. Finally, the set S− contains the remaining vertices undominated by S.

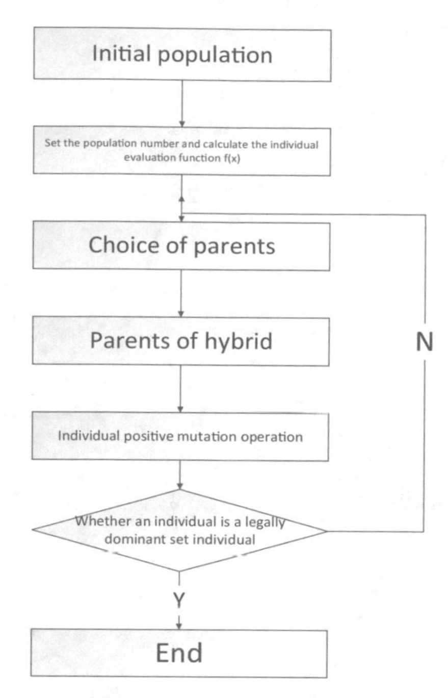

Fig. 1. Idea of algorithm design

In order for the dominating set S to dominate the whole undirected graph G, we know that the S− must be an empty set, such that the dominating set is feasible. We can define objective function $\varphi(x) = |S-|$ as their size of S−, i.e., we minimize the number of vertices in S−. When $\varphi(x) = 0$, it indicates that the current dominating set is feasible.

Taking Fig. 2 as an example, the undirected graph, as a dominant set configuration, is internally divided into three parts. The blue vertex is set S, the gray vertex is set S+, and the orange vertex is set S−. And one can see that $\varphi(x) = 6$, which is the number of orange vertices. The MA described below tries to minimize the value of $\varphi(x)$.

3.3 Population Parameters and Evaluation Function Settings

Given the undirected graph $G = (V, E)$ to be solved, this paper transforms the problem into a population optimization problem. The specific approach is to initialize a population, set the number of individuals in the population as a constant value of 5, and each individual represents a dominant set problem configuration. Hybridization is carried out among population individuals to produce offspring and make their variation maintained for individual optimization. The algorithm adopts the method of survival of the fittest and maintains a constant number of population individuals to restrict the population size. As one can see that the value of the individual evaluation function $\varphi(x)$ is used to represent the merits of an individual, the smaller the better. When $\varphi(x) = 0$, it is the feasible dominating set individual, indicating that the current population has completed its goal. A new round of population initialization can be conducted to solve the individuals with a smaller k-dominating set problem.

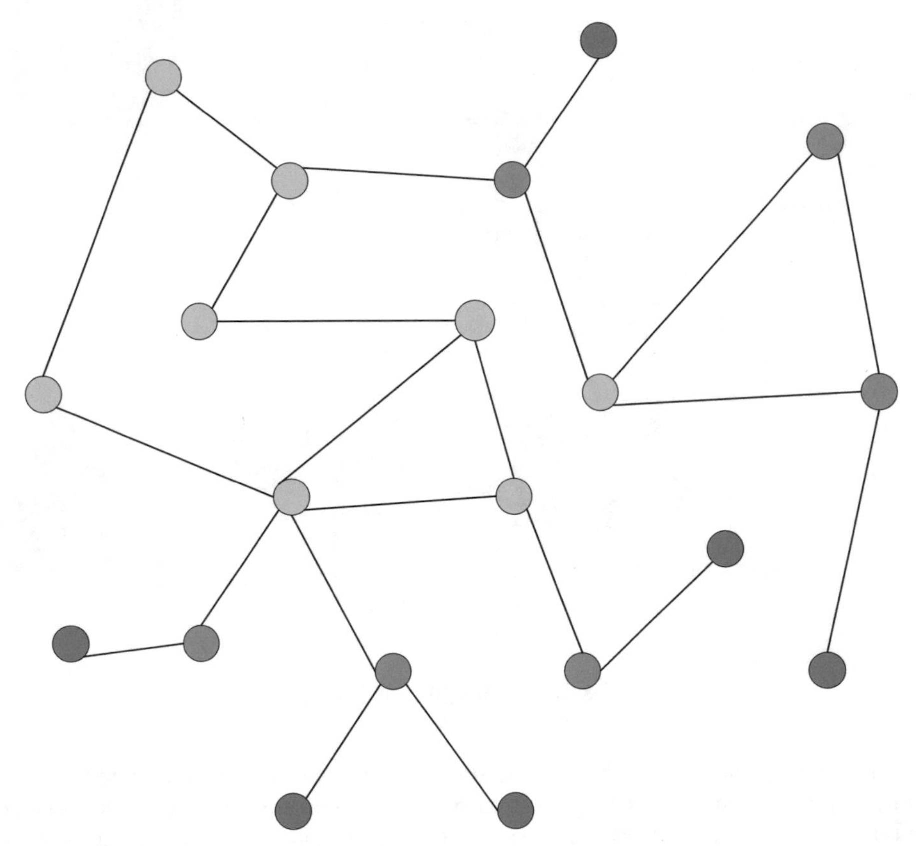

Fig. 2. Dominant set individual set partition diagram

3.4 Parental Selection and Hybridization

For the selection of parents, the algorithm adopts the method of random extraction to increase the diversity of the population. Each time, two individuals are randomly selected from five individuals in the population for hybridization, and the genes of both parents are fused to produce new offspring individuals. The vertices in the progeny individuals set S are the intersection of the set S of the parents. If the number of vertices in the progeny individuals set S does not reach k, the vertices will be randomly selected from the parents' individuals set S for a supplement such that the number of vertices reaches k.

3.5 Mutation of the Individual Offspring

After the successful generation of the offspring, the mutation process requires the modification of the genes of the individual progeny. This paper will set an evolution time h, within which individual genes will be continuously optimized. The mutation part is designed as a local search process which continuously exchanges vertices in sets S and S+ of this individual randomly and tries to reduce the S- size. The change operation exchanges single vertex to single vertex, rather than multiple vertices. For each iteration, the algorithm chooses the exchange operator that improves the objective function most, i.e., the one that reduces $\varphi(S) = |S - |$ most. The process terminates if the time limit is reached. If the mutated individual offspring became feasible dominating sets, indicating the current population has achieved the goal. The algorithm conducts a new round of population evolution. Otherwise, the offspring individual is judged. If it is better than the worst individual in the current population, it will replace the worst in the population. As shown in Fig. 3, the vertex elements in the set S of the current offspring individuals are B, D, and C, which is, the number of the dominant vertices of the population k = 3, the vertex elements in the set S+ are A, G, E, and the vertex elements in the set S− are F, where $\varphi(S) = 1$. After exchanging vertex D in set S with vertex G in set S+, all vertices of the undirected graph are dominated by the vertex of set S, therefore $\varphi(S) = 0$. After this exchange, the population has found the feasible dominating set.

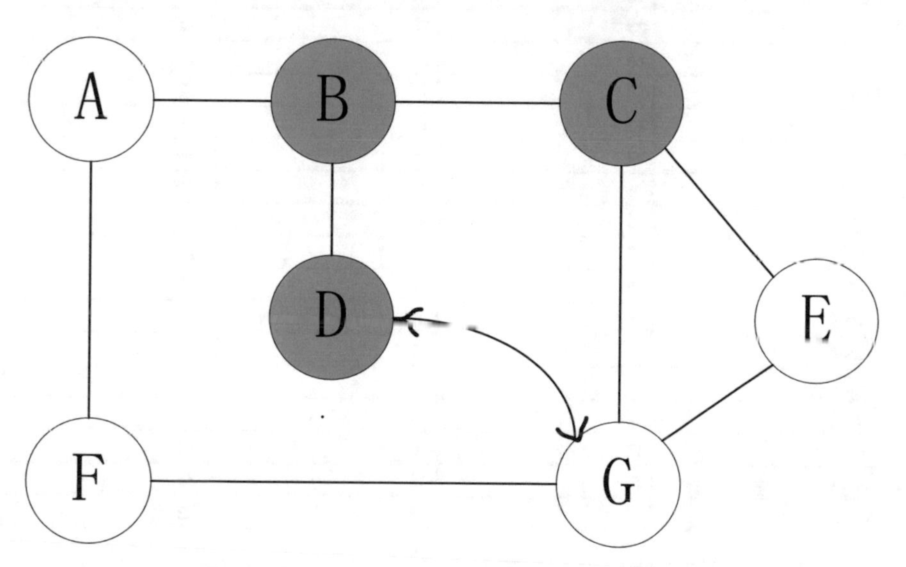

Fig. 3. An evolutionary strategy for vertex swapping

4 Experiment

In this section, a comparative experiment is conducted on the effectiveness of the memetic algorithm in solving the minimum dominant set problem. The algorithm was implemented in Java with JDK8. The experimental platform was Intel Core i5-7300HQ 2.5 GHz, 8 GB memory PC, and the operating system was Windows 10.

In this section, the data set is formatted in the DIMACS standard. This data set is often applied to undirected graphs with the number of vertices within 20000 for graph-related problems. The tabu search algorithm (TSA), local search algorithm (LSA), and MA were compared. Table 1 below records the optimal values of the results obtained by the three algorithms in 20 independent operations. TSA uses a tabu strategy that prevents

Table 1. Experiment of contrast

Instance	MA	TSA	LSA
le450_5a	30	32	58
le450_5b	29	33	52
le450_5c	20	22	34
le450_5d	20	23	35
le450_15a	25	26	61
le450_15b	26	27	82
le450_15c	11	12	20
le450_15d	11	12	21
le450_25a	37	36	136
le450_25b	33	32	196
le450_25c	13	14	23
miles500	9	9	15
miles750	6	6	7
miles1000	4	4	4
miles1500	2	2	2
mulsol.i.1	61	61	61
mulsol.i.2	17	17	17
mulsol.i.3	12	12	12
mulsol.i.4	12	12	12
mulsol.i.5	12	12	12
queen1	3	3	3
queen2	3	3	4
queen3	4	4	5
queen4	5	5	5
queen5	6	6	7
queen6	5	6	8
queen7	6	7	8
queen8	6	7	10
queen9	7	8	11
queen10	7	8	12
queen11	8	8	13
queen12	9	9	14
queen13	9	11	15
zeroin.i.1	87	87	87
zeroin.i.2	56	56	56
zeroin.i.3	51	51	51

the algorithm from falling into a locally optimal optimization prematurely. LSA is a basic local search framework algorithm, which uses a simple gradient descent solution to solve the minimum-dominating set problem.

According to the observation and analysis in Table 1, TSA's solution result is better than LSA, indicating that the tabu search process is better than the local search process, and TSA prevents the program from falling into the local optimal trap early by using tabu strategy. The result of MA is better than that of LSA, which indicates that MA, as an algorithm with parallel structure, improves the program evacuation in a certain length. Compared with TSA, MA and TSA have their own advantages and disadvantages, which indicates that both algorithms are more optimized than TSA in solving the minimum dominant set problem.

5 Conclusion

This paper proposes a memetic algorithm to solve the minimum dominating set problem. The algorithm breaks the original problem into several k-dominating set problems and solves these subproblems consecutively using a genetic algorithm-based scheme. The genetic algorithm consists of a specially defined hybridization and a mutation strategy which help it to produce reliable results. Compared with several other algorithms for the minimum dominating set problem, the proposed one performs better. Future work may lay on introducing more sophisticated hybridization or mutation strategies to improve further the effectiveness of the algorithm.

References

1. Balasundaram, B., Butenko, S.: Graph domination, coloring and cliques in telecommunications. In: Handbook of Optimization in Telecommunications, pp. 865–890. Springer, Boston, MA (2006). https://doi.org/10.1007/978-0-387-30165-5
2. Slavik, M., Mahgoub, I.: Spatial distribution and channel quality adaptive protocol for multihop wireless broadcast routing in VANET. IEEE Trans. Mob. Comput. **12**(4), 722–734 (2012)
3. Li, D., Jia, X., Liu, H.: Energy efficient broadcast routing in static ad hoc wireless networks. IEEE Trans. Mob. Comput. **3**(2), 144–151 (2004)
4. Approximation algorithms for NP-hard problems. ACM Sigact News **28**(2), 40–52 (1997)
5. Baaj, M.H., Mahmassani, H.S.: Hybrid route generation heuristic algorithm for the design of transit networks. Transp. Res. Part C: Emerg. Technol. **3**(1), 31–50 (1995)
6. Wang, D., Tan, D., Liu, L.: Particle swarm optimization algorithm: an overview. Soft. Comput. **22**(2), 387–408 (2018)
7. Gallego, R.A., Romero, R., Monticelli, A.J.: Tabu search algorithm for network synthesis. IEEE Trans. Power Syst. **15**(2), 490–495 (2000)
8. Dorigo, M., Birattari, M., Stutzle, T.: Ant colony optimization. IEEE Comput. Intell. Mag. **1**(4), 28–39 (2006)
9. Mirjalili, S.: Genetic algorithm. In: Evolutionary Algorithms and Neural Networks, pp. 43–55. Springer, Cham (2019). https://doi.org/10.1007/978-3-319-93025-1
10. Vince, A.: A framework for the greedy algorithm. Discret. Appl. Math. **121**(1–3), 247–260 (2002)

11. Dagdeviren, Z.A., Aydin, D., Cinsdikici, M.: Two population-based optimization algorithms for minimum weight connected dominating set problem. Appl. Soft Comput. **59**, 644–658 (2017)
12. Pinacho-Davidson, P., Blum, C.: BARRAKUDA: a hybrid evolutionary algorithm for minimum capacitated dominating set problem. Mathematics **8**(11), 1858 (2020)
13. Hedar, A.R., Ismail, R.: Hybrid genetic algorithm for minimum dominating set problem. In: International Conference on Computational Science and Its Applications. Springer, Berlin, Heidelberg, pp. 457–467 (2010). https://doi.org/10.1007/978-3-642-12189-0_40

Exploration of Neural Network Imputation Methods for Medical Datasets

Vivatchai Kaveeta[1](✉), Prompong Sugunnasil[3], and Juggapong Natwichai[2]

[1] Faculty of Engineering, Chiang Mai University, Chiang Mai, Thailand
vivatchai.k@cmu.ac.th

[2] Data Science Consortium, Chiang Mai University, Chiang Mai, Thailand
juggapong@eng.cmu.ac.th

[3] Department of Software Engineering, College of Arts, Media and Technology, Chiang Mai University, Chiang Mai, Thailand
prompong.sugunnasil@cmu.ac.th

Abstract. Datasets, especially those related to medicine, commonly suffer from missing data. The missing data originates from various sources. Examples include ever-changing medical diagnosis and treatment techniques, the absence of lab results, or even data collection errors. Most machine learning methods are trained on dense datasets. The sparse samples are either discarded or filled in with imputation. Imputation methods generate missing data by examining the variables in the relevant samples. Therefore, the performance of subsequence prediction models might be impacted by these methods. In this study, we explore neural network-based imputation methods to generate the missing data in medical datasets. The experimental results show that compared with traditional imputation methods, neural network imputation can be more effective in the classification and prediction tasks. We discuss some of the method's differences and assess their suitability for the dataset's specific characteristics.

1 Introduction

1.1 Missing Data

Medical data plays a critical role in helping the clinician take care of the patients. Experimental data was likely gathered with a collection protocol. However, clinical data usually collected upon contact with patients. The data may not be regularly collected and not strictly guided. The collection may suffer from various limitations such as different or changing treatment diagnoses or protocols, equipment availability or failure, lack of survey response, etc. Though the missing data may not be significant for clinical planning, machine learning algorithms suffer from this incompleteness. Some models perform worse in sparse datasets, while others are strictly unable to handle them.

In the definition of missing data in [23], the missingness pattern and mechanism are represented as the relationship between the missing data and other

L. Barolli (Ed.): EIDWT 2023, LNDECT 161, pp. 441–450, 2023.
https://doi.org/10.1007/978-3-031-26281-4_46

variables within the dataset. The actual missing data are the combinations of these types of mechanisms. Difference kinds of missingness are including [2, 19]

- Missing completely at random (MCAR). This term describes the missingness which not related to the studied person.
- Missing at random (MAR). This term describes the missingness that relates to persons but can be predicted from the person's other information.
- Missing not at random (MNAR). This term describes the missingness as directly related to the value. For example, seriously ill patients are likely to have missing questionnaires.

As mentioned, addressing the missing value is crucial to improve data utilization. In general, there are two different ways of processing the missing data. The simplest method is to exclude incomplete samples from the data. But this could introduce bias into the data and eventually affect the model performance. Alternatively, other methods try to predict the missing values using statistical or machine learning techniques. Numerous imputation algorithms are available for generating missing values. Statistical and machine learning imputation methods consider the input data, model the relationships, and impute data to restore the underly true distribution. Some basic statistical methods include mean, median, and most frequent. These techniques are mostly used method and successfully utilized in various fields.

The recent trend in machine learning studies leads to interest in adopting machine learning models for data imputation tasks. Machine learning models can recognize complex data patterns and make predictions based on them. Therefore, machine learning imputation methods can capture the data characteristics and fill in the missing values. Multiple models were proposed for imputation. The performance depends on the dataset characteristics such as the type of missingness.

The rest of this paper is organized as follows. Section 2 reviews the imputation methods. Section 5 discusses the differences between these methods. Section 3 explains the experimental detail. Section 4 shows and discusses the experimental results. Section 5 concludes the study.

2 Imputation Methods

Multiple imputation methods can be utilized for medical data. In this study, we look into some popular methods shown in Table 1. The detail of each method is as follows.

1. Mean imputation
   The mean of each variable is calculated and the missing values are imputed by this mean. This method can lead to very biased impute data [11].
2. Median imputation
   Similarly, the missing values are replaced by the median of the observed variables.

Table 1. Imputation methods in the experiment

Method	Algorithm
Mean	–
Median	–
kNN	K-Nearest Neighbors
ICE	Chained Equation
MICE [3]	Chained Equation
MissForest [25]	Random Forest
SoftImpute [8]	Alternating Least Squares
Sinkhorn [22]	Optimal Transport
GAIN [27]	Generative Adversarial Network
MIRACLE [15]	Neural Network
MIWAE [21]	Autoencoder
MIDAS [16]	Autoencoder

* Bold names indicates neural imputation methods

3. kNN (k Nearest Neighbor) imputation
   A type of Hot Deck imputation where the missing value is replaced with a value contributed from similar units. When k=1, values from the most similar neighbor are used. With k neighbors, the mean or the weighted mean of neighbors is imputed. Multiple neighbor distance calculation and weighting methods are proposed [13,17,18]
4. ICE (Imputation By Chained Equations) imputation
   The univariate version of MICE
5. MICE (Multivariate Imputation by Chained Equations) imputation [3]
   A multiple imputation method which accounts for statistical uncertainty in imputation [1]. The chained equation can apply to both continuous and binary data types.
6. MissForest imputation [25]
   Use the random forest algorithm to iteratively improve the imputed values. It works with mixed-type data and can impute continuous and categorical data.
7. SoftImpute imputation [8]
   Combining two matrix completion approaches, nuclear-norm-regularized matrix approximation [4] and maximum-margin matrix factorization [24]. This produces an efficient algorithm that outperforms the originals.
8. Sinkhorn imputation [22]
   Use optimal transport distance as the loss function for imputation. With the assumption that two random batches from the dataset should share the same distribution.
   The following methods are neural network-based imputation methods.

9. GAIN (Generative Adversarial Imputation Nets) imputation [27]
   Based on the Generative Adversarial Nets (GAN) framework, The pair of the generator (G) and discriminator (D) observe components of the observed data and output the completed vector. The discriminator force generator to learn the data distribution.
10. MIRACLE (Missing data Imputation Refinement And Causal LEarning) imputation [15]
   This method is a casually-aware imputation algorithm that iteratively learns the missingness graph. The underlying imputation algorithm can be any imputation method. The result shows significant improvement compared to traditional methods.
11. MIWAE (Generative Adversarial Imputation Nets) imputation [21]
   This method is based on the importance-weighted autoencoder (IWAE). It fit a deep latent variable model (DLVK, [14]) to the incomplete data and uses it for imputation.
12. MIDAS (Generative Adversarial Imputation Nets) imputation [16]
   Use unsupervised neural networks, Denoising Autoencoder, which was originally designed for dimensional reduction. Repurposing the autoencoders for multiple imputations.

In this study, we focus on the imputation method for numerical data. Time series datasets can benefit from recurrent neural networks-based imputation methods [5,7,20]. In the next section, we discuss the experimental designs.

3 Experiment

At least two performance comparison methods can be used to evaluate imputation methods. First, we take a fully complete dataset and delete part of the data to simulate the missing values. Then, the imputation algorithms can be used to impute the simulated dataset. When compared with the complete original data, the various matrices such as error rate can be measured. This directly measures the imputation method performance. Alternatively, we can machine learning experiments on the actual incomplete datasets and measure the accuracy of the final predictive models. In this study, we perform the later evaluation scenario. The machine learning task is survival prediction of prostate cancer patients, to classify whether the patient is alive 3 years after the observed data collection. Next, we describe the datasets and experimental methods.

3.1 Dataset

In this section, we show the detail of the datasets and pre-processing methods. The dataset used in this study is a publicly available prostate cancer datasets. As the datasets are publicly anonymous datasets, the ethical consideration is weaved. Four datasets are as follows.

1. Metastatic castration-sensitive prostate cancer (MSK, Clin Cancer Res 2020) Metastatic castration-sensitive prostate cancer patient dataset from [26]. This dataset will be referred to as "MCSPC 2020" in the following section.
2. Prostate Adenocarcinoma (TCGA, PanCancer Atlas) Data from The Cancer Genome Atlas Program (TCGA) PanCancer [10]. This dataset will be referred to as "PAN 2018" in the following section.
3. Prostate Adenocarcinoma (TCGA, Firehose Legacy) Data from The Cancer Genome Atlas Program (TCGA) source from GDAC Firehose [6]. This dataset will be referred to as "TCGA 2016" in the following section.
4. Prostate Adenocarcinoma (MSKCC, PNAS 2014) Data from prostate cancer patient cohort with long-term clinical outcome [9]. This dataset will be referred to as "MSKCC 2014" in the following section.

Before proceeding with the training, the datasets are pre-processed. Only related clinical parameters are kept. Unrelated variables were removed, the imaging and genomics data are excluded. Categorical features are encoded with one hot encoding. A requirement for the selected sample is to contain an overall survival time parameter. This value shows the survival period after the data collection. The target feature is generated by considering the overall survival time and alive status. When the survival time is at least 36 months, we keep the patients in the alive group. The patients who decease under 36 months are in the death group. The alive patients whose recorded survival times are under 36 months are excluded from this study.

Datasets are randomly split into 4-to-1 training-to-test sets. In the experiment, the training samples are imputed with imputation methods as shown in Table 1. The imputations are done on numerical features. For categorical features, the most frequent values are used. The distributions of missing values in the datasets are shown below (Figs. 1 and 2).

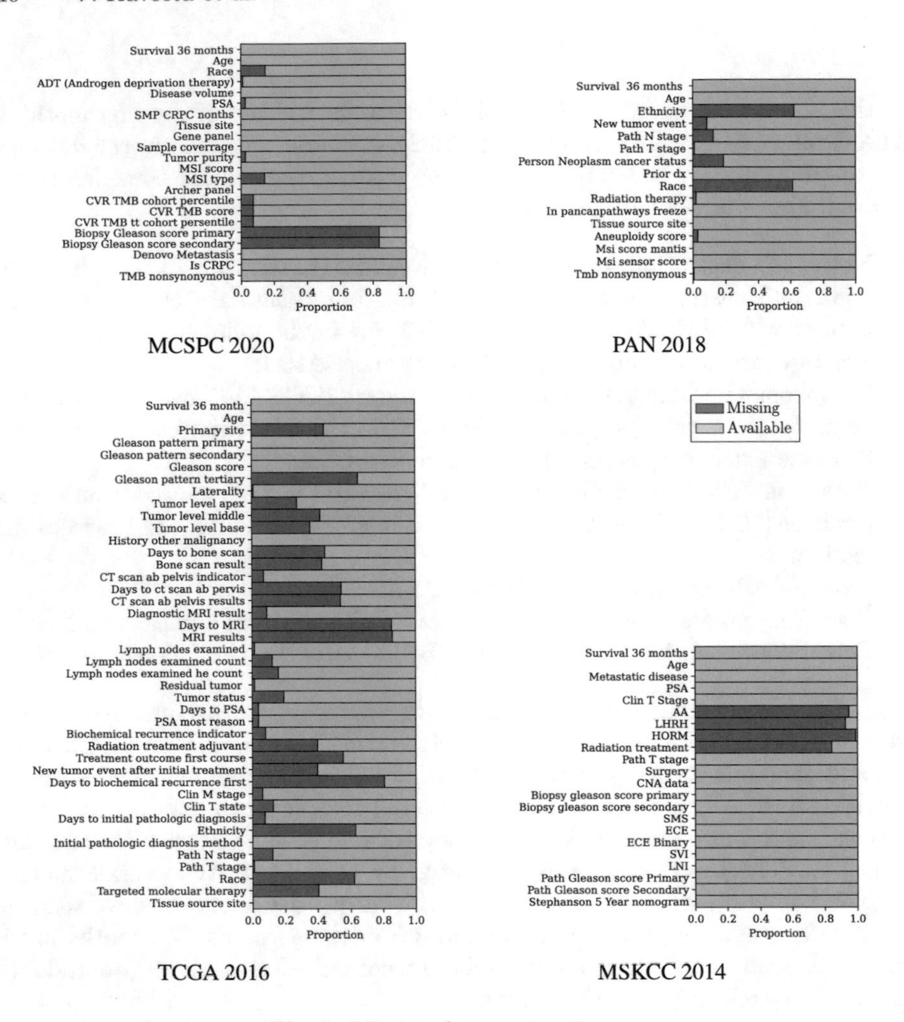

Fig. 1. Missing data proportions.

3.2 Setup

The experiment's training and test process are displayed in Fig. 3. In the training process, the training set is imputed by the methods. The imputation model with the same parameters is used to impute each test set sample. This process ensures that knowledge in the test set is not used for the imputation process. For machine learning, the SVM classifiers are trained on the imputed training set. Finally, we evaluate the model accuracy on the imputed test set. The imputation methods tested are shown in Table 1. The detail of each method was previously explained in Sect. 2. The actual imputation implementation utilizes Hyperimpute [12].

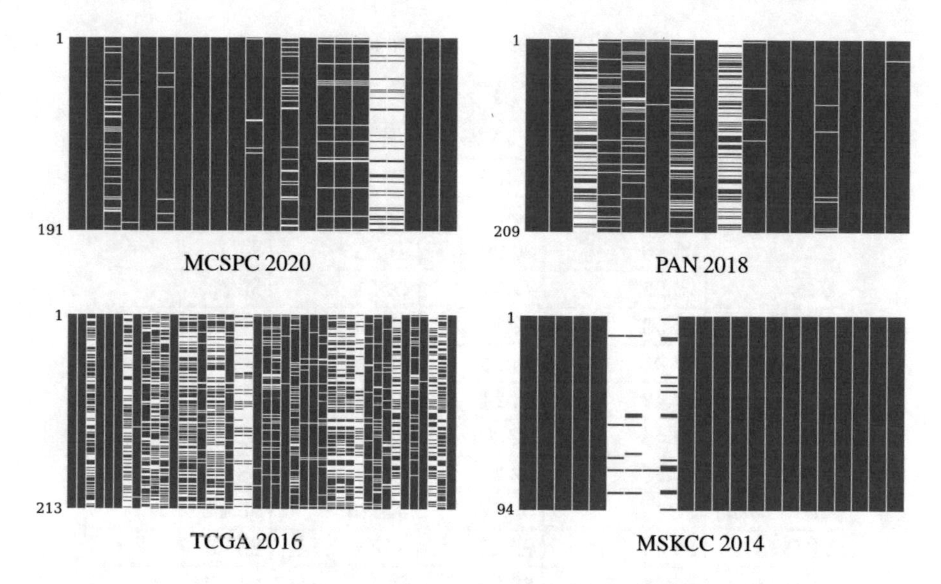

Fig. 2. Missing data map. (White shows missing values)

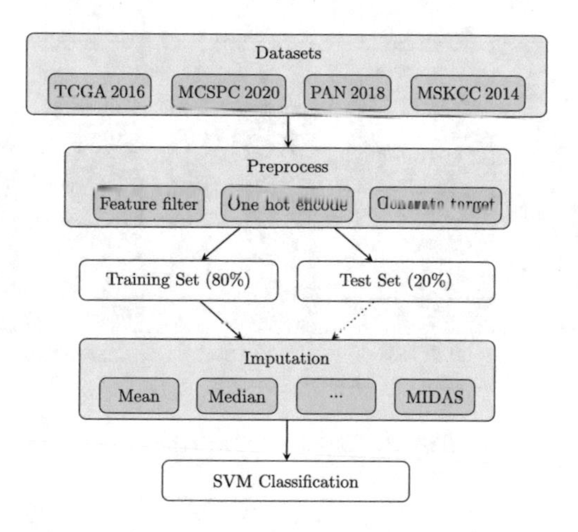

Fig. 3. Experimental pipeline.

4 Result

The experimental results are shown in Table 2 and Fig. 4. The results show that MIRACLE performs best in two datasets (MCSPC 2020, TCGA 2016). Sinkhorn performs best on PAN 2018 and MSKCC 2014. In general, machine learning especially neural network-based imputation methods perform better than traditional statistical methods. Sinkhorn performs exceptionally well with the result equal to or exceeding machine learning methods. Additionally, it has a significantly shorter runtime than machine learning methods.

Table 2. Experimental results.

Method	Prediction accuracy (%)			
	MCSPC 2020	PAN 2018	TCGA 2016	MSKCC 2014
Mean	76.96	72.25	77.83	75.53
Median	75.92	73.21	78.33	74.47
kNN	85.86	86.60	81.28	87.23
ICE	76.96	77.51	79.31	80.85
MICE	78.53	82.78	83.25	77.66
MissForest	86.91	83.25	82.76	84.04
SoftImpute	87.43	84.21	81.77	85.11
Sinkhorn	88.48	**93.78**	91.13	**93.62**
GAIN	90.58	92.34	92.61	88.30
MIRACLE	**93.19**	91.87	**93.60**	90.43
MIWAE	85.34	90.43	86.21	90.43
MIDAS	88.48	89.95	86.7	91.49

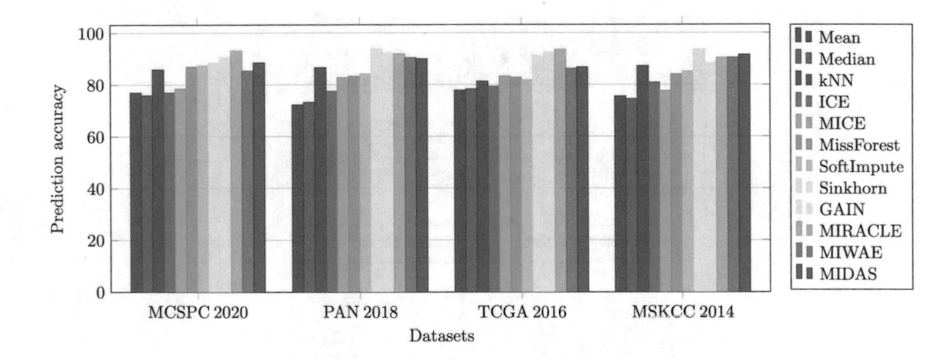

Fig. 4. Experimental result.

kNN imputation performs significantly better than other traditional methods. Some datasets even surpass machine learning techniques. This method is a time-efficient alternative. Neural network imputation methods perform closely to each other. Their runtimes are highly dependent on the architecture and training parameters. One possible challenge for the neural network methods is the limited size of the datasets. We expect better prediction improvement over the traditional methods on larger dimension and sample size datasets.

5 Conclusion

In this study, we study the missing data in medical datasets. Missingness types are described. The methods to address the missing data are shown. Imputation algorithms that we used in this study are listed. The methods are generally divided into traditional and machine-learning approaches. In the experiment section, we test the imputation methods by training binary classification models on four prostate cancer datasets. The SVM classification models are trained with imputed data by the methods. The experimental results show that neural network methods perform well on the datasets. Sinkhorn and kNN imputation are also performing similarly to the machine learning techniques. Lastly, we discuss the possibility of the neural network imputation performing better on larger datasets.

Acknowledgments. This research was partially supported by Chiang Mai University.

References

1. Azur, M.J., Stuart, E.A., Frangakis, C., Leaf, P.J.: Multiple imputation by chained equations: what is it and how does it work? Int. J. Methods Psychiatr. Res. **20**(1), 40–49 (2011)
2. Bland, M.: An Introduction to Medical Statistics. Oxford University Press, Oxford (2015)
3. Buck, S.F.: A method of estimation of missing values in multivariate data suitable for use with an electronic computer. J. Roy. Statist. Soc.: Ser. B (Methodol.) **22**(2), 302–306 (1960)
4. Candès, E.J., Tao, T.: The power of convex relaxation: near-optimal matrix completion. IEEE Trans. Inf. Theory **56**(5), 2053–2080 (2010)
5. Cao, W., Wang, D., Li, J., Zhou, H., Li, L., Li, Y.: Brits: bidirectional recurrent imputation for time series. In: Advances in Neural Information Processing Systems, vol. 31 (2018)
6. cBioPortal: Prostate Adenocarcinoma (TCGA Firehose Legacy). www.cbioportal.org/study/summary?id=prad_tcga
7. Du, W., Côté, D., Liu, Y.: Saits: Self-attention-based imputation for time series. arXiv preprint arXiv:2202.08516 (2022)
8. Hastie, T., Mazumder, R., Lee, J.D., Zadeh, R.: Matrix completion and low-rank SVD via fast alternating least squares. J. Mach. Learn. Res. **16**(1), 3367–3402 (2015)
9. Hieronymus, H., et al.: Copy number alteration burden predicts prostate cancer relapse. Proc. Nat. Acad. Sci. **111**(30), 11139–11144 (2014)
10. Hoadley, K.A., et al.: Cell-of-origin patterns dominate the molecular classification of 10,000 tumors from 33 types of cancer. Cell **173**(2), 291–304 (2018)
11. Jamshidian, M., Bentler, P.M.: Ml estimation of mean and covariance structures with missing data using complete data routines. J. Educ. Behav. Stat. **24**(1), 21–24 (1999)
12. Jarrett, D., Cebere, B.C., Liu, T., Curth, A., van der Schaar, M.: Hyperimpute: generalized iterative imputation with automatic model selection. In: International Conference on Machine Learning, pp. 9916–9937. PMLR (2022)
13. Jiang, C., Yang, Z.: CKNNI: an improved KNN-based missing value handling technique. In: Huang, D.-S., Han, K. (eds.) ICIC 2015. LNCS (LNAI), vol. 9227, pp. 441–452. Springer, Cham (2015). https://doi.org/10.1007/978-3-319-22053-6_47
14. Kingma, D.P., Welling, M.: Auto-encoding variational Bayes. arXiv preprint arXiv:1312.6114 (2013)

15. Kyono, T., Zhang, Y., Bellot, A., van der Schaar, M.: Miracle: causally-aware imputation via learning missing data mechanisms. Adv. Neural Inf. Proces. Syst. **34**, 23806–23817 (2021)
16. Lall, R., Robinson, T.: The MIDAS touch: accurate and scalable missing-data imputation with deep learning. Polit. Anal. **30**(2), 179–196 (2022)
17. Lee, J.Y., Styczynski, M.P.: NS-kNN: a modified k-nearest neighbors approach for imputing metabolomics data. Metabolomics **14**(12), 1–12 (2018)
18. Liang, C., Zhang, L., Wan, Z., Li, D., Li, D., Li, W.: An improved kNN method based on spearman's rank correlation for handling medical missing values. In: 2022 International Conference on Machine Learning and Knowledge Engineering (MLKE), pp. 139–142. IEEE (2022)
19. Little, R.J., Rubin, D.B.: Statistical Analysis with Missing Data, vol. 793. John Wiley, Hoboken (2019)
20. Luo, Y., Zhang, Y., Cai, X., Yuan, X.: E2gan: end-to-end generative adversarial network for multivariate time series imputation. In: Proceedings of the 28th International Joint Conference on Artificial Intelligence, pp. 3094–3100. AAAI Press (2019)
21. Mattei, P.A., Frellsen, J.: Miwae: deep generative modelling and imputation of incomplete data sets. In: International Conference on Machine Learning, pp. 4413–4423. PMLR (2019)
22. Muzellec, B., Josse, J., Boyer, C., Cuturi, M.: Missing data imputation using optimal transport. In: International Conference on Machine Learning, pp. 7130–7140. PMLR (2020)
23. Rubin, D.B.: Inference and missing data. Biometrika **63**(3), 581–592 (1976)
24. Srebro, N., Rennie, J., Jaakkola, T.: Maximum-margin matrix factorization. In: Advances in Neural Information Processing Systems, vol. 17 (2004)
25. Stekhoven, D.J., Bühlmann, P.: Missforest-non-parametric missing value imputation for mixed-type data. Bioinformatics **28**(1), 112–118 (2012)
26. Stopsack, K.H., et al.: Oncogenic genomic alterations, clinical phenotypes, and outcomes in metastatic castration-sensitive prostate cancer. Clin. Cancer Res. **26**(13), 3230–3238 (2020)
27. Yoon, J., Jordon, J., Schaar, M.: Gain: Missing data imputation using generative adversarial nets. In: International Conference on Machine Learning, pp. 5689–5698. PMLR (2018)

Applying BBLT Incorporating Specific Domain Topic Summary Generation Algorithm to the Classification of Chinese Legal Cases

Qiong Zhang[1] and Xu Chen[2](✉)

[1] School of Management Information and System, Zhongnan University of Economics and Law, Wuhan 430073, China
zqiong@stu.zuel.edu.cn

[2] School of Information and Safety Engineering, Zhongnan University of Economics and Law, Wuhan 430073, China
chenxu@zuel.edu.cn

Abstract. In response to the challenge that most existing case retrieval platforms can not effectively extract feature information of Chinese legal cases, and thus perform unsatisfactorily in terms of indicators such as relevance and accuracy of retrieval results. We propose to apply LBBT model incorporating domain-specific topic-based text summary generation algorithm to the classification of Chinese legal cases. In our proposed LBBT model, we use LDA to extract subject keywords for each type of legal documents separately, and then the TextRank algorithm is introduced to generate abstract for each legal document by combining the extracted subject words. BERT is used to vectorize the generated abstracts adopted as the inputs of BiLSTM to implement the task of classification on Chinese legal documents. The experimental result on the data set of 2500 single charge Chinese legal judgment documents obtained from CAIL2022 shows that our proposed LBBT model can effectively remove the redundant information in legal documents and improve the ability of LSTM to grasp the global key semantic information of long texts.

1 Introduction

Artificial intelligence technologies, represented by deep learning and natural language processing, have made great breakthrough in recent years, driving the development of intelligent approaches to justice [1]. However, the processing of legal texts is more challenging than other fields due to the complexity of legal sources and legal language [2]. Although deep learning has been applied to improve text classification performance, its effectiveness in the legal domain needs to be further explored [3].

The Supreme People's Court of China has opened the Judicial Documents Network since 2013, which has made almost all Chinese judicial decision documents public. The number of these documents has exceeded 100 million. At the same time, as the size of the documents keeps growing, the content of the cases contained in them also grows in a manner. Therefore, how to extract and manage the knowledge contained therein has become a key problem to be solved [4].

L. Barolli (Ed.): EIDWT 2023, LNDECT 161, pp. 451–459, 2023.
https://doi.org/10.1007/978-3-031-26281-4_47

Specifically, this paper conducts Chinese case classification based on the dataset of publicly available legal judgment documents from the class case search race track of CAIL (Challenge of AI in Law) in 2022. The research objectives of this paper are as follows. (1) Try to propose a model incorporating text summary generation algorithm with legal domain topics to efficiently solve the problem of classification of Chinese legal cases based on charges. (2) Try to verify whether some existing text classification models still have better effectiveness on the task of Chinese legal document classification.

To achieve the above research objectives, we extract eight categories of single charge with a total of 2500 Chinese legal judgments as dataset from the publicly available datasets from the class case search race track of CAIL 2022. Firstly, we used the TF-IDF-based LDA model to extract thematic keywords for each of the above eight categories of accusations datasets separately. Subsequently, we use word2vec to represent the feature of each legal verdict, and introduce TextRank algorithm to generate legal text summary with the extracted subject keywords to remove the redundant information in the case texts and improve the correlation between the case classes. Finally, the classification of the generated summary texts is completed by BERT-BiLSTM. The contributions of this paper are as follows.

- We combine the concept of domain and apply a legal domain-based subject keyword extraction algorithm to legal text classification.
- We adopt a text summary generation algorithm for data pre-processing, which effectively reduces the memory and time costs of the model training on excessively long legal text sequences in the real world.
- This paper completes the feature representation of text sequences based on the pre-training task of BERT and combines it wsith BiLSTM to achieve text classification.
- We use some typical text classification models for comparative analysis and explored their applicability in legal domain.

The rest of the this paper mainly consists of four sections. Some previous works related to text classification are presented in Sect. 2. The details of implementing the proposed method are elaborated in Sect. 3. Section 4 contains the experimental results and comparative analysis conclusions. The final section summarizes the shortcomings and challenges of our proposed method in this paper and then start a discussion on the future work.

2 Related Works

Legal text classification is the determination of a legal text category based on the association between the legal text and that text category [5], which is essentially similar to the variable-length text similarity measurement problem. This section examines previous work in terms of domain topic extraction, text vectorized representation and model selection and comparison included in the task of variable-length text classification in the legal domain.

2.1 Field Theme Extraction

[6] pointed out that extracting the key concepts of the domain to correctly represent the key information of the corpus of domain documents is the main and crucial step in entity learning. The aim is to construct entities by identifying relevant domain concepts and their semantic relations from the text corpus. Since keywords are the smallest unit to express the meaning of text topics, the existing methods for domain topic extraction are mainly focused on the extraction of keywords or phrases. There are traditional word statistics-based domain keyword extraction methods such as TF-IDF, word frequency, word co-occurrence, etc. There are domain keyword extraction methods based on topic models such as LDA, TextRank, etc. [7].

2.2 Text Feature Extraction

In the field of natural language processing, converting ordinary text into a computer-computable form has been a hot research problem. Text vectorization is the process of converting text into a form that can be correctly recognized and processed by computer. Traditional bag-of-words models such as one-hot, TF-IDF, etc. are based on word frequency mapping, which have the drawback of not effectively capturing text order and semantic information [8]. Tomas Mikolov et al. [9] proposed a neural network-based word embedding model word2vec to train word vectors with contextual semantic information. Zu Cheng [10] proposed to apply a pre-trained task-based language model BERT to the vectorized representation of Chinese text, and the experimental results showed that BERT also can effectively solve the problem of multiple meanings of a word that cannot be solved by word2vec on the basis of obtaining semantic information in the text.

2.3 Model Selection and Comparison

It's difficult for traditional text classification methods to mine implicit features from unstructured data corpus and it requires a lot of labor to build a feature lexicon suitable for a specific text classification task. Fortunately, with the development of deep learning techniques, many effective neural network-based text classification methods have been designed by researchers at home and abroad, such as BERT [10, 12, 13] and BiLSTM [11, 14].

Gao et al. [11] applied an attention mechanism-based BiLSTM model to Chinese text classification and verified the stability and effectiveness of the model through several comparative analyses. Khadhraoui Mayara et al. [12] applied BERT to an automated multi-class classification task for scientific texts. However, the maximum positional embedding of BERT is 512, so its performance is not promising in some text tasks that exceed its length limitation. Ding M et al. [13] proposed a CogLTX framework based on the core assumption that for most NLP tasks, a few key sentences in the text store sufficient and necessary information to complete the task. Specifically by dividing the original text into blocks of utterances from which key sub-sentence blocks are selected to form a new text. CogLTX aims to compress the text by removing redundant clause blocks from the text without changing the core concepts of the original text. This framework clearly outperforms SVM, FastText, TextCNN, etc. in news text classification tasks.

3 Proposed Method

In this section, we elaborate the implementation details of the proposed method in this paper. The overall architecture of our method is shown in Fig. 1, which mainly contains threee steps: Chinese legal case data preprocessing, case text semantic feature extraction, and case text classification.

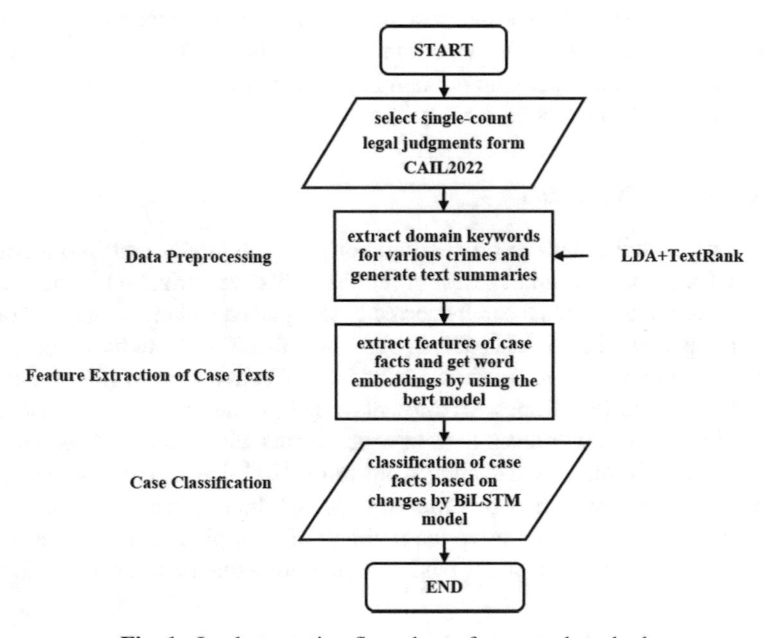

Fig. 1. Implementation flow chart of proposed method.

3.1 Legal Case Data Preprocessing

The data set of this paper is derived from the verdict documents published on China Judicial Documents Network, with a total of 8 charges and 2500 Chinese legal judgments. Because the judgment documents are a variable-length long text sequences. If we use BERT directly for text contextual semantic feature extraction, problems such as forced truncation of case text and loss of semantics at a distance will inevitably occur [13].

Therefore, we first use the TF-IDF-based LDA model for keyword extraction for different categories of legal cases in the data preprocessing stage. But in practice, the extracted keywords are not all domain keywords [6]. Therefore, we perform a manual screening to transform the extracted keywords to domain keywords, and the domain keywords for each category of legal cases are shown in Table 1.

Subsequently, as shown in Algorithm 1, we use the word2vec-based TextRank model to generate text summaries for each judgment document based on the core assumption that the key information conveyed by the case text can be fully expressed by the fused sub-sentence blocks in it, and since the generated text summaries still contain much noisy information, we further combine the domain keywords to reduce them in text summaries. Thus, we obtain a compressed text summary data set with fused domain keywords that can replace the original judgment documents.

Table 1. Results of topics keywords extracted in each category of judgment documents

Category	Crime 0	Crime 1		Crime 6	Crime 7
Keywords	Transport	Intoxication	…	Smuggle	Illegal operation
	Drive	Hazardous driving		Trafficking	Verify
	park	Wine testing	…	Seizure	Sale
	……	……		……	……
	road	Vehicle		Drug	Permit
	VEHICLE	Driving licence		Tablet	Finance

Algorithm 1:Pseudo codes for Text Summary Extraction Process

```
1. input X,
//input case text X
2. output X=[x[0],....,x[n]],
//slice the case text X into blocks of clauses in order
3. for x in X:
    words+=word for word in jieba.cut(x) and not in stopwords
    //subclause participles and removal of stopwords
    word2vec_embedding(words)
    //comput feature representation for each clause
4. cosine(x[i],x[j]),
//i≠j.construct the cosine similarity matrix
5. graph(x,cosine),
//the subclause and cosine similarity are used as the node and
transfer probability, respectively
6. output X'=[x[s1],....,x[sn]],
//s1<,...,<sn and len(s1+,...,+sn)<max_length:guarantee relative
position of clause blocks and the length of the fused case text is
less than the maximum length tolerated by the model training
7.end
```

3.2 Pre-trained Word Embeddings

The method proposed in the paper will use the pre-trained BERT model to extract features from the case texts, and the obtained vector representation of the word/sentence embedding will be used as the input of the downstream BiLSTM model. BERT compared to traditional models such as TF-IDF and word2vec, can effectively capture the location information and semantic information of the word/sentence [10]. As shown in Fig. 2, the input of the BERT model consists of the superposition of three parts: tokens embedding, segment embedding and position embedding, and the embedding representation of the same word/sentence in different or the same text will be different, which can effectively solve the problem of multiple meanings of a word and make the training of the downstream BiLSTM model obtain better performance. The token embedding represent the encoding of the current word, the segment embedding represents the encoding of the position of the sentence in which the current word is located, the position embedding represents the encoding of the position of the current word, and each sentence is marked using [CLS] and [SEP] as the beginning and the end of the sentence.

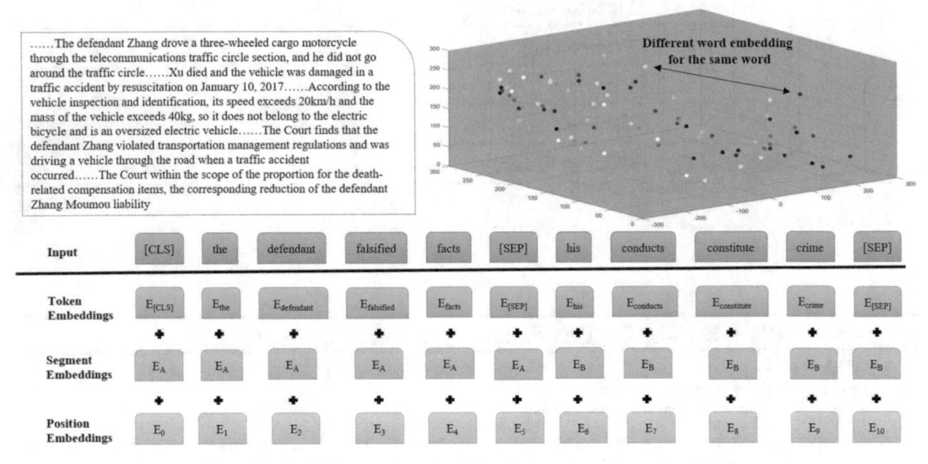

Fig. 2. BERT word embedding structure diagram. [10]

3.3 Legal Case Classification

We use BiLSTM to train the input of text feature representation and use softmax function for case text classification. As shown in Fig. 3, BiLSTM consists of a combination of a forward LSTM and a backward LSTM, where the forward LSTM is responsible for the forward operation of the input information and the backward LSTM is responsible for the reverse operation of the input information. Compared with models such as TextCNN and LSTM [11], BiLSTM can fully capture the larger span of contextual information contained in the case text by combining LSTMs in both forward and backward directions.

4 Experimental Setup and Results

4.1 Experimental Setup

We use python language to complete the experiment in a computer environment with 2.10 GHz processor and 16 GB RAM. The model-related parameters are set as shown in Table 2.

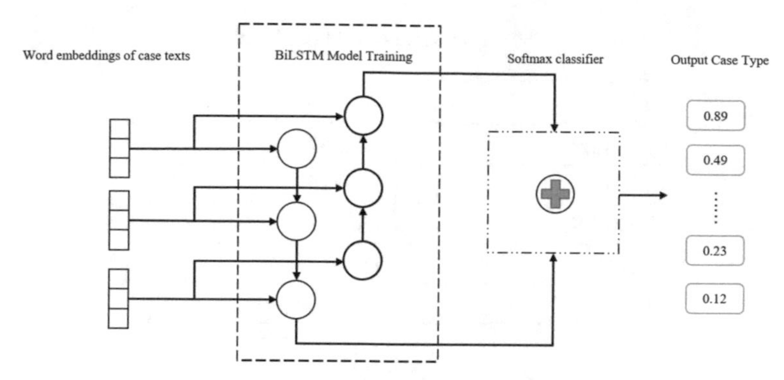

Fig. 3. BiLSTM model training schematic

Table 2. Model hyperparameters and experimental environment

Environment	Configuration parameters
Python version	3.7
Deep learning framework	1.12.1
Learning_rate	2e−5
Max_seq_length	300
Epoch	10
Batch_size	16
Output_size	8

4.2 Experimental Results

We measure the performance effectiveness of each model by comparing the accuracy on test sets, as they are more commonly used evaluation metrics in text classification [2], and the accuracy is defined as shown in Eqs. (1): where TP is the number of correctly classified cases and FP is the number of incorrectly classified cases.

$$\text{accuracy} = \frac{TP + TN}{TP + FP + TN + FN} \tag{1}$$

As shown in Table 3, our proposed method obtains 0.87 classification accuracy in the legal document classification task, which is significantly better than FastText, LSTM, TextCNN and BERT. This illustrates the feasibility of text summary generation algorithms incorporating keywords from the legal domain for legal case classification.

Table 3. Model performance and model comparison

Model	Accuracy
FastText	0.485573
LSTM	0.734160
TextCNN	0.829436
BERT	0.849174
Our	0.871726

5 Conclusion and Future Work

In this paper, we propose to use a fusion model for the classification task of legal documents. And we compare and analyze the proposed approach with some traditional text classification models such as FastText, TextCNN, etc. on our data set. The experimental result show that our method can not only effectively remove the noisy information from the text and improve the relevance of the text between classes, but also better extract the contextual semantic information of the Chinese legal case text for a better classification performance.

However, we still have the following two main problems. (1) Our data set is small due to the limitations of the experimental setting. (2) Different judges have different scales of decision and trial factors for similar cases [15], which requires us to consider more practical factors for Chinese legal case classification in the future.

Acknowledgement. This research was funded by National Funds of Social Science (21BXW076), National Natural Science Foundation of China (61602518), Philosophy and Social Science Research Project of Hubei Provincial Department of Education (20G026), Innovation Research of Young Teachers of Central Universities in 2021 (2722021BZ040), The Key Social Science Projects in Wuhan in 2021 (2021010), Prof. Liu Yaqi's Outstanding Youth Innovation team Construction Project (Big Data Intelligent Information Processing and Application Technology Innovation Team) and School-level reform project of Zhongnan University of Economics and Law (YB202158).

References

1. Aman, F., Yanchuan, W.: An intelligent adjudication method for multitasking legal cases based on BERT model. Microelectronics and Computers **39**(09), 107–114 (2022). https://doi.org/10.19304/J.ISSN1000-7180.2022.0217
2. Chen, H., Wu, L., Chen, J., Lu, W., Ding, J.: A comparative study of automated legal text classification using random forests and deep learning. Inf. Process. Manag. **59**(2) (2022)
3. Hongshui, S.: Research on judicial big data text mining and sentencing prediction model. Jurisprudence **07**, 113–129 (2020)
4. Yu, H., Li, H.: A knowledge graph construction approach for legal domain. Tehnički vjesnik **28**(2), 357–362 (2021)

5. Kang, Y.-B., Haghighi, P.D., Burstein, F.: CFinder: an intelligent key concept finder from text for ontology development. Expert Syst. Appl. **41**(9), 4494–4504 (2014). https://doi.org/10.1016/j.eswa.2014.01.006
6. Mao, L.Q., Shi, T., Wu, L., Ma, T.A.: Unsupervised text keyword extraction model based on domain adaption: an example of "artificial intelligence risk" domain text. Intell. Theory Pract. **45**(03), 182–187 (2022). https://doi.org/10.16353/j.cnki.1000-7490.2022.03.025
7. Zhou, N., Shi, W., Liang, R., Zhong, N.: Textrank keyword extraction algorithm using word vector clustering based on rough data-deduction. Comput. Intell. Neurosci. **2022**, 1–19 (2022). https://doi.org/10.1155/2022/5649994
8. Yuxuan, J.: Summary and analysis of text vectorization representation methods. Electron. World **22**, 10–12 (2018). https://doi.org/10.19353/j.cnki.dzsj.2018.22.003
9. Mikolov, T., Chen, K., Corrado, G., Dean, J.: Efficient Estimation of word representations in vector space. CoRR, abs/1301.3781 (2013)
10. Cheng, Z.: BERT-based vectorized representation of Chinese text. Technol. Innov. **21**, 107–108 (2021). https://doi.org/10.15913/j.cnki.kjycx.2021.21.046
11. Gao, C.L., Xu, H., Gao, K.: Combining lexical information for Chinese text classification based on attention mechanism of bidirectional LSTM. J. Hebei Univ. Sci. Techn. **39**(05), 447–454 (2018)
12. Khadhraoui, M., Bellaaj, H., Ammar, M.B., Hamam, H., Jmaiel, M.: Survey of BERT-base models for scientific text classification: COVID-19 case study. Appl. Sci. **12**(6), 2891 (2022). https://doi.org/10.3390/app12062891
13. Ding, M., et al.: Cogltx: applying bert to long texts. Adv. Neural. Inf. Process. Syst. **33**, 12792–12804 (2020)
14. Li, G., Wang, Z., Ma, Y.: Combining domain knowledge extraction with graph long short-term memory for learning classification of Chinese legal documents. IEEE Access **7**, 139616–139627 (2019)
15. Sun, H.: Rediscovering the "same case": constructing the criteria for judging the similarity of cases. Chinese Jurisprudence **06**, 262–281 (2020). https://doi.org/10.14111/j.cnki.zgfx.2020.06.014

Implementation of a Fuzzy-Based Testbed for Coordination and Management of Cloud-Fog-Edge Resources in SDN-VANETs

Ermioni Qafzezi[1](✉), Kevin Bylykbashi[2], Elis Kulla[3], Makoto Ikeda[2], Keita Matsuo[2], and Leonard Barolli[2]

[1] Graduate School of Engineering, Fukuoka Institute of Technology (FIT), 3-30-1 Wajiro-Higashi, Higashi-Ku, Fukuoka 811–0295, Japan
bd20101@bene.fit.ac.jp

[2] Department of Information and Communication Engineering, Fukuoka Institute of Technology (FIT), 3-30-1 Wajiro-Higashi, Higashi-Ku, Fukuoka 811-0295, Japan
kevin@bene.fit.ac.jp, makoto.ikd@acm.org, {kt-matsuo,barolli}@fit.ac.jp

[3] Department of System Management, Fukuoka Institute of Technology, 3-30-1 Wajiro-Higashi, Higashi-Ku, Fukuoka 811-0295, Japan
kulla@fit.ac.jp

Abstract. The integration of Cloud-Fog-Edge computing in Software-Defined Vehicular Ad hoc Networks (SDN-VANETs) brings a new paradigm that provides the needed resources for supporting a myriad of emerging applications. While an abundance of resources may offer many benefits, it also causes management problems. In this work, we implement a testbed to flexibly and efficiently manage resources in SDN-VANETs. We used our testbed and carried out some experiments. The results demonstrate the feasibility of the proposed approach in coordinating and managing the available SDN-VANETs resources.

1 Introduction

According to World Health Organization, around 1.3 million people die every year because of road traffic crashes [20]. The key risk factors come from human error (speeding, wrong decisions, etc.), irresponsible behavior (drinking, distracted driving, fatigue, etc.), unsafe road infrastructure, and bad weather conditions (e.g., inadequate visibility and slippery roads) [3,5,17]. Vehicular Ad hoc Networks (VANETs) have emerged as a solution to alleviate all these factors by means of different applications [11,16,18]. For example, implementing an accident prevention system in VANETs that considers velocity, weather condition, risk location, nearby vehicles density, and driver fatigue can reduce the number of road crashes and consequently the number of deaths [2]. Other applications, on the other hand, can improve traffic management and the driving experience [9,12,19]. VANETs goal is to benefit all road users, drivers, passengers and walkers, without any exception.

L. Barolli (Ed.): EIDWT 2023, LNDECT 161, pp. 460–470, 2023.
https://doi.org/10.1007/978-3-031-26281-4_48

Vehicular networks applications depend on vehicle-to-vehicle (V2V) communications, Vehicle-to-Infrastructure (V2I), Vehicle-to-Network (V2N) and Vehicle-to-Pedestrian (V2P) communications. Through these communication links they exchange important information that come from other sources such as data about traffic lights, public safety information, weather conditions, the state of other vehicle and its surrounding environment, and so on. Such data improves the accuracy and decisions taken in the network.

However, there are still many challenges that are yet to be addressed. One of these challenges is the management of the abundant information and resources available in these networks [1,4,16,19]. The volume generated data from VANETs keeps increasing as the number of vehicles and the sensors incorporated keeps increasing. On the other hand, many new applications have strict requirements for more resources, leading to increased complexity in network management. Dealing with resource management problems while still satisfying application requirements is the focus on our proposed approach.

In a previous work [15], we deal with the resource management problem and propose an integrated intelligent architecture based on Fuzzy Logic (FL) and Software Defined Networking (SDN) approach that can efficiently manage cloud-fog-edge storage, computing, and networking resources in VANETs, from a bottom-up perspective by exploiting the resources of edge layer first and then the fog and cloud resources. The integrated system is composed of three subsystems, namely Fuzzy-based System for Assessment of QoS (FS-AQoS), Fuzzy-based System for Assessment of Neighbor Vehicle Processing Capability (FS-ANVPC), and Fuzzy-based System for Cloud-Fog-Edge Layer Selection (FS-CFELS), each having a key role in the proposed approach.

In order to evaluate the simulation results of the aforementioned integrated system, in this work we implement a testbed and carry out experiments. We consider FS-CFELS system and analyze it experimental results. The experimental results show the feasibility of FS-CFELS system in coordinating and managing the available resources.

The rest of this paper is organized as follows. In Sect. 2, we present an overview of Cloud-Fog-Edge SDN-VANETs. The proposed approach is presented in Sect. 3. The details of testbed implementation are given in Sect. 4. Section 5 discusses the evaluation results. The last section, Sect. 6, gives some concluding remarks and ideas for future work.

2 Cloud-Fog-Edge SDN-VANETs

The integration of Cloud-Fog-Edge computing in VANETs is the solution to handle complex computation, provide mobility support, low latency and high bandwidth. While they offer scalable access to storage, networking and computing resources, SDN provides higher flexibility, programmability, scalability and global knowledge. In Fig. 1, we give a detailed structure of SDN-VANET architecture. It includes the topology structure, its logical structure and the content distribution in the network. Specifically, the architecture consists of cloud computing data centers, fog servers with SDN Controllers (SDNCs), Road-Side Units (RSUs), RSU Controllers (RSUCs), Base Stations (BSs) and vehicles.

The implementation of this architecture can enable and improve the VANET applications such as road and vehicle safety services, traffic optimization, video surveillance, telematics, commercial and entertainment applications.

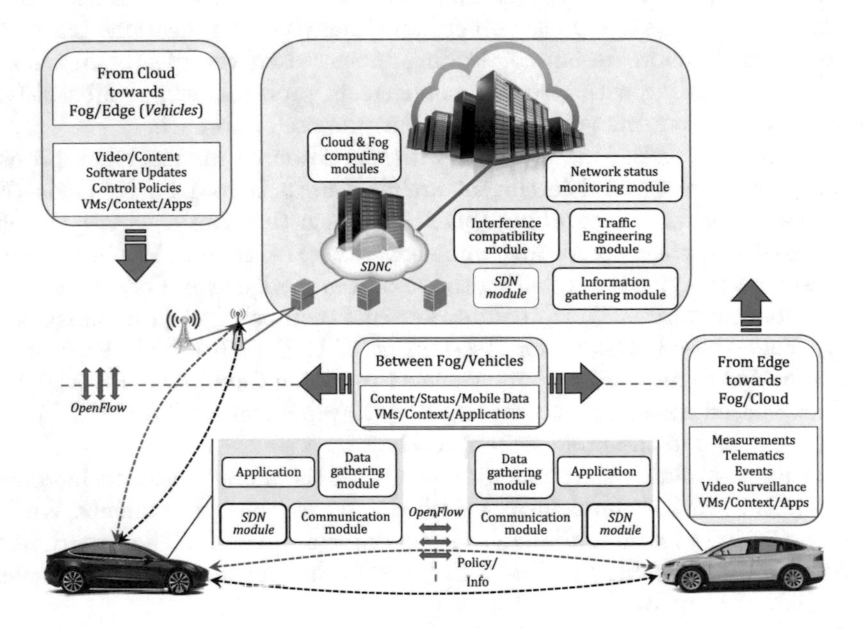

Fig. 1. Logical architecture of Cloud-Fog-Edge SDN-VANET with content distribution.

3 Description of Fuzzy-based Integrated System

This section presents the architecture of our proposed approach for coordination and management of VANETs resources. The proposed approach, called Integrated Fuzzy-based System for Coordination and Management of Resources (IFS-CMR), considers a layered Cloud-Fog-Edge SDN architecture that is coordinated by a fuzzy system implemented in the SDNC and the vehicles which

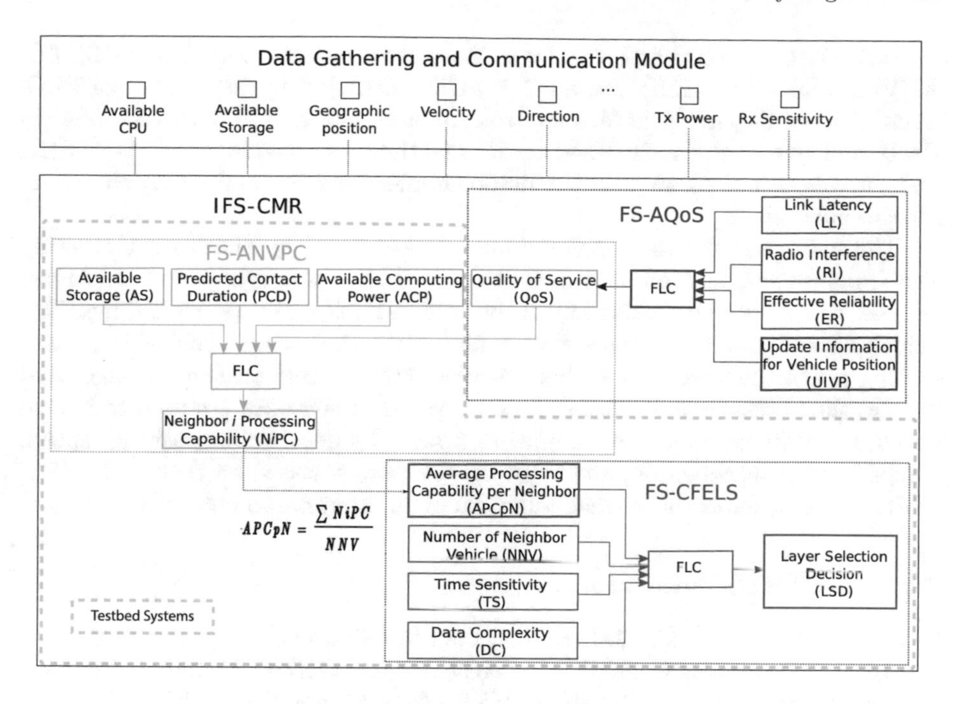

Fig. 2. Structure of fuzzy integrated system.

are equipped with an SDN module. SDNC manages the resources of the edge, fog, and cloud layer and based on the output of the fuzzy system determines the most suitable layer for storing and processing data application.

We consider this architecture from a bottom-up approach, which implies that the edge layer is the first layer selected based on the available connections and service requirements. If the application requirements are not fulfilled, then the fog layer is selected. The last layer to be considered is the cloud layer. Through this approach, all network resources are utilized effectively and massive traffic flow in the core network is avoided.

In the next subsections, we give details of the composition of FS-CFELS proposed system, describe the input and output parameters and present the design and implementation of FS-CFELS testbed.

3.1 IFS-CMR Design

The input parameters of IFS-CMR subsystems do not correlate to one another, leading to an NP-hard problem. FL can deal with these problems. Moreover, we want our systems to make decisions in real time and fuzzy systems can give very good results in decision making and control problems [8,10,13,14,21,22].

IFS-CMR is comprised of three integrated subsystems (FS-AQoS, FS-ANVPC and FS-CFELS). Each of them is controlled by its respective FLC. Each of the subsystems have a key role in the system. The structure of IFS-CMR is shown in Fig. 2. However for this testbed we consider only FS-CFELS, and for this reason we explain the design implementation and parameters only for this subsystem.

The data used for the input parameters are taken by the Data Gathering and Communication Module, which consists of the sensors implemented in the vehicles to gather information about the vehicle itself and its surroundings. The vehicle share this information gathered also with other nearby vehicle by broadcasting beacon messages. Neighbor vehicles receive these beacon messages and extract the necessary information. In our system, these data are used to inform about the vehicles available computing power, its geographic position, speed, available storage, direction, and transmission power. Based on these data, IFS-CMR then calculates the current condition of all input parameters of the system.

3.2 FS-CFELS Parameters

The parameters of FS-CFELS are described in following.

Average Processing Capability per Neighbor (APCpN): This parameter is the average of the Processing Capability (PC) of all neighboring vehicles within *the vehicle's* communication range. It represents the processing and storage capability of the edge layer, which is comprised by the total number on vehicles that are able to communicate and share their storage and processing capabilities with one-another. It is calculated as the sum of the PC of each neighbor vehicle divided by the number of neighboring vehicles.

Number of Neighboring Vehicles (NNV): Nearby neighbors are always changing in such a dynamic environment like VANETs. Vehicles change their speed and direction constantly, leading to fluctuations between a dense environment which is a better environment for sharing the vehicles capabilities with each-other, and not crowded environment in which vehicle-to-vehicle interactions are less likely to happen.

Time Sensitivity (TS): Different applications have different requirements in terms of latency. Safety applications in particular can not tolerate delays more than a few milliseconds [6,7]. Whereas, entertainment applications are considered delay tolerant and tolerate longer delays.

Data Complexity (DC): There are many factors that dictate the data complexity, its volume, its processing requirement and data analyzes to name a few. On the other hand, the heterogeneous data generated in VANETs make them different in type and structure. Even a single sensor might generate different types of data, which have different requirements in terms of processing, analyzing and extracting important information.

Table 1. Testbed setup.

Vehicles	5 RPi Model 3 B+
Mobility trace generator	sumo
Area size	200 m × 200 m
Communication range	50m
TS	[0.1, 0.5, 0.9]
DC	[0.1, 0.5, 0.9]
Acceptable PCD	2 s
Experimental time	1500 s

Layer Selection Decision (LSD): This is the output parameter value which decides the best layer to run applications. We consider three intervals [0, 0.3], [0.3, 0.7] and [0.7, 1] which indicate the selection of edge, fog and cloud layer, respectively.

4 Testbed Design

To evaluate the feasibility of IFS-CMR, we have designed and implemented a small-scale testbed using Raspberry Pis (RPi). We use five RPis with 50 m communication range that represent the vehicles moving for about 25 min in a 200 m × 200 m urban area. One of the vehicles is *the vehicle* in need of resources, whereas the other four are the vehicles that could turn into potential neighbors for sharing their resources with the vehicle in need. The setup of the testbed is summarized in Table 1. The movement of vehicles are generated using the *sumo* simulator and the layout is given in Fig. 3.

The position of vehicles changes every second. Vehicles current position, together with vehicle previous position, vehicle id, speed, direction and timestep are shared with the neighbors though beacon messages which are broadcasted every second. Neighbor vehicles receive these beacon messages and extract the necessary information. In our system, these data are used to inform about vehicles geographic position, their speed, direction, transmission power, available storage, available computing power. Based on this data, IFS-CMR then calculates the current condition of all input parameters of the system. In FS-CFELS, these data are used for calculating APCpN. NNV is decided by the number of neighbors which response to help beacons broadcasted by *the vehicle*. The neighbors response message indicate that these neighbors are inside *the vehicle* communication range and are willing to help it. Based on the calculated APCpN and NNC and the application requirement, which decide TS and DC, the fuzzy system FS-CFELS implemented in *the vehicle* decides then the appropriate layer for processing data. This process is shown in Fig. 4. One snapshots from the command line of the *vehicle* is shown in Fig. 5.

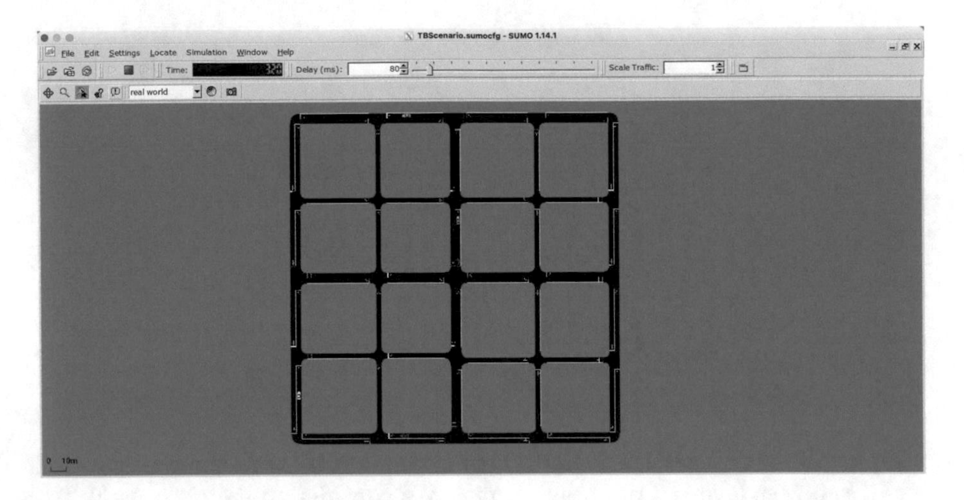

Fig. 3. A screenshot of the vehicles moving around the considered area in sumo simulator.

5 Experimental Results

The testbed results are shown in Fig. 6, in which we keep DC and TS constant while changing NNV and APCpN. In the testbed, the results are not easily distinguished because the single red line represents the results for different NNV and APCpNV at the same time.

In Fig. 6(a) are shown the results for non-complex applications that are delay tolerant. We can see that LSD values fall into the interval of [0.24, 0.78]. LSD values more than 0.7 indicate that the most appropriate layer to be used by *the vehicle* to run its applications is the cloud layer. Many of the LSD values fall between [0.3, 0.7] interval, which show the selection of fog layer for running applications. Whereas, for LSD values less than 0.3, the most suitable layer is decided the edge layer.

With the increase of data complexity in Fig. 6(b), LSD values are increased which show that fog and cloud layer are selected. Whereas in Fig. 6(c), as a results of having very complex data, the cloud layer is most appropriate. In both cases, the edge layer is rarely selected because we are dealing with very complex data and real-time processing is not a requirement.

Time sensitivity applications are shown in Fig. 6(d) – 6(f). For low data complexity (see Fig. 6(d)), edge layer and fog layer are suitable to handle these applications, because LSD values are less than 0.6. The increase of data complexity (see Fig. 6(e) and Fig. 6(f)) increases LSD value, however never reaches the cloud layer. Instead, it chooses the fog layer, as this layer provides powerful computing capabilities and real-time processing at the same time.

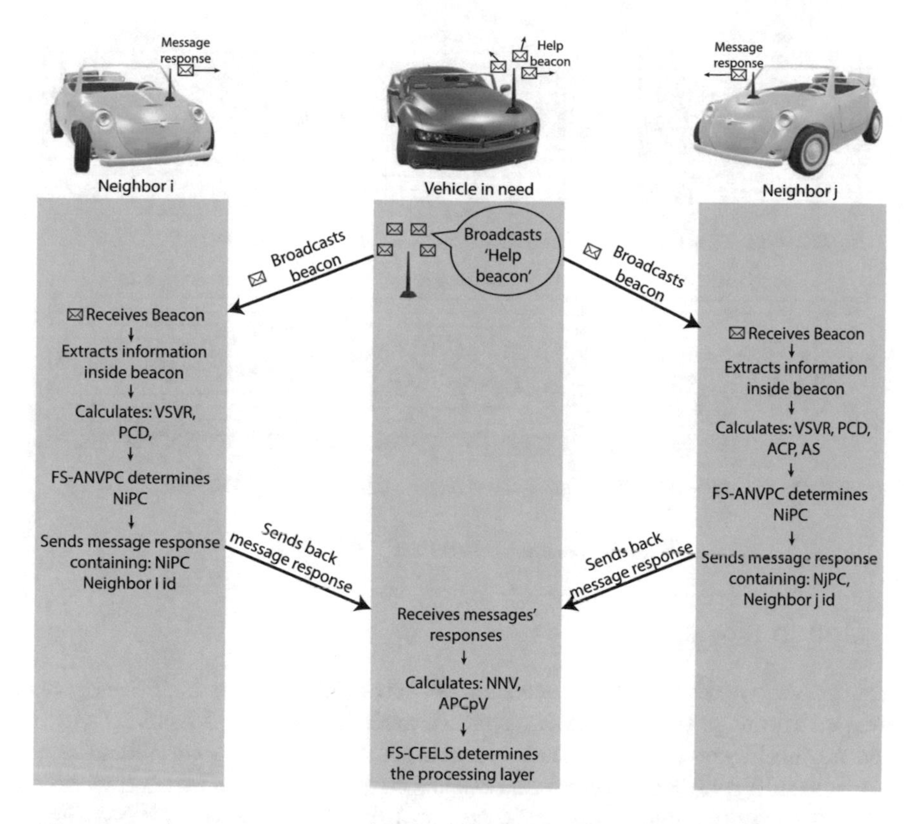

Fig. 4. A scheme of the communication between the *vehicle* and its neighbors.

```
pi@raspberrypi: ~/Desktop/Testbed-Final-Version-Vehicle
File  Edit  Tabs  Help

Received a message from neighbor with vehicle id: t_2!
Received a message from neighbor with vehicle id: t_1!
Received a message from neighbor with vehicle id: t_4!
Received a message from neighbor with vehicle id: t_3!
At this time there are 4 neighboring vehicles!
The average processing capability per neighbor stands at 0.762118!
The current values for all parameters are: , DC = 0.1, TS = 0.1, NNV = 1, APCpN = 0.762118!
Determining the processing layer for this application by using FS-CFELS ...!

Processing Layer recommended by FS-CFELS --> EDGE


Received a message from neighbor with vehicle id: t_2!
Received a message from neighbor with vehicle id: t_1!
Received a message from neighbor with vehicle id: t_4!
Received a message from neighbor with vehicle id: t_3!
At this time there are 4 neighboring vehicles!
The average processing capability per neighbor stands at 0.731865!
The current values for all parameters are: , DC = 0.1, TS = 0.1, NNV = 1, APCpN = 0.731865!
Determining the processing layer for this application by using FS-CFELS ...!

Processing Layer recommended by FS-CFELS --> EDGE
```

Fig. 5. Snapshots from the command line of the *vehicle*.

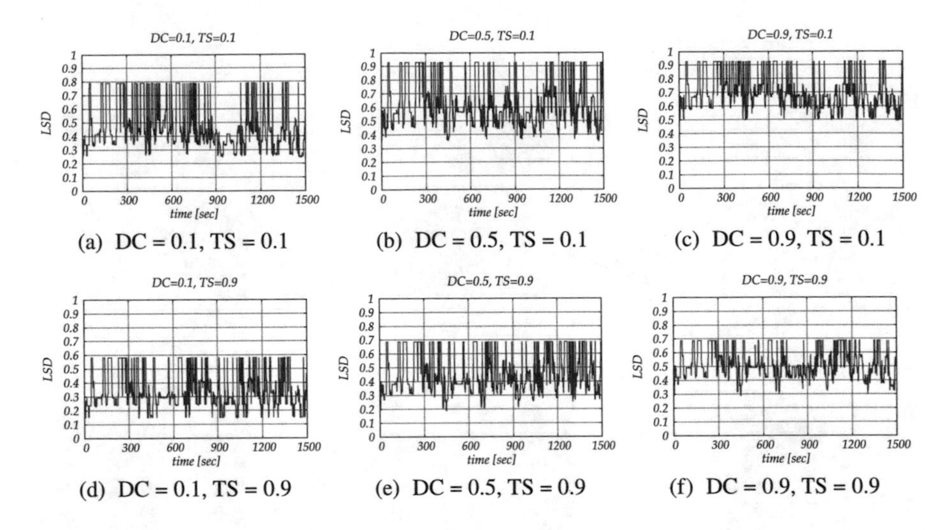

Fig. 6. Testbed results for FS-CFELS.

6 Conclusions

In this paper, we discussed the need for new strategies that can efficiently coordinate and manage the abundant resources available in SDN-VANETs and proposed an intelligent approach that can achieve this goal in a very flexible way. The proposed FS-CFELS system decides the best resources to be used by a vehicle that needs additional resources to accomplish different tasks. We have shown through experimental results by using the implemented testbed the effect of considered parameters on our system and its feasibility. From the experimental results we conclude as follows.

- In a dense environment, moderate complex data can be processed in the edge only if there are many potentially helpful neighbors in the vicinity.
- Time-sensitive applications are run either in edge or fog layer and never in the cloud.
- With the increase of data complexity less data is processed in the edge layer even if vehicles stay connected to the same potentially helpful neighbors for a long time.

In the future, we would like to improve IFS-CMR by implementing a testbed with mobile RPis moving randomly and not using sumo simulator for extracting location of vehicles.

References

1. Al-Heety, O.S., Zakaria, Z., Ismail, M., Shakir, M.M., Alani, S., Alsariera, H.: A comprehensive survey: benefits, services, recent works, challenges, security, and use cases for SDN-VANET. IEEE Access **8**, 91028–91047 (2020). https://doi.org/10.1109/ACCESS.2020.2992580

2. Aung, N., Zhang, W., Dhelim, S., Ai, Y.: Accident prediction system based on hidden Markov model for vehicular ad-hoc network in urban environments. Information **9**(12), 311 (2018). https://doi.org/10.3390/info9120311
3. Colagrande, S.: A methodology for the characterization of urban road safety through accident data analysis. Transp. Res. Procedia **60**, 504–511 (2022). https://doi.org/10.1016/j.trpro.2021.12.065
4. Fadhil, J.A., Sarhan, Q.I.: Internet of vehicles (IoV): a survey of challenges and solutions. In: 2020 21st International Arab Conference on Information Technology (ACIT), pp 1–10 (2020). https://doi.org/10.1109/ACIT50332.2020.9300095
5. Wu, G.F., Liu, F.J., Dong, G.L.: Analysis of the influencing factors of road environment in road traffic accidents. In: 2020 4th Annual International Conference on Data Science and Business Analytics (ICDSBA), pp 83–85 (2020). https://doi.org/10.1109/ICDSBA51020.2020.00028
6. Ge, X., Li, Z., Li, S.: 5G software defined vehicular networks. IEEE Commun. Mag. **55**(7), 87–93 (2017)
7. Hartenstein, H., Laberteaux, K.: Cooperative vehicular safety applications, John Wiley & Sons., pp 21–48 (2010). https://doi.org/10.1002/9780470740637.ch2
8. Kandel, A.: Fuzzy Expert Systems. CRC Press Inc, Boca Raton (1992)
9. Karagiannis, G., et al.: Vehicular networking: a survey and tutorial on requirements, architectures, challenges, standards and solutions. IEEE Commun. Surv. Tutorials **13**(4), 584–616 (2011)
10. Klir, G.J., Folger, T.A.: Fuzzy Sets, Uncertainty, and Information. Prentice Hall, Upper Saddle River, NJ, USA (1988)
11. Lu, N., Cheng, N., Zhang, N., Shen, X., Mark, J.W.: Connected vehicles: solutions and challenges. IEEE Internet Things J. **1**(4), 289–299 (2014). https://doi.org/10.1109/JIOT.2014.2327587
12. Luan, T.H., Cai, L.X., Chen, J., Shen, X.S., Bai, F.: Engineering a distributed infrastructure for large-scale cost-effective content dissemination over urban vehicular networks. IEEE Trans. Veh. Technol. **63**(3), 1419–1435 (2014). https://doi.org/10.1109/TVT.2013.2251924
13. McNeill, F.M., Thro, E.: Fuzzy Logic: A Practical Approach. Academic Press Professional Inc, San Diego, CA, USA (1994)
14. Munakata, T., Jani, Y.: Fuzzy systems: an overview. Commun. ACM **37**(3), 69–77 (1994)
15. Qafzezi, E., Bylykbashi, K., Ampririt, P., Ikeda, M., Matsuo, K., Barolli, L.: An intelligent approach for cloud-fog-edge computing SDN-VANETs based on fuzzy logic: effect of different parameters on coordination and management of resources. Sensors **22**(3), 878 (2022). https://doi.org/10.3390/s22030878
16. Raza, S., Wang, S., Ahmed, M., Anwar, M.R.: A survey on vehicular edge computing: architecture, applications, technical issues, and future directions. Wirel. Commun. Mob. Comput. **3159**, 762 (2019). https://doi.org/10.1155/2019/3159762
17. Rolison, J.J., Regev, S., Moutari, S., Feeney, A.: What are the factors that contribute to road accidents? An assessment of law enforcement views, ordinary drivers' opinions, and road accident records. Accid. Anal. Prev. **115**, 11–24 (2018). https://doi.org/10.1016/j.aap.2018.02.025
18. Seo, H., Lee, K.D., Yasukawa, S., Peng, Y., Sartori, P.: LTE evolution for vehicle-to-everything services. IEEE Commun. Mag. **54**(6), 22–28 (2016). https://doi.org/10.1109/MCOM.2016.7497762
19. Shrestha, R., Bajracharya, R., Nam, S.Y.: Challenges of future VANET and cloud-based approaches. Wirel. Commun. Mob. Comput. **2018** (2018)

20. World Health Organization (2018) Global status report on road safety 2018: summary. World Health Organization, Geneva, Switzerland, (WHO/NMH/NVI/18.20). Licence: CC BY-NC-SA 3.0 IGO)
21. Zadeh, L.A., Kacprzyk, J.: Fuzzy Logic for the Management of Uncertainty. John Wiley & Sons Inc, New York, NY, USA (1992)
22. Zimmermann, H.J.: Fuzzy control. In: Zimmermann, H.J. (ed.) Fuzzy Set Theory and Its Applications, pp. 203–240. Springer, Cham (1996)

A Consistency Maintenance Method Integrating OT and CRDT in Collaborative Graphic Editing

Chen Weijie, Xiong Caiquan(✉), and Wu Xinyun

School of Computer Science, Hubei University of Technology, Wuhan 430068, China
x_cquan@163.com, xinyun@hbut.edu.cn

Abstract. Operational Transformation (OT) algorithms and Commutative Replicated Data Type (CRDT) are methods commonly used to solve the problem of consistency maintenance for text objects in collaborative editing, while rarely studied in collaborative graphics editing. This paper proposes a consistency maintenance method for collaborative graphics editing by integrating OT and CRDT. Firstly, the conflict detection algorithm is given for three basic operations. Secondly, a data structure combining a hash table and a double linked list is used to store graph object operations, and operation conflicts are solved by operation transformation. At the same time, the algorithms of integrating local operation and integrating remote operation are given respectively to make the operation sequence consistent in the double linked list. Finally, an example and experiment verify the effectiveness and correctness of the method.

1 Introduction

Collaborative graphic editing takes graphics as the object of operation, and supports various collaborators in different locations to jointly edit shared graphic documents through the network [1]. Graphics such as rectangles, circles, and lines in shared graphics document are operation objects that can be edited [2]. Each object has position, size, color and other attributes. Users can create, modify, and delete graphics objects. During the collaborative process, multiple operations issued by multiple users on the same graph may be generated at the same time. Due to the uncertainty of the network environment, after each operation is generated at the local site and sent to the remote site, the execution order of the operations on each site may be inconsistent [3], resulting in inconsistent shared graphics documents, and the user's operation intention cannot be fully guaranteed [4]. Therefore, Consistency maintenance is an important issue in collaborative graph editing.

At present, the idea of operation transformation (OT) and commutative replicated data type (CRDT) are two main types of consistency maintenance methods [5]. The essence of two methods is to control the execution of operations and solve the concurrent conflicts between operations to achieve the purpose of consistency maintenance. The core idea of operation transformation is to obtain the correct operation position by correcting the parameters in the conflicting operation when the operation conflicts. The operation transformation algorithm is mostly used for text object, and the algorithm can be improved accordingly to solve the problem of graphic objects consistency

L. Barolli (Ed.): EIDWT 2023, LNDECT 161, pp. 471–482, 2023.
https://doi.org/10.1007/978-3-031-26281-4_49

maintenance. For example, Xu [6] extended the dOPT algorithm and proposed a concurrency control algorithm based on graphic objects. The algorithm effectively maintains the consistency of shared graph documents through transformation functions, and achieves semantic maintenance of operations. Sun [7] improved the context-based operation transformation (COT) algorithm by preprocessing the propagation and reception of operations. The improved algorithm can effectively reduce the repetition and number of transformations between operations. Gao [8] transformed two-dimensional graphics operations into linear operations, improved the ABST algorithm to make it suitable for collaborative graphics editing, and developed a Web-based collaborative graphics editing system CoWebDraw.

CRDT is a novel consistency maintenance method that maintains the user's operation attention. The CRDT algorithm has great scalability, and can achieve consistent maintenance of text, graphics, CAD systems, file management, and even mobile terminals [9]. The performance of the CRDT algorithm depends on the underlying data structure [10]. In the research of graphics, Gao [11] proposed an efficient CRDT algorithm, which can be applied to large-scale collaborative graphics editing systems, which not only solves the inconsistency of concurrency conflicts in basic operations, but also uses dynamic rule bases. The operational semantics of each site are effectively guaranteed.

The OT-CRDT method proposed in this paper combines the idea of operation transformation on the basis of CRDT. The method adopts the data structure of hash table and double linked list to store graphics object operations. When a concurrent operation is generated, the operation transformation is performed on the concurrent operation, so that the execution of the operation between the sites is consistent, so as to realize the consistency maintenance of the collaborative graphics editing system.

The content of this paper is organized as follows. Section 2 describes some basic definitions in collaborative graph editing. Section 3 describes the general idea of the OT-CRDT method. Section 4 provides a case to verify the correctness of the method. Section 5 compares the OT-CRDT method with the CRDT method, and the paper is summarized in Sect. 6.

2 Basic Definitions

Definition 1: Graphic Object Operation. A graph object operation can be defined as a 9-tuple <type, l, t_key, c_key, pre_key, next, prior, link, s>. Among them, type is the type of graphic object operation; including create, update, delete operations; l is a boolean variable, the values true or false represent local operations and remote operations; t_key is the identifier list of the target operations; c_key is the identifier of the operation; pre_key representing the identifier of the previous executed operation; next is the pointer which points the next operation in the double linked list; prior is the pointer which points the prior operation in the double linked list; link is the pointer used for chaining in the hash table; s is the current state of the execution of the operation.

Definition 2: Oa $\prec$ O_b. Given any two operations O_a and O_b in the double linked list, the position of O_a is donated as pos(O_a), and the position of o_b is donated as pos(O_b). If: pos(O_a) < pos(o_b), then $O_a \prec O_b$.

Definition 3: State Vector (SV). Assuming that the Number of collaborative sites in the system is N, the sites are identified by natural numbers 1, 2, ..., N, respectively. SV requires each site to maintain an N-dimensional state vector, each item corresponding to a site state. The operation rules of site SV_k are: (1) In the initial state, $SV_k = 0$, $i \in \{0, 1, ..., N-1\}$; (2) Every time an operation generated by site i is executed on Site k: $SV_k[i] = SV_k[i] + 1$.

Definition 4: Identifier (ID). The identifier of each graphic object operation is a three-tuple <s, ssv, site>: (1) s is the identifier of the session; (2) ssv is the sum of the state vectors of an operation; (3) site is the unique identifier of the site.

Definition 5: Given any two graphics object operations O_a and O_b, the identifiers of the two operations are ID_{Oa} and ID_{Ob} respectively. When any of the following conditions are met, $ID_{Oa} \prec ID_{Ob}$: (1) $ID_{Oa}[s] < ID_{Ob}[s]$; (2) $ID_{Oa}[s] = ID_{Ob}[s]$, $ID_{Oa}[ssv] < ID_{Ob}[ssv]$; (3) $ID_{Oa}[s] = ID_{Ob}[s]$, $ID_{Oa}[ssv] = ID_{Ob}[ssv]$, $ID_{Oa}[site] < ID_{Ob}[site]$.

3 OT-CRDT Method

3.1 The Main Framework

The main framework of the OT-CRDT method is shown in Fig. 1. In the collaborative graphic editing system, all graphic object operations are stored and linked through a data structure combined with a hash table and a double linked list, and the interactive view provides an interactive interface for collaborators. This method combines the idea of operation conversion on the basis of CRDT to solve the concurrency conflict between operations. The specific control procedure is as follows: when a collaboration site generates a local operation O_i, the operation O_i is executed and directly linked to the tail of the double linked list. When the remote operation O_j is received, it is assumed that its target operation O_3 is the same as the O_4. If the operation O_j conflicts with the O_4, the ConflictResolve function is called to compare the two identifiers to perform operation transformation on one of the operations to resolve the conflict. Otherwise, the CompatibleResolve function is called to determine the order of execution by their identifiers.

3.2 Conflict Detection

In the collaborative graphics editing system, there are three basic operation types, namely New (N) operation, Delete (D) operation and Update (U) operation. The conflict and compatible relationship between operations are represented by $\otimes$ and $\odot$, respectively, where A(O) represents the attribute identifier of the graphic object operated by operation O, and V(O) represents the attribute value of the graphic object operated by operation O. Suppose operation O_1 is the operation that has been executed and operation O_2 is the remote operation to be executed. The algorithm for judging whether two operations conflict is shown in Algorithm 1.

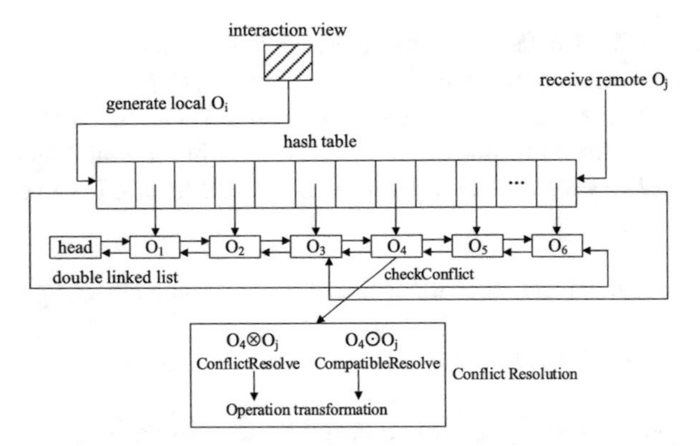

Fig. 1. The main framework.

```
Algorithm 1: checkConflict(O1,O2)
Input: O1,O2
Output: O1⊗O2 or O1⊙O2
  begin
    if O1.type = D then
      if O2.type = D then
        O1⊗O2
      else if O2.type = U then
        O1⊗O2
      else
        O1⊙O2
      end if
    else if O1.type = U then
      if O2.type = D then
        O1⊗O2
      else if O2.type = U then
        if A(O1) = A(O2) then
          if V(O1) = V(O2) then
            O1⊙O2
          else
            O1⊗O2
          end if
        else
          O1⊙O2
        end if
      else
        O1⊙O2
      end if
    else
      O1⊙O2
    end if
end.
```

It can be known from Algorithm 1 that, assuming two operations O_a and O_b are given, $O_a \parallel O_b$ and the target object graphic object identifiers of O_a and O_b operations are the same, the two operations have the relationship shown in Table 1.

Table 1. Conflict/Compatibility relationship between graphic object operations

	N	D	U
N	⊙	⊙	⊙
D	⊙	⊗	⊗
U	⊙	⊗	⊗/⊙

3.3 Conflict Resolution

Suppose operation O_1 is the operation that has already been executed, and operation O_2 is the remote operation to be executed. Algorithm 2 describes how to resolve conflicts between operations. The priority of the operation is determined by comparing the IDs of the two operations. The lower priority operation is transformed into null and not stored in the double-linked list. E(O) represents operation O is transformed.

```
Algorithm 2: ConflictResolve(O1⊗O2)
Input: O1⊗O2
  begin
    if ID(O1)<ID(O2) then
      E(O2) ← NULL
    else
      E(O1) ← NULL, O2 is executed and linked after
O1.prior
    end if
end.
```

Algorithm 3 describes how to resolve the compatibility relationship between two operations. Operations with higher priority are executed first, and operations with lower priority are executed later.

```
Algorithm 3: CompatibleResolve(O₁⊙O₂)
Input: O₁⊙O₂
  begin
    if ID(O₁)≺ID(O₂) then
      O₂ is executed and linked after O₁
    else
      E(O₁) ← O₂, O₁ is redone and linked after O₂
    end if
end.
```

3.4 Integrating Local/Remote Graphic Object Operations

Algorithm 4 shows how to integrate a local operation. If the double linked list is empty, operation O is executed and linked after head. If the graphics object operation from the previous session is reused, operation O is executed and linked at the tail of the double linked list. Otherwise, operation O is executed and linked after the operation (Pre) that has just been executed in the current session.

```
Algorithm 4: IntegrateLocal(O)
Input: O
  begin
    if O.pre_key = NULL then
      if head.next = NULL then
        O is executed and linked next to head
      else
        O is executed and linked after tail
      end if
    else
      Pre ← hash(O.pre_key), O is executed and linked
after Pre
    end if
O is broadcast to remote sites
end.
```

Algorithm 5 shows how to integrate a remote operation. When the remote operation O is received, the method Target(O) is invoked to find its target operation. The target operation is generally a graphics object operation generated by its NEW operation. After finding the target operation in the double linked list, operation O compares the identifier with the operation behind the target operation. If a conflict occurs, the function ConflictResolve is called to resolve the conflict operations. Otherwise, the function CompatibleResolve is called to resolve the compatible operations.

```
Algorithm 5: IntegrateRemote(O)
Input: O
  begin
    Tar ← Target(O), N ← Tar.next
    if Tar = NULL then
      O is executed and linked after tail
    else
      if N = NULL then
        O is executed and linked after Tar
      else
        while N.c_key.s < O.c_key.s do
          N ← N.next
        end while
        while N != NULL && ID(N)≺ID(O) do
          if N.s = O.s && checkConflict(N, O) = N⊗O
then
            ConflictResolve(N⊗O), Break
          else
            N ← N.next, Continue
          end if
        end while
        if N = NULL then
          O is executed and linked after N.prior
        end if
        if ID(O)≺ID(N) then
          if N.s = O.s && checkConflict(N, O) = N⊗O
then
            ConflictResolve(N⊗O)
          else if N.s = O.s && checkConflict(N, O) =
N⊙O then
            CompatibleResolve(N⊙O)
          else
            O is executed double linked before N
          end if
        end if
      end if
    end if
end.
```

4 Case Analysis

In order to verify the correctness and effectiveness of the proposed method, it is assumed that there are two collaborative sites Site1 and Site2 in the collaborative graphics editing system to complete a design task together, and this case is carried out on the Microsoft Windows 10 operating system using the IntelliJ IDEA platform simulation. ID[sitel] < ID[site2]. L stands for double linked list. We use "–" to connect all graphics object operations, described in two session phases.

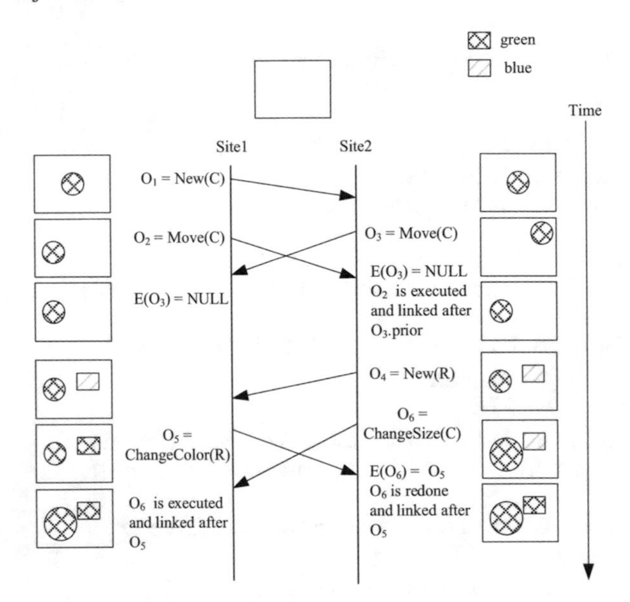

Fig. 2. The first collaborative process.

Figure 2 shows the first collaborative process between Site1 and Site2. Site1 generates three local operations, O_1, O_2, and O_5. Site2 also generates three operations, O_3, O_4, and O_6. The state vectors for all operations are: $SV(O_1) = (1, 0)$, $SV(O_2) = (2, 0)$, $SV(O_3) = (1, 1)$, $SV(O_4) = (2, 2)$, $SV(O_5) = (3, 2)$, $SV(O_6) = (2, 3)$. The IDs for all operations are $ID_{O1} = (1, 1, 1)$, $ID_{O2} = (1, 2, 1)$, $ID_{O3} = (1, 2, 2)$, $ID_{O4} = (1, 4, 2)$, $ID_{O5} = (1, 5, 1)$, $ID_{O6} = (1, 5, 2)$.

At Site1, the local operation O_1 is generated and executed immediately. At this time, the double linked list L_1 is empty and directly linked after the head node of L_1, $L_1 = O_1$. Operation O_2 is executed after generation and linked after O_1, $L_1 = O_1 - O_2$. When the remote operation O_3 is received, the operation O_2 conflicts with the operation O_3, and their target operation is O_1. Since $ID_{O2} \prec ID_{O3}$, the operation O_3 is transformed, and $E(O_3) = NULL$, $L_1 = O_1 - O_2$. When a remote operation O_4 is received, O_4 is directly linked at the tail of the double linked list, $L_1 = O_1 - O_2 - O_4$. The local operation O_5 is executed after it is generated, and O_5 is directly linked at the tail of the double linked list, $L_1 = O_1 - O_2 - O_4 - O_5$. When remote operation O_6 is received, the target operation of operation O_6 is O_1. After operation O_1, there are operations O_2, O_4, and O_5. Among them, the state vectors of O_2 and O_4 are smaller than O_6, and $ID_{O5} \prec ID_{O6}$, therefore, the operation O_6 is linked after operation O_5, $L_1 = O_1 - O_2 - O_4 - O_5 - O_6$.

At Site2, when remote operation O_1 is received, it is executed immediately, $L_2 = O_1$. The local operation O_3 is executed after generation and linked after O_1, $L_2 = O_1 - O_3$. When the remote operation O_2 is received, the operation O_2 conflicts with the operation O_3. Since $ID_{O2} \prec ID_{O3}$, the operation O_3 is transformed, $E(O_3) = NULL$, the operation O_2 is executed, $L_2 = O_1 - O_2$. The local operation O_4 is executed after it is generated, and O_4 is directly linked at the tail of the double linked list, $L_2 = O_1 - O_2 - O_4$. The local operation O_6 is the same as O_4. $L_2 = O_1 - O_2 - O_4 - O_6$. Finally, when the remote operation O_5 is received, the target operation of O_5 is O_4. Since $ID_{O5} \prec ID_{O6}$ and operation O_5 and O_6 are compatible, $E(O_6) = O_5$, O_6 is redone and linked after O_5, $L_2 = O_1 - O_2 - O_4 - O_5 - O_6$.

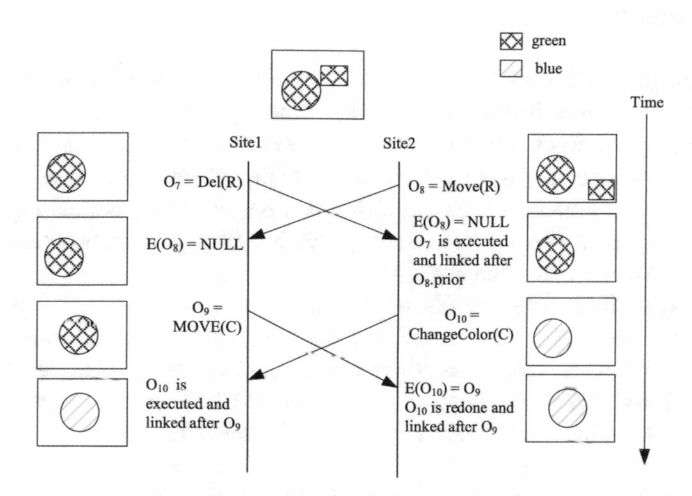

Fig. 3. The second collaborative process.

Figure 3 shows the second collaborative process between Site1 and Site2. Site1 generates two local operations, O_7 and O_9. Site2 also generates two operations, O_8 and O_{10}. The state vectors for all operations are: $SV(O_7) = (4, 3)$, $SV(O_8) = (3, 4)$, $SV(O_9) = (5, 4)$, $SV(O_{10}) = (4, 5)$. The IDs for all operations are $ID_{O7} = (2, 7, 1)$, $ID_{O8} = (2, 7, 2)$, $ID_{O9} = (2, 9, 1)$, and $IDO_{10} = (2, 9, 2)$.

At Site1, the local operation O_7 is executed after generation and linked after O_6, $L_1 = O_1 - O_2 - O_4 - O_5 - O_6 - O_7$. After receiving the operation O_8, the target operation of O_8 is O_4, and the operation O_7 is conflict with O_8. Since $ID_{O7} \prec ID_{O8}$, the operation O_8 is transformed, $E(O_8) = NULL$, $L_1 = O_1 - O_2 - O_4 - O_5 - O_6 - O_7$. The local operation O_9 is executed immediately after it is generated. O_9 is directly linked at the tail of the double linked list and $L_1 = O_1 - O_2 - O_4 - O_5 - O_6 - O_7 - O_9$. The remote operation O_{10} is finally received, the target operation of the operation O_{10} is O_1. Since $ID_{O9} \prec ID_{O10}$, operation O_{10} is linked after the O_9, $L_1 = O_1 - O_2 - O_4 - O_5 - O_6 - O_7 - O_9 - O_{10}$.

At site 2, the local operation O_8 is generated after it is generated and linked after O_6, $L_2 = O_1 - O_2 - O_4 - O_5 - O_6 - O_8$. After receiving the operation O_7, the target operation of O_7 is O_4, and the operation O_7 is conflict with O_8, so the operation O_8 is transformed, $E(O_8) = NULL$, and operation O_7 is executed, $L_2 = O_1 - O_2 - O_4 - O_5 - O_6 - O_7$. The local operation O_{10} is executed after generation and linked after O_7, $L_2 = O_1 - O_2 - O_4 - O_5 - O_6 - O_7 - O_{10}$. The remote operation O_9 is finally received, O_1 is the target operation of the O_9. Since $ID_{O9} \prec ID_{O10}$ and operation O_{10} is compatible with O_9, $E(O_{10}) = O_9$, O_{10} is redone and linked after O_9, $L_2 = O_1 - O_2 - O_4 - O_5 - O_6 - O_7 - O_9 - O_{10}$.

Ultimately, the double linked lists of both sites are: $L = O_1 - O_2 - O_4 - O_5 - O_6 - O_7 - O_9 - O_{10}$, and the results of the collaborative graphic document are consistent.

5 Experiment

In this section, the CRDT method is compared with the OT-CRDT method in this paper through simulation experiments. The merits of the algorithm can be measured by the comparison times between operations. In the CRDT algorithm, the number of comparisons between operations is mainly determined by the integration of remote operations. Therefore, the performance of the algorithm can be judged by comparing the operation comparison times of integrated remote operation during the comparison simulation experiment between the two methods.

When solving the problem of consistency maintenance of cooperative graphics editing, The CRDT method resolves conflicts between graphic object operations by using tombstone operations to make operations invisible. $\ominus$ indicates that operation $\ominus$ is an invisible tombstone operation. For example, when the CRDT method solves the preceding cases, the consistency maintenance result is as follows: $L = O_1 - O_2 - \ominus_3 - O_4 - O_5 - O_6 - O_7 - \ominus_8 - O_9 - O_{10}$.

In the collaboration scenario as shown in Fig. 4, the operation set consists of 9 operations, and the operation objects and target operations of all operations are consistent and meet the conditions: $O_1 \otimes O_2 \otimes O_3$ 、 $O_4 \otimes O_5 \otimes O_6$ 、 $O_7 \otimes O_8 \otimes O_9$.

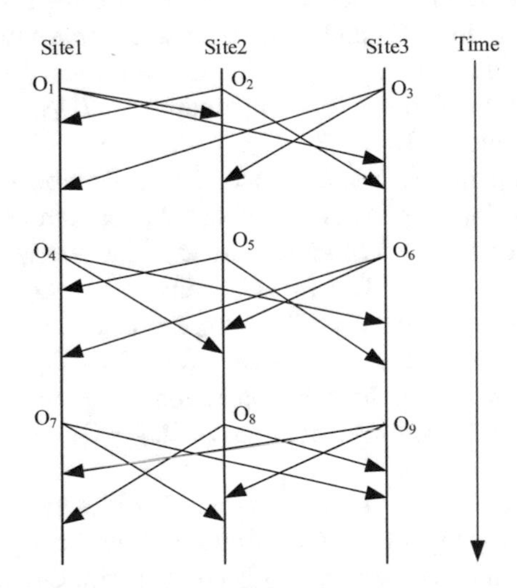

Fig. 4. The collaboration scenario.

The experiment counts the comparison times of remote operations, and the results are shown in Fig. 5. As can be seen from the experimental comparison results, when integrating remote operations, with the increase of cooperative operations, the comparison times between operations of CRDT method is higher than OT-CRDT method. Due to the high responsiveness of operation transformation, the method in this paper effectively reduces the number of comparisons between operations by performing an operation transformation and not storing the tombstone operation. It is proved that the OT-CRDT method can effectively reduce the response time of remote operation and increase the efficiency of real-time collaborative graphics editing system.

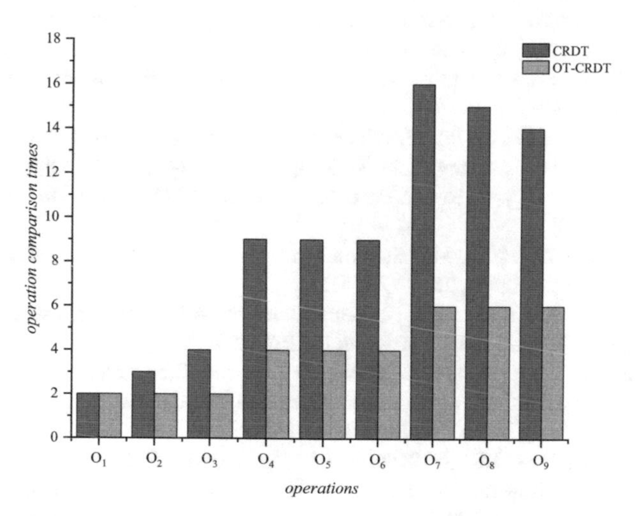

Fig. 5. The experimental results of the OT-CRDT method and CRDT method.

6 Conclusion

In this paper, a consistency maintenance method OT-CRDT integrating OT and CRDT is proposed for collaborative graph editing system. First, a conflict detection is given for the conflict compatibility relationship between operations. Secondly, the overall process of the method is described in detail, including resolving conflict/compatible operations and integrating local and remote operations. Thirdly, the method is verified by a case. Finally, the effectiveness of the method is proved by the comparison of experiments.

In the future work, on the basis of integrating OT and CRDT, we will explore the consistency maintenance algorithm suitable for Undo/Redo operation in collaborative graphics editing. For the CRDT algorithm, how to design a more favorable data structure and explore a real-time collaborative graph editing algorithm with high response and strong scalability is also one of the future research contents.

Acknowledgments. This research is supported by Science and Technology Plan Project of Hubei Province under grant number 2021BLB171, National Natural Science Foundation of China under grant number 61902116, and Green Industry Technology Leading Project (product development category) of Hubei University of Technology under grant number CPYF2017008.

References

1. Sun, C.Z., Chen, D.: Consistency maintenance in real-time collaborative graphics editing systems. ACM Trans. Comput. Hum. Interact. (TOCHI) **9**(1), 1–41 (2002)
2. Wu, C., Li, L., Peng, C., Wu, Y., Xiong, N., Lee, C.: Design and analysis of an effective graphics collaborative editing system. EURASIP J. Image Video Process. **2019**(1), 1–21 (2019). https://doi.org/10.1186/s13640-019-0427-6
3. Yang, J., Dou, W.F.: A new multiple versions incremental creation algorithm. Chin. J. Comput. **31**(4), 702–710 (2008)
4. He, F.Z., Lv, X., Cai, W.W.: Survey of real-time collaborative editing algorithms supporting operation intention consistency. Chin. J. Comput. **41**(4), 840–867 (2018)
5. Sun, C.Z., Sun, D., Ng, A.: Real differences between OT and CRDT for co-editors. arXiv: 1810.02137 (2018)
6. Xu, J., Jiang, X.F., Zhang, K.: Research on consistency maintenance based on graphic object. Comput. Appl. Softw. **29**(2), 261–265 (2012)
7. Sun, M., Wang, R.H.: Research on concurrent control algorithm based on operational transformation. Comput. Appl. Softw. **34**(1), 263–269 (2017)
8. Gao, L., Gao, D., Xiong, N., Lee, C.: CoWebDraw: a real-time collaborative graphical editing system supporting multi-clients based on HTML5. Multimed. Tools Appl. **77**(4), 5067–5082 (2017). https://doi.org/10.1007/s11042-017-5242-4
9. Lv, X., Yuan, J.C., Ben, K.R.: A CRDT-based sequence transformation algorithm for supporting collaborative editing on mobile device. J. Huazhong Univ. Sci. Technol. (Natural Science Edition) **50**(02) 130–135 (2022)
10. Cai, W., He, F., Lv, X.: Multi-core accelerated CRDT for large-scale and dynamic collaboration. J. Supercomput. **78**(8), 10799–10828 (2022). https://doi.org/10.1007/s11227-022-04308-7
11. Gao, L.P., Xu, X.F.: CRDT-based consistency maintenance in large-scale real-time graphics editing systems **40**(7) 1361–1367 (2019)

Data Pipeline of Efficient Stream Data Ingestion for Game Analytics

Noppon wongta(✉) and Juggapong Natwichai

Department of Computer Engineering, Faculty of Engineering, Chiang Mai University, Chiang Mai, Thailand
noppon.w@cmu.ac.th, juggapong@eng.cmu.ac.th

Abstract. This paper explores approaches to optimize micro-services that manage game analytics in terms of stream analysis. Typically, the micro-services parameters should be adjusted to suit the streaming data efficiently. We focus on the important data pipeline's issues, i.e. the throughput and velocity of the data stream generated by extraction information modules. We investigate the existing technologies employed by the empirical studies as well as the architecture of the micro-services that make up the end-to-end data pipeline. The findings are reported as conclusion.

1 Introduction

Game analytics gains the interests increasingly. Important methods, e.g. data mining, business intelligence, statistics, and usability testing methods, are employed to extract information from the game data. There is a trend in several studies to collect spatial information on every player's activity in the games. Due to its effectiveness and usability in identifying "who" is carrying out "what" activity ", when", and "where" in your game, this approach is rapidly gaining popularity within the industry. In [15], game data pipeline analytics is used to capture in-game data by integrating game state on the client at nearly real-time analytics, demonstrating how to operate a prediction model and show the result of prediction during game-play. In which, it possibly affects the performance of their clients. To efficiently manage spatial information that should be sent via back-end services, the game client would not use owner resources and could immediately use the data for analytics. It is necessary to design a transmission structure that considers the appropriate amount of data flow, throughput, and transmission method. The data pipeline that serves as the foundation for this analysis is made to operate concurrently with the player's gaming system. In addition, to avoid affecting how to use the least amount of space in the data lake, which could be focused on storing and compressing data.

This work studies the performance characteristics of these streaming sources under various loads, indicates a performance trade-off interaction, and identifies the limits of acceptable performance for data pipeline and streaming source integration. In our work, the data pipeline first processes and extracts information

L. Barolli (Ed.): EIDWT 2023, LNDECT 161, pp. 483–490, 2023.
https://doi.org/10.1007/978-3-031-26281-4_50

from binary log data from the data lake before moving on to the pre-processing and analytical modules by continuous stream data. In this case, following on Dota2 game replay processing [5] provides an open-source of the parser a replay and open access archive replay server [7]. A workflow is customized by sending a log interval to message protocol as remote procedure calls (RPCs). The Extracting information module called "Parser" is based on Java language and Supports running on micro-service such as docker and Kubernetes. This helps explore the scale-up of a loading ingest through the data pipeline. The data ingestion module use messaging system service such as flume [11], Kafka [18], ActiveMQ [22], and RabbitMQ [13]. This service is meant to be built to deliver reliable data analysis throughput and efficient data transmission. There is a requirement for compatibility between extracting information module and the analytic module. Finally, to study this domain as a continuum, the analytic module leveraged the Spark ecosystem [21] to manage batch and stream processing to analytic process with configurable message size and CPU burden per message.

The remaining sections of this paper are organized as follows. Section 2 presents the literature review. Section 3 describes the system to be studied. Section 4 provides the experiment results of the study. Finally, the proposed work is concluded and discussed in Sect. 5.

2 Literature Review

Analytics in games demands understanding the reasoning behind game engagement, monetization, fraud, and error reporting. In competitive e-sports games [8], the process of analyzing player behaviour has been applied in various issues such as win prediction [9,16,17,20,24,26], encounter detection [23] or identifying patterns in combat [25], and Classify player roles [14]. Typically, the data flow used in academia and industry would have similar techniques. Batch and stream processes are the two different forms of data flow utilized in computation. Employing a micro-service module that covers the same process approach generates interest in the field of data engineering that produces practical workflows.

The character of the process of extracting information module is brief period and thick log data. The messaging service would connect and handle the traffic data [19]. In between processes [10] of data-flow would have considered throughput, CPU Cost, and Streaming latency. Then the data flows into the analysis service, and the ingest technique must be handled by scheduling jobs [12] to pass continuing data through the analytical process successfully. The system experiment would apply virtualization to create an end-to-end data pipeline. To Investigation data pipeline with the objective of throughput, CPU Cost Streaming latency, and other issues. Because of the flexibility of the data pipeline design and the extension of monitoring tools like Prometheus [6] and Grafana [2], it is decided, in our work, to employ virtualization to manage the exploration.

3 System Architecture

Infrastructure based on hypervisor supports guest virtual machines by coordinating calls for central processing unit (CPU), memory, disk, network, and other resources through the physical host's OS. In our studies, the virtual environment has the following characteristics: Intel(R) Core(TM) i7-4770 CPU with a clock frequency of 3.40 GHz and eight-core, 15 GB of RAM, Solid State Drive size of 430 GB with raid 1 (120 GB per drive). In this section, the settings for the research experiment's network, micro-services, and problems encountered when executing data flows are discussed.

3.1 Virtualization

The process of virtualization makes it possible to use physical computer hardware more effectively. To enable the division of a single computer's hardware components, processors, memory, storage, and more into several virtual computers, also known as virtual machines (VMs). This work uses Proxmox as the base operating system. Because the design base is based on Debian GNU/Linux [1] and uses a custom Linux Kernel that is a Kernel-based Virtual Machine (KVM). Comparatively with to other container solutions, Kubernetes [3] offers a simple method for scaling your application. Many resource management and provisioning operations can be automated using the Kubernetes API. This research chooses Microk8s [4] from the Ubuntu server to run Kubernetes. The central cluster node is setting virtualization by 8 core CPU and 10 GB memory, and virtual hard disk size of 60 GB.

3.2 Micro-service

The work of the data pipeline has four micro-services that are related to each other as shown in Fig. 1:

1. Dota2parser to extract the replay data and send log messages in small packets for every recording interval.
2. Flume which has collected and moved substantial amounts of log data from Dota2parser via RPCs protocol and forwards it to the Message broker as Kafka.
3. Kafka to take the data and allocates Splitting data to make real-time processing work efficiently and adequately ingest to analytics module as Spark.
4. Spark to receive streaming data and processes analytics data continuously.

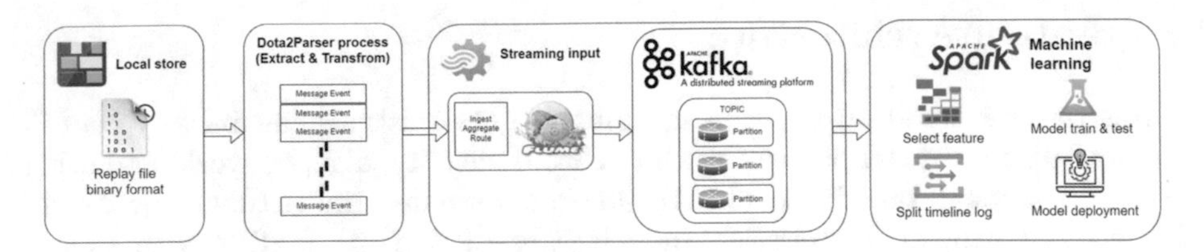

Fig. 1. Analytics system architecture

3.3 Data Flow Issues

The extraction information and collected log data module show adaptable of a channel to stream data which issues occur from the CPU cost of distributed messages in extraction processes and message buffer in each channel. Firstly, to scale up throughput from a single extraction module would desire to split each log data to each channel collected log data module. But the cost of split data must be notation as $O(n)$ if running more channels, and the CPU cost will grow linear. Therefore, the channel and throughput can be raised, but the CPU cost must also be considered. Second, in collected log data, can adjust the memory buffer in each channel would affect handling copious amounts of log data. As a result, there is less need for channel extension, yet the capacity limit on data transmission will impact the amount of delayed data.

4 Experiment Results

In this section, the result from the empirical studies are presented. Experiments are set to evaluate the effects of parameters to the performance indicators, e.g. throughput of data packets or CPU workload.

For the first experiment, Fig. 2 shows an increased target at dota2 parser to flume, effected to the data transfer rate. It can be seen that the result when the capacity is increased, the data transmission rates are higher.

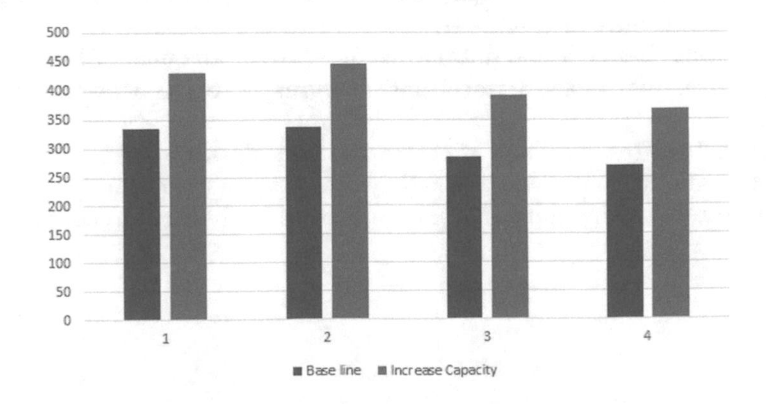

Fig. 2. Change capacity in Flume module

When we focus more on each individual microservice, i.e. Dota2parser, which will send out to multiple Flume agents. Figures 3 and 4 show the result when we increase in multiple targets sending. It can be seen such effect as the result of network bandwidth. Obviously, throughput cannot improve by multiple targets increasing. Although, capacity and transaction per capacity would increase.

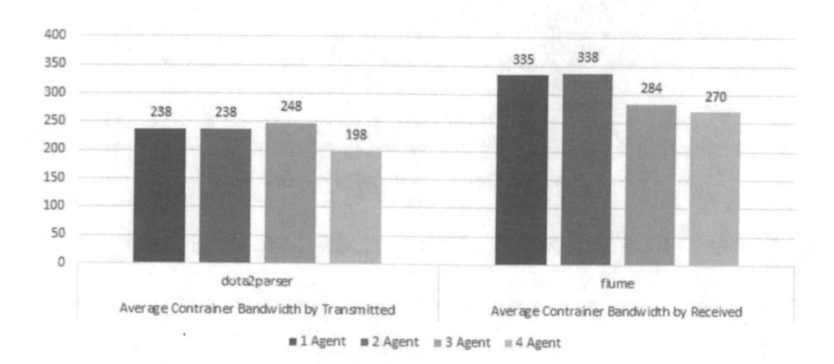

Fig. 3. Average container bandwidth by transmitted in dota2parser module in case baseline

Figure 5 shows how the data pipeline operates and extracts information to send the messages to Kafka through the time.

Figure 6 shows the increased transactions per capacity affected on data transfer rate. It can be seen that the increments does not affect the overall uptake and transmission through the flume.

In summary, the increased capacity showed a better performance than the default baseline. It also resulted in a more noticeable when the number of agents in the system increased, particularly, when more than three agents was set. In general, the difference between a single agent and the dual agents could be

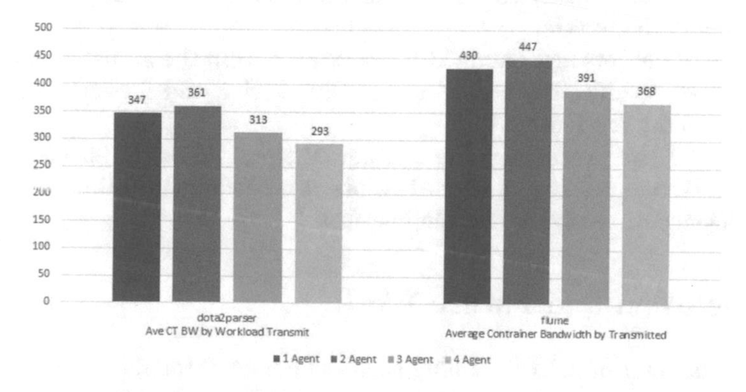

Fig. 4. Average container bandwidth by transmitted in dota2parser module in case baseline

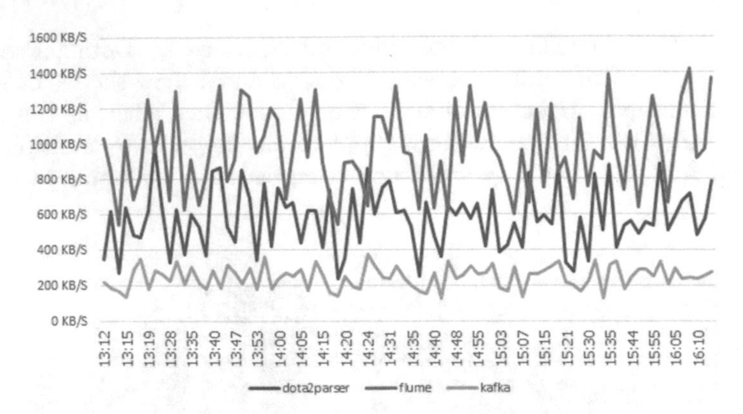

Fig. 5. Rate of transmitted packets in flume module through the time

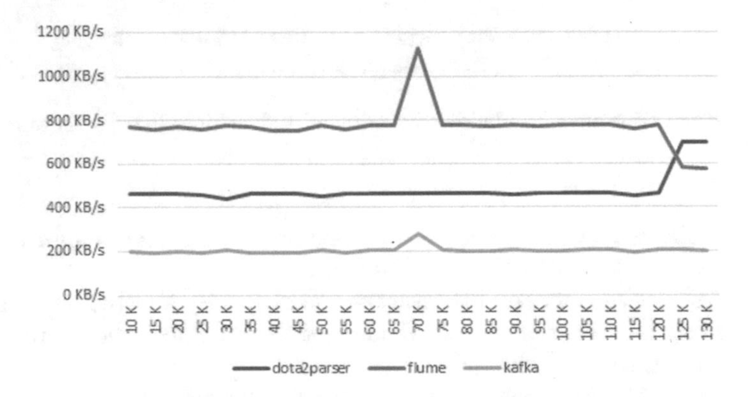

Fig. 6. Average container bandwidth by transmitted in dota2parser module in case baseline

seen since the dual worker can share the workload. As a result, receiving and transmitting data is better.

In terms of packets and bandwidth, the results from the adjusted parameters were increased significantly as the agents can receive and forward data to the next service more efficiently.

In terms of the processing, the number of agents directly affects the CPU workload, so the increasing number of agents also needs to be adjusted according to the efficiency of receiving and sending data.

5 Conclusion and Future Work

With the character of data streaming from the parser through the data pipeline, in-game analytics would frequently send after extracting information altogether, but the packet size is not large or complex. With our architecture, the micro-services can handle messages from the parser via RPCs protocol. Primarily, the protocol RPCs could operate well and is simple to implement. In addition, the micro-services can benefit from the optimizer to adjust further, e.g. the memory channel to improve the efficiency of throughput.

In the future work, further experiments will focus on constraints determination theoretically, and improve the data pipeline on Kafka and Spark modules to achieve an end-to-end data pipeline.

References

1. Debian – the universal operating system. https://www.debian.org/index.en.html
2. Grafana: The open observability platform— grafana labs. https://grafana.com/
3. Kubernetes. https://kubernetes.io/
4. Microk8s - zero-ops kubernetes for developers, edge and IoT. https://microk8s.io/
5. odota/parser: Replay parse server generating JSON log events from Dota 2 replay files. https://github.com/odota/parser
6. Prometheus - monitoring system & time series database. https://prometheus.io/
7. SteamPipe - Valve Developer Community. https://developer.valvesoftware.com/wiki/Replay, https://developer.valvesoftware.com/wiki/SteamPipe
8. Dax: Data-driven audience experiences in esports, pp. 94–105 (2020). https://doi.org/10.1145/3391614.3393659
9. Agarwala, A., Pearce, M.: Learning dota 2 team compositions, pp. 2–6 (2014). https://cs229.stanford.edu/proj2014/Atish%20Agarwala,%20Michael%20Pearce,%20Learning%20Dota%202%20Team%20Compositions.pdf
10. Blamey, B., Hellander, A., Toor, S.: Apache spark streaming, Kafka and HarmonicIO: a performance benchmark and architecture comparison for enterprise and scientific computing. In: Gao, W., Zhan, J., Fox, G., Lu, X., Stanzione, D. (eds.) Bench 2019. LNCS, vol. 12093, pp. 335–347. Springer, Cham (2020). https://doi.org/10.1007/978-3-030-49556-5_30
11. Chambers, C., et al.: FlumeJava: easy, efficient data-parallel pipelines (2010). https://doi.org/10.1145/1809028.1806638, https://research.google/pubs/pub35650/
12. Cheng, D., Chen, Y., Zhou, X., Gmach, D., Milojicic, D.: Adaptive scheduling of parallel jobs in spark streaming (2017). https://doi.org/10.1109/INFOCOM.2017.8057206
13. Dobbelaere, P., Esmaili, K.S.: Industry paper: Kafka versus RabbitMQ: a comparative study of two industry reference publish/subscribe implementations. In: DEBS 2017 - Proceedings of the 11th ACM International Conference on Distributed Event-Based Systems, pp. 227–238 (2017). https://doi.org/10.1145/3093742.3093908
14. Eggert, C., Herrlich, M., Smeddinck, J., Malaka, R.: Classification of player roles in the team-based multi-player game dota 2. In: Chorianopoulos, K., Divitini, M., Hauge, J.B., Jaccheri, L., Malaka, R. (eds.) ICEC 2015. LNCS, vol. 9353, pp. 112–125. Springer, Cham (2015). https://doi.org/10.1007/978-3-319-24589-8_9
15. Hodge, V.J., Devlin, S., Sephton, N., Block, F., Cowling, P.I., Drachen, A.: Win prediction in multiplayer esports: Live professional match prediction. IEEE Trans. Games **13**, 368–379 (2021). https://doi.org/10.1109/TG.2019.2948469, https://ieeexplore.ieee.org/document/8906016/
16. Kalyanaraman, K.: To win or not to win? A prediction model to determine the outcome of a DotA2 match (2014). https://cseweb.ucsd.edu/jmcauley/cse255/reports/wi15/Kaushik_Kalyanaraman.pdf

17. Kinkade, N., Jolla, L., Lim, K.: DOTA 2 win prediction. Univ. Calif. **1**, 1–13 (2015)
18. Kreps, J., Narkhede, N., Rao, J.: Kafka: a distributed messaging system for log processing. ACM SIGMOD Workshop on Networking Meets Databases, p. 6 (2011). http://research.microsoft.com/en-us/um/people/srikanth/netdb11/netdb11papers/netdb11-final12.pdf
19. Maarala, A.I., Rautiainen, M., Salmi, M., Pirttikangas, S., Riekki, J.: Low latency analytics for streaming traffic data with apache spark, pp. 2855–2858 (2015). https://doi.org/10.1109/BigData.2015.7364101
20. Makarov, I., Savostyanov, D., Litvyakov, B., Ignatov, D.I.: Predicting winning team and probabilistic ratings in "dota 2" and "counter-strike: global offensive" video games. In: van der Aalst, W.M.P., et al. (eds.) AIST 2017. LNCS, vol. 10716, pp. 183–196. Springer, Cham (2018). https://doi.org/10.1007/978-3-319-73013-4_17
21. Meng, X., et al.: Mllib: machine learning in apache spark. J. Mach. Learn. Res. **17**, 1–7 (2016)
22. Sarhan, Q.I., Gawdan, I.S.: Java message service based performance comparison of apache ActiveMQ and apache Apollo brokers. Sci. J. Univ. Zakho **5**(4), 307–312 (2017). https://doi.org/10.25271/2017.5.4.376
23. Schubert, M., Drachen, A., Mahlmann, T.: Esports analytics through encounter detection. In: MIT Sloan Sports Analytics Conference, pp. 1–18 (2016)
24. Seif El-Nasr, M., Drachen, A., Canossa, A.: Game Analytics. Springer, London (2013). https://doi.org/10.1007/978-1-4471-4769-5
25. Yang, P., Harrison, B., Roberts, D.L.: Identifying patterns in combat that are predictive of success in MOBA games, pp. 1–8 (2014)
26. Yu, L., Zhang, D., Chen, X., Xie, X.: MOBA-slice: a time slice based evaluation framework of relative advantage between teams in MOBA games. Commun. Comput. Inf. Sci. **1017**, 23–40 (2019). https://doi.org/10.1007/978-3-030-24337-1_2

IPT-CFI: Control Flow Integrity Vulnerability Detection Based on Intel Processor Trace

Zhuorao Yang, Baojiang Cui(✉), and Can Cui

Beijing University of Posts and Telecommunications, Beijing, China
{alex,cuibj,kidic}@bupt.edu.cn

Abstract. Control flow integrity has always been the focus of security research, but how to achieve complete and accurate control flow graph construction and how to achieve transparent and efficient control flow detection has always been a current research difficulty. In this paper, we design a static construction and optimization method of control flow graph for programs, and propose a transparent and accurate extraction technology of program execution flow using Intel Processor Trace under Windows, and design a control flow integrity detection strategy based on IPT (IPT-CFI). The experimental results show that, in terms of security, the system can successfully detect the control flow hijacking attack. In terms of system performance, the static CFG code coverage of this system is over 99%, and the IPT running overhead is about 5%.

1 Introduction

To prevent software damage caused by vulnerabilities, the new operating system adopts methods such as Data Execution Prevention (DEP) and Address Space Layout Randomization (ASLR), which can effectively prevent the exploitation of most software vulnerabilities attack. However, the attacker can still hijack the control flow of the process by constructing a sophisticated payload to achieve the attack purpose [1]. For example, Return-Oriented Programming (ROP) in control flow hijacking attacks can reuse existing code fragments in the process to bypass DEP protection.

In 2005, the University of California and Microsoft jointly proposed a Control Flow Integrity (CFI) mechanism to defend against control flow hijacking attacks [2]. CFI technology generates a Control Flow Graph (CFG) by static and dynamic analysis of the program, and then restricts the running control flow of the process to only run in the limited CFG [3]. If the target address of the indirect branch instruction is not in the CFG, the process is considered to be attacked by the control flow hijacking vulnerability.

However, there are two main challenges that hinder the practicality of CFI. First, how to construct a complete and sound control flow graph. The second challenge is how to achieve efficient, transparent and accurate CFI detection. To overcome these challenges, we propose a static control flow graph optimization method for binary programs, which realizes the optimization of direct function calls, indirect function calls, indirect jumps and tail calls, and improves the static CFG code coverage rate to more than 99%. And we use Intel Processor Trace (IPT) to efficiently and transparently extract the running program control flow, and design the CFI detection system (IPT-CFI) under Windows.

L. Barolli (Ed.): EIDWT 2023, LNDECT 161, pp. 491–499, 2023.
https://doi.org/10.1007/978-3-031-26281-4_51

Experiments show IPT-CFI can detect control flow hijacking vulnerabilities with 100% accuracy and only 5% running overhead.

2 Related Work

Control-Flow Hijacking Detection. Control flow hijacking is a new type of vulnerability exploitation method. Attackers use stack overflow, heap overflow and other vulnerabilities to illegally hijack the program running control flow, and finally make the program perform malicious functions. According to the types of payload, Wang divided control flow hijacking attacks into code injection and code reuse attacks [4].

For the code injection attack, Microsoft proposed software DEP (data execution prevention) and hardware DEP [5]. Software DEP, alias SafeSEH, judges whether an exception handling function is located on a non-executable page, and thus judges whether to execute the exception handling function. Hardware DEP identifies whether instructions are executed on the page by adding a special attribute (NX/XD) to the page table. DEP improves the OS memory management mechanism and effectively prevents code injection attacks.

For the code reuse attack, it is mainly divided into static defense before running and dynamic defense at runtime. The former modifies the compiler before running a program to protect memory such as stack space boundary checking and address randomization. StackGuard attached the detection byte canary at the low address before the return address of the function stack frame [6]. If the canary has been modified when the function returns, it is considered a hijacking control flow attack has occurred.

Control-Flow Integrity Detection. Control-Flow Integrity (CFI) is used to solve Control-flow hijacking. Control flow graph construction is the key to CFI. Construction methods are divided into two categories: source-based compilation and binary-based analysis [7]. Joern extracts the Abstract Syntax Tree (AST) from the source code based on island grammar, so as to construct the control flow graph of the program [8]. Because the source code of a real-world program is hard to obtain, most control flow graphs are constructed with binary programs. BINCOA provides a framework for analyzing binary programs [9]. It uses DBA (Dynamic Bitvector Automata) to obtain the abstract representation of binary programs. IDA Pro is a disassembly software that provides binary program analysis and generates static control flow graphs [10]. It obtains program jump instructions by recursively scanning the binary program to construct CFG.

CFI has two challenges: complex processing before running and high runtime consumption. In recent years, several solutions have been proposed to address these two problems. BinCFI classifies indirect transfer instructions so that the same target set only contains the same type of transfer instructions, and uses a coarse-grained detection strategy to improve the running efficiency of programs [11]. However, binCFI needs to generate CFG, and its detection accuracy for ROP is low.

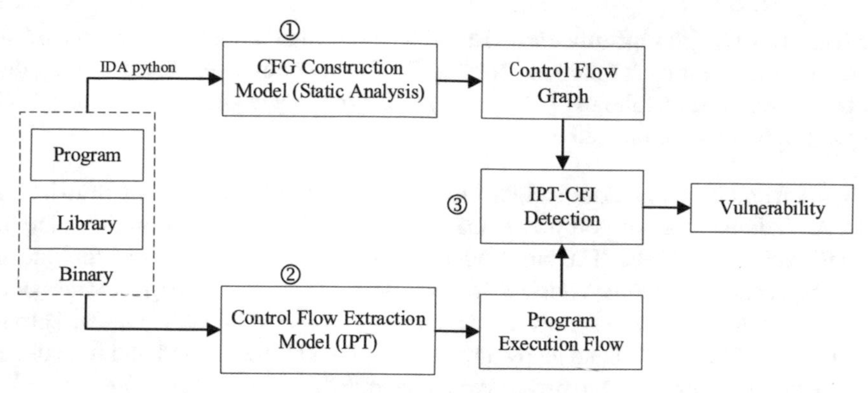

Fig. 1. IPT-CFI Overview.

3 IPT-CFI Overview

The IPT-CFI overview is shown in Fig. 1, which includes the CFG construction module, IPT execution flow extraction module and CFI detection module.

Firstly, the protected process is parsed by the CFG construction module to generate static CFG (step 1), which realizes the optimization of direct function calls, indirect function calls, indirect jumps and tail calls. Then we use the IPT execution flow extraction module to monitor the protected program (step 2), which automatically extracts the execution flow of the runtime program in real time. And the program execution flow will be input to the CFI detection module (step 3), which will judge whether the execution flow violates the static CFG according to the CFI detection policy.

3.1 Control Flow Graph Construction

CFG is a graph consisting of nodes and directed edges, representing all possible paths that a program's control flow can reach during execution [12].

Control Flow Graph Construction Based on Static Analysis. We divide instructions into direct branch instructions, indirect branch instructions, and non-branch instructions. Direct branch instruction means that the destination address of the call and jump can be determined by static analysis (eg, *call printf, jmp puts*). The indirect branch instruction means that the destination address of the call and jump cannot be determined by static analysis (eg, *call eax*). The remaining instructions are all considered non-branch instructions.

Obtain all function start addresses in binary files through recursive descent static analysis. Then parse the instruction from the entry of each function. If the instruction is a non-branch instruction, add it directly to the current basic block. If the instruction is a direct branch instruction, record the destination address of the instruction. Then use this instruction as the basic block end address. Finally, construct the edge from the current basic block to the destination basic block.

Control Flow Graph Optimization. Based on the method above, we can create basic blocks and edges for each function. But this CFG is not complete, the main limitations can be summarized as follows: 1. Not analyze the indirect branch instructions; 2. Not analyze the tail-call optimization.

Indirect branch instructions include indirect function calls and indirect jump instructions. An indirect function call means that its target address depends on the value of a register, such as call %eax. The target address is obtained during dynamic runtime and cannot be accurately obtained during static analysis. However, the target address of the indirect call instruction is generally a function pointer or a virtual table pointer. Through referring to ITC-CFG [13] and analyzing the function calling method and the vulnerability exploitation method, we regard the target address of the indirect function call as a legal address when it meets the following requirements: 1. The target address is the function start address of the.text section of the program; 2. The target address is the function address stored in the program.data segment and.rodata segment. According to the above method, we create edges for indirect function call basic blocks.

The indirect jump means that the target address of the jump instruction is dependent on the value of the register, such as jmp %eax. Indirect jumps can generally be divided into indirect function jumps and switch-case conditional jumps. This paper deals with these two situations as follows: 1. The processing method of indirect function jump is the same as that of indirect function call; 2. Get the jump table for this switch-case from the.rodata section, and obtain the set of possible target basic blocks of switch-case according to the jump table, then create the edge from the basic block of an indirect jump instruction to the target basic block.

Due to compilation optimization, some function calls are processed with tail call optimization. For example, function 1 calls function 2. At the end of function 2, use the jump instruction to call function 3. At this point, function 3 will directly return to function 1 when returning, instead of returning to function 2. This paper uses cross-reference to handle tail-call optimization. Firstly, traverse the function instructions(e.g., *function 2*) to find if there is a jump instruction using an indirect function (e.g., *jmp function 3*). If the instruction is found, find the basic block (e.g., *function 1*) that calls or jumps to function (*function 2*) by cross-reference. Then build the edge from the return function (*function 3*) basic block back to call basic block ().

3.2 Control Flow Integrity Detection Based on IPT

Extraction of Program Control Flow Based on IPT. IPT is a new processor trace technology used by Intel in its core M and 5th Intel processors [12]. IPT will perform an execution flow trace of the running program and generate highly compressed information in real time. Table 1 shows the information types and scenarios related to the control flow transfer generated by IPT. IPT mainly generates four types of packets: TNT, TIP, FUP and MODE. Among them, TNT and TIP will record the jump of program control flow, which is also the data analyzed in this paper.

The extraction system is divided into two parts: IPT extraction driver and decoding recovery module. The IPT extraction driver monitors the IPT record buffer. When the

Table 1. IPT output and associated control flow type

IPT output	Control flow type	Scenarios
TNT	Conditional branch	Jcc, J*CXZ, LOOP
TIP	Indirect branch and near return	JMP\CALL and return TIP (indirect)
FUP	Far transfers	Interrupts, traps, etc

buffer overflows, it outputs the buffer content and clears the buffer. At the same time, the driver records the dynamic link library name and base address required for decoding recovery. The decoding recovery module is mainly aimed at TIP and TNT packets. When the module reads the TIP packet, it recovers the target address of the control flow transfer from the TIP packet. When the module reads the TNT package, it will combine the loaded library information and use the xed (The X86 Encoder Decoder,) to get the disassembly code of the library file to get the target address of the TNT package jump.

Control Flow Integrity Detection. IPT-CFI divides the execution flow generated by IPT into three categories for detection: deterministic instructions, non-deterministic instructions and safe jump instructions. For direct function calls and conditional jump instructions, we have a set of definite target addresses through CFG optimization technology, which IPT CFI calls definite instructions. For indirect function calls and unconditional jump instructions, although the target address cannot be determined during static analysis, we can obtain the set of possible target addresses through CFG optimization technology, which IPT-CFI calls non-deterministic instruction. In addition, for the system library function call and the dynamic library function call, although the destination address cannot be determined, we can get a set of library function addresses through static analysis, which is called the safe jump instruction.

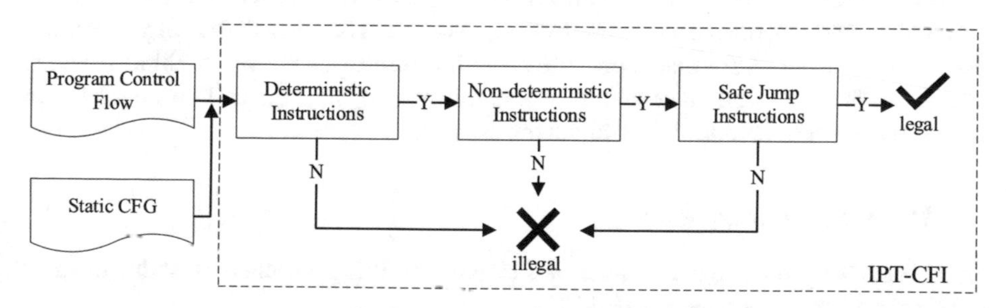

Fig. 2. The process of IPT-CFI.

The Process of IPT-CFI is shown in Fig. 2. First, check whether the target address of the execution flow is in the library function address set, and if so, it means that the execution flow is a safe address. If not, check whether the destination address is in the set of target addresses of the determined instruction, and if so, regard the execution flow as a safe address. If not, check whether the target address is in the set of non-deterministic instruction target addresses, and if so, the execution flow can be considered safe. If the above three steps of detection do not pass, the execution flow is considered to be an illegal execution flow, and IPT-CFI will warn that a control flow hijacking attack has occurred.

4 Evaluation

In this section, we evaluate the accuracy of IPT-CFI for detecting the control flow hijack attacks, and its efficiency in terms of runtime performance. Our testing platform is based on the Windows 10 with an Intel(R) Core(TM) i7-7700HQ CPU @ 2.80 GHz and 16G memory.

4.1 Security Evaluation

To verify the security of the IPT-CFI. According to the opinions of Yan Wang [14], we select the Adobe Reader real vulnerability to construct samples for security test.

Table 2. The results of security evaluation.

	Root cause	Sample number	Detection number	Accuracy
CVE-2018–4990	UAF	20	20	100%
CVE-2013–3346	UAF	10	10	100%
CVE-2010–2883	Stack overflow	30	30	100%
CVE-2010–0188	Stack overflow	10	10	100%

First construct a CFG for the 4 versions of Adobe Reader. Then use IPT to monitor the vulnerable program, run the vulnerability samples and extract the vulnerability execution flow, and use IPT-CFI for detection. The detection results are shown in Table 2 below. The IPT-CFI successfully detected all vulnerability samples, which shows our system can accurately detect control flow hijacking attacks.

4.2 Performance Evaluation

To evaluate the performance, we select 4 CTF vulnerability programs to test the integrity of CFG and the performance of IPT-CFI.

The integrity of the CFG depends on the CFG generation process for all executable code of the binary program as much as possible. To measure the processing performance of static CFG on binary program executable code, this paper proposes the concept of static CFG code coverage. A binary program file can contain.text section,.data section,.rsrc section and.reloc section, etc. Usually executable code is stored in the.text execution code segment. However, the.text section is not entirely code, there are alignments (such as memory alignment, file alignment, etc.) and resource data, and this part of non-code is not executable. So the static CFG code coverage calculation method is as follows:

$$\text{Coverage} = \frac{\text{Length}_{\text{CFG}}}{\left(\text{Length}_{\text{text}} - \text{Length}_{\text{not-ex}}\right)} \tag{1}$$

$\text{Length}_{\text{text}}$ indicates the data size of the binary program.text section. $\text{Length}_{\text{not-ex}}$ Indicates the size of non-executable code in the.text segment. $\text{Length}_{\text{CFG}}$ represents all basic block sizes of a static CFG.

When the code coverage of the static CFG reaches more than 99.0%, it can be considered that the static CFG is sound and complete. The CFG code coverage of the 4 CTF programs selected in this paper is shown in Table 3.

Table 3. The results of the CFG code coverage.

Program name	Lengthtex	Lengthn	$\text{Length}_{\text{CFG}}$	Coverage
Ret2shellcode	0x18E000	0x82D	0x18C4BB	99.692%
Ret2Text	0x14B00	0x56BC	0xF2A4	99.335%
Ret2Lib	0x14A00	0x578C	0xF166	99.565%
SEHHack	0x167A0	0x6C3D	0xF940	99.181%

It can be seen from Table 3 that the static CFG code coverage of each program has reached more than 99.0%, indicating that our static CFG construction is complete and sound.

We evaluate the running overhead of IPT-CFI from two aspects: one is the overhead of IPT itself; the other is second is the impact of IPT runtime on the monitored program. The running overhead of this paper adopts the average of the CPU utilization at three time points when the calculation program starts to run, during the running and before the end of the running.

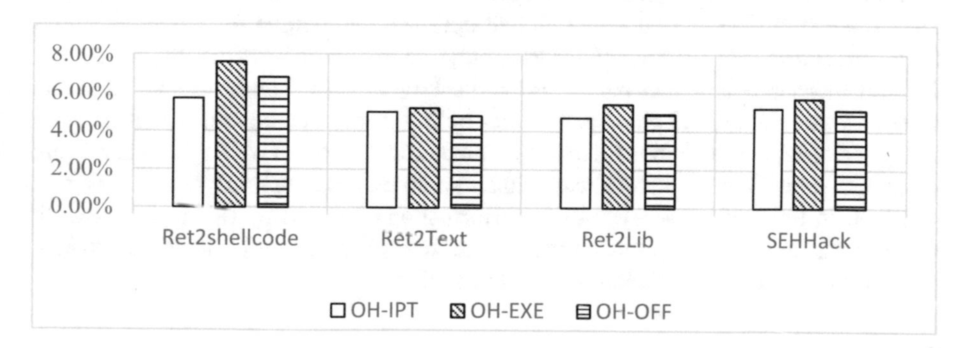

Fig. 3. The Performane of IPT-CFI

The experimental results are shown in Fig. 3, where OH-IPT represents the overhead of the IPT runtime itself, OH-EXE represents the runtime overhead of the program monitored by IPT, and OH-OFF indicates the impact of IPT monitoring on the program. It can be seen from the table that the overhead of IPT itself is about 5%, and the overall impact on the load of the program is relatively low, and the average overhead is less than 10%.

Compared with the existing control flow integrity detection technology, the vulnerability exploitation detection method implemented in this paper has a great improvement in practicability, as shown in the following Table 4:

Table 4. Performance comparison of CFI detection systems.

Name	Modification	Integrity	OS	Multi-thread
ROPecker [15]	N	N	Linux	Y
PathArmor [16]	N	N	Linux	Y
CCFIR [17]	Y	N	Linux	Y
FlowGuard [13]	N	Y	Linux	Y
IPT-CFI	N	Y	Windows	Y

IPT-CFI is a reproducible inspection of the complete instruction stream without modifying the executable file, which is more complete than the existing related technologies. And the current researches are mainly on the Linux operating system platform, but the number of users and attack events on the Windows platform is much larger than that on the Linux platform. Therefore, our IPT-CFI is more practical and has the value of in-depth research.

5 Conclusion

Control flow hijacking attacks are currently widely used by attackers to break through traditional security mechanisms, causing great damage. To prevent control flow hijacking attacks, we design a CFG construction and optimization method for binary execution programs, and use IPT to efficiently extract program execution flow. Finally, we propose IPT-CFI to accurately detect control flow hijacking attacks. The experimental results show that our static CFG construction and optimization method can make the static CFG code coverage reach more than 99%. The running overhead of IPT itself is 5%, and it only increases the running overhead of the monitored program by 10%, which is more practical. In the future, we will use the information extracted by IPT to dynamically optimize the control flow graph, and reduce the size of the control flow graph, so as to improve the security and efficiency of CFI detection.

References

1. Zhang, J., Chen, C., Zheng, H.: Fuzzing sample optimization method for software vulnerability detection. Journal of Shandong University (Natural Science) (09), 1–8+35 (2019)

2. Erlingsson, M.A.M.B.U., Jigatti, J.: Control-flow integrity. In: CCS'05: Proceedings of the 12th ACM conference on Computer and Communications Security, pp. 340–353 (2005)
3. Huang, Y., Fully Context-Sensitive Control-Flow Integrity. Huazhong University of Science and Technology (2017)
4. Wang, F., Zhang, T., Xu, W., Sun, M.: Overview of control-flow hijacking attack and defense techniques for process. Chin. J. Netw. Inf. Secur. **5**(6), 10–20 (2019)
5. Gao, Y., Zhou, A., Liu, L.: Data-execution prevention technology in windows system. Chinese Journal of Information Security and Communication Confidentiality (07), 77–79+82 (2013)
6. Cowan, C., et al. :Stackguard: automatic adaptive detection and prevention of buffer-overflow attacks. In: USENIX Security Symposium, vol. 98, pp. 63–78 (1998)
7. Tan, G., Jaeger, T.: CFG construction soundness in control-flow integrity. In: Proceedings of the 2017 Workshop on Programming Languages and Analysis for Security (PLAS '17), pp. 3–13. Association for Computing Machinery, New York, NY, USA (2017)
8. Joern. https://joern.io/docs/
9. Bardin, S., Herrmann, P., Leroux, J., Ly, O., Tabary, R., Vincent, A.: The BINCOA framework for binary code analysis. In: Gopalakrishnan, G., Qadeer, S. (eds.) CAV 2011. LNCS, vol. 6806, pp. 165–170. Springer, Heidelberg (2011). https://doi.org/10.1007/978-3-642-22110-1_13
10. IDAPro disassembler. http://www.datarescue.com/idabase/
11. Zhang, M., Sekar, R.: Control flow integrity for{COTS)binaries. In: Presented as Part of the 22nd Security Symposium, pp. 337–35 (2013)
12. Wang, X., Liu, Y., Chen, H.: Transparent protection of kernel module against ROP with intel processor trace. Chin. J. Soft. **29**(05), 1333–1347 (2018)
13. Liu, Y., Shi, P., Wang, X., Chen, H., Zang, B., Guan, H.: Transparent and efficient CFI enforcement with Intel processor trace. In: 2017 IEEE International Symposium on High Performance Computer Architecture (HPCA), Austin, TX, pp. 529–540 (2017)
14. Wang, Y., et al.: Revery: from proof-of-concept to exploitable. In: Proceedings of the 2018 ACM SIGSAC Conference on Computer and Communications Security (CCS '18), pp. 1914–1927. Association for Computing Machinery, New York, NY, USA (2018). https://doi.org/10.1145/3243734.3243847
15. Cheng, Y., et al.: ROPecker: a generic and practical approach for defending against ROP attack (2014)
16. Van der Veen, V., et al.: Practical context-sensitive CFI. In: Proceedings of the 22nd ACM SIGSAC Conference on Computer and Communications Security (2015)
17. Zhang, C., et al.: Practical control flow integrity and randomization for binary executables. In: 2013 IEEE Symposium on Security and Privacy. IEEE, pp. 559–573 (2013)

Business Intelligence: Alternative Decision-Making Solutions on SMEs in Indonesia

Agustina Fitrianingrum[1(✉)], Maya Indriastuti[2], Andi Riansyah[3], Abdul Basir[4], and Dedi Rusdi[2]

[1] Department of Management, Faculty of Economics, Universitas Islam Sultan Agung, Semarang, Indonesia
agustinafitrianingrum@unissula.ac.id
[2] Department of Accounting, Faculty of Economics, Universitas Islam Sultan Agung, Semarang, Indonesia
{maya,dedirusdi}@unissula.ac.id
[3] Department of Informatics, Faculty of Industrial Technology, Universitas Islam Sultan Agung, Semarang, Indonesia
andi@unissula.ac.id
[4] Department of Mathematics Education, Faculty of Education, Universitas Islam Sultan Agung, Semarang, Indonesia
abdulbasir@unissula.ac.id

Abstract. In this modern and digital era, digital transformation is echoed as one of the organization's efforts to survive through Business Intelligence (BI). BI has become a buzzword even among business actors or organizations, not least for Small and Medium Enterprises (SMEs). SMEs are one of the sectors affected by the COVID-19 pandemic, namely the number of SME players who have lost their income and are finally forced to go out of business. BI is a combination of techniques and methods in terms of fulfilling access to information and a concise data management mechanism to be able to have a positive influence on SME business activities. It is because the strength of BI significantly impacts strategic decision-making using processing tools from Microsoft, namely SQL Server Integration Services (SSIS) and SQL Server Reporting Services (SSRS). This study aims to see the extent of BI as an alternative solution in decision-making by all SMEs in Indonesia. This research contributes to SMEs through the implementation of BI; SMEs get explicit knowledge about the factors that affect the performance of SMEs to help SMEs in making decisions.

Keywords: Business intelligence · SQL server integration services · SQL server reporting services · SMEs Indonesia

1 Introduction

This The development of information technology integrated with work processes is a necessity that an organization must carry out. It helps to improve the ability to analyze

L. Barolli (Ed.): EIDWT 2023, LNDECT 161, pp. 500–507, 2023.
https://doi.org/10.1007/978-3-031-26281-4_52

problems faced by the organization and the decision-making process by management. Nowadays, the application of the Business Intelligence (BI) system assists the data collection, information, and analysis. It also opens access to sources of information and company performance. It also opens the possibility of measuring the strength of competitor growth.

BI is defined as a set of theories, methodologies, processes, architectures, and technologies that transform raw data into meaningful and useful information for business purposes [1]. As BI system can provide meaningful information, it can assist a company in designing a strategic plan. Therefore, BI system implementation is significant in building a competitive advantage and a company's sustainability.

BI combines techniques and methods to fulfill access to information and a concise data management mechanism [1]. The BI approach will facilitate data access collection, storage, analysis, and provision. Then, the information will assist the company in making decisions on problems or even predicting the long-term possibility of the company's mission [2]. The data generally collected in one access to information are transaction data, warehouses, reports, and visualization tools (prediction analysis and modeling) [2]. With the development of the times, existing technology provides convenient opportunities for various aspects in answering the problems that arise. For every choice that is based on processed data in SMEs, which manage complex and uncertain condition for their operational business needs, an information system is required as a reference. BI approach can be used as an alternative solution to meet these needs. The urgency of BI for SMEs can be seen from BI's function as a decision support tool. More specifically, the BI application in an SME is very helpful in managing SME business management through the fulfillment of access to information.

Simplicity is the key success factor of BI model [3]. It enables the provision of more flexible solutions that aligns with the client's business objectives [4]. Regarding [5] BI as a software platform can deliver 13 critical capabilities across three categories: enabling, producing, and consuming, as well as supporting four use cases. There are several empirical research pieces of evidence [5–7] shown that customer loyalty is increased by BI model implementation and able to elevate relationships to a new level. Customers are educated and expect higher performance and quality from this software suite as stated by [2, 8], hence BI is the answer to managing today's data overload. The segmentation of BI market according to [9] traditional, mobile, cloud, and social business intelligence, depending on the product architecture and user interface. However, the software industry is having fierce competition. It needs a more customer-centered [10] and vendor [11] approach. This statement is also supported by [12] which correctly assesses this perception that it is rigorous for companies in the high-tech and software business. From the point of view of using data to gain business advantage, BI tools can help in information management, helping executives to make more informed decisions [13].

The implementation of BI provides several advantages for SMEs. First, information and data are the basic things that support arguments in making decisions and convince several interested parties. Second, easy access to information by users impacts increasing the value of SME data and information. Third, increasing the value of organizational data and information makes it easier for an organization to monitor its performance. BI can show an organization's Key Performance Indicators (KPIs). Fourth, the application of

BI can become an investment in information technology that has existed before because BI does not replace technology but as a function of scale-up services on the system so that the information is more comprehensive. Overall, BI can make companies work more efficiently to save time because access to find information is easy to understand to minimize costs for employee training [1].

With the benefits obtained from BI, SMEs can determine the steps for implementing BI using a method that suits the company's needs. BI implementation begins with analyzing the problem you want to simplify in the company by evaluating the existing business cases, then planning a solution model. Next, planning results need to be tested for accuracy with in-depth analysis. Then if the analysis phase meets the terms and conditions, the BI system can enter the design phase divided into the design phase (information design) and the construction phase (prototyping). With the complexity of the stages and the benefits of BI implementation, it is hoped that the company can coherently understand the above steps so that BI can become a provider of information that facilitates the company's organizational operations.

The contribution of this research for SMEs are encouraging SMEs to implement BI system by providing model implementation of BI for SMEs. This also explores explicit knowledge about the factors that affect SMEs business performance. BI model implementation assists SMEs in decision-making. In the long term, this will enable

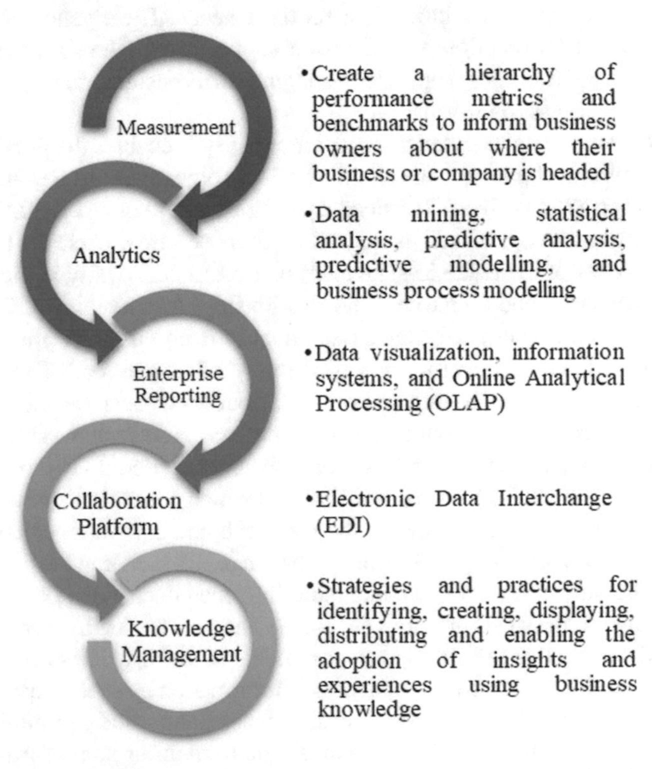

Fig. 1. Business intelligence implementation in Indonesia SMEs

SMES to build its competitive advantage by having the ability to analyze market trends. Understanding the trends will help SMEs to determine strategies required to anticipate changes (Fig. 1).

2 Paper Preparation

For 65% of SMEs in Indonesia, technology is an essential investment priority. It is because investment in technology can be used to build digital capabilities in terms of sales and services (85%), digital marketing and social media (76%), and network and technology management (64%). Furthermore, 88% of SMEs in Indonesia believe that this technology will guarantee a more sustainable business model in the long term. Technology can help them manage cash flow better and reduce pressure on cash flow by reducing costs related to marketing and delaying the replacement of equipment such as laptops, desktops, and machines [14].

SMEs shall apply 5 (five) stages in implementing BI system. The first stage is measurement application. It creates a hierarchy of performance metrics and benchmarking processes. The result of this stage is to inform business owners about the direction of the business or company goals which is essential for the business. It makes it easier for management to understand market needs and take the necessary steps from a business outlook. By doing the first step, business owners or management boards can choose a method that suits the company's business in its application.

The second is Analytics, an application that builds quantitative processes: data mining, statistical analysis, predictive analysis, predictive modeling, and business process modeling. The result of analytics optimizes the decision in making business knowledge discovery. Analytics helps analyze obstacles, advantages, and solutions and assists companies in determining cost planning.

The Third stage is enterprise reporting for building a strategic infrastructure for business reporting. It is to fulfill strategic business management and not operational reporting. This report produces data visualization, information systems, and Online Analytical Processing (OLAP). Furthermore, enterprise reporting is helpful for the company to determine business decisions because it helps read and interpret data. For example, the BI application collects information on market conditions and processes it into a valuable report.

The fourth stage is the collaboration platform, which enables an organization to create distinct scopes inside and outside the business using Electronic Data Interchange (EDI). Collaboration platforms have a reasonable budget because they are easier to use. It also accelerates business stakeholders' work. The time to acquire data from inside and outside the company is faster. Special knowledge is not required.

Fifth, knowledge management enables the adoption of business knowledge insights and experiences by identifying, creating, displaying, and disseminating business knowledge insights and experiences. Businesses will receive more prepared data forecasts for long-term decision-making with this business intelligence tool. Business intelligence is essential for any organization. Planning, data management, and giving final results in the form of information for all parties involved in a business or business are the responsibilities of business intelligence [15] (Fig. 2).

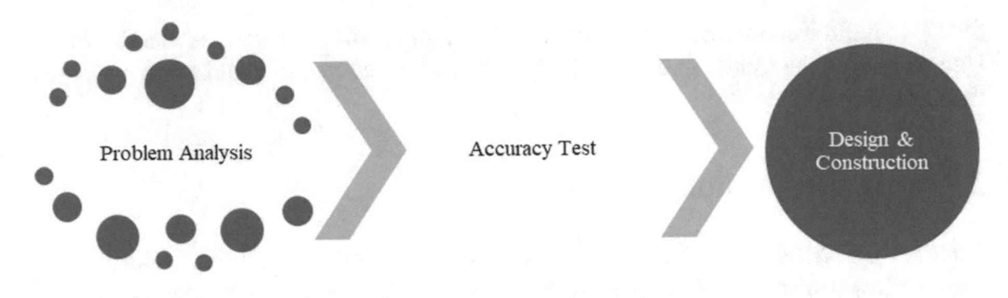

Fig. 2. Stages of the business intelligence implementation process in Indonesia SMEs

3 Literature Review

3.1 Business Intelligence

Business intelligence (BI) is a system and application that converts data (operational data, transactional data, or other data) within a business or organization into knowledge. This program examines historical data, analyzes it, and then employs that understanding to enhance organizational planning and decision-making so that it can be used in crucial corporate decision-making processes or decisions to achieve business goals [1].

In order to aid an organization in decision-making and, at the same time, boost its competitive advantage, business intelligence can also help a firm or organization get specific understanding of the elements that influence organizational performance. An organization can use BI to examine changing trends and identify the tactics required to foresee changing trends. BI can be defined as knowledge derived from the outcomes of data analysis derived from an organization's operations [4].

3.2 Microsoft Processing Tools

3.2.1 SQL Server Integration Services (SSIS)

A Microsoft SQL Server database software component called SQL Server Integration Services (SSIS) is capable of carrying out a variety of data migration activities. SSIS is a platform for workflow applications and data integration. Data warehousing tools with this capability are used for data extraction, transformation, and loading (ETL). This program can also update data in multidimensional cubes and automate SQL Server database maintenance [15].

3.2.2 SQL Server Reporting Service (SSRS)

Microsoft's SQL Server Reporting Services (SSRS) is a server-based reporting tool. It is a component of the Microsoft SQL Server suite of services, which is handled using a Web interface and includes SSAS (SQL Server Analysis Services) and SSIS (SQL Server Integration Services). It can be used to create and deliver a variety of interactive reports that are printable. SQL administrators and developers can connect to SQL databases using the SSRS service's interface for Microsoft Visual Studio and utilize the SSRS tools to format SQL reports in intricate ways. Additionally, a "Report Generator" tool is available for less sophisticated users to produce simpler SQL reports [15].

3.2.3 Decision-Making

Decision-making is universally defined as the choice among various alternatives. This understanding includes both choice-making and problem-solving. Rational decision-making is needed for every organization where managers must take more rational action in dealing with every situation. Managers define issues, weigh potential solutions, and select the best ones for the organization through the systematic process of rational decision-making [1] and [8] (Figs. 3, 4 and 5).

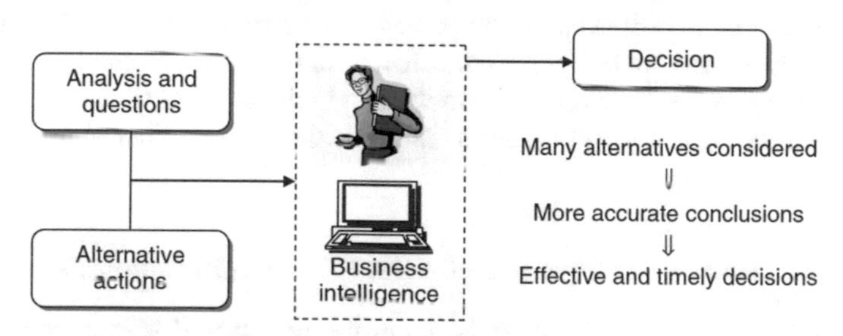

Fig. 3. Benefits of a business intelligence system Source: [1]

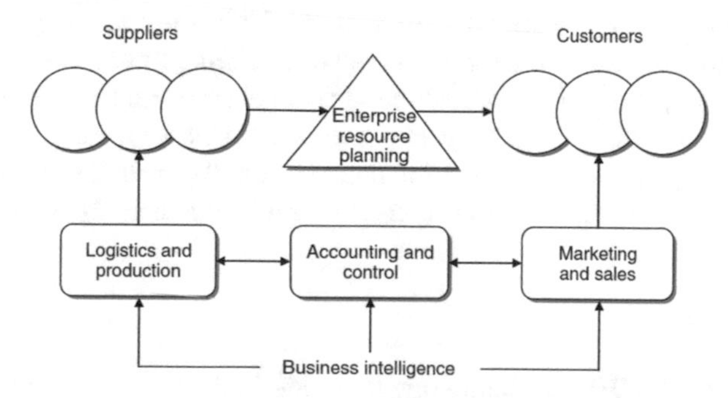

Fig. 4. Departments of an enterprise concerned with business intelligence systems Source: [1]

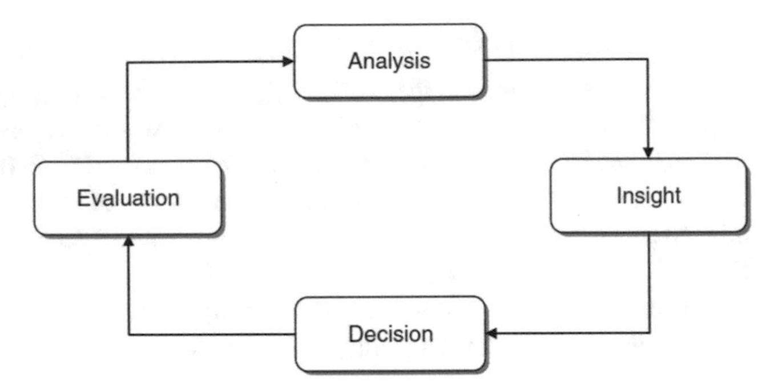

Fig. 5. Cycle of a business intelligence analysis Source: [1]

4 Research Method

4.1 Population and Sample

The population in this study is all SMEs in Indonesia that still exists today with the sampling technique using the purposive sampling. Only SMEs who have used a business intelligent system are eligible to participate in this study as respondents. The rules of thumb are referenced to [16] because the population's size is unknown. For the majority of studies, sample sizes more than 30 and lower than 500 are suitable. Additionally, the sample size in multivariate research (including multiple regression analyses) should be at least 10 times (ideally 100 times) as large as the total number of variables being examined.

4.2 Stages of Business Intelligence Activities for SMEs in Indonesia

a. Observation Stage: The first stage is planning, which will start by preparing any tools/software needed in making this application and conducting research using interview methods to find out the needs of application users.
b. Literature Study: Discussing the problems in developing this application service, books and articles from the internet are needed relating to business intelligence.
c. Data Processing Stage: The data obtained from observations will then measure the perspective errors that generate questions and will be analyzed.
d. Testing and Implementation: After the business intelligence discussion is complete, then carry out testing or test cases on the system later and apply them to the system.

5 Conclusion

Business intelligence (BI) is a technology that aids in the analysis and visualization of business data by businesses, particularly SMEs. A characteristic of BI allows for the conversion of raw data into valuable information that can aid in decision-making and give the business a competitive edge. Businesses can better serve their customers by evaluating the data they own. High levels of client loyalty and retention must come after rising sales.

References

1. Vercellis, C.: Business Intelligence: Data Mining and Optimization for Decision Making. Wiley, United Kingdom (2009). www.microsoft.com
2. Swoyer, S.: Analytics in the cloud: the challenges and benefits, TDWI Report (2013). http://tdwi.org/research/2013/11/tdwi-ebook-analytics-in-the-cloud-the-challenges-and-benefits.aspx
3. Luoma, E.: Examining business models of software-as-a-service firms. In: Altmann, J., Vanmechelen, K., Rana, O.F. (eds.) GECON 2013. LNCS, vol. 8193, pp. 1–15. Springer, Cham (2013). https://doi.org/10.1007/978-3-319-02414-1_1
4. Thompson, W., Van der Walt, J.: Business intelligence in the cloud. SA J. Inf. Manag. **12**(1), 1–5 (2010)
5. Sallam, R.L., Hostmann, B., Schlegel, K., Tapadinhas, J., Parenteau, J., Oestreich Thomas, W.: Magic quadrant for business intelligence and analytics platforms (2015). www.gartner.com/technology/reprints.do?id=1-2AEKLJ3&ct=150223&st=sb
6. Business-Software Report: Top 10 business intelligence software report – comparison of the leading business intelligence software vendors, Business-Software (2013). http://c3330831.r31.cf0.rackcdn.com/top_10_bi.pdf
7. Dresner, H.: Wisdom of crowds embedded intelligence market study. Dresner Advisory Services (2013). www.dundas.com/wisdom-of-crowds-embedded-bi-study
8. Chen, H., Chiang, R.H.L., Storey, V.C.: Business intelligence and analytics: from big data to big impact. MIS Q. **36**(4), 1165–1188 (2012)
9. Redwood Capital: Sector report on business intelligence (2014). http://www.redcapgroup.com/media/98e342dd-420c-4716-be25-f21a14f46691/Sector%20Reports/2014-04-09_Business_Intellegence_Report_April_2014_pdf
10. Van der Lans, R.F.: Data virtualization for business intelligence systems: revolutionizing data integration for data warehouse, Morgan Kaufmann, Waltham, MA (2012). http://store.elsevier.com/Data-Virtualization-for-Business-Intelligence-Systems/Rick-van-der-Lans/isbn-9780123978172/
11. Baur, A.W., Bühler, J., Bick, M.: How pricing of business intelligence and analytics SaaS applications can catch up with their technology. J. Syst. Inf. Technol. **17**(3), 229–246 (2015). https://doi.org/10.1108/jsit-03-2015-0024
12. Mohr, J.J., Sengupta, S., Slater, S.F.: Marketing of High-Technology Products and Innovations, 3rd edn. Prentice Hall, Upper Saddle River (2010)
13. Muntean, M.I., Muntean, C.: Evaluating a business intelligence solution: feasibility analysis based on Monte Carlo method. Econ. Comput. Econ. Cybern. Stud. Res. **2**(47), 85–102 (2013). www.mediaindonesia.com
14. Kadas, K.A.: How is Business Intelligence Implemented in Companies? (2022). https://myskill.id/blog/dunia-kerja/penerapan-business-intelligence/
15. www.microsoft.com
16. Sekaran, U.: Research Methods for Business: A Skill Building Approach. Wiley, New York (2003)

Author Index

L. Barolli (Ed.): EIDWT 2023, LNDECT 161, pp. 509–510, 2023.
https://doi.org/10.1007/978-3-031-26281-4

Printed in the United States
by Baker & Taylor Publisher Services